CATACLYSMIC VARIABLES

ASTROPHYSICS AND SPACE SCIENCE LIBRARY

VOLUME 205

CATACLYSMIC VARIABLES

Proceedings of the Conference
held in Abano Terme, Italy, 20–24 June 1994

Edited by

A. BIANCHINI

M. DELLA VALLE

Dipartimento di Astronomia,
Università di Padova, Italy

and

M. ORIO

Osservatorio Astronomico di Torino, Italy

SPRINGER SCIENCE+BUSINESS MEDIA, B.V.

Library of Congress Cataloging-in-Publication Data

Cataclysmic variables / edited by A. Bianchini and M. Della Valle and
M. Orio.
       p.   cm. -- (Astrophysics and space science library , v. 205)
    Papers of a congress held in Padova.
    ISBN 978-94-010-4148-5   ISBN 978-94-011-0335-0 (eBook)
    DOI 10.1007/978-94-011-0335-0
    1. Cataclysmic variable stars--Congresses.   I. Bianchini, A.
(Antonio) II. Della Valle, M. (Massimo) III. Orio, M. (Marina)
IV. Series.
    QB837.5.C37 1995
    523.8'446--dc20   95-34348

ISBN 978-94-010-4148-5

Printed on acid-free paper

TABLE OF CONTENTS

1. OBSERVATIONAL OVERVIEW

2. INDIVIDUAL OBJECTS

3. THERMONUCLEAR RUNAWAYS

4. SPECTRAL EVOLUTION OF NOVAE AND PROBLEMS IN NEBULAR PHYSICS

5. ACCRETION INDUCED PHENOMENA

6. THE ROLE OF MAGNETIC FIELDS

7. SUPERSOFT X-RAY SOURCES AND THEIR RELATION TO CVs

8. EVOLUTION AND EVOLUTIONARY LINKS

PREFACE

Cataclysmic Variables (CVs) are *well furnished laboratories* in which one can investigate a number of important physical processes. These processes include: thermonuclear reactions, nebular physics, winds, mass accretion phenomena on compact objects, magnetohydrodynamics, plasmas, thermal and non thermal radiative processes, relativistic phenomena, dust formation, etc.. In the recent year the most exciting development in our opinion was that the small, cold and invisible secondary components of CVs were found to be the prime motors for the evolution of close binary systems, disclosing new scenarios for research. All branches of astronomy and astrophysics often take advantage of the new ideas brought by the study of CVs. It is for this reason, we believe, that meetings on CVs are important for most of the astronomical community.

The idea of a Congress on CVs in Padova has been fluttering in the air for a few years. After all – we thought – a large part of the scientific heritage of Padova and Asiago Observatories is represented by the study of variable stars and, in particular, of CVs, as Prof. L. Rosino reminded us at the very beginning of the Congress. We finally chose Abano Terme as the site of the Conference, not merely because Abano is one of the most famous centers in Europe for mud therapy, and astronomers easily suffer from rheumatism, but also because we knew that a very hearty welcome from the Abano Municipality and touristic associations awaited us. Another reason was because this small town, at the end of the XIII century, was the birthplace of *Peter of Abano*. It was he, who first started in the just born University of Padova, the tradition of the studies in astronomy, as Prof. Armando Gennaro stated when greeting us.

The first day of the conference was completely dedicated to an observational overview and presentation of new outstanding results. On the second and third day we talked about classical and recurrent novae, from the observational and theoretical point of view, dedicating half a day to the problems of nebular physics. On the fourth day we discussed statistical properties of CVs and the possible connection with Super Soft X-Ray Sources and Type I Supernovae. Finally, on the fifth day accretion induced phenomena were discussed together with the evolutionary scenario, and some final conclusions were drawn. Three Podium Discussions were also organized: by Bob Williams, on *The Nebular Physics of Novae*, by

Hakki Ögelman, on *The Nature of the Supersoft X-Ray Sources*, and by Mario Livio and Alvio Renzini, about *The Progenitors of SNe-I and AIC borne neutron stars.*

After the enterprise of organizing a conference, one looks back at all the hectic activities undertaken in a few days and tries to draw a balance. We are aware now that we could have organized several practical things better and more efficiently. However, we feel very satisfied with the large number of participants and the scientific content of the conference. Actually, one of our goals was to guarantee participation of colleagues from a large number of countries. An effort has been made to allow participation of graduate students and young post-doctoral fellows. They brought fresh enthusiasm and at the same time had the occasion to learn a lot. We also felt the need to bring together scientists with different backgrounds, not only the "experts" of this field, but also experts of related areas of stellar astrophysics. In addition, work of amateur astronomers' organizations, which contribute very much to the continuous monitoring of CVs, was presented to the scientific community.

We must thank the Scientific Organizing Committee for having designed the scientific program so as to achieve all these goals. Some of the "exhausted" participants might have thought was too full a program for five days, but it was successful and the scientific debate was lively and extremely interesting. Thanks are due to all members of the Local Organizing Committee for their assistance in the organization of the Conference, and to Ms. Marisa Zanon, Ms. Daniela Trevisan and Ms. Gilda Rota for their collaboration in the administration of the conference. We are particularly indebted to Ms. Paola Orio for her help in the administration and in the assistance to the participants. Finally, we are grateful to Thomas Zwitter and to the very young and very energetic Pippo Bianchini for their endless work in the conference room every day of the conference! Financial support was given by the following institutions that we thank: the Italian CNR (Consiglio Nazionale delle Ricerche), the Italian space industry Alenia, the Banca Antoniana, the Community Councils of the towns Abano e Padova, the University and the Astronomical Observatory of Padova, the International Science Foundation. Travel grants for several participants were awarded by means of a generous grant of the Directorate for Science,

Research and Development of the European Community Council. Finally, we wish to thank the mayor of Abano Terme, Mr. Cesare Pillon, the President of the Azienda di Promozione Turistica, Prof. Armando Gennaro, and also Mr. Gianpietro Bano, Mr. Mario Sbalchiero, Mr. Fabio Stecca and the Association of Abano Hotels for offering accommodation in very good hotels at a very convenient price.

Padova,Torino   Antonio Bianchini
January 1995   Massimo Della Valle
                                          Marina Orio

Scientific Organizing Committee

A. Bianchini (Chairperson)	S. Starrfield
M. Kato	P. Szkody
M. Friedjung	A. Tutukov
J. Mattei	N. Vogt
H. Ögelman	B. Warner
G. Shaviv	R. Williams

Local Organizing Committee

M. Orio (Chairperson)	U. Munari
A. Bianchini	P. Rafanelli
M. Della Valle	M. Hack
R. Claudi	P.L. Selvelli
G.C. Favero	

CONFERENCE PARTICIPANTS

1) Tim ABBOTT, European Southern Observatory, La Silla, Chile

2) Ivan ANDRONOV, Dept. of Astronomy, Odessa, Ukraine

3) G.C. ANUPAMA, IUCAA, Pune, India

4) Thomas AUGUSTEIJN, University of Amsterdam, The Netherlands

5) Solen BALMAN, Physics Dept., University of Wisconsin, Madison WI, USA

6) Raymundo BAPTISTA, Space Telescope Science Institute, Baltimore MD, USA

7) Cesare BARBIERI, Dipartimento di Astronomia, Università di Padova, Italy

8) Paul BARRETT, Goddard Space Flight Center, Greenbelt, MD, USA

9) Corrado BARTOLINI, Dip. di Astronomia, Università di Bologna, Italy

10) A. P. BEARDMORE, The Open University, Milton Keynes, U.K.

11) Paul BENNIE, Univesity of St. Andrews, Scotland, U.K.

12) Klaus BEUERMANN, Universitäts-Sternwarte Goettingen, FRG

13) Antonio BIANCHINI, Dipartimento di Astronomia, Università di Padova, Italy

14) Angela BRAGAGLIA, Osservatorio Astronomico di Bologna, Italy

15) Nina BRESKOVNAJA, Pulkovo Observatory, St Petersburg, Russia

16) David BUCKLEY, Cape Town University, South Africa

17) Vadim BURWITZ, MPI für Theor. Physik, Garching bei München, FRG

18) John CANNIZZO, Goddard Space Flight Center, Greenbelt, MD, USA

19) Maria Sandi CATALAN, Astronomy Centre, University of Brighton, U.K.

20) Ganesh CHANMUGAM, Baton Rouge University, Louisiana, USA

21) Cesare CHIOSI, Dipartimento di Astronomia, Università di Padova, Italy

22) Lidia CHINAROVA, Dept. of Astronomy, Odessa, Ukraine

23) Drakomir CHOCHOL, Astronomical Institute, Slovak Academy, Slovakia

24) Riccardo CLAUDI, Osservatorio Astronomico di Padova, Italy

25) Marcella CONTINI, Dept. of Phys. and Astron., Tel Aviv University, Israel

26) Massimo DELLA VALLE, Dip. di Astronomia, Università di Padova, Italy

27) Guido DE MARCHI, Space Telescope Science Institute, Baltimore, MD, USA

28) Vickram DHILLON, Royal Greenwhich Observatory, Tenerife, Spain

29) Domitilla DE MARTINO, ESA-Vilspa, Villafranca del Castillo, Spain

30) Marcos DIAZ, Sao Paulo University, Brazil

31) Stefan DIETERS, University of Amsterdam, The Netherlands

32) Danuta DOBRZYCKA, Center For Astrophysics, Cambridge, MA, USA

33) Simon DUCK, Department of Astrophysics, Oxford, U.K.

34) Hilmar DUERBECK, Institute of Astronomy, Münster, FRG

35) Hasan H. ESENOĞLU, Osservatorio di Asiago, and Istanbul University, Turkey

36) S.D.S. EYRES, Dept. of Astrophysics, Oxford, U.K.

37) Giancarlo FAVERO, Dipartimento di Chimica, Università di Padova, Italy

38) Michael FRIEDJUNG, Institute d' Astrophysique, Paris, France

39) William F. FRY, Physics Dept., University of Wisconsin, Madison, WI, USA

40) John GALLAGHER, University of Wisconsin, Madison, WI, USA

41) Boris GÄNSICKE, MPI für Extr. Physik, Garching bei München, FRG

42) Mark GARLICK, University of Sussex, Astronomy Centre, UK

43) Silvia GAUDENZI, Padova University, Italy

44) Bob GEHRZ, Physics Dept., University of Minnesota, Mimneapolis, USA

45) Patrick GODON, Physics Department, Technion, Haifa, Israel

46) Rosario GONZALEZ-RIESTRA, IUE, ESA-Vilspa, Spain

47) Jochen GREINER, MPI für Extr. Physik, Garching bei München, FRG

48) Jem GUNTHER, Amateur Astronomer, France

49) Peter KAHABKA, MPI für Extr. Physik, Garching bei München, FRG

50) Mariko KATO, University of Sussex, Astronomy Centre, UK

51) Taichi KATO, Kyoto University , Japan

52) Andrew KING, Department of Astronomy, University of Leicester, UK

53) Christian KNIGGE, Physics Department, University of Oxford, U.K.

54) Ulrich KOLB, MPI für Astrophysik, Garching bei München, FRG

55) Sergej KOLESNIKOV, Astronomy Dept., Odessa, Ukraine

56) Attay KOVETZ, Department of Geophysics, Tel Aviv University, Israel

57) Joachim KRAUTTER, Landessternwarte, Heidelberg, FRG

58) Chris JOMARON, University of St. Andrews, Scotland, U.K.

59) Pasi HAKALA, University of Helsinki, Finland

60) Margaret HARROP-HALLIN, Astronomy Dept., Cape Town, South Africa

61) Peter HAUSCHILDT, Arizona State University, Tempe AZ, USA

62) Frederick HESSMAN, MPI für Astronomie, Königstuhl,Heidelberg, FRG

63) Masahito HIROSE, Department of Astronomy, University of Tokyo, Japan

64) Robert HJELLMING, National Radio Astronomical Obs., Socorro, USA

65) Kent HONEYCUTT, Department of Astronomy, Bloomington, Indiana, USA

66) Keith HORNE, Astronomical Institute, Utrecht, The Netherlands

67) Steve HOWELL, Planetary Science Institute, Tucson, AZ, USA

68) Min HUANG, Physics Department, Villanova University, PA, USA

69) Icko IBEN, Astronomy Dept., University of Urbana-Champaigne, IL, USA

70) Irit IDAN, Physics Department, Technion, Haifa, Israel

71) Nazar IKHSANOV, Pulkovo Observatory, St. Petersburg, Russia

72) Manabu ISHIDA, Institute for Space Science, Kanagawa, Japan

73) Laurits LEEDJARV, Tartu Astrophysical Observatory, Estonia

74) Elia LEIBOWITZ, Dept. of Phys. and Astron., Tel Aviv University, Israel

75) Mario LIVIO, Space Telescope Science Institute, Baltimore MD, USA

76) Knox LONG, Space Telescope Science Institute, Baltimore MD, USA

77) Rosemary MARDLING, Dept. of Mathematics, Monash University, Australia

78) Frederich MEYER, MPI für Astrophyisk, Garching bei München, FRG

79) Emmi MEYER-HOFFMEISTER, MPI für Astrophyisk, Garching bei München

80) Peter MARKS, Astronomy Centre, Brighton, U.K.

81) Nicola MASETTI, Dipartimento di Astronomia, Università di Padova, Italy

82) Janet MATTEI, AAVSO, Boston, MA, USA

83) Ronald MENNICKENT, Universidad Catolica, Santiago, Chile

84) Joanna MIKOLAJEWSKA, Copernicus Astronomical Center, Warsaw, Poland

85) Maciej MIKOLAJEWSKI, Inst. of Astronomy, Torun, Poland

86) Joe MONAGHAN, Department of Mathematics, Monash University, Australia

87) Claudio MORENO, Royal Greenwhich Observatory, Tenerife, Spain

88) Joseph MORRIS, Department of Mathematics, Monash University, Australia

89) Martine MOUCHET, DAEC, Observatoire de Meudon, Paris, France

90) Koji MUKAI, Goddard Space Flight Center, Greenbelt, MD, USA

91) Ulisse MUNARI, Osservatorio di Padova-Asiago, Asiago, Italy

92) Urs MUERSET, Institute of Astronomy, Zurich, Switzerland

93) J. MURRAY, Department of Mathematics, Monash University, Australia

94) Andy NORTON, The Open University, U.K.

95) Darragh O'DONOGHUE University of Cape Town, South Africa

96) Hakkı ÖGELMAN, Physics Dept., University of Wisconsin, Madison, WI, USA

97) Gabriela OPRESCU, University of Bucarest, Romania

98) Marina ORIO, Osservatorio Astronomico di Torino, Pino Torinese, Italy

99) Yoji OSAKI, University of Tokyo, Japan

100) M. Türker ÖZKAN, Istanbul University Observatory, Turkey

101) Elena PAVLENKO, Crimean Astrophysical Observatory, Ukraina

102) Daniel PEQUIGNOT, Observatoire de Meudon, France

103) Villpu PIIROLA, Tuorla University, Finland

104) Mike POLITANO, Physics Dept. Arizona State University, Tempe AZ, USA

105) Dina PRIALNIK, Department of Geophysics, Tel Aviv University, Israel)

106) Daniel PROGA, Center for Astrophysics, Cambridge, MA, USA

107) Piero RAFANELLI, Dipartimento di Astronomia, Università di Padova, Italy

108) Enikö REGÖS, Institute of Astronomy, Cambridge, U.K.

109) Klaus REINSCH, Sternwarte, Göttingen, FRG

110) Alvio RENZINI, Dipartimento di Astronomia, Università di Bologna, Italy

111) Alon RETTER, Dept. of Physics and Astronomy, Tel Aviv University, Israel

112) Gerold RICHTER, Sternwarte Sonneberg, FRG

113) Fred RINGWALD, Department of Physics, Keele, U.K.

114) Hans RITTER, MPI für Astrophysik, Garching bei München, FRG

115) Craig ROBINSON, Astronomy Dept., University of Texas at Austin, USA

116) Leonida ROSINO, Dipartimento di Astronomia, Università di Padova, Italy

117) Samar SAFI-HARB, University of Wisconsin, Madison, WI, USA

118) Pedro SAIZAR, University of Pennsylvania, Philadelphia, USA

119) Nikolai SAMUS, Sternberg Astronomical Institute, Moscow, Russia

120) Marek SARNA, Copernicus Astronomical Center, Warsaw, Poland

121) Eric M. SCHLEGEL, Goddard Space Flight Center, Greenbelt, MD, USA

122) Axel SCHWOPE, Zentralinstitut für Astrophysik, Postdam, FRG

123) Pierluigi SELVELLI, CNR, Trieste, Italy

124) K. SEKIGUCHI, SAA Observatory, Cape Town, South Africa

125) Weltraut SEITTER, Astronomy Dept., University of Münster, FRG

126) Allen SHAFTER, San Diego State University, CA, USA

127) Lea SHANLEY, Astron. Dept., University of Wisconsin, Madison WI, USA

128) Michael SHARA, Space Telescope Science Institute, Baltimore MD, USA

129) Giora SHAVIV, Physics Department, Technion, Haifa, Israel

130) Sergej SHUGAROV, Sternberg Astronomical Instutute, Moscow, Russia

131) Andrew SILBER, Centre for Astrophysics, Boston, USA

132) Zladislav ŜIMA, Astronomical Institute, Praga, Chec Republic

133) Edward SION, Physics Dept., Villanova University, Villanova, PA, USA

134) Joe SMAK, Copernicus Astronomical Center, Warsaw, Poland

135) Robert SMITH, Astronomy Centre, Brighton, Sussex, U.K.

136) Giovanni SOSTERO, Ass. Friulana di Astronomia e Meteorologia, Italy

137) Warren SPARKS, Los Alamos National Laboratories, LOS Alamos,NM, USA

138) Lee SPROATS, Mullard Space Science Laboratory, U.K.

139) Sumner STARRFIELD, Arizona State University, Tempe AZ, USA

140) Rudolf STEHLE, MPI für Extr. Physik, Garching bei München, FRG

141) Martin STILL, Astronomy Centre, Brighton, Sussex, U.K.

142) Paula SZKODY, University of Washington, Seattle, USA

143) Enrico G. TANZI, Istituto di Fisica Cosmica del CNR, Milano

144) Peter TAYLOR, The Open University, Milton Keynes, U.K.

145) Christopher TOUT, Institute of Astronomy, Cambridge, U.K.

146) Gaghik TOVMASSIAN, UNAM, Mexico City, Mexico

147) Jim TRURAN, University of Chicago, IL, USA

148) Ytzchak TUCHMAN, Hebrew University, Jerusalem, Israel

149) Alexander TUTUKOV, Sternerg Astronomical Institute, Moscow, Russia

150) Nikolaus VOGT, Universita Catolica de Chile, Chile and MPA, Garching

151) Irina VOLOSHINA, Sternberg Astronomical Institute, Moscow, USA

152) Sonja VRIELMANN, Astronomisches Institut, Universität Muenster, FRG

153) Brian WARNER, Cape Town University, South Africa

154) William WELSH, Department of Physics, Keele University, U.K.

155) Dayal WICKRAMASINGHE, University of Sidney, Australia

156) Robert WILLIAMS, Space Telescope Institute, Baltimore, MD, U.S.A.

157) Janet WOOD, Department of Physics, Keele University, U.K.

158) Graham WYNN, Department of Astronomy, University of Leicester, U.K.

159) Kinwah WU, Department of Astrophysics, University of Sidney, Australia

160) Boris YUDIN, Sternberg Astronomical Institute, Moscow, Russia

161) Lev YUNGELSON, Sternberg Astronomical Institute, Moscow, Russia

162) Luca ZANGRILLI, Dipartimento di Astronomia, Università di Padova, Italy

163) Tomaz ZWITTER, Osservatorio di Padova-Asiago, Asiago (VI), Italy

1. OBSERVATIONAL OVERVIEW

JURISDICTIONAL OVERVIEW

50 YEARS OF CV'S AT THE LOIANO AND ASIAGO OBSERVATORIES

LEONIDA ROSINO
Dipartimento di Astronomia, Universitá di Padova
vicolo Osservatorio 5, 35122, Padova, Italy.

I have been requested by Antonio Bianchini to say a few words at the beginning of this Conference to recall the work on cata clysmic variables carried out for about fifty years, at first at Loiano (Bologna), and after some years, at the Astrophysical Observatory at Asiago.

It is known that under the name of 'cataclysmic variables' are classified close binary stars which display rapid brightness The active component is a collapsed object, generally a white dwarf, and the outbursts are due to the flux from gas ejected from the companion of a later spectral type towards the white dwarf or the accreting disk surrounding it.

Very little was known about CV's when, in 1939, I began to work as assistant astronomer at the Observatory of Bologna. My predecessor in this office, Dr.Luigi Jacchia, before leaving Italy in consequence of the iniquitous racial laws, had started a programme of photographic observations of U Gem variables with the 0.6 m telescope of Loiano in the Apennines. I was asked by the Director of the University Observatory of Bologna to complete the survey and reduce the nearly 900 photographs collected from 1937 to 1939.

At that time the Observatory of Bologna did not possess a photometer. I examined therefore the plates by eye with an enlarging lens, determining visually the magnitudes of the variables by comparing nearby sequences of stars of known magnitude. The limiting photographic (pg) magnitude on the Ferrania plates used for the survey (the only ones then available in Italy) was about 17.5. Ten U Gem variables were included in the programme. Of these stars we derived (Table 1) : pg magnitudes, amplitudes, light curves and average periods (days elapsed between two consecutive maxima). Some peculiarities in the light curves were also pointed out, e.g. the occasional alternation of narrow and wide maxima, the latter brighter than the former, and the occurrence at minimum of irregular brightness fluctuations, a sort of

3

A. Bianchini et al. (eds.), Cataclysmic Variables, 3–8.

flickering,which is well apparent in the U Gem stars with a long permanence at minimum, like EY Cygni.

The paper (Rosino 1942) in which these preliminary results were given, was practically ignored out of Italy. I felt, however, a certain satisfaction to see, after the war, that most of the results obtained at Loiano had been independently confirmed abroad with much better equipments. I returned back to U Gem stars some years later. Seventyfive U Gem variables, moderately concentrated towards the galactic equator, were known in 1945. The fact that all these stars were extremely weak, even at maximum, could indicate that their brightness was mainly depending on their low intrinsic luminosity. A statistical discussion of all the available data based on this assumption, and a comparison with the distribution of other objects of known luminosity led us to conclude that very likely the apparent magnitude of the U Gem variables at minimum was from +6 to + 9 or even more (Rosino 1946).

In 1953 I moved from Bologna to the University of Padua, where in the mountain Station of Asiago, was available a telescope of 1.22 m , at that time the largest in Europe, and well equipped with spectrographs, photometers and other auxiliary instruments. I was busy at that time with other research (globular clusters, nebular variables, flare stars, etc). The discovery by Merle Walker (1954) of the binary nature of the old nova DQ Her, and the fact that also some U Gem stars were suspected to be close binaries, called again our attention to the CV's. At Asiago we observed, with a Lallemand photoelectric photometer, SS Cyg (Pinto and Rosino, 1959). We did not find eclipses, but what was remarkable was the continuous flickering of this star at minimum, with irregular fluctuations of 0.2–0.3 magnitudes, like those observed in EY Cygni.

A series of investigations carried out by Kraft, Krzeminski, Mumford, et al., from 1963 to 1969, left no doubt that all CV's (U Gem stars, recurrent and classical novae, symbiotic objects) were close binaries in which a white dwarf was accompanied by a late type star filling its Roche lobe and that flickering and outbursts were an effect of the stream of gas rich of hydrogen falling through the through the accretion disk on the white dwarf. Continuing at Asiago the programme of researches on the CV's , we decided to pass from the U Gem variables to the recurrent novae, considering the strict affinity between these two classes of stars. We were rather lucky. On 1958 July 14 the recurrent nova RS Oph, which had already shown maxima in 1898 and 1933 rose suddenly from visual magnitude 11.5 to 4.8 . The observations of this object began at Asiago on July 15 and were regularly continued during its decline until October when the nova reached its minimum. And as it was previously observed in T CrB, another famous recurrent nova, in RS Oph, during its nebular phase, were present strong coronal lines of [FeX], [AX] and even [FeXIV]. We had the occasion of observing at Asiago two other maxima of this star, in 1967 and 1985 . The light curves and the circumstances of the spectral evolution were strictly the same as observed in the previous outbursts (Rosino 1985): very rapid decline and high expansion velocity (4000 km/s and even more) in the first days after maximum, strong HeI and HeII emissions, forbidden nebular and coronal lines, with [FeX] even stronger than H. The suspected presence of a red giant (M2-III) in symbiosis with the collapsed blue star, suspected since 1958, was confirmed by a series of observations at minimum, carried out at Asiago after 1959.

Three other recurrent novae, T Pyx, U Sco and WZ Sge were observed at Asiago in the following years. Nova T Pyx, which had shown outbursts in 1890, 1902, 1920 and 1944, brightened again on January 1967, passing from magnitude 14 to 6.9. In spite of its difficult position (declination :– 32°) Chincarini obtained from January to April with the 1.22 m telescope of Asiago an excellent series of spectra. The decline from maximum was relatively slow and the degree of ionization somewhat lower than in RS Oph, although the forbidden lines of [OIII] and [FeX] emerged during the nebular phase. U Sco was somewhat peculiar for its relatively high amplitude ≈ 10.4. Outbursts occurred in 1863, 1906, 1936 and 1979 and in 1987. The spectrum, although of relatively high excitation with emission lines of HeI, HeII, NIII, NIV, NV did never display forbidden nebular lines. WZ Sge, was observed at Asiago in the course of its third outburst, in 1978 (the previous outbursts occurred in 1913 and 1946) but the absence of nebular lines led us to conclude that the star was rather a peculiar U Gem variable than a recurrent nova.

In the last thirty years the work carried out at Asiago on CV's was concentrated on the classical novae. Eighteen novae were observed from 1960 to 1994 (Table 1). The observations were performed at Asiago soon after the announcement of the discovery and continued as long as possible, at first with the 1.22 m telescope and then, after 1975, with the 1.82 m telescope of Ekar. B and V magnitudes were also derived mostly from photographs taken with the Schmidt telescope. In conclusion, for each nova wavelengths and identifications of the lines, their profiles, and radial velocities, were given and were determined the intensities relative to H, at first in a semiempirical scale, but later, after 1985, as soon as Reticons and CCD devices became available at Asiago, with absolute measures of the fluxes. Some spectral observations were extended to the near infrared, to about one micron. In general light curves were also obtained from the Schmidt material. The study of 18 galactic novae , observed with the same instruments and methods, was sufficient to show that, although all of them display common characteristics, each nova has its own personality. The

TABLE 1 . U Gem variables, recurrent and classical novae (Loiano - Asiago)

U Gem variables			Recurrent Novae		
Loiano (1937-40)		P(days)	Asiago (1953-1985)		
EY Cyg	(11.4) 15.5pg	240	RS Oph	1958	4.3- 12.5 V
AB Dra	11.0 15.3	13	(1898,1933)	1967	Nebular and
AY Lyr	12.8 17.0	24		1985	coronal lines
CY Lyr	13.7 16.5	12-17			(Rosino&Iijima 1986)
CZ Ori	12.4 16.5	25-4	T Pyx	1967	7.0- 15.8 B
RU Peg	11.0 13.0	70	(1890,1902,		Nebular and
TZ Per	12.0 15.6	17-21	1920,1944)		weak coronal
SU UMa	12.5 15.1	13-20			lines
TW Vir	11.8 15.8	25			(Chincarini&Rosino 1968)

WZ Sge   1978   7.0- 15.5 B
(1913,1946)   U Gem ?
                    (Ortolani et al. 1988)

U Sco   1987   8.7-19.3 V
(1863,1906)   no nebular lines
(1935,1979)   (Rosino&Iijima 1988)

Classical Novae

V444 Her (1960)	2.8 -18 P	Chincarini,Rosino :1963, Asiago Co. 139
V533 Her (1963)	3 -16 P	Chincarini, Rosino: 1964, Ann.d'Astroph. 27,449
LV Vul (1968-1)	5.1 -16.9 B	Mammano,Rosino: 1969,Wien Annalen, 29,2
LU Vul (1968-2)	9.2 -(21 P	Rosino et al.:1969, ApSS ,4,392,1969
V1229 Aql (1970)	6.7- 18 B	Rosino, Ciatti, 1974, A&AS 16,305
V1330 Cyg (1970)	7.5- 18.1 B	
V368 Sct (1970)	7.7- 19.3 P	
IV Cep (1971)	7.0- 19.3 B	Rosino: 1975, Var.stars and stellar evol.,pp347
		Sherwood & Plaut,pp.347.
V 400 Per (1974)	8 - 19.5 B	Rosino: 1978, ApSS 55,383
V373 Sct (1975)	7.1- 18.5 B	
R Del (1967)	3.9- 12.4 V	Rafanelli,Rosino: 1978,A&AS,31,337
FH Ser (1970)	4.4- 16 V	Rosino, Ciatti, Della Valle: 1986, A&A,158,34
V1500 Cyg (1975)	1.7-(20.5 V	Rosino, Tempesti: 1977,Soviet Astr.21(3)
N Aql 1982	6.5- 20 V	Rosino,Iijima, Ortolani: 1983, MNRAS 205,1069.
N Sgr 1982	8 - 18.5 B	Iijima, Rosino :1983,PASP 95,506
V443 Sct (1989)	7.5- (19 V	Rosino et al.: 1991, AJ,101,1807
QU Vul (1984-2)	5.5- 19 V	Rosino et al., 1992, A&A,257,603
V1974 Cyg (1992)	4.4- 18 V	Rafanelli,Rosino,Radovic: 1995,A&A 294,488

----

novae listed in Table 1, displayed in fact different light curves, spectra, expansion velocities, composition and physical properties In the Archives of the Observatory at Asiago are collected hundreds of spectra of the novae listed in Table 1. I think that a re-examination of these spectra with modern instruments (PDS) would really be of some interest.

A fundamental point in the study of the CV's, and in particular novae, is the knowledge of their absolute magnitudes. About fifty years ago, these magnitudes were still rather uncertain, because of the difficulty of obtaining reliable parallaxes and acceptable values of the interstellar absorption in different regions of the sky. It was known, however, that the intrinsic luminosity of the novae at maximum was about as high as that of the brightest stars of our Galaxy. The best way to overcome these difficulties was therefore that of determining magnitudes and characteristics of the novae appearing in nearby galaxies, like M 31 or M33, where, at less than one constant, apparent and absolute magnitudes are the same. So, following the pioneer work oh Hubble (1929) and about at at the same time of Arp (1956), I began a search of novae in M31, at first at Loiano with the 60 cm telescope, but without success, and later in 1953, at Asiago with the 1.22 m telescope, and this time our expectations were successful. We decided therefore to continue the survey. In 1964 I published a list of 46 novae discovered in M31, with finding charts, positions , magnitudes and lightcurves (Rosino 1964); a second list on other 44 novae (Rosino 1973) and a third of 52 novae (Rosino et al.1989) concluded the survey. The novae discovered in M31, by examining at the blink-comparator 922 plates distributed over 32 years, were 142, 40 of which were discovered on plates obtained with the Ekar 1.82 m telescope. The Asiago survey showed with great evidence the large variety of the light curves of novae. In M31, as in our Galaxy, there were fast and slow novae, some with linear decline, others with a double maximum , and possibly also two recurrent novae. The nova rate per year in M 31 was estimated of 30 ± 4 . From a thorough discussion of the Asiago material in conjunction with other published results it was found that the absolute magnitudes at maximum (MM) of the novae were distributed, with a few exceptions, between -9 and -6.5. Moreover, as it was previously suggested by Arp (1956), there was a very strict relation with their rate of decline (RD), the brightest novae being in general those with the highest RD (MMRD relation, Capaccioli et al.,1989). Assuming that the same MMRD relation is followed also by the novae in other galaxies , the novae become a powerful distance indicator, even more effective than the Cepheids.

Until 1988 the only existing estimates of the nova rate in M33 were based on the works of Hubble and Sandage (1954) and Arp (1956), who agreed on a rate of 1 nova per year. However, in the 60's, a nova survey of M33 was started at Asiago Observatory. Although sparser than the contemporary survey of M31 there were significant results. Analysis of 41 plates obtained with the 1.22m telescope allowed Rosino and Bianchini (1973) to conclude that the nova rate in M33 was probably not '...*so low as previously assumed* '. This first result was afterwards confirmed by Della Valle et al. (1994) who estimated ≈ 5 novae per year through the analysis of 72 plates taken with the Ekar 1.82m telescope.

Finally, I should like to spent a few words on the symbiotic variables, which were also included in the past years in the Asiago programmes. But the time is getting short. I wish only to remember the pioneer study of V 1016 Cyg by Ciatti and Mammano and the studies by the same persons on the enigmatic star SS 433, which led to an acceptable model of this unique star. That of the symbiotic variables is still an open field in which are operating, with the modern techniques, other colleagues of Asiago, who will speak later on their researches. So I like to close at this point my relation, with the best wishes, to my youngest colleagues who are continuing, much better than me, the study of the Cataclysmic variables, and to all the participants in this Conference.

References

Arp,H.,1956,AJ, 61,15
Capaccioli, M., Della Valle, M., D'Onofrio, M., Rosino,L.,1989,AJ, 97,1622
Chincarini, G., Rosino, L. 1968, IAU Coll., Budapest
Della Valle, M., Rosino, L., Bianchini, A., Livio, M. 1994, A&A, 287, 403
Hubble, E. 1929,ApJ, 69,103
Hubble, E., Sandage, A. 1954,Ann. Rep. Mount Wilson and Palomar Obs.,53,23
Ortolani et al.:1988, A&A, 87, 31
Pinto, G., Rosino, L., 1959, Mem. Acc.Patavina, 57,1
Rosino,L., 1942, Mem.SAIt., 14-2
Rosino,L., 1946, Mem.SAIt., 18-2
Rosino,L., 1964, Ann.d'Astrophys.,27,498
Rosino,L., 1973, A&As,9,347
Rosino, L., Bianchini, A. 1973, A&A, 22, 461
Rosino,L.,1986, Proc.Manchester Co. Ed.Bode,VNU Science Press, 1
Rosino, L., Iijima, T. 1986, Proc.Manchester Co. Ed.Bode,VNU Science Press
Rosino, L., Iijima, T. 1988, A&A, 201,89
Rosino,L., Capaccioli,M., D'Onofrio,M., Della Valle,M.,1989,AJ ,97,83

1a. CLASSICAL NOVAE

AN UPDATE ON THE X-RAY OBSERVATIONS OF CLASSICAL NOVAE

HAKKI ÖGELMAN AND MARINA ORIO

Department of Physics, University of Wisconsin Madison, Wisconsin 53706, U.S.A.

AND

Osservatorio Astronomico di Torino I-10025 Pino Torinese (TO), Italy

1. Introduction

Sitting down at a sidewalk cafe at Abano, basking in the June sun and going through the transparencies, composing your talk as you sip your favorite drink is a very pleasant experience. Some months later, when you are reminded by the editors that you are late in writing up your talk that stress sets in and you can not recover that pleasant Abano experience. Here is an attempt to convince ourselves that writing can be fun too. This contribution was largely written at a Madison fast-food joint with a winter background of drifting snow and sub zero (C) temperatures outside, looking at the transparencies and freely associating.

The aim of this contribution is to present an update on the X-ray observations of classical novae since the *EINSTEIN - EXOSAT* era of late 70's to mid 80's, i.e. the *ROSAT* results. In order to achieve this we shall first motivate the reader by showing the several ways novae can emit X-ray during its outburst and post outburst stage; i.e. the interval between the outburst and the stage at which it appears to have returned to its pre-outburst, normal accreting Cataclysmic Variable (CV) state. Then we shall show how limited the observations prior to *ROSAT* were, and how much more information is available now. New observations answer some of the old questions and raise more new ones, as is the case here. We shall try to outline these advances and the new challenges.

11

A. Bianchini et al. (eds.), Cataclysmic Variables, 11–19.

2. Why X-rays from novae?

Novae have two main components where we can look for X-ray emission. One is the circumnova environment where some $10^{-5} - 10^{-4} M_\odot$ of material are ejected from the white dwarf at velocities $\sim 10^8$ cm s^{-1}; the other is the central engine, a catacylismic variable (CV) with the white dwarf at the seat of the explosion.

In the **ejecta** we can imagine a number of shock fronts forming either where it confronts the circum-nova environment or internal shocks in the post-nova wind interacting with the first ejecta (e.g. O'Brien et al, 1994). Independently of the details we expect strong shocks with associated temperatures of keV's and hence X-ray emission. One can also estimate the volume and density of the emission region from the expansion and ambient medium parameters finally arriving at a luminosity of 10^{33} to 10^{35} erg s^{-1}. The cooling time associated with this plasma would be some decades.

The CV, on the other hand can emit X-rays in a variety of ways. The **white dwarf** itself is very hot after the explosion and any part of the unejected envelope will return back on to this hot surface and provide the hydrogen rich fuel for further thermonuclear burning. One can intuitively see that the luminosity will be close to the Eddington luminosity of the white dwarf ($\sim 10^{38}$ erg s^{-1}) since anything less will allow more mass to fall on to the white dwarf and increase its luminosity, anything more will stop the envelope from accreting. Hence, we expect the $\sim 10^{38}$ luminosity to be radiated from a white dwarf photosphere of $\sim 10^9$ cm radius with an effective temperature of 20 to 50 eV. How long can this envelope burning go on is one of the puzzling aspects revealed by the X-ray data. The expected 10 to 100 year timescale ($\sim M_{env} \times 3 \times 10^{18}$erg/g/$L_{Edd.}$) seem to conflict with the experimentally observed timescales which appear to be a factor of ten less. The recent introduction of the *OPAL* opacities (Kato & Hachisu 1994) partially explains the puzzle, but it is likely that also other factors are at work in getting rid of the remnant envelope (Orio *et al.* 1992; Livio et al. 1990).

Finally, if accretion is re-established, we would again expect X-ray emission from the CV powered by the gravitational energy release of the accreted material onto the white dwarf. Depending on the accretion rate, the magnetic field of the white dwarf one would get the typical X-ray emission of a CV: hard X-rays with luminosity of 10^{32} to 10^{34} erg s^{-1} from the boundary layer (corresponding to $\dot{M} \sim 10^{-11} - 10^{-9}\ M_\odot$ yr^{-1}) and maybe a soft component of comparable luminosity as found in polars.

There is another important point we should consider when we are discussing re-established accretion from the companion. If, due to some mechanism like radiation induced (self-excited) mass transfer is established in the early stages of the post-nova while the white dwarf is still hot, then the accreted hydrogen rich mass will nuclear burn and release ~ 30 times more energy per gram in comparison to the gravitational energy release. These sources will also look like the envelope consuming post nova hot white dwarfs or the supersoft sources recently brought to our attention by the *ROSAT* discoveries.

3. Pre-*ROSAT* data

Despite the expectations listed above, early hard X-ray observations of classical novae during outburst yielded only upper limits, indicating that these objects were not copious emitters in the 2 to 6 keV range (Hoffman et al. 1976; Cruise 1977). In the soft X-ray region (0.1-4.0 keV), sufficient sensitivity was available after *EINSTEIN* and *EXOSAT* . Since most of the currently discovered galactic novae are at distances of few kiloparsecs and in the galactic plane, absorbtion by the cold interstellar gas further complicated the soft X-ray detection problem.

The only classical nova observed by *EINSTEIN* close to the outburst was V1500 Cyg. In spite of an earlier claim of non-detection (Hutchings 1970), Kałużny and Chlebowsky (1988) have re-analyzed the data and shown that V1500 Cyg was seen at a 3σ level of 0.01 cts s^{-1} , 1390 days after its optical maximum. The first discovery of soft X-rays from classical novae during the outburst stage was accomplished with the *EXOSAT* satellite where Nova GQ Mus was detected at about 460 days after outburst (Ögelman, Beuermann and Krautter 1984). With further observations of the three brightest novae whose outburst stage coincided with the lifetime of the *EXOSAT* satellite, namely Nova GQ Mus , Nova PW Vul and Nova QU Vul the soft X-ray light curve of classical novae were sampled from optical maximum to ~ 900 days after (Ögelman, Krautter and Beuermann 1987). In Figure 1, the first 4 points of GQ Mus, 3 points of PW Vul and 2 points of QU Vul are all *EXOSAT* data.

Since the *EXOSAT* low-energy telescope did not have any energy resolution except for some filters which were difficult to use for low count rate sources, we basically had no spectral evidence about the origin of the x-rays. At the time, the last data point for Nova GQ Mus was interpreted as a possible signal of the permanent turn-off of the X-ray flux from a hot white dwarf which had just consumed or ejected its envelope. There was however, evidence from optical spectra of Nova GQ Mus suggesting that the central source has continued to increase in temperature (Krautter and Williams 1989), thus contradicting the turnoff interpretation of the X-ray data; this was later clearly demonstrated to be the case by *ROSAT* .

4. *ROSAT* data

ROSAT has given us information about the X-ray emission from novae in outburst and post-outburst stage with unprecedented detail. Few times larger area, better angular and energy resolution, longer exposures and above all, a systematic program of monitoring a number of novae through their outburst stage have been the key factors in making *ROSAT* better than its predecessors. Despite these key improvements we should make it clear that the proportional counter type soft X-ray telescopes have poor energy resolution compared to what optical astronomers are used to. The energy range of *ROSAT* is 0.1 to 2.4 keV and the resolution $\Delta E/E$ is of the order of 50% with the Position Sensitive Proportional Counter (PSPC), implying that we have about 5 independent energy samplings over the sensitive range. The Wien peak of the radiation in the temperature range of interest of a hot white dwarf at Eddington luminosity is around 60 to 150 eV which lies at the lower sensitivity range of the detector. Then there is the additional complication of interstellar absorption which has an energy dependent photoelectric cross section that goes like $\sigma \sim \sigma_o E^{-3}$ and gives rise to an absorption $exp(-\sigma_o N_H E^{-3})$ where N_H is the equivalent column density of hydrogen in the line of sight. The consequences of the above considerations are that for column densities $N_H \gtrsim 3 \times 10^{20}$ cm^{-2} the Wien peak gets absorbed and the temperature and the flux has to be determined from the exponential tail of the black body spectrum. In these circumstances the spectral parameters become very sensitive to the value of N_H, which has to be fitted along with the other parameters; the whole process leads to a poorly restricted parameter space.

Up to the date of this review (summer 1994), the accomplishments of *ROSAT* can be summarized under several subsections including the *ROSAT* All-Sky-Survey (RASS) results and some pointed observation of recent novae in outburst. We proceed along these lines.

4.1. *ROSAT* ALL-SKY-SURVEY (RASS) RESULTS

The first mission of *ROSAT* after launch was to survey the whole sky in a scanning mode where, depending on the ecliptic lattitude, every part of the sky would get an exposure of few hundred seconds or more. During this survey, some 283 positions of known nova or nova-like object positions were examined for X-ray emission. The result was a meager detection of 7 classical novae where 6 were CV like emission from old novae with established accretion and only one, GQ Mus (Nova Muscae 1983) was a recent nova (Orio *et al.* 1993). In fact out of the 26 novae which had their outburst in the last 10 years before the *ROSAT* all-sky-survey, GQ Mus was the only one detected.

Although the sensitivity of the RASS is low (~ 300 s of exposure) compared to the pointed observations ($\sim 10,000$ s of exposure), it is clear from the above results that most novae are not bright in X-rays at approximately Eddington luminosity after 10 years or less; Nova GQ Mus appears to be an exception.

4.2. GQ MUS: SUPERSOFT AND BRIGHT A DECADE AFTER OUTBURST

GQ Mus (Nova Muscae 1983) was the first classical novae to be detected close to the outburst (Ögelman et al. 1984, 1987) by EXOSAT . The data were limited by the small effective area and very poor energY resolution of the detector and could not distinguish between a hot thermonuclear burning white dwarf remnant emitting $3.0\text{-}3.5 \times 10^5$K black body radiation at 10^{37} to 10^{38} erg s^{-1} , or shocked circumstellar material emitting 10^7K thermal bremsstrahlung radiation with 10^{35} erg s^{-1} luminosity. GQ Mus was observed with the ROSAT Position Sensitive Proportional Counter (PSPC) during 25-26 February, 1992 as a very soft black-body-like source with effective temperature $\sim 3.5 \times 10^5$K and a luminosity near Eddington luminosity $\sim 10^{38}$ erg s^{-1}) from a $1 M_\odot$ white dwarf (Ögelman et al. nature). This was like the long anticipated constant luminosity envelope burning phase (refs). As mentioned before, GQ Mus appears to be an exceptional nova since it was the only such nova detected in the All-Sky-Survey (Orio) forcing the conclusion that steady nuclear burning of the remaining hydrogen rich envelope is indeed a rare phenomenon that occurs in less than few percent of the classical novae. Ögelman et al. (1993) have suggested that the source of the hydrogen rich fuel of GQ Mus may be fresh material, currently being accreted from the companion at a rate of $\sim 10^{-7}$ $M_\odot$ y^{-1} and that the short orbital period of 85.5 min (12) maybe the key difference in developing this irradiation induced mass transfer.

Finally, GQ Mus surprised us again by turning off shortly after the first ROSAT measurements; by Jan 1993, the X-ray count rate declined had by a factor 17. In September 1993, the soft X-ray flux was below the ROSAT threshold limit, implying a decrease of a factor ≥ 30 in the count rate (Shanley et al. 1995; also see this volume). We still do not know if the envelope was consumed or the radiation induced mass transfer stopped.

4.3. NOVA V1974 CYGNI (1992)

There is a contribution in this volume by Krautter et al. on the exciting story of Nova V1974 Cygni. The X-ray light curve of this nova is by far the best studied. Furthermore, the nova lived up to its observations and displayed a spectacular Super-Soft-Spectrum with count rates that almost reached 10^2 c/s with the ROSAT PSPC. The rise and fall of the soft emission within ∼2 years of the outburst, appears to be a "text-book" example of the constant bolometric luminosity phase of envelope burning (Krautter *et al.* 1995). The orbital period $P_{orb} \simeq 1.95$ hr (De Young and Schmidt 1994) is also on the short side like GQ Mus; does short orbital period imply Super-Soft-Spectrum ?

4.4. OTHER NOVAE THAT SHOWED ONLY HARD FLUX

Two other novae have been detected by *ROSAT* during their outburst stage, namely Nova Herculis 1991 and Nova Pup 1991. Nova Herculis 1991 was observed 5 days, a year and 19 months after the outburst (Lloyd *et al.* 1991, Szkody & Hoard 1994); it showed X-ray flux after 5 days and turned into a very faint source after a year ($L_x \lesssim 10^{31}$ erg/s). Nova Pup 1991 was observed by us 16 months after the outburst and it appeared in the hard *ROSAT* energy band with a luminosity $L_x \simeq 10^{34}$ ergs/s assuming a distance of 5 kpc that can be inferred from the maximum magnitude vs. rate of decline relationship (Orio *et al.* , 1995, in preparation).

In pointed observations and archival data available up to now we found that *ROSAT* was also pointed at a few more novae in the Galaxy (including V1500 Cyg 18 years after the outburst) and 10 novae in the LMC. No flux was detected from these sources, thus yielding only upper limits.

5. Putting everything together

We finally attempted to construct one figure, Figure 1, where all the data mentioned above could be summarized. The hope was that if one stared long enough to this figure one would be enlightened on the X-ray ways of novae. We chose the time axis to be logaritmic in order to show the data in the range of days to 10's of years after the outburst. The luminosity axis was chosen as what *ROSAT* would count if the source was at a distance of 1 kpc and the interstellar column density N_H was 10^{21} cm^{-2}. These corrections could only be done very approximately since neither the distance nor the N_H is accurately known for any source. In the case of sources with hard spectra (novae Her 91 Pup 91 and CP Pup) the N_H correction was not that critical. However, in the case of sources with known supersoft spectra like GQ Mus (Nova Musca 83) and V1974 Cyg (Nova Cyg 92) the N_H correction is very inaccurate and can easily be off by a factor of 10. The normalization for the novae observed with *EXOSAT* (GQ Mus, PW Vul and QU Vul) was done assuming that they had a supersoft spectra. The fact that V2974 Cyg was about a factor 30 higher than GQ Mus at the peak of the light curve despite their expected 10^{38} erg s^{-1} Eddington luminosity is due to the inability of blackbody models to give appropriate fits to the detailed data of a source like V2974 Cyg. The unlabeled down-pointing arrows are the upper limits for a number of LMC novae that were in the field of view of *ROSAT* observations and were derived under the assumption of a hard spectrum. These points on the graph together with the other hard spectrum points in the normalized *ROSAT* count rates of 0.1 to 3 counts/s can be considered as hard spectra CV like luminosities around 10^{33} to 3×10^{34} erg s^{-1} .

Below are some conclusions we can draw from this figure 2 and questions we may raise ara listed below:

- X-ray light curves of different novae show a large variety. Are the main differences determined by the white dwarf, the companion, the orbit or the circum-nova environment?
- On the time scale of *days* to *months* after outburst we have seen hard X-rays from novae Her 91, V1974 Cyg and probably from QU Vul and PW Vul (as estimated from *EXOSAT* data). The emission is weak and at a luminosity level of 10^{33} to 3×10^{34} erg s^{-1} . The most likely explanation of this is some shock heating due to the interaction of the ejected envelope or the subsequent wind with the environment. There is *no nova for which the data conflict with the above luminosity and hardness estimate.*
- On the time scale of $\frac{1}{2}$ *year* to *several years* we have detected for sure two Super-Soft-Sources (SSS): GQ Mus and V1974 Cyg. In the case GQ Mus, the supersoft phase lasts almost a decade, and for V1974 Cyg $\sim$ 2 years. It is interesting to note that both of these novae have short orbital periods (85.5 m for GQ Mus and and 117 min for V1974 Cygni). We also know from the survey data that out of the 26 nova which had their outburst in the last 10 years before the *ROSAT* all-sky-survey, GQ Mus was the only one detected as a SSS source (Orio *et al.* 1993). One more nova that showed no supersoft emission at the appropriate time was Nova Pup 91 which had only hard emission 1 yr after outburst (Orio *et al.* 1995, in preparation).
- On the time scale of *year* to *decades* after outburst we still see hard sources ($L_x \sim 10^{33} - 10^{34}$ erg s^{-1}) like Pup 91, QU Vul, CP Pup. This stage probably represents the re-establishment of accretion and the associated luminosities that are determined by the gravitational energy release.

We hope that by the time someone writes another update on the X-ray emission of novae, we would have solved some of these mysteries listed above and in the process, added other new and challenging puzzles.

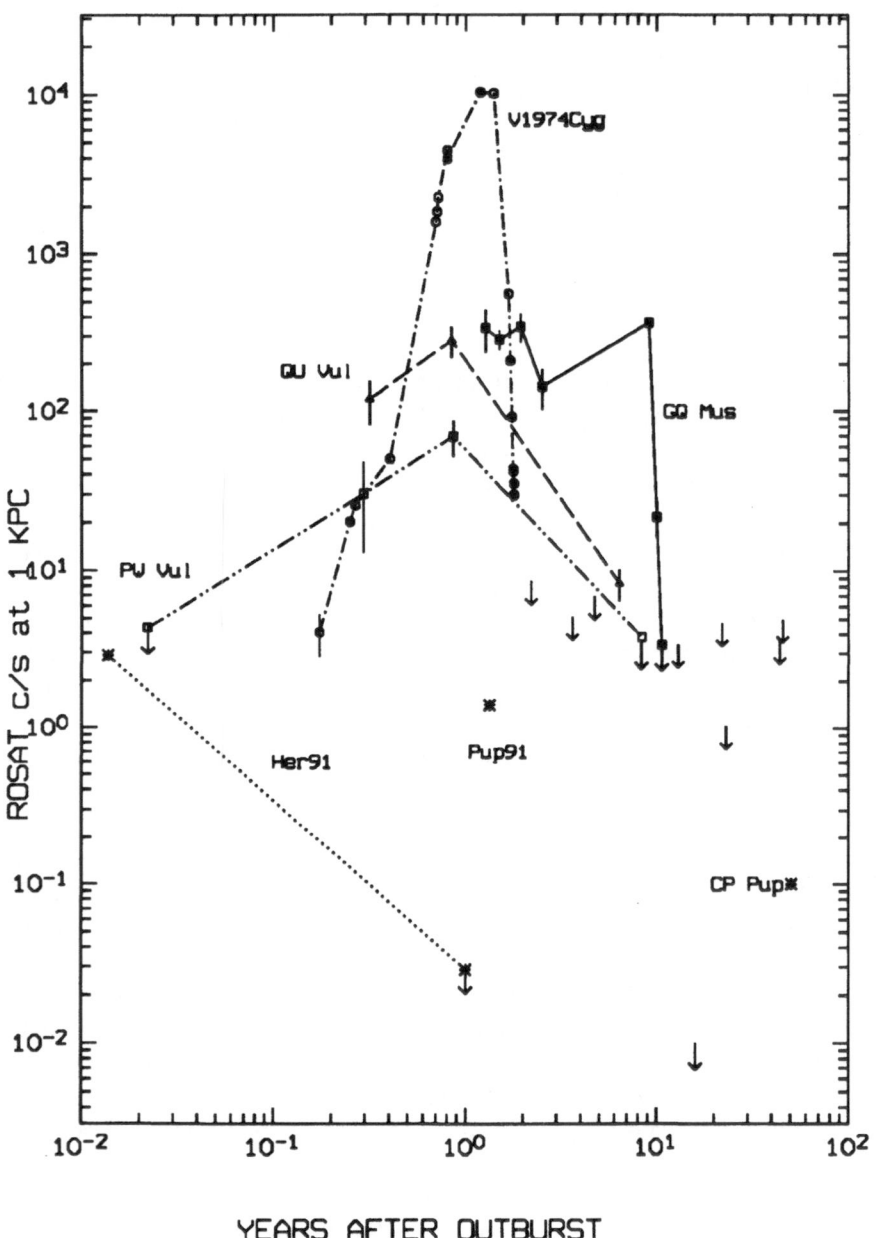

Figure 1. The X-ray light curves of various novae as a function of time elapsed since the outburst. The luminosity has been expressed as *ROSAT* count rate at a distance of 1 kpc with an intervening column density of $N_H = 10^{21}$ cm^{-2}. See the text for details.

References

Cruise, A.M., 1977, *Nature*, **267**, 685

De Young, J. A., Schmidt, R.E., 1993, IAU Circ. 5814

Hoffman, J.A., et al., 1976, *Nature*, **261**, 208

Hutchings, J.B., 1980, it PASP, **92**, 458

Kałużny, J., Chlebowsky, T., 1988, *Acta Astr.*, **38**, 329

Kato, M., Hachisu, I., 1994, *Ap J*, **347**, 802

Krautter,J., Williams, R. E., 1988, *Ap J*, **341**, 968

Krautter, J., Ögelman, H., Wachmann, R., Starrfield, S., Trümper, J., 1995, *preprint*

Livio, M., Shankar, A., Burkert, A., Truran, J.W., 1990, *Ap J*, **356**, 250

Lloyd, H.M., it et al., 1992, *Nature*, **356**, 222

Ögelman, H., Beuermann, K., Krautter, J., 1984, *Ap J*, **287**, L31

Ögelman, H., Krautter, J., Beuermann, K., 1987, *A & A*, **177**, 110

Ögelman, H., Orio. M., Krautter, J. Starrfield, S., 1993, *Nature* **361**, 331

O' Brien, T.J., Lloyd, H.M., Bode, M.F., 1994, *MNRAS*, **271**, 155

Orio, M., et al., 1992, *Adv. Space Res.*, Vol. 13, No. 12, 351

Orio, M., Trussoni, E., Ögelman, H., 1993, *A & A*, **257**, 548

Shanley, L., Ögelman, H. Gallagher, J., Orio, M., Krautter, J., 1995, *Ap J*, **438**, L95

Szkody, P., Hoard, D.W., 1994, *Ap J*, **429**, 857

THE EVOLUTION OF CLASSICAL NOVAE IN THE ULTRAVIOLET

R. GONZALEZ-RIESTRA
IUE Observatory - VILSPA
P.O. Box 50727, 28080 Madrid, Spain

1. Introduction

During the more than sixteen years since its launch the IUE satellite has observed the outbursts of a large number of Classical Novae. An examination of the large database built up during these years shows that the number of spectra available for each of the objects is highly variable, depending on the brightness, the rate of decline, spacecraft constraints, etc. Attending to the number and time distribution of the existing spectra, the CNe observed with IUE can be grouped into four categories:

- Objects which have been well followed during and after outburst. Belonging to this group are: V1668 Cyg 1978 (the first nova observed with IUE, Stickland et al.1981), GQ Mus 1983 (the nova which has been observed with IUE for the longest period –11 years–, Krautter et al. 1984), PW Vul 1984 (Andreä et al. 1991), QU Vul 1984 (Saizar et al. 1992), OS And 1986 (Starrfield et al. 1988) and V838 Her 1991 (Starrfield et al. 1992). Three extragalactic novae, LMC 1988 No. 1 (Sonneborn et al. 1988), LMC 1988 No.2 (González-Riestra and Cassatella 1988) and LMC 1991 (González-Riestra et al. 1991) have been observed during the first year after the outburst.
- Other objects were well observed during the first stages of the outburst (weeks or months), but there is not any observation at later phases. In this category are V694 CrA 1981 (Williams et al. 1985), V1370 Aql 1982 (Snijders et al. 1987), LMC 1990 No. 1 (Shore et al. 1990) and LMC 1992 (Shore et al. 1992)
- For some objects there are very few observations, either during the outburst and/or at later stages (e.g. V842 Cen 1986 or QV Vul 1987).

21

A. Bianchini et al. (eds.), Cataclysmic Variables, 21-28.

  – At the time of writing (June 1994) three objects are being regularly
    observed with IUE: V351 Pup 1991, V1974 Cyg 1992 (Shore et al.
    1994a) and Nova Cas 1993 (Shore et al. 1994b).

A preliminary analysis of the UV evolution of classical novae has already
been presented by Cassatella and González-Riestra (1990). This review will
describe the behaviour of the objects which have been observed since the
first phases of the outburst until well into the nebular phase. Some characteristics of the systems included in this work are given in Table 1.

2. General characteristics of the UV evolution of a Classical Nova

The general trends of the development of the outburst of a classical nova
in the UV are schematically summarized in the next sections, although it
should be taken into account that each object shows features which make
it unique. A sketch of the UV evolution of a nova during the first two
years after the outburst is shown in Figure 1. The light curve in the figure
corresponds to the short-wavelength (1230–1950 Å) integrated flux of V1974
Cyg (Shore et al. 1994a). Each of the spectra shown belongs to a different
object, as specified, in a different evolutive stage which corresponds to the
part of the light curve marked. Not all the phases have necessarily been
observed in all the objects.

2.1. PHOTOMETRIC EVOLUTION

The objects in the sample show UV light curves that, although similar in
many respects, show significant differences in aspects like the rates of raise
and/or decline, duration of the maximum, presence of secondary outbursts
or dips, etc (UV light curves of some of the objects in the sample can be
found in Cassatella & González-Riestra 1990).

Thanks to the flexibility of the scheduling of IUE, the first UV spectrum
of a nova is generally obtained in coincidence, or very close, to optical
maximum. This first spectrum usually shows a very low flux, specially in
the short–wavelength range. Only in one case was a nova caught at an
earlier stage, the so-called "fireball" phase: V1974 Cyg, whose first IUE
spectrum was obtained less than one day after the announcement of the
discovery and well before the optical maximum (see Shore et al. 1994a).
In this case the UV flux dropped in a few hours by more than a factor
of 10 due recombination of the iron peak elements and the corresponding
opacity increase. During the first weeks after the outburst the UV flux
gradually increases. The general trend is that the maximum is reached
later at the shortest wavelengths, i.e. the maximum of the flux distribution

TABLE 1. The sample of objects

Nova	t_3 (opt)	t_{10} (SWP) days	delay SWP-opt	E(B-V)	distance kpc	Max UV[†] Lum 10^4 $L_\odot$
			Galactic Novae			
V1668 Cyg	30 [1]	120	27	0.40 [1]	3.6 [1]	2.8
V694 CrA	12 [2]	50*	≤11**	≤0.15 [3]	9.0 [2]	≤11
V1370 Aql	10 [4]	20*	≤26**	0.60 [4]	5.0 [4]	1.9
GQ Mus	40 [5]	550*	≤35**	0.45 [5]	4.8 [5]	0.87
PW Vul	147 [6]	230	66	0.55 [6]	1.3 [6]	0.67
QU Vul	31 [7]	•••	•••	0.61 [8]	3.5 [8]	0.59
OS And	20 [9]	75	19	0.26 [9]	4.3 [9]	2.1
V838 Her	3 [10]	15	5	0.53 [10]	3.4 [10]	2.9
V351 Pup	45 [11]	290	30	0.70 [12]	3.1 [12]	3.2
V1974 Cyg	42 [13]	240	22	0.20 [14]	3.0 [14]	3.0
			LMC Novae			
1988 No. 1	33 [15]	150	35	0.05		3.8
1988 No. 2	10 [16]	160	14	Galactic		5.4
1990 No. 1	6: [11]	34*	≤3*	+	50.1 [18]	5.8
1991	6 [17]	50:	3	0.10		15
1992	25 [11]	>70	22	LMC		3.9

[†]: Integrated luminosity in the range 1230–3200 Å
*: From observed maximum
**: Maximum in first observation
***: Maximum not observed
References:
1.- Stickland et al. 1981   7.- Rosino et al. 1992   13.- Chochol et al 1993
2.- Caldwell 1982   8.- Saizar et al. 1982   14.- Shore et al. 1994a
3.- Williams et al. 1985   9.- Kikuchi et al. 1988   15.- Cappacioli et al. 1990
4.- Rosino et al. 1983   10.- Lynch et al. 1992   16.- Sekiguchi et al. 1989
5.- Krautter et al. 1984   11.- From IAU Circulars   17.- Della Valle 1991
6.- Andreã et al. 1991   12.- This work   18.- Panagia et al. 1991

moves toward shorter wavelengths. Once reached the maximum, that can last up to several weeks, the flux starts to decrease.

The involved timescales can be quantified in a similar way as done in the optical by defining a characteristic UV decay time. In this work we have used t_{10} SWP, defined as the time needed for the flux in the IUE

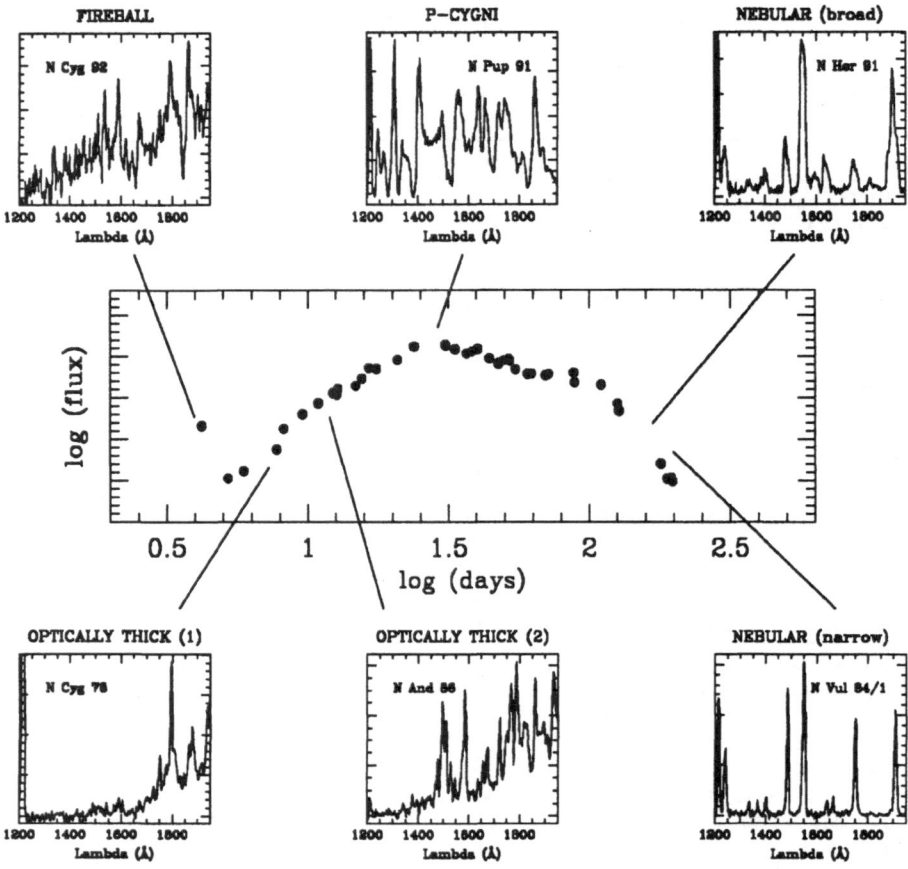

Figure 1. Schematic evolution of a classical nova in the UV during the first two years after the outburst. The light curve is that of V1974 Cyg and the spectra correspond to different objects chosen as representative examples, as labeled.

short–wavelength range to decrease by a factor of 10 from the observed maximum. Values of this decay time for the different objects are given in Table 1, together with the delay between optical and UV (IUE short-wavelength range) maxima. In Figure 2 we show the relation between the optical decay time t_3 and these two UV characteristic times. Open circles in the figure correspond to uncertain values of t_{10} SWP and refer to those cases in which the actual UV maximum has not been observed and the decay time has been measured from the observed maximum flux. The most discrepant points in the $t_3 - t_{10}$ SWP relation correspond to PW Vul and GQ Mus. In the first object the wide oscillations in the optical light curve make the measurement of t_3 very inaccurate. GQ Mus has an unusually slow UV evolution most likely due to its peculiar nature (Díaz and Steiner 1994). A similar comparison can be performed by using the flux in the IUE long–wavelength band (1950–3200 Å), with the result that in this range the maximum never occurs later than in the short–wavelength range, and that the decay is faster (t_{10} SWP is roughly twice t_{10} LWP).

During the decay phase some objects show dips in the light curve which are related to the condensation of dust in the envelope. This is for instance the case of V1370 Aql (Snijders et al. 1987), V842 Cen (Gerhz 1990) or Nova LMC 1988 No. 1 (Whitelock 1988). In Nova Cas 1993 the episode of dust formation took place near UV maximum (Shore et al. 1994b). The transition to the nebular stage coincides in some case with a change of slope in the light curve which becomes steeper.

2.2. SPECTROSCOPIC EVOLUTION

The different phases of the photometric evolution above outlined are closely related to the spectroscopic changes. During the "fireball" stage the nova shows a strong emission all over the UV range (see the first spectrum of V1974 Cyg shown in Figure 1, taken from Shore et al. 1994a). Once the iron peak elements recombine the opacity increases, blocking all the radiation shortward 1700 Å. This is stage in which the first IUE spectra are generally taken. Later on, the optical depth in the UV lines decreases and the flux starts to raise, with numerous spectral features appearing (what has been called the "lifting of the iron curtain", see Hauschildt et al. 1994). Near UV maximum the usual nebular lines start to appear and the resonance lines develop strong P-Cygni profiles. A peculiar case is that of GQ Mus in which, about 150 days after the outburst, the continuum increased and the emission lines became wider appearing strong P-Cygni profiles. This phenomenon has not been observed in any other object and is most likely due to the ejection of a second shell (Hassall et al. 1990). The next phase in the spectral development is the "nebular" phase, in which there is only a very faint continuum with the typical nebular emission lines superimposed on it. Most of the novae have narrow emission lines (i.e. not resolved in the IUE low dispersion mode) during this stage, but novae with broad lines during the outburst (V694 CrA, LMC 1991, V838 Her) still have them in the nebular phase.

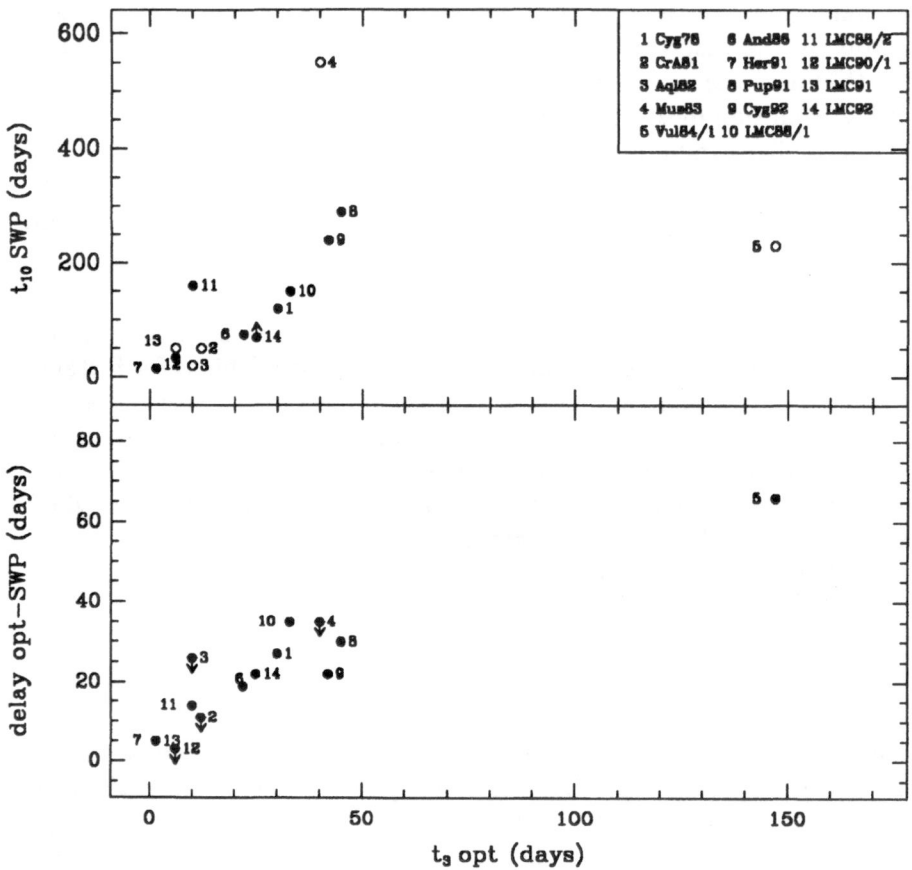

Figure 2. Relation between the optical decay time t_3 and the characteristic UV times. The upper panel shows the UV decay time t_{10} SWP *vs.* t_3. The relation between t_3 and the delay between optical and UV maxima is shown in the lower panel.

3. The UV Luminosity of Classical Novae

One of the predictions of the TNR theory of classical novae outbursts is the existence of a phase of constant bolometric luminosity. This was first confirmed first by the OAO-2 observations of FH Ser (Gallagher and Code 1974) and later on by the detailed multiwavelength observations of V1668 Cyg (Stickland et al. 1981) and V1370 Aql (Snijders et al. 1987). During this period the effective temperature increases and the peak of the energy distribution shifts toward shorter wavelengths, from the UV to the soft X-rays region. Although the maximum UV luminosity is not by itself representative of the bolometric luminosity the outburst it can provide a useful lower limit. Listed in Table 1 are the maximum UV luminosities

(1230–3200 Å) of the objects in the sample. Reddenings and distances used to compute the luminosities are also listed. The uncertainty in the distance is very large in most cases. For V351 Pup the reddening was derived by comparison of the UV spectra and light curve with those of V1974 Cyg given their similar evolution. The result of the comparison is that the color excess of V351 Pup is approximately 0.5 magnitudes larger than that of V1974 Cyg. The distance has been computed by assuming that both novae reach a similar brightness at maximum. For all LMC novae a distance of 50.1 kpc has been used (Panagia et al. 1991), and the reddening has been assumed to be the same for all of them: a galactic foreground reddening of $E(B-V)_{GAL} = 0.05$ – with the Savage and Mathis (1987) extinction law – plus a LMC internal reddening of $E(B-V)_{LMC} = 0.10$ – with the non–30 Dor extinction law (Fitzpatrick 1986) –. As shown in the Table, *in the UV* all novae radiate at maximum close to the Eddington limit for a 1 $M_\odot$ white dwarf. The LMC novae represent an especially interesting subset since it is possible to perform a direct comparison of the objects without the usual large uncertainties in reddening and distance. Although the LMC sample is too reduced to draw any conclusion, there seems to exist a correlation between the optical decay time t_3 and the maximum UV luminosity, except for Nova LMC 1990 No.1 which is abnormally bright (in fact it is the brightest object in the sample), in agreement with the suggestion made by Della Valle (1991) that it belongs to a "super-bright" nova population.

Some of the objects listed in Table 1 have been detected by IUE at late stages of the outburst (González-Riestra et al. 1994). The system with the slowest evolution in the UV is GQ Mus, for which it has been possible to obtain IUE spectra eleven years after the outburst. Spectra of PW Vul, QU Vul, V842 Cen and OS And have been acquired approximately five years after the outburst. In other cases (e.g. V838 Her) the decrease has been very fast. The lack of relation between the UV turn–off time and other characteristics such as the optical speed class or the nova type seems to indicate that there are parameters other than the WD mass or the mass accretion rate which play an important role in the development of the outburst.

Acknowledgments

Angelo Cassatella is acknowledged for many interesting discussions and for his helpful comments on the first versions of this paper.

References

Andreä, J. Drechsel, H., Snijders, M.A.J., Cassatella, A. (1991), *A&A*, **244**, p. 111
Caldwell, J.A.R. (1982), *IBVS*, **2147**

Cappacioli, M., Della Valle, M., D'Onofrio, M. and Rosino, L. (1990), *ApJ*, **360**, p.63
Cassatella, A. and Gonsáles-Riestra, R. (1990), in *The Physics of Classical Novae*, eds.
    A. Cassatella and R. Viotti, Springer-Verlag, p. 115
Chochol, D., Hric, L., Urban, Z., Komsik, R., Grygar, J. and Papousek, J. (1993), *A&A*,
    277, p.103
Della Valle, M. (1991), *A& A*, **252**, L9
Días, M.P. and Steiner, J.E. (1994),*ApJ*, **425**, p.252
Fitspatrick, E.L. (1986), *AJ*, **92**, p. 1068
Gallagher, J.S. and Code, A.D. (1974), *ApJ*, **189**, 303
Gehrs, R.D. (1990), in *The Physics of Classical Novae*, eds. A. Cassatella and R. Viotti,
    Springer-Verlag, p. 138
Gonsáles-Riestra, R. and Cassatella, A. (1988), *IAU Circular 4669*
Gonsáles-Riestra, R., Clavel, J., Cassatella, A., Krautter, J. (1991), *IAU Circular 5253*
Gonsáles-Riestra, R., Orio, M. and Gallagher, J.S. (1994), *submitted to A&A*
Hassall, B.J.M. et al. (1990), in *The Physics of Classical Novae*, eds. A. Cassatella and
    R. Viotti, Springer-Verlag, p. 202
Hauschildt, P.H$_i$, Starrfield, S., Austin, S., Wagner, R.M., Shore, S.N., Sonneborn, G.
    (1994), *ApJ*, **422**, p. 831
Kikuchi, S., Kondo, M. and Mikami, Y. (1988), *PASJ* **40** , p. 491
Krautter, J., Beuermann, K., Leitherer, C., Oliva, E., Moorwood, A.F.M., Deul, E.,
    Wargau, W., Klare, G., Kohoutek, L., van Paradijs, J., Wolf, B. (1984), *A&A*, **137**,
    p. 307
Lynch, D.K., Hackwell, J.A. and Russell, R.W. (1992), *ApJ*, **398**, p. 632
Panagia, N., Gilmossi, R., Macchetto, F., Adorf, H.M., Kirshner, R.P. (1991), *ApJ*, **380**,
    L21
Rosino, L., Iijima, T. and Ortolani, S. (1983), *MNRAS*, **205**, p. 1069
Rosino, L., Iijima, T., Benetti, S., D'Ambrosio, V., Di Paloantonio, A. and Kolotilov,
    E.A. (1992), *A&A*, **257**, p.603
Saizar, P., Starrfield, S., Ferland, G.J., Wagner, R.M., Truran, J. W., Kenyon, S.J.,
    Sparks, W.M., Williams, R.E., Stryker, L.L. (1992), *ApJ*, **398**, p. 651
Sekiguchi, K., Kilkenny, D., Winkler, H. and Doyle, G. (1989), *MNRAS*, **241**
Shore, S.N., Sonneborn, G., Shrader, C., Starrfield, S. (1990), *IAU Circular 4949*
Shore, S.N., Starrfield, S., Gonsáles-Riestra, R., Sonneborn, G. (1992), *IAU Circular*
    5657
Shore, S.N., Sonneborn, G., Starrfield, S., Gonsáles-Riestra, R., Ake, T. (1993), *AJ*, **106**,
    p. 2408
Shore, S.N., Sonneborn, G., Starrfield, S., Gonsáles-Riestra, R., Polidan, R. (1994a), *ApJ*,
    421, p. 344
Shore, S.N. Starrfield, S., Gonsáles-Riestra, R., Hauschildt, P.H. and Sonneborn, G.
    (1994b), *Nature*, **369**, p. 539
Snijders, M.A.J., Batt, T.J., Roche, P.F., Seaton, M.J., Morton, D.C., Spoelstra, T.A.T,
    Blades, J.C. (1987), *MNRAS*, **228**, p. 329
Sonneborn, G., Starrfield, S., Sparks, W. (1988), *IAU Circular 4610*
Starrfield, S., Stryker, L.L., Sonneborn, G., Sparks, W.M., Ferland, G., Wagner, R.M.,
    Williams, R.E., Gehrs, R.D., Ney, E.P., Kenyon, S., Wade, R., Truran, J.W. (1988),
    ESA SP-281, p. 159
Starrfield, S., Shore, S.N., Sparks, W. M., Sonneborn, G., Truran, J.W., Politano, M.
    (1992), *ApJ*, **391**, L71
Stickland, D.J., Penn, C.J., Seaton, M.J., Snijders, M.A.J., Storey, P.J. (1981), *MNRAS*,
    197, p. 107
Whitelock, P. (1988), *IAU Circular 4601*
Williams, R.E., Ney, E.P., Sparks, W.M., Starrfield, S., Wycoff, S., Truran, J.W. (1985),
    MNRAS, **212**, p. 753

INFRARED OBSERVATIONS OF CLASSICAL NOVAE: PHYSICAL PARAMETERIZATION AND CONTRIBUTIONS TO THE ISM

R. D. Gehrz
Department of Astronomy, School of Physics and Astronomy, University of Minnesota,
116 Church Street, S. E., Minneapolis, MN 55455

Abstract

Astrophysical parameters of classical nova outbursts that can be quantified by infrared (IR) measurements are reviewed. Recent observational results on V1974 Cyg (a neon nova) and Nova Cas 1993 (an extreme DQ Her-type CO nova) are reported. The data imply that the ejecta of these novae were extremely overabundant in C, Ne, Mg, Al, and Si. We suggest that the abundance anomalies inferred for V1974 Cyg from IR forbidden emission line measurements apparently require that the thermonuclear runaway (TNR) excavate substantial quantities of material from the white dwarf in ONeMg nova systems. Analysis of the visual and infrared light curves of Nova Cas 1993 suggests that the grains in the dust shell grew to radii of ≈ 0.55 μm. We argue that when CO novae form visually optically thick dust shells, the dust shells probably effectively cover the central engine in almost all cases.

1. Introduction

Infrared (IR) observations of classical novae have proven useful for determining the physical characteristics of the thermonuclear runaways (TNRs) on white dwarfs (WDs) accreting matter in close binary systems, and for quantifying the physical parameters of the resulting ejecta (Bode and Evans 1989, Gehrz 1988, 1990; 1993; Starrfield 1988). Recent studies have particularly emphasized evaluations of the abundances of elements such as C, N, O, Ne, Na, Mg, Al, Si, and S in the ejecta to determine the degree to which classical novae participate in the chemical evolution of the Galaxy (see Gehrz, Truran, and Williams 1993). Two fundamentally different types of nova events can be distinguished by IR measurements. CO novae, resulting from TNRs on low mass CO WDs, produce carbon-rich ejecta in which dust condenses after 30-80 days. In many cases, typified by DQ Her, enough carbon dust condenses to produce a visually optically thick carbon dust "cocoon" that acts as a calorimeter by completely obscuring the central engine at short wavelengths while reradiating the entire luminosity of the central engine in the thermal IR. The ejected shells of the most extreme CO novae tend to be high mass and have expansion velocities at the low end of the range observed in novae. TNRs on more massive ONeMg WDs produce lower mass high velocity shells in which the shell density is too low at the base of the condensation zone to enable the production of appreciable amounts of dust. These novae are distinguished in the IR by the formation soon after outburst of strong infrared forbidden emission lines that include a number in the near IR as well as those from [NeVI] at 7.62 μm and [NeII] at 12.8 μm. The exceptionally powerful [NeII] emission from the prototype coronal nova QU Vul led Gehrz et al. (1985) to term these "neon" novae. We refer to the two fundamentally different classes of novae described above as CO and ONeMg (neon) novae hereafter. Both CO and neon novae appear similar in the infrared immediately following the outburst with an initial optically thick

29

A. Bianchini et al. (eds.), Cataclysmic Variables, 29-37.
© 1995 Kluwer Academic Publishers.

expanding gas fireball giving way to optically thin thermal bremsstrahlung (free-free emission) a few days later as the shell cools and thins. The free-free phase is followed either by a dust formation episode in the CO novae and the coronal phase in neon novae. This review summarizes the physical parameters of the nova outburst that can be quantified by infrared measurements and reports the latest results on the recent bright novae V1974 Cyg (a neon nova) and Cas 1993 (an extreme DQ Her-type CO nova).

2. Physical Parameters of Nova Outbursts Calculable from Infrared Measurements

Infrared measurements of classical novae can be used to determine many astrophysical quantities that describe the outburst and subsequent development of the expanding ejecta . Table 1 summarizes the properties of novae that are directly measured by infrared observations and the astrophysical quantities that can be calculated by applying these parameters to a uniformly expanding constant velocity ejected shell. Quantitative relationships given in column 3 are gleaned from many studies of the infrared temporal development of novae that have been conducted during the past 20 years (see the discussions presented by Gallagher and Ney 1976; Ennis et $al.$ 1977; Ney and Hatfield 1978; Bode and Evans 1989; Gehrz et $al.$ 1980a, 1980b; Gehrz, Grasdalen , and Hackwell 1985; Gehrz and Ney 1987, 1990; Gehrz 1988, 1990, 1993; Hayward et $al.$ 1992). Optically thick outburst fireballs of all novae and optically thick dust shells formed by CO novae produce blackbody spectral energy distributions (SED's). The apparent luminosity of the SED is particularly straight forward to estimate in these cases, since the integrated flux F under a blackbody energy distribution with temperature T_{BB} source is measured by the quantity $(\lambda F_\lambda)_{max} = (\nu F_\nu)_{max}$ where $F = 1.36(\lambda F_\lambda)_{max}$ (Gehrz and Ney 1992). Blackbody angular radii can be combined with Doppler expansion velocities inferred from optical/infrared spectra to yield the distance to a nova and therefore its absolute luminosity. In cases where this method has been successfully applied, the outburst luminosity is always seen to be greater than or equal to the Eddington luminosity $L_{Edd} = 4\pi GMHc\kappa_T^{-1} = 3.5\times10^4 (M_{WD}/M_\odot)L_\odot$ for a 1-1.4 $M_\odot$ WD. The dust that forms in nova ejecta is mainly composed of carbon, although small amounts of silicon carbide (SiC), oxygen-rich silicate minerals based on SiO_2 molecules (such as olivine, enstatite, and pyroxene), and hydrocarbons (PAHs) have been observed to form in the ejecta of a small number of novae (Gehrz 1988, 1990, 1993; Gehrz et $al.$ 1992). The strengths of dust emission features and dust continuum emission can be used to determine abundances of the condensable elements in nova ejecta.

3. Observations of the Coronal Phase of the Neon Nova V1974 Cygni

Early observations of V1974 Cyg (Nova Cygni 1992) with the Cornell University imaging infrared spectrometer SpectroCam-10 showed this nova to be the latest example of the neon nova phenomenon (Hayward et $al.$ 1992). It was the brightest northern hemisphere nova since V1500 Cyg in 1975, and IR observers were able to construct a rather complete picture of the temporal development of a neon nova. Strong [NeII] emission at 12.8μm and hydrogen emission dominated the 7-14μm spectrum 54 days after the outburst, and hydrogen and helium lines were the principal

Table 1: Physical Parameters of Classical Novae Calculable from Infrared Measurements

PARAMETERS (units)	DEFINITION	RELATIONSHIPS (units of variables as specified in column 1)
t (days)	Time elapsed since outburst	Analysis of V and IR light curves
V_o (km s^{-1})	Outflow velocity from lines	Visual and infrared line widths, doppler shifts
T_{BB} (K)	Blackbody temperature	V/IR measurements of pseudophotospheres and dust shells
$(\lambda F_\lambda)_{max}$ (W cm^{-2})	Apparent flux at blackbody maximum	V/IR measurements of pseudophotospheres and dust shells
θ_r (milliarcseconds)	Blackbody angular radius	$\theta_r = 1.01\times10^{14}[(\lambda F_\lambda)_{max}]^{1/2}T_{BB}^{-2}$
D (Kpc)	Distance by blackbody parallax	$D = 5.75\times10^{-4}V_o t\theta_r^{-1} = 5.69\times10^{-18}T_{BB}^2 V_o t[(\lambda F_\lambda)_{max}]^{-1/2}$
λ_c (μm)	Wavelength for $\tau_{\text{free-free}} = 1$	Inspection of IR spectral energy distribution (SED)
M_{gas} (M$_\odot$)	Ejected mass from Thompson scattering	$M_{gas} = 3.3\times10^{-13}V_o^2 t^2$
M_{gas} (M$_\odot$)	Ejected mass from free-free model	$M_{gas} \approx 4.27\times10^{-14}V_o^{5/2}t^{5/2}\lambda_c^{-1}$
L_o (L$_\odot$); L_{IR} (L$_\odot$)	Outburst luminosity; IR luminosity	$L_o; L_{IR} = 4.11\times10^{17}D^2(\lambda F_\lambda)_{max} = 1.33\times10^{-17}V_o^2 T_{BB}^4 t^2$
T_d (K)	Grain condensation temperature	Observed to be $\approx$1000K in CO novae from IR SED
t_d (days)	Time of dust emission maximum	Inspection of IR light curves
R_c (cm)	Base of the dust condensation zone	$R_c = [L_o/16\pi\sigma T_c^4]^{1/2} = 1.18\times10^{12}L_o^{1/2}$
t_c (days)	Time for ejecta with V_o to reach R_o	$t_c = 137 L_o^{1/2}V_o^{-1}$
a_{gr} (μm)	Dust grain radius at IR maximum	$a_{gr} = 1.87\times10^{22}L_o V_o^{-2}t_c^{-2}T_{BB}^{-6}$
M_d (M$_\odot$)	Dust mass in CO nova ejecta	$M_d = 1.17\times10^6\rho_{gr}L_{IR}T_{BB}^{-6} = 4.81\times10^{23}\rho_{gr}D^2(\lambda F_\lambda)_{max}T_{BB}^{-6}$ $= 1.56\times10^{-11}\rho_{gr}V_o^2 t^2 T_{BB}^{-2}$
n_H (cm^{-1})	Gas density from free-free continuum	$n_H = 4.67\times10^{21}e^{7195/\Delta T}T^{1/4}D\Delta\lambda^{1/2}(\lambda F_\lambda)_{\text{free-free}}^{1/2}V_o^{-3/2}t^{-3/2}$
$I_{12.8}$ (W cm^{-2}s^{-1})	12.8 μm [NeII] intensity for ONeMg novae	Analysis of IR spectra
n_{NeII}/n_H	Abundandance of NeII in ONeMg novae	$n_{NeII}/n_H = 3.48\times10^{-11}I_{12.8}(\lambda F_\lambda|_{\lambda=12.8\mu m})^{-1}n_H$
n_Y/n_X	Abundandance of Y to X from IR lines	Analysis of IR spectra (see e.g. Greenhouse et al. 1994)

emission features in the near infrared for the first 100 days (Dinerstein *et al.* 1993). Nova Cyg entered a coronal phase several hundred days after outburst (see Gehrz 1993 and the references therein). By day 270, the temperature of the central engine had increased sufficiently to drive the NeII to higher ionization states. A series of observations with SpectroCam-10 (Hayward *et al.* 1995) and the HIFOGGS IR spectrometer operating on the NASA Kuiper Airborne Observatory (KAO) and at Mt. Lemmon, Arizona (Gehrz *et al.* 1994) showed that by day 270 the 12.8μm [NeII] emission line had decreased sharply while 7.6 μm emission from [NeVI] had grown very strong. V1974 Cyg is still sufficiently bright that IR observers are continuing to monitor its post-eruptive spectral development.

Excellent temporal spectroscopic coverage of the development of the coronal phase has facilitated a number of estimates of the abundances of the coronal species (see Table 2) using the relationships summarized in Table 1. In general, the rather large abundances of Ne, Al, and Mg with respect to H and Si tend to confirm the models of TNRs on ONeMg WDs proposed by S. Starrfield and J. Truran and their collaborators. The conformance between the observations and the abundance elevations predicted by these models requires that a considerable amount of material be dredged up from the surface layers of the WD during the TNR.

Table 2: Abundances in the Ejecta of V1074 Cyg and Nova Cas 1993

NOVA	ABUNDANCE	INSTRUMENT	INVESTIGATORS
V1974 Cyg	$\frac{(Ne/H)_{V1974}}{(Ne/H)_\odot} \geq 4$	SpectroCam-10	Hayward *et al.* 1992
V1974 Cyg	$\frac{(Ne/H)_{V1974}}{(Ne/H)_\odot} \geq 10$	HIFOGGS	Gehrz *et al.* 1994
V1974 Cyg	$\frac{(Ne/Si)_{V1974}}{(Ne/Si)_\odot} \approx 35$	HIFOGGS	Gehrz *et al.* 1994
V1974 Cyg	$\frac{(Al/Si)_{V1974}}{(Al/Si)_\odot} \approx 5$	CRSP; CGAS	Woodward *et al.* 1995
V1974 Cyg	$\frac{(Mg/Si)_{V1974}}{(Mg/Si)_\odot} \geq 3$	CRSP; CGAS	Woodward *et al.* 1995
Cas 1993	$\frac{(C/H)_{Cas93}}{(C/H)_\odot} \geq 12.5$	UM Bolometer	This paper

4. Dust Formation in the Ejecta of the Extreme DQ Hercules-type Nova Cas 1993

Nova Cas 1993 has recently provided a classic example of the CO nova phenomenon. The visual extinction event(transition) caused by the formation of its dust shell was the largest observed since the dust formation episode in DQ Her 60 years ago (Figure 1*a*). When blackbody thermal IR emission from the dust shell was first observed by Kidger *et al.* (1994) on day 71, the dust formation phase was already well underway and the shell temperature was about 1220K (Figure 1*c*). It appears from University of Minnesota measurements (Gehrz *et al.*, 1995) that the dust had begun to form as early as Day 58 (Figure 1*b*). The visual extinction and IR dust emission both reached a maximum around day 106. By day 137, the dust shell was becoming optically thin again due to combined effects of cessation of grain growth and expansion of the ejecta. The thinning of the dust shell is clearly seen in the visual light curve (Figure 1*a*) by recovery of the intensity after day 120.

We can apply the parameters described in Table 1 to determine the physical characteristics of the dust that formed in the ejecta of Nova Cas. Presuming that the

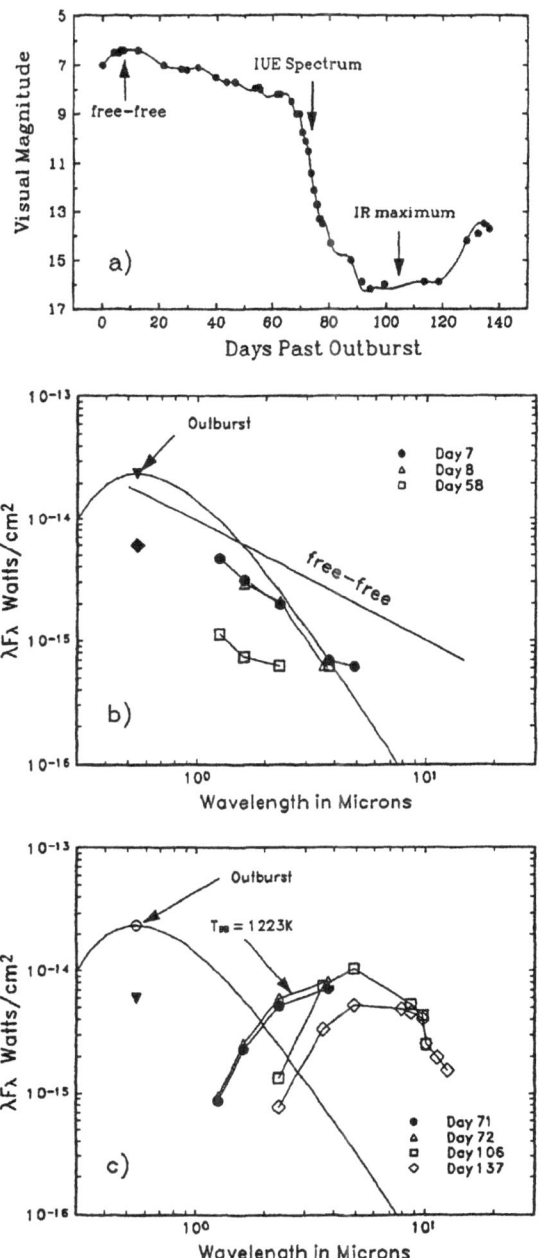

Figure 1: The visible/infrared temporal development of the DQ Her-type CO nova Cas 1993 showing a) the visible (0.55μm) light curve, the optically thin free-free emission phase, and c) the evolution of the thermal emission from the optically thick (at V) dust shell. The depth of the transition in the visible light curve and the thermal emission from the dust shell can be used as described in Table 1 to estimate that the grains grew to radii as large as ≈ 0.55 μm. Visual data in panel a) are gleaned from the I. A. U. Circulars, IR data for day 71 are from Kidger *et al.* (1994), and all other IR data are from Gehrz *et al.* (1995).

interstellar visual extinction to Nova Cas is $A_v \approx 1.5$ mag as implied by the value of $E(B-V) = 0.5$ given by Hauschildt *et al.* (1995), the dust shell reradiated only about 50% of the outburst luminosity. It appears unlikely that the IR maximum was missed since the shell absorption and IR emission both continued to increase between days 71 an 106. Two other alternatives are that the shell had large enough voids in it to allow 50% of the radiation from the central source to escape, or that the luminosity of the central engine declined by a factor of two shortly after the cessation of the TNR. Following the arguments given in section 5. below, we presume that the latter is most likely, and presume that the luminosity of the central engine is $L_o \approx L_{IR}$ for day 106 in applying the relationships in Table 1. IR spectroscopy by Woodward and Greenhouse (1993) gives an expansion velocity for the principal ejecta of $V_o \approx 840$ km s^{-1}. On day 106, the dust shell had an angular radius of $\theta_r \approx 22$ milliarcseconds, a temperature of $T_{BB} \approx 690K$, and a peak apparent flux of $\lambda F_\lambda \approx 1.1 \times 10^{-14}$ W cm^{-2}. The resulting distance of 2.33 Kpc and luminosity of $L_o \approx 2.5 \times 10^4 L_\odot$, which is at the Eddington Luminosity for a 0.7 $M_\odot$ WD. This is entirely consistent with the picture that the most extreme CO novae occur on rather low mass WDs (see Gehrz, Truran, and Williams (1993).

The dust grain radius required to produce the observed IR maximum luminosity on day 106 is $a_{gr} \approx 0.55$ μm. Independent confirmation that large grains formed in the ejecta of Nova Cas 1993 comes from an analysis by Shore *et al.* (1994) of the UV circumstellar dust extinction using IUE spectra taken shortly after the dust began to form. They found that the grains had already grown to radii of about 0.2 μm by day 76.5. Gehrz *et al.* (1980a, 1980b) found evidence for large grains in the ejecta of LW Ser and V1668 Cyg. We note that these grains are significantly larger than the grains that cause the general interstellar extinction, but are similar in size to the small grains that make up interplanetary dust particles (IDPs - see Brownlee *et al.* 1987). The total dust mass implied by the IR luminosity on day 106 is $M_d \approx 6 \times 10^{-7} M_\odot$, and the fact that the fireball had become optically thin by day 7.5 puts an upper limit on the gas mass of $M_{gas} \leq 1.3 \times 10^{-5} M_\odot$. Thus, the gas to dust mass ratio is $\approx 22/1$ and given the solar value of 275 for the mass ratio of hydrogen to carbon, we conclude that carbon is at least a factor of 12.5 overabundant in Nova Cas 1993. Hauschildt *et al.* (1995) report that CNO enhancements of 100 or more are implied by IUE UV spectra taken before the dust formation episode.

5. The Physical Characteristics of CO Nova Dust Shells

Photographic studies have shown that the expanded shells of very old novae are very clumpy, with evidence for ejecta moving over a wide range of velocities. Elementary calculations such as those presented in Table 1 are based on a model in which the ejecta uniformly cover the central engine. They refer to times in the expansion very shortly after the outburst, and will be valid if the clumps are still well merged or are superimposed so as to completely cover the central engine. Data on 15 recent bright novae (see Figure 2) provide evidence that the use of such a model is probably justified for at least the first few hundred days following the outburst. The degree to which the central engine is covered by the ejecta in CO novae is measured by the ratio L_{IR}/L_o. Presuming that L_o stays constant for at least a few hundred days, this ratio will be unity for a shell that covers completely. Given that the luminosity of

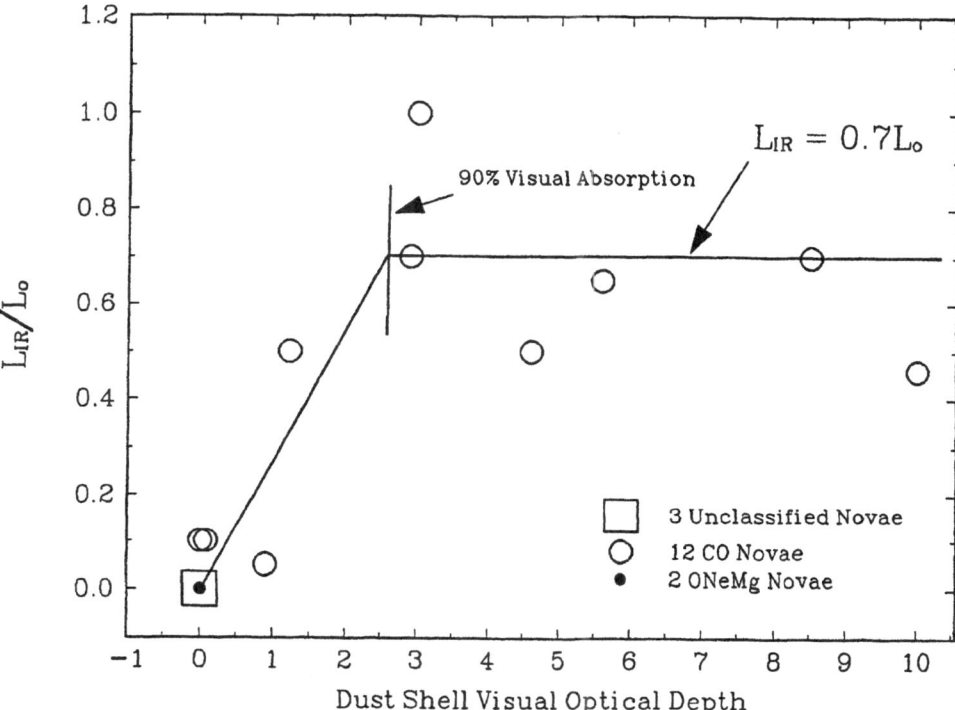

Figure 2: The luminosity re-radiated by nova shells is plotted as a function of shell visual absorption optical depth for seventeen classical novae. The latter is determinated from the depth of the transition in the visible light curve at the time of the dust formation episode. No novae that formed optically thick dust shells (those for which L_{IR}/L_o = 0.5 to 1.0) exhibited small visual transitions, suggesting that the ejecta cover the central engine fairly efficiently for several months following the eruption even though visual photographic observations of nova shells made many years after the outburst demonstrate that the ejecta are clumpy.

the central engine drops quickly by a factor of 2 or so before entering the constant luminosity phase, one would expect that values of L_{IR}/L_o in the range 0.5 to 1.0 indicate complete coverage of the central engine. Figure 2 demonstrates that this condition is met for all CO novae with visual dust extinctions of greater than 2.5 magnitudes (absorption of $\approx$ 90% of the visual light for a uniformly covering shell). If there were large holes in these dust shells one would expect to find 1) examples of very small values of L_{IR}/L_o accompanied by large visual extinction (a low-mass dust shell with a dust clump lying by chance along the line of sight to the central engine), and 2) examples of large values of L_{IR}/L_o accompanied by virtually no visual extinction (a high-mass dust shell in which the central engine is viewed through a hole).

We conclude that a visual transition deeper than 2.5 magnitudes indicates that the central engine of the novae is covered uniformly by the ejecta, and that the model used to derive the entries in Table 1 can be used to estimate the physical properties of the event to first order. The shells of novae that do not produce dust remain very transparent after the initial fireball expansion, and therefore clump about the origin in Figure 2 as expected.

Acknowledgments

The author was supported for this work by the National Science Foundation, NASA, the US Air Force, and the Graduate School of the University of Minnesota. His attendance at the Abano Conference was supported in part by a travel grant from the American Astronomical Society.

References

Benjamin, B., and Dinerstein, H. L. 1990, *Astron. J.*, **100**, 1588

Benjamin, B., Dinerstein, H. L. 1991, in *Symposium in Celebration of the 50th Anniversary of MacDonald Observatory of the University of Texas at Austin: Frontiers of Stellar Evolution.*, p91.

Bode, M. F., and Evans, A. 1989, in *Classical Novae*, eds. M. F. Bode and A. Evans, John Wiley and Sons Ltd.: London, p. 163.

Brownlee, D. E., Pilachowski, L., Olszewski, E., and Hodge, P. W. 1987, in *Solid Particles in the Solar System*, eds. I. Halliday and B. A. McIntosh, Reidel: Dordrecht, p. 333.

Cameron, A. G. W. 1982, in *Essays in Nuclear Astrophysics*, Eds. C. A. Barnes, D. D. Clayton, and D. N. Schramm, Cambridge University Press: Cambridge, pp. 23-43.

Cameron, A. G. W. 1991, in *Protostars and Planets III*, eds. E. Levy and J. Lunine (University of Arizona Press: Tucson), p. 47

Dinerstein, H. L. *et al.* 1993, *Astrophys. J.*, to be submitted.

Ennis, D., Becklin, E. E., Beckwith, S., Elias, J., Gatley, I., Matthews, K., Neugebauer, G., and Willner, S. P. 1977, *Astrophys. J.*, **214**, 478.

Gallagher, J. S., and Ney, E. P. 1976, *Astrophys. J. Lett.*, **204**, L35.

Gehrz, R. D. 1988, *Ann. Rev. Astron. Astrophys.*, **26**, 377-412.

Gehrz, R. D. 1990, in *Physics of Classical Novae*, eds. A.Cassatella and R. Viotti, (Springer Verlag: Berlin), p. 138.

Gehrz, R. D. 1993, *Annals of the Israel Physical Society*, **10**, 100.

Gehrz, R. D., and Ney, E. P. 1987, *Proc. Nat. Acad. Sci. (USA)*, **84**, 6961.

Gehrz, R. D., and Ney, E. P. 1990, *Proc. Nat. Acad. Sci. (USA)*, **97**, 4354.

Gehrz, R. D., Grasdalen, G. L., and Hackwell, J. A. 1985, ,*Astrophys. J. Lett.*, **298**, L47.

Gehrz, R. D., Truran, J. W., and Williams, R. E. 1993, in *Protostars and Planets III*, eds. E. Levy and J. Lunine (University of Arizona Press: Tucson), p. 75.

Gehrz, R. D., Grasdalen, G. L., Hackwell, J. A., and Ney, E. 1980a, *Astrophys. J.*, **237**, 855.

Gehrz, R. D., Hackwell, J. H., Grasdalen, J. A., Ney, E. P., Neugebauer, G., and Sellgren, K. 1980b. *Astrophus. J.*. **239**, 570.

Gehrz, R. D., Hackwell, J. H., Grasdalen, J. A., Ney, E. P., Neugebauer, G., and Sellgren, K. 1980b, *Astrophys. J.*, **239**, 570.

Gehrz, R. D., Woodward, C. E., Greenhouse, M. A. Starrfield, S. G., Wooden, D. H., Witteborn, F. C., Sandford, S. A., Allamandola, L. J., Bregman, J. D., and Klapisch, M. 1994, *Astrophys .J.*, **421**, 762.

Gehrz, R. D., Jones, T. J., Woodward, C. E., Greenhouse, M. A., Wagner, R. M., Harrison, T. E., Hayward, T. L., Benson, J. 1992, *Astrophys. J.*, **400**, 671

Gehrz, R. D., Mason, C., Mergen, J., and Woodward, C. E. 1995, *Astrophys. J.*, in preparation.

Hauschildt, P. H., Starrfield, S., Shore, S. N., Gonzalez-Riestra. R., Sonneborn, G., and Allard, F. 1995, *Astrophys. J.*, in press.

Hayward, T. L., Gehrz, R. D., Miles, J. W., Houck, J. R. 1992, *Astrophys. J. Lett.* **401**, L101.

Kidger, M., Devany, N., Sahu, K., and López, S. 1994, *I. A. U. Circ.*, No. 5936.

Livio, M., and Truran, J. W. 1990, in *Nonlinear Astrophysical Fluid Dynamics*, eds. J. R. Buchler and S. T. Gottesman, Journal of the New York Academy of Sciences, **617**, 126.

Ney, E. P., and Hatfield, B. F. 1978, *Astrophys. J. Lett.*, **219**, L111.

Saizar, P., Starrfield, S., Ferland, G. J., Wagner, R. M., Truran, J. W., Kenyon, S. J., Sparks, W. M., Williams, R. E., Stryker, L. L. 1992, *Astrophys. J.*, **398**, 651.

Starrfield, S. G. 1988, in *Multiwavelength Studies in Astrophysics*, ed. F. Cordova, Cambridge: Cambridge Univ. Press, p. 159.

Shore, S. N., Starrfield, S., Gonzales-Riestra, R., Hauschildt, P. H., and Sonneborn, G. 1994, *Nature*, **369**, 539.

Starrfield, S., Truran, J. W., Sparks, W. M., and Krautter, J. 1986, *Astrophys. J. Lett.*, **303**, L5.

Woodward, C. E., and Greenhouse, M. A. 1993, *I. A. U. Circ.*, No. 5910.

Woodward, C. E., Greenhouse, M. A., Gehrz, R. D., Pendleton, Y. J., Joyce, R. R., Van Buren, D., Fischer, J., Jennerjohn, N. J., and Kaminskii, C. D. 1995a, *Astrophys. J.*, **438**, in press. .

Woodward, C. E., Gehrz, R. D., Jones, T. J., Lawrence, G. R., and Skrutskie, M. 1995b, *Astrophys. J.*, in preparation.

PHOTOMETRIC PROPERTIES OF CLASSICAL NOVAE AT MINIMUM

HILMAR W. DUERBECK
Astronomisches Institut, Universität Münster
D-48149 Münster, Germany, and
Grupo de Astrofísica, P. Universidad Católica de Chile
Santiago 22, Chile

Abstract. Our present knowledge of the number of Galactic classical novae is briefly reviewed, and recurrent nova candidates are listed. The photometric behaviour of novae at minimum is studied, and the existence of a general decline of about $dm/d\log t = 1.3$ (decline in magnitudes for each ten-fold temporal separation from the outburst, measured in years) is established. A small group of novae has post-outburst magnitudes which are noticeably brighter than their pre-outburst magnitudes, and are identified as short-period (magnetic) systems. Both groups show a similar brightness decline. Number counts $N(m)$ of cataclysmic binaries may be helpful for solving the problem of recurrence times and hibernation properties of novae. The magnitude limit of existing complete samples, however, is too bright to draw firm conclusions.

1. A census of novae

Our knowledge of minimum properties of novae has increased substantially during the last few years. Triggered by the predictions of the hibernation hypothesis (for one of its incarnations, see, Shara et al. 1986, Livio 1993), optical and infrared photometric surveys were carried out (Harrison 1992, Szkody 1994, Weight et al. 1994, Duerbeck et al. 1995), and spectroscopic surveys are under way (Duerbeck and Seitter 1987, Bianchini et al. 1992).

In order to study the behaviour of classical novae at minimum, they should be clearly discriminated against related, but different types of variables, such as dwarf novae with long cycle times, symbiotic stars, x-ray no-

39

A. Bianchini et al. (eds.), Cataclysmic Variables, 39-46.
© *1995 Kluwer Academic Publishers.*

vae, or possible gravitational lensing objects. Since the compilation of the
Reference Catalogue and Atlas of Galactic Novae (Duerbeck 1987), considerable progress has been made. The most straightforward discrimination is
based on spectroscopic observations near maximum, which is usually carried out for newly discovered objects, but not available for about a quarter
of the ≈ 200 well documented older objects. For 24 suspected novae, the
very existence of the objects is still not clarified, although sometimes a late
solution of the problem emerges – V360 Her is a good example. Last year, I
inspected the discovery plate taken 1892 at the Observatoire de Paris, and
immediately recognized that it must be a plate carrying two exposures of
different star fields, but before I could get a scan of the plate and identify
the underlying field, Webbink (1993) discovered marginalia in the USNO
copy of the Astrographic Catalogue, which showed that this fact had been
known a long time ago, but was forgotten when the 'nova' was announced
in 1927!

In recent years, monitoring has revealed a number of recurrent novae –
V394 CrA (1949,1987), V3890 Sgr (1962,1990), V745 Sco (1937,1989) – as
well as dwarf novae with long outburst cycles (CI Gem, V592 Her, SS LMi,
GW Lib, HV Vir), in many cases first recognized as such by the Sonneberg
astronomers Richter and Wenzel. Table 1 gives an up-to-date census of
classical Galactic novae (confirmed (N), possible (N:) and dubious (N??))
and related types (recurrent novae (NR) and x-ray novae (XN)).

TABLE 1.   Numbers of known
classical novae and related objects

Type	N(1987.0)	N(today)
N	136	156
N:	52	52
N??	24	24
NR	6	8
XN	6	9

There remains the lottery of *predicting* recurrent novae, either on the
basis of their small amplitude, their red colour at minimum, or other criteria. Since amateurs are interested in monitoring such objects, we should
strive to define *reliable* candidates.

My own list, based on novae with small amplitude, comprises LS And,
MU Ser, V794 Oph and V909 Sgr. The first one, which has no spectroscopic record, may also be a dwarf nova with long cycle times. Harrison's

(1992) red objects, AR Cir and V723 Sco, are based on incorrect minimum identifications in Duerbeck (1987 – see Duerbeck and Grebel 1993 for a revision), while Weight et al.'s (1994) and Szkody's (1994) red objects, EU Sct, V3965 Sgr, LW Ser and V368 Sct, are more promising, although existing outburst spectra show neither strong coronal lines nor anything else unusual. My actual best guess would be V840 Cen, a novalike variable discovered by Liller in early 1986, which shows a symbiotic-like spectrum at minimum (Duerbeck and Seitter 1989). With an amplitude of 7^m and a light curve decay time of about a week, it qualifies as a recurrent nova similar to RS Oph.

Let us now focus on our knowledge of photometric properties at minimum. Szkody (1994) has provided an extremely valuable inventory of present-day optical and infrared nova magnitudes. We have focussed our attention: (1) on nova magnitudes determined at different times after outburst; (2) magnitudes determined both before and after outburst. Some results have been published (Duerbeck 1992,1993); a detailed paper is in preparation (Duerbeck et al. 1995). Here we give an overview of the available results.

2. Additional evidence for hibernation

Photometric data of a nova taken at different times after outburst shows that, on the average, novae tend to evolve into fainter stages as time goes by. Duerbeck (1992) collected old visual photometry by Steavenson and by members of the RASNZ, to compare them with new CCD observations, as well as additional evidence based on photoelectric photometry.

TABLE 2. Nova decline rates $dm/d\log t$ (t in years). A ten-fold increase in time leads to the listed magnitude decline.

No.	Method	Author	$dm/d\log t$
1	Theoretical	(Kovetz 1988)	0.70
2	Statistical	(Vogt 1990)	2.28 ± 0.62
3	16 systems	(Duerbeck 1992)	1.00 ± 0.32
4	22 systems	(this paper)	1.35 ± 0.27
5	Statistical	(this paper)	1.22 ± 0.43

Recently, visual photometry by members of the AFOEV has been analyzed, and old POSS red magnitudes were compared with new CCD red magnitudes. A summary is given in Table 2, where the new result of the average decline rate is given as No. 4. It is in good agreement with the previous result based on the same method (No. 3).

3. A revision of the hypothesis $m\,(\text{prenova}) = m\,(\text{postnova})$

In a seminal paper published almost 20 years ago, Robinson (1975) stated that "with the possible exception of BT Mon, the pre-eruption and post-eruption magnitudes are the same for all 18 stars for which both magnitudes are known." As it turned out later, BT Mon is a system with deep eclipses, and most pre-outburst observations had been made when it was near minimum, giving even more credibility to Robinson's statement. Nevertheless, any hibernation scenario – even one based on marginal evidence as indicated in Table 2 – is in strict contradiction to Robinson's statement.

We have repeated Robinson's analysis by comparing old POSS magnitudes with modern CCD magnitudes, collected at the 1.0 m telescope of Wise Observatory and at the 1.5 m Danish telescope at ESO (Table 3). Additional data were taken from the recent literature (Table 4). The results are shown in Figure 1. In view of the large photometric variability of some exnovae found by Honeycutt (1995) the results only have statistical value, and should not be used for individual systems.

Figure 1 shows that most novae some years after outburst are about 2^{m} brighter than before outburst, reaching pre-maximum brightness after about $10 - 30$ years. However, a group of novae exists which is $2^{\mathrm{m}} - 4^{\mathrm{m}}$ brighter than the others (the magnitude difference is noticeably larger than the fluctuations found by Honeycutt). The novae of this 'bright group' are clearly separated in Fig. 1. A cursory study indicates that this behaviour is explained by the fact that they erupted from a *fainter* stage, while their post-outburst luminosities seem to resemble those of the other group.

Identifying objects of both groups in a histogram of orbital periods (Fig. 2), it is seen that the 'bright group' is obviously confined to short-period systems. We summarize:

- Short-period (magnetic?) systems erupt from fainter prenova stages, so that they appear $2^{\mathrm{m}} - 4^{\mathrm{m}}$ brighter for at least several decades, as shown in Fig. 1. Examples are CP Pup, V1500 Cyg and GQ Mus, and possible candidates for magnetic systems are RW UMi and HZ Pup.
- Long-period systems follow an exponential decay which leads to equality between pre-eruption and post-eruption magnitude about 20 years after outburst. Whether such *novae become systematically fainter about 100 years after outburst*, as suggested in Fig. 1, is a matter of debate. Examinations of the old magnitude scales, or re-investigations of old photographic records are necessary.

The individual points of different novae of long orbital period available at different times after outburst can also be interpreted as presenting the behaviour of a 'typical' nova. We have determined the slope $dm/d\log t$ of the decay, by using 'long-period' systems with $\Delta t > 1$. Its value, 1.22 ± 0.43, is

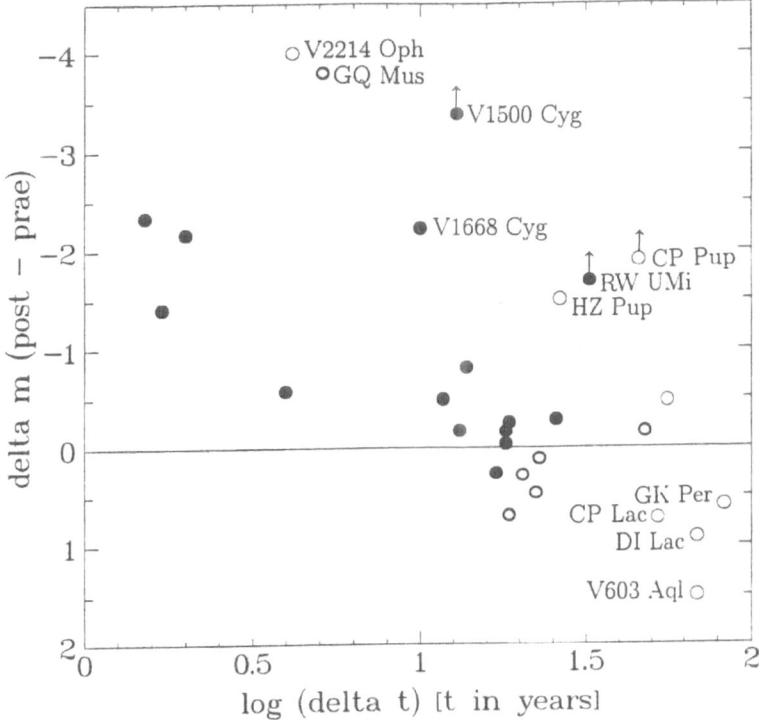

Figure 1. $m(\text{pre}) - m(\text{post})$ as a function of time. Data from Table 3 are marked by filled circles. Data from Table 4, based on the same magnitude sequence, are marked by heavy open circles, data based on two different magnitude sequences are marked by light circles. Symbols with arrows indicate lower limits. Some important objects are identified.

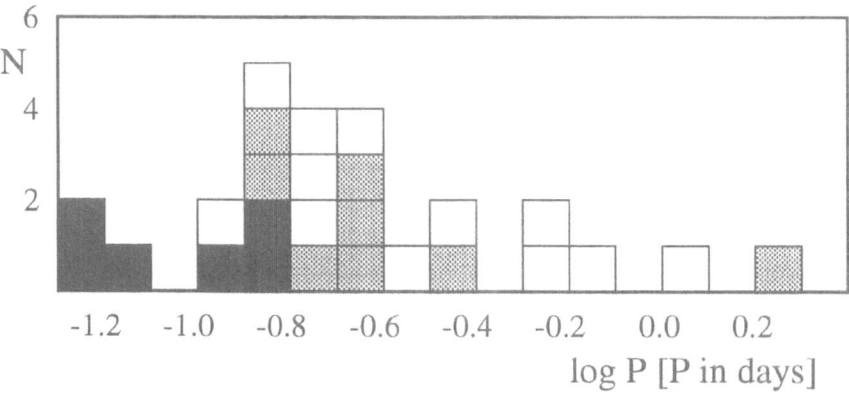

Figure 2. Period distribution of novae. Novae belonging to the 'bright group' (= faint prenovae) are marked in black, and populate the histogram at short periods. Novae of the 'faint group' are marked in gray, they are medium- and long-period systems. White boxes indicate novae for which the photometric behaviour is not yet known.

TABLE 3. Pre- and post-outburst magnitudes of novae (Duerbeck et al. 1995)

object	m(pre)	m(post)	Δm	Δt
OS And	18.29	16.88	−1.41	1988.6 − 1986.9 = 1.7
V1229 Aql	17.40	17.36	−0.04	1988.6 − 1970.3 = 18.3
IV Cep	16.18	16.44	+0.26	1988.6 − 1971.5 = 17.1
V1330 Cyg	17.02	16.86	−0.16	1988.6 − 1970.4 = 18.2
V1500 Cyg	[20.0	16.62	]−3.38	1988.6 − 1975.6 = 13.0
V1668 Cyg	20.64	18.41	−2.23	1988.6 − 1978.7 = 9.9
V1819 Cyg	18.49	16.32	−2.17	1988.6 − 1986.6 = 2.0
V533 Her	15.76	15.48	−0.28	1988.6 − 1963.1 = 25.5
V827 Her	19.38	17.04	−2.34	1988.6 − 1987.1 = 1.5
V400 Per	19.10	18.29	−0.81	1988.6 − 1974.8 = 13.8
V373 Sct	18.42	18.25	−0.17	1988.6 − 1975.5 = 13.1
FH Ser	16.80	16.55	−0.25	1988.6 − 1975.6 = 18.5
RW UMi	[20.0	18.31	]−1.69	1988.6 − 1956.6 = 32.0
NQ Vul	16.88	16.39	−0.49	1988.6 − 1976.8 = 11.8
PW Vul	16.63	16.05	−0.58	1988.6 − 1984.6 = 4.0

Notes: V1229 Aql is blend of three stars on POSS. The magnitudes of these have been merged for the CCD magnitude.
IV Cep was variable between 16.6 and 18.1 (blue) before outburst.

TABLE 4. Pre- and post-outburst magnitudes of novae (from various sources)

object	m(pre), pg	m(post)	Δm	Δt
V603 Aql	10.5	12.00 B	1.5	1988.5 − 1918.4 = 70.1
HR Del	12.17	12.28 B	0.11	1990.6 − 1967.5 = 23.1
DQ Her	14.65	14.49 B	−0.16	1983.0 − 1934.9 = 48.1
CP Lac	15.32	16.05 B	0.73	1988.7 − 1936.5 = 52.5
DI Lac	13.82	14.74 B	0.92	1980.4 − 1910.9 = 69.5
BT Mon	15.32	15.79 pg	0.47	1962.1 − 1939.7 = 22.4
GQ Mus	21.	17.2 B	−3.8	1988.2 − 1983.1 = 5.1
V2214 Oph	20.5j	16.5 V	−4.0	1992.4 − 1988.2 = 4.2
GK Per	13.34	13.94 B	0.60	1984.1 − 1901.1 = 83.0
RR Pic	12.75	12.28 B	−0.47	1981.4 − 1925.2 = 56.2
CP Pup	[17.	15.14 B	]−1.9	1988.2 − 1942.9 = 45.3
HZ Pup	18.5	17.00 B	−1.5	1989.3 − 1963.1 = 26.2
V368 Sct	19.3	20.00 B	0.7	1989.3 − 1970.6 = 18.7
LV Vul	16.25	16.53 B	0.28±0.2	1988.7 − 1968.3 = 20.4

Notes: BT Mon − both magnitudes on the scale of Wachmann
V368 Sct − Kosai gives 18.6 for the POSS B magnitude.

in good agreement with the average values of the slopes of nova decline light curves, as listed in Table 2. We have taken it as an independent approach of determining decline rates of nova light curves, the result is included as entry 5 in Table 2. The number of short-period systems certainly follows a similar trend, but the number of data points is too scarce to attach any significance to the value of the slope.

4. Observed and calculated number counts

On the basis of observed trends in postnova brightness evolution, we assumed several 'typical' brightness evolutions of novae in the interval between outbursts, as suggested by the hibernation scenario. Furthermore, we assumed a schematical model of the Galaxy for the solar neighbourhood with a homogeneous absorption layer, and with the known z-distribution and outburst density of novae (Duerbeck 1984). Using a Monte-Carlo method, we distributed the novae at *statistically chosen times between outbursts*, and assigned to them absolute magnitudes according to pre-selected specific 'hibernation scenarios' (various combinations of long and short phases of high and low states), and determined the apparent magnitudes as seen from Earth. This leads to number counts in different magnitude intervals, which can be compared with the observed number counts of exnovae and other types of cataclysmic variables, as listed in the CV catalogue of Downes and Shara (1993).

A very preliminary comparison of observed and calculated number counts favours:

- relatively short intervals between nova explosions, of the order 5000 – 10000 years,
- not very extended times in hibernation, typically 2500 – 5000 years, very similar to the 'empirical hibernation model' suggested by Duerbeck (1993).
  However,
- marked differences between different models occur only at magnitudes fainter that magnitude $11^m - 12^m$, which indicates that we need better magnitude- and distance limited samples to discriminate between different models!

Acknowledgements

I gratefully acknowledge support from the German-Israeli Foundation for Scientific Research and Development (GIF) under contract I-133-109.7/89. The nova photometry is carried out in collaboration with E. Leibowitz, W. Seitter and M. Shara. I also acknowledge support from the Deutscher

Akademischer Austausch-Dienst, which partly financed my stay at P.U.C. in Santiago, where this paper was brought into its final form.

References

Bianchini, A., Della Valle, M., Duerbeck, H.W., Orio, M. 1992, ESO Messenger 69, 42.

Downes, R., Shara, M.M. 1993, Publ. Astr. Soc. Pacific 105, 127.

Duerbeck, H.W. 1984, Astrophys. Space Sci. 99, 363.

Duerbeck, H.W. 1987, Space Sci. Rev. 45, 1.

Duerbeck, H.W. 1992, Mon. Not. R. astr. Soc. 258, 629.

Duerbeck, H.W. 1993, 2nd Technion CV Conf., Annals Israel Physical Society 10, p. 77.

Duerbeck, H.W. Grebel, E.K. 1993, Mon. Not. R. astr. Soc., 265, L9.

Duerbeck, H.W., Leibowitz, E.M., Seitter, W.C., Shara, M.M. 1995, in preparation.

Duerbeck, H.W., Seitter, W.C. 1987, Astrophys. Space Sci. 131, 467.

Duerbeck, H.W., Seitter, W.C. 1989, Publ. Astr. Soc. Pacific 101, 673.

Harrison, T.E. 1992, Mon. Not. R. astr. Soc. 259, 17p.

Honeycutt, R.K. 1995, these proceedings.

Livio, M. 1993, in Interacting Binary Stars, 22nd Saas-Fee Advanced Course, ed. H. Nussbaumer and A. Orr, Berlin: Springer, p. 135.

Robinson, E.L. 1975, Astr. J. 80, 515.

Shara, M.M., Livio, M., Moffat, A.F.J., Orio, M. 1986, Astrophys. J. 311, 163

Szkody, P. 1994, Astr. J. 108, 639.

Webbink, R.F. 1993, Inform. Bull. Var. Stars, 3910.

Weight, A., Evans, A., Naylor, T., Wood, J.H., Bode, M.F. 1994, Mon. Not. R. astr. Soc. 266, 761.

1b. RECCURENT AND SYMBIOTIC NOVAE

THE GENERAL TRENDS OF RECURRENT NOVAE

G.C. ANUPAMA
*Inter-University Centre for Astronomy and Astrophysics,
Post Bag 4, Ganeshkhind, Pune 411007, India*

Abstract.
The observational properties of recurrent novae are reviewed in this talk. The class of recurrent novae may be divided into subgroups as (a) systems with a dwarf secondary and (b) systems with a giant secondary. The primary in these systems is a white dwarf, accreting at rates $\sim 10^{-8}\ M_\odot$ yr^{-1}. The outbursts are powered by thermonuclear runaway reactions on the surface of the white dwarf.

1. Introduction

Recurrent novae (RNe) constitute a small class of objects which bear many similarities to the classical novae. These systems undergo outbursts with a recurrence period of $\sim 8-80$ years. Webbink et al (1987) proposed that the designation of RNe be restricted to those systems which have two or more recorded outbursts (a) reaching an absolute magnitude comparable to those of classical novae ($M_V \leq -5.5$) and (b) accompanied by the ejection of a high velocity shell with $V_{\rm exp} \gtrsim 300$ km s^{-1}. The first criterion distinguishes RNe from dwarf novae, while the second distinguishes them from symbiotic novae. Till date, there are eight systems known which satisfy the above criteria — TPyx, USco, V394CrA, Nova LMC 1990#2 (LMC#2), RSOph, TCrB, V3890Sgr and V745Sco.

The observational properties of these systems, both at quiescence and at outburst indicate heterogeneity of the group. The infrared colours of these systems at quiescence (Harrison et al 1993) clearly indicate they may be further sub-classified as (a) systems with dwarf secondary — TPyx, USco, V394CrA, LMC#2 and (b) systems with giant secondary — RSOph, TCrB, V3890Sgr, V745Sco.

A Bianchini et al (eds), Cataclysmic Variables, 49-57
© *1995 Kluwer Academic Publishers*

The outbursts in RNe are powered by thermonuclear runaway (TNR) reactions on white dwarfs, similar to classical novae. For the recurrence of the outbursts to be of the order of a few decades as observed, the models require a massive white dwarf ($M_{\mathrm{WD}} \gtrsim 1.3\ M_\odot$) accreting at rates $M_{\mathrm{acc}} \gtrsim 10^{-8}\ M_\odot\ \mathrm{yr}^{-1}$ (Starrfield et al 1985). Prialnik and coworkers (these proceedings) have however been able to reproduce recurrent nova outbursts on white dwarfs not necessarily massive. Webbink et al (1987) and Livio (1988) proposed an alternative outburst mechanism for RSOph and TCrB, i.e. systems with a giant secondary. According to their model, the outbursts are powered by accretion events by a burst of mass transfer from the giant secondary onto the primary, which is a main-sequence star. However, recent outbursts of RSOph, V3890Sgr and V745Sco show no evidence for the presence of a main sequence accretor. Further, Kato (1991) has successfully been able to model the outbursts of RSOph as TNR events.

In what follows, the observational properties of these systems will be discussed. Also, the observational evidences which indicate a TNR powered outburst in the systems with giant secondary will be discussed.

2. Systems with Dwarf Secondary

This class has four members; TPyx, USco, V394CrA, LMC#2. Of these four, except TPyx, the other three members are similar in their observational properties.

2.1. T PYXIDIS

TPyx is distinctly different from the other members of the class of recurrent novae, both during outburst and at quiescence. At outburst, this nova has an extremely slow development of the light curve with a rate of decline 0.034 mag day^{-1}. The spectral development (Catchpole 1969) during outburst is similar to classical novae, with the development of absorption systems. The broad band $B - V$ and $U - B$ colors become redder during rise to visual maximum, characteristic of an expanding photosphere, as in the case of classical novae. The outburst development is remarkably similar during each outburst.

At quiescence, TPyx has a typical disk spectrum. The broad band colors are extremely blue $(B - V)_0 = -0.26$, $(U - B)_0 = -1.25$ and $(V - R)_0 = -0.11$, indicating a high mass transfer rate ($\dot{M} \sim 5 \times 10^{-8}\ M_\odot\ \mathrm{yr}^{-1}$) on a short period system. The UV continuum energy distribution is remarkably constant in slope and intensity, while the emission lines show substantial changes in their intensity (see Selvelli et al, these proceedings).

The spectral and light curve development during outburst, and the colors during quiescence indicate the outbursts are powered by TNR reactions on the white dwarf, accreting at a high rate. Starrfield et al (1988) and Kato (1990) have modeled the outbursts of TPyx. Kato also gives a prediction of the light curve energetics for the next outburst.

Photometry during the years 1966–1990 (Shaefer et al 1992) indicates a roughly sinusoidal modulation ($\Delta m = 0.09$ mag). The orbital period varies between 2.5–1.5 hrs with the best estimate for the period as 0.07616 ± 0.00017 days (1.82784 ± 0.00408 hrs). The sinusoidal motion does not appear tied with the orbital period. Vogt et al (1990) obtain a spectroscopic period $P = 3.439$ hrs, with $K = 30$ km s^{-1}, a mass function $f(m) = 0.00028$ $M_\odot$, and $i = 27°$.

TPyx is unique among recurrent novae in having a discernable nebulosity around it. This nebulosity consists of an inner shell ~ 10 arcsec in diameter, with a faint diffuse envelope almost 20 arcsec in diameter surrounding the inner shell (Duerbeck & Seitter 1979, Williams 1982, Shara et al 1989). These shells are formed by the outflowing gas from previous eruptions. The shell is clumpy in nature similar to the shell of GK Per (Shara; these proceedings). The distribution of nitrogen in the shell is asymmetric while the distribution of oxygen is symmetric (Anupama 1990). The shell is a slowly expanding ($V_{exp} = 350$ km s^{-1}), photoionised gas with roughly solar abundances (Shara et al 1989, Williams 1982).

2.2. U SCORPII ET AL

This class of recurrent novae have an extremely fast development of the outburst. The light curve declines at a rate ~ 0.6 mag day^{-1}. The outburst is accompanied by the ejection of matter at extremely high initial velocities. No absorption systems are present in the outburst spectrum. The emission lines are extremely broad (FWZI ~ 10000 km s^{-1}). He II lines are extremely strong, and the helium abundance He/H~ 2 indicate helium enhancement. Nitrogen is enhanced compared to carbon and oxygen, while the CNO abundance is nearly solar. This indicates a TNR processing. The outburst luminosity is super Eddington. The continuum spectrum during the later stages of outburst is extremely flat, with $S_\nu \propto \nu^{0.4}$. The outburst characteristics are similar at each outburst.

The spectrum at quiescence is unusual in the absence of hydrogen lines, and the presence of strong helium lines. The absence of hydrogen lines indicate an under abundance of hydrogen (Duerbeck et al 1993, Johnston & Kulkarni 1992, Duerbeck & Seitter 1990). The secondary is of spectral type F8–G6 in U Sco (Johnston & Kulkarni 1992, Schaefer 1990), and of type G (later than USco) in V394CrA (Duerbeck et al 1993).

Photometry of U Sco (Schaefer 1990) show eclipses with an amplitude of ~ 1.5 mag. The duration of the eclipses is 0.17 in phase with flickering outside the eclipses. The estimated period is $P = 1.2344 \pm 0.0025$ days. V394CrA also shows a large and significant variablity in its quiescence magnitude (Schaefer 1990), which are roughly sinusoidal with a period $P = 0.7577$ days.

USco is an ideal test for TNR models of RNe since both radial velocity curves, that of the white dwarf and the cool component can be measured. Further, masses of the individual components can also be estimated. Johnston & Kulkarni (1992) estimate the following parameters: secondary spectral type F8$\pm$2, $P = 1.225806$ day, $K_{WD} = 35 \pm 17$ km s^{-1}, $K_s = 156 \pm 19$ km s^{-1}, $f_a(M_{WD} = 0.0054 \, M_\odot$, and $M_{WD} = 0.23 \pm 0.12 \, M_\odot$ for $M_s = 1.3 \, M_\odot$. They obtain a 3σ limit to the mass of the white dwarf as $M_{WD} = 0.9 \, M_\odot$. This poses a problem for the TNR powered models of outburst, which require a massive white dwarf.

Duerbeck et al (1993) have remeasured the radial velocities for USco, and obtain the following parameters: $P = 1.234518$ day with a sine fit scatter of 74 km s^{-1}, $K_{WD} = 164 \pm 333$ km s^{-1}, $K_s = 116 \pm 35$ km s^{-1}, $M_s = 1.64 \pm 0.83 \, M_\odot$, and $M_{WD} = 1.16 \pm 0.69 \, M_\odot$. This estimate of the mass of the white dwarf is compatible with TNR models. It should be noted here that both USco and V394CrA are extremely faint at quiescence, and the data on which the radial velocity measurements are based are fairly noisy with large uncertainities. Better quality data are required for an accurate estimate of the binary parameters.

3. Systems with Giant Secondary

This class consists of four members, RSOph, TCrB, V3890Sgr and V745Sco. These four systems are quite similar in their properties. They are fast novae with a rate of decline ~ 0.3 mag day^{-1}. Matter during outburst is ejected with extremely high initial velocities (~ 4000 km s^{-1}), which decrease with time. Also, the spectrum develops high excitation coronal lines and OI 8446 Å line enhanced by Lyβ fluorescence (see for eg. Williams et al 1991). RSOph was detected as a strong X-ray source about 55 days following the 1985 outburst maximum, and also as a synchrotron emission source in the radio at a similar period (Mason et al 1987, Hjellming et al 1986). The development of the strong coronal lines, the narrowing of the emission lines and the development of non-thermal X-ray and radio radiation in these systems is interpreted as arising from a region shock heated as the fast moving nova shell interacts with the slow moving pre-outburst circumstellar material from the stellar wind of the giant companion (Gorbatskii 1972, Bode & Kahn 1985, O'Brien & Kahn 1987, O'Brien, Bode & Kahn 1992).

The optical spectrum at quiescence is composite, with strong emission lines superposed over the spectrum of the late type giant secondary, similar to the symbiotic stars. The origin of the emission lines is uncertain. The spectral type estimate for the giant secondary is:

T CrB: M3

RS Oph: K5–M1

V745 Sco: M4–M7

V3890 Sgr: M5–M8

In the next two sections, the systems TCrB and RSOph at quiescence are discussed in detail.

3.1. T CRB

At quiescence, TCrB is the brightest recurrent nova $V \approx 10.2$ mag. Sanford (1949) first detected radial velocity variations with a period of 230.5 days and velocity amplitude $K_g = 21$ km s^{-1}. The orbit was later refined by Kraft (1958), by Paczyński (1965) and most recently by Kenyon & Garcia (1986). The orbital parameters as estimated by Kenyon & Garcia are: $P = 227.53$ days, $K_g = 23.32$ km s^{-1}, $K_h = 33.76$ km s^{-1}, $i \approx 68°$, $M_g = 3.34 \pm 0.73$ $M_\odot$, $M_h = 2.31 \pm 0.29$ $M_\odot$.

The mass estimate for the hot component implies a main sequence accretor, and poses a problem for TNR powered outburst models. Webbink (1976) proposed an accretion powered outburst model, with sporadic mass transfer onto the main sequence accretor. It should be noted here that the origin of the emission lines is uncertain, and is probably not associated with the hot component. In this case, the velocity amplitude K_h which is based on the emission lines does not represent the hot component, and hence the estimated mass cannot be attributed to the hot component.

The optical quiescence spectrum is dominated by that of the giant secondary, while the UV spectrum is that of an accretion disk. Selvelli et al (1992) describe the UV spectrum in detail and from their estimates of the UV luminosity and mass accretion rates give arguments for the presence of a white dwarf accretor. The slope and luminosity of the UV continuum energy distribution is found have variations. The mean UV luminosity of ~ 40 $L_\odot$ implies a mean transfer rate of $\sim 5 \times 10^{-8}$ $M_\odot$ yr^{-1}. Similar rates are implied by the HeII 1640 Å luminosity. The EUV luminosity estimated based on the HeII line flux, implies an inner disk temperature $T_* = 3 \times 10^5$ K.

The emission line strengths in the UV region are variable and correlated with the UV continuum variations. Variability is also present in the optical emission lines. Hα emission line shows a long term variation with orbital phase dependant variations superposed (Anupama & Prabhu 1991). Maxima in intensity occurs around phases 0 and 0.5, which is anti-correlated with broad band photometric variations, which show a minima around these phases (Peel 1990), but correlated with UV variations. These variations probably arise due to a variation in the mass transfer rate.

3.2. RS OPH

The 1985 outburst of RSOph was extremely well studied in all wavelength regions, from the X-rays to radio, the details of which may be obtained from Bode (1987).

At quiescence, this star resides at $m_{\mathrm{vis}} \approx 11.5$ mag, with variations of the order $\Delta m \sim 0.5$ mag. Oppenheimer & Mattei (1994) analyzed the visual magnitudes of RSOph from 1930–1993 in the AAVSO archives. They find several significant periods in each interval between outbursts. No single period repeated from one interval to the next. Analysis of a similar data by Dobrzycka & Kenyon (1994) show the light curve to be a superposition of a long period variation (2178 ± 160 days) and a short period variation (508 ± 46 days). Analysis of visual magnitude estimates between 1979–1994 (Anupama & Gilmozzi 1994) reveals periods of $P \sim 1500$ days and $P \sim 470$ days.

Garcia (1986) obtained the orbital parameters for this star, which has recently been revised by Dobrzycka & Kenyon (1994) as: $P = 460 \pm 10$ days, $K_{\mathrm{g}} = 12.8 \pm 2.0$ km s^{-1}, $K_{\mathrm{h}} = 4.6 \pm 1.4$ km s^{-1}, $i \lesssim 40°$. As in the case of TCrB, the origin of the emission lines and hence the value of K_{h} is uncertain. Dobrzycka & Kenyon treat RSOph as a single lined spectroscopic binary, and assuming $M_{\mathrm{h}} \sim 1.4\ M_{\odot}$, they obtain $M_{\mathrm{g}} \lesssim 1.4\ M_{\odot}$. The spectroscopic period is similar to the short period detected in photometric data.

Similar to TCrB, RSOph also shows variations in the UV (1150–3000 Å) continuum slope and intensity (Anupama & Gilmozzi 1994). The variations appear correlated with the visual magnitude variations. For eg. the UV luminosity reached a maximum in 1982 October, when the visual magnitude also recorded an increase to 9.2 mag. The mean UV luminosity of $L_{\mathrm{UV}} \sim 10^{36}$ erg s^{-1} implies a mass transfer rate of $\dot{M} \sim 6 \times 10^{-8}\ M_{\odot}$ yr^{-1}. The He II 1640 Å flux also implies a similar mass transfer rate. The inner disk temperature is $T_{*} \sim 7 \times 10^{5}$ K. Corresponding to this temperature, the outer radius of the optically thick disk is $R_{\mathrm{out}} \sim 1\ R_{\odot}$. The presence of a large accretion disk is consistent with the extremely blue continuum in the optical region.

The optical spectrum is a combination of an extremely blue continuum superposed over which are strong emission lines of the hydrogen Balmer series, FeII, HeI, CaII and OI 8446 Å, and that of the giant secondary. The presence of the TiO bands indicate a spectral type K5–M1 for the secondary. The CaII and FeII emission lines originate in a region with $n_{e} \sim 10^{12}$ cm^{-3}

and $T_e \sim 6000 - 7000$ K (Anupama & Gilmozzi 1994), possibly in the stellar wind photoionized by the hot white dwarf.

The hydrogen lines show a variable broad component which is double peaked either in the blue or red side of the narrow component (Iijima et al 1994). Variations have also been detected in the secondary spectral type, from K5–M2 (Anupama & Gilmozzi 1994), and these are possibly correlated with the variations in the UV flux and the visual magnitudes.

RSOph has also been detected as a soft X-ray source at quiescence by ROSAT (Orio 1993). Assuming an optically thin model of thermal plasma, Orio obtains an X-ray luminosity of $L_X \simeq 2.8 \times 10^{31} - 1.64 \times 10^{32}$ erg s^{-1}. This implies a mass transfer rate $\dot{M} < 10^{-9}$ $M_\odot$ yr^{-1}, a value much less that that estimated by UV data.

3.3. NATURE OF THE HOT COMPONENT

There are no observational evidences for the presence of a main sequence accretor from the recent outbursts of RSOph (1985), V745Sco (1989) and V3890Sgr (1990). We present in this section the evidences for the presence of a white dwarf accretor.

RSOph 1985 outburst
• Detection of remnant X-ray radiation 250 days after outburst maximum, implying a temperature $T = 3.5 \times 10^5$ K, luminosity $L = 10^{37}$ erg s^{-1} and a blackbody radius $R \sim 10^9$ cm.
• Radius and temperature of the central ionizing source inferred from hydrogen and helium emission lines, 204 days after outburst maximum (Anupama & Prabhu 1989): $T_* = 3.6 \times 10^5$ K, $R_* = 0.03$ $R_\odot$.

V3890Sgr 1990 outburst
• Radius and temperature of the central ionizing source infrerred from hydrogen and helium emission lines, 18 days after outburst maximum (Anupama & Sethi 1994): $T_* = 3 \times 10^5$ K, $R_* = 0.06$ $R_\odot$.

TCrB at quiescence
• Bulk of the disk luminosity arises in the UV with negligible contribution in the optical. The observed UV luminosity (~ 40 $L_\odot$) is incompatible with a main sequence accretor as this would imply extremely high accretion rates ($\sim 10^{-3}$ $M_\odot$ yr^{-1}) and hence significant contribution in the optical region, which is not observed (Selvelli et al 1992).
• Presence of strong HeII 1640 Å line, implying temperatures $\sim 10^5$ K (Selvelli et al 1992).
• Presence of broad wings (velocities $\sim$few 100 km s^{-1}) in the emission lines of CIV and HeII (Selvelli et al 1992).
• Flickering in U band photometry (Walker 1977).

OI 8446 line

• Presence of strong OI 8446 Å line in the quiescent spectrum of RSOph and V745Sco. This line which is enhanced by Lyβ fluorescence indicates the presence of a hot source of UV photons.

3.4. OUTBURST PROCESS

Observational evidences exist for the outbursts in these systems to be powered by TNR reactions on the surface of the white dwarf. We list here the evidences.

RSOph

• Sustained bolometric luminosity plateau lasting over 57 days from outburst (Snijders 1987, Evans et al 1988).

• Increase in temperature of the ionizing source with a decrease in the radius, with increasing time, following the 1985 outburst maximum (Anupama & Prabhu 1989).

• Outburst luminosity of $1.3 \times 10^5\, L_\odot$ (Harrison et al 1993).

• Elemental abundances indicating enhancement of helium and nitrogen (Snijders 1987, Bohigas et al 1989, Anupama & Prabhu 1989).

V3890Sgr

• UV maximum was reached ~ 20 days after visual maximum, similar to what happens in classical novae (Gonzalez-Riestra 1992).

• Estimated IUE luminosity 18 days after the outburst: $L_{UV} = 1.3 \times 10^5\, L_\odot$, implying $L_{bol} >> L_{Edd}$ (Gonzalez-Riestra 1993).

• An enhanced helium abundance of He/H= 0.2 (Anupama & Sethi 1994).

Mass accretion rates

• The mass accretion rates estimated from the UV luminosity for both TCrB and RSOph at quiescence, which have values $\gtrsim 10^{-8}\, M_\odot\, yr^{-1}$.

4. Conclusions

In conclusion, the class of recurrent novae may be divided into two different subclasses based on the nature of the secondary: (a) systems with dwarf secondary and (b) systems with giant secondary. The first subclass is further ditinguished as the USco type or the TPyx type. The outbursts in recurrent novae are powered by TNR reactions on the white dwarf accreting at rates $\sim 10^{-8}\, M_\odot\, yr^{-1}$.

References

Anupama G.C. 1990 Ph.D Thesis, Bangalore University
Anupama G.C., Gilmozzi, R. 1994 (in preparation)
Anupama G.C., Prabhu T.P. 1989 *JA&A*, **10**, 237
Anupama G.C., Prabhu T.P. 1991 *MNRAS*, **253**, 605

Anupama G.C., Sethi S. 1994 *MNRAS*, **269**, 105

Bode M.F. 1987 ed. *RS Ophiuchi (1985) and the Recurrent Nova Phenomenon*, VNU Sci. Press, Utrecht.

Bode M.F., Kahn F.D. 1985 *MNRAS*, **217**, 205

Bohigas J., Echevarria J., Diego F., Sarmiento J.A. 1989 *MNRAS*, **238**, 1395

Catchpole R.M. 1969 *MNRAS*, **142**, 119

Dobrzycka, D., Kenyon, S.J. 1994 *AJ*, (in press)

Duerbeck H.W., Duemmler R., Seitter W.C., Leibowitz E.M., Shara M.M, 1993 *ESO Messenger*, **71**, 19

Duerbeck H.W., Seitter W.C. 1979 *ESO Messenger*, **17**, 1

Duerbeck H.W., Seitter W.C. 1990 in Cassatella, A., Voitti, R., eds, *The Physics of Classical Novae*, Springer-Verlag, Heidelberg p.425

Evans A., et al 1988 *MNRAS*, *234*, 755

Garcia M., 1986 *AJ*, **91**, 1400

Gonzalez-Riestra R., 1992 *A&A*, **265**, 71

Gorbatskii V.G., 1972 Sov. Astr. **16**, 32

Harrison T.E., Johnson J.J., Spyromilo J., 1993 *AJ*, **105**, 320

Hjellming et al 1986 *ApJL*, **305**, L71

Iijima T., et al 1994 *A&A*, **285**,

Johnston H.M., Kulkarni S. 1992 *ApJ*, **396**, 267

Kato M 1990 *ApJ*, **355**, 277

Kato M 1991 *ApJ*, **369**, 471

Kenyon S.J., Garcia M. 1986 *AJ*, **91**, 125

Kraft R.P. 1958 *ApJ*, **127**, 625

Livio M 1988 in J. Mikolajewska et al, eds., *The Symbiotic Phenomenon*, Kluwer Academic Pub., Dordrecht, p.323

Mason K.O., Córdova F.A., Bode M.F., Barr P., in Bode, M.F., ed, *RS Ophiuchi (1985) and the Recurrent Nova Phenomenon*, VNU Sci. Press, Utrecht, p.167

O'Brien, T.J., Bode, M.F., Kahn, F.D. 1992 *MNRAS*, **255**, 683

O'Brien, T.J., Kahn, F.D., 1987 *MNRAS*, **228**, 277

Oppenheimer, Mattei 1994

Orio M. 1993 *A&A*, **274**, L41

Paczynski B. 1965, Acta Astr. **15**, 197

Peel M. 1990 J.Br. Astr. Assoc., **100**, 136

Sanford R.F., 1949 *ApJ*, **109**, 81

Selvelli P.L., Cassatella A., Gilmozzi R., 1992 *ApJ*, **393**, 289

Schaefer B.E. 1990 *ApJL*, **335**, L39

Schaefer B.E., Landolt A.U., Vogt N., Buckley D., Warner B., Walker A.R., Bond H.E., 1992 *ApJS*, **81**, 321

Shara M.M., Moffat A.F.J., Williams R.E., Cohen J.G., 1989 *ApJ*, **337**, 720

Snijders M.A.J. 1987 in Bode, M.F., ed, *RS Ophiuchi (1985) and the Recurrent Nova Phenomenon*, VNU Sci. Press, Utrecht, p.51

Starrfield S., Sparks W.M., Truran J.W., 1985 *ApJ*, **291**, 136

Starrfield et al 1988 in Longdon, N., & Rolfe E.J., eds. *A Decade of UV Astronomy with the IUE Satellite*, ESA SP-281, 1, p.167

Vogt N., Barrera L.H., Barwig H., Mantel K.-H., 1990 in ed. Mauche C.W., *Accretion Powered Compact Binaries*, Camb. Univ. Press, Cambridge (UK), p.391

Walker A.R., 1977 *MNRAS*, **179**, 587

Webbink R.F., 1976 Nature, **262**, 271

Webbink R.F., Livio M., Truran J.W., Orio M., 1987 *ApJ*, **314**, 653

Williams R.E. 1982 *ApJ*, **261**, 170

Williams R.E., Hamuy M., Phillips M.M., Heathcote S.R., Wells L, Navarrete M., 1991 *ApJ*, **376**, 721

SYMBIOTIC NOVAE

URS MÜRSET & HARRY NUSSBAUMER

Institut für Astronomie
ETH-Zentrum
CH-8092 Zürich
Switzerland

Abstract. The outbursts of symbiotic novae are physically related to classical novae. After attempting a definition of symbiotic novae and describing their spectral evolution we review some fundamental properties and the spectral evolution. Then we focus on three issues that link observations and theory: i) the evolution of surface temperature, radius, and luminosity; ii) the chemical composition; iii) mass-loss.

Among interacting binaries, symbiotic stars are the systems with the widest orbits, with periods ranging from years to decades or even centuries. They contain a red giant and a very hot white dwarf. The stellar components are embedded in an emission nebula which is due to the cool giant's wind partly photo-ionized by the hot star. Symbiotic novae are a small subclass of the symbiotic stars. They are distinct by a strong, ongoing outburst with the following observational characteristics (cf. Mürset & Nussbaumer 1994):

i) the outburst amplitude is large, up to $\Delta V \approx 7^{\mathrm{m}}$.

ii) the outburst is of long duration (many decades). No known symbiotic nova has completely recovered from the outburst, although the outburst of one of them began in the middle of the 19th century.

iii) no secondary outbursts occur.

iv) the object may have had no typical symbiotic spectrum before outburst; possibly, the system has symbiotic characteristics only during outburst.

v) in atmospheres and ejecta of symbiotic novae, nitrogen is over-abundant.

The outburst energy and the elemental composition prove that symbiotic novae are powered by nuclear burning. The typical outburst energy amounts to $E \sim 10^{47}$ erg (requiring $M_{\mathrm{H}} \sim 10^{-5}$ M$_\odot$ of hydrogen to be burnt).

1. The spectral evolution

Three phases can be distinguished in the evolution of the visual spectrum of a symbiotic nova (cf. Figur 2):

i) THE 'GIANT' PHASE: When the outburst started, half of the objects expanded strongly, in one case up to a radius $R_{\mathrm{max}} \approx 100$ R$_\odot$ (see Table). The photosphere cooled down, and the object spectroscopically resembled an A- or F-type giant or supergiant. No heavy mass-loss is observed in that phase. – Other objects skipped the expanded phase.

59

A. Bianchini et al. (eds.), Cataclysmic Variables, 59-64.

TABLE I

Basic data on the symbiotic nova systems and their outbursts. The first 7 rows characterize the outburst, row 8 gives the orbital period, and the last rows give the pulsation period of the cool component, if it is a mira variable, or else its spectral type. References: Most of the information is derived or cited in Mürset & Nussbaumer (1994). Parameters for SMC 3 are from Morgan (1992) and Vogel & Morgan (1994). Orbital periods are from Nussbaumer et al. (1989), Schmutz (1995; AG Peg), Schmid & Schild (1995; V1016 Cyg), Vogel (personal communication; PU Vul), and the period of RT Ser is roughly estimated form line profile variations. Spectral types are from Kenyon & Fernández-Castro (1987), Schild et al. (1992), and Belyakina et al. (1982), and mira periods from Harvey (1974) and Feast et al. (1983).

	AG Peg	RT Ser	RR Tel	HBV 475	V1016 Cyg	HM Sge	PU Vul	SMC 3
outburst	≈1850	1909	1944	1963	1964	1975	1978	1981
ΔV [mag]	3	6	8	3	4	6	6	1.5
M_V^{max} [mag]	−3	−6	−5	−3	−3	−3	−3	−4
T_{min} [K]	8000	8000	7000	40 000	60 000	40 000	6000	
R_{max} [$R_\odot$]	20	100	90	3	0.6	2	50	
L_{max} [$L_\odot$]	4000	30 000	20 000	20 000	40 000	30 000	4000	
E [10^{46} erg]	3	10	4	4	8	4	0.4	
P_{orbit} [yr]	2.2	≈ 6		2.6	≈40		≈13	
P_{mira} [d]			387		450	540		
Spec. type	M3	M6		M6			M4	M0

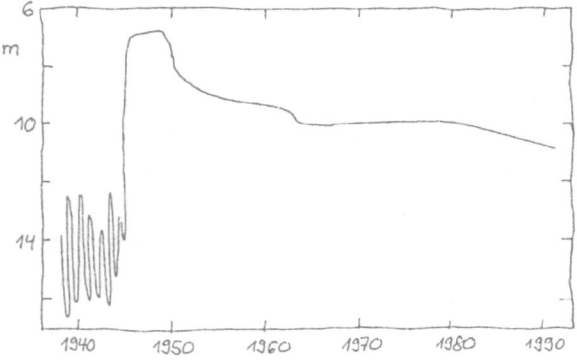

Fig. 1. Light curve of the symbiotic nova RR Tel

ii) THE 'WN' PHASE: At some point, for whatever reason, heavy mass-loss starts. We observe broad emission bands of He II, N III, N IV, and N V, as in the spectra of N-rich Wolf Rayet stars. Whithin a few years the extended atmosphere is pealed off; the photosphere recedes and grows hotter. The star starts ionizing its surroundings.

iii) THE NEBULAR PHASE: Finally the WN features disappear, and strong emission lines from an ionized nebula begin to dominate the spectrum. The degree of ionization gradually increases, see Figure 2.

Though the hot star becomes much fainter in the visual light than the nebula and the cool component, it remains luminous. This is seen indirectly

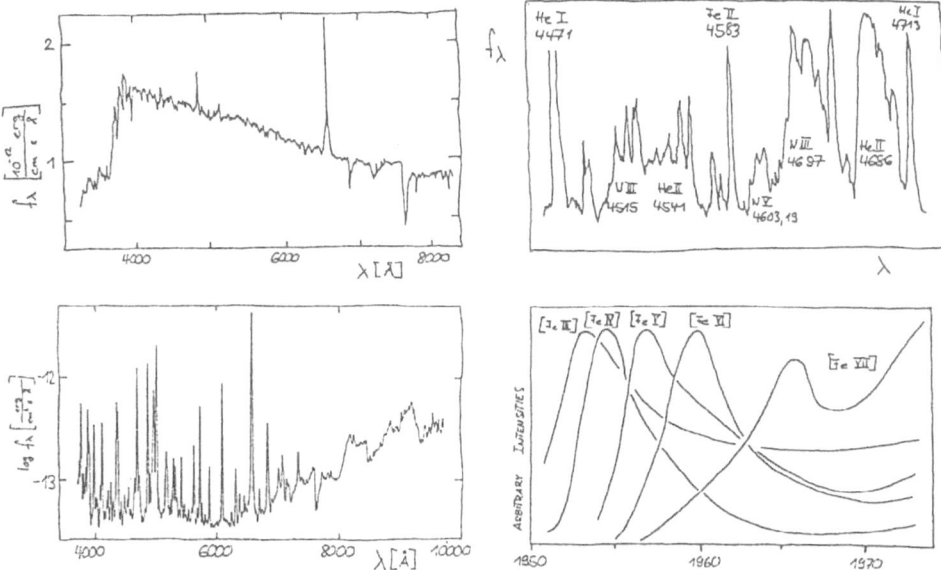

Fig. 2. Spectra of symbiotic novae in the giant (PU Vul; upper left panel; Kenyon 1986), WN (RR Tel; upper right; Thackeray & Webster 1974), and nebular phase (RR Tel; lower left; observed in 1993 at ESO). Lower left panel: increasing ionisation stages of iron in RR Tel (from Thackeray 1977).

through the nebular emission and a discernible contribution to the UV spectrum. It can even be observed directly in the soft X-ray range. Jordan et al. (1994) analyzed a spectrum recorded with ROSAT. In their model (Figure 3), the bulk of the emission emerges from a hot NLTE model atmosphere.

2. Evolutionary tracks in the HR diagram

Unlike in classical novae, nebular matter is continuously supplied by the red giant. It converts the unobservable EUV emission of the hot remnant into visual nebular radiation. From that radiation we can derive fundamental parameters of the hot stars, as demonstrated by Nussbaumer & Schild (1981), Kenyon & Webbink (1984), and Mürset et al. (1991). Hence, the evolution of temperature, luminosity, and radius of symbiotic novae can be tracked! This was done for all commonly accepted galactic symbiotic novae by Mürset & Nussbaumer (1994), and for AG Peg also by Kenyon et al. (1993). Figure 4 displays three examples of HR diagram tracks.

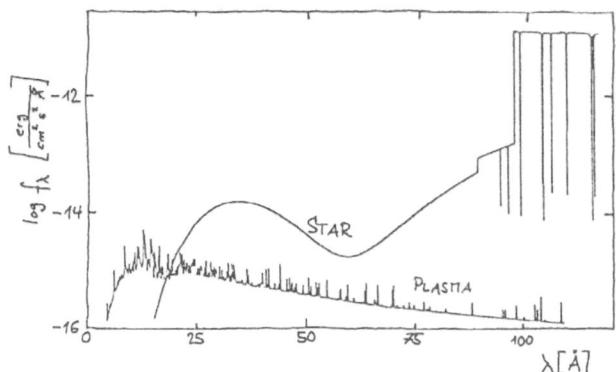

Fig. 3. Model for the X-ray spectrum of RR Tel consisting of the emissions from a hot NLTE model atmosphere (T_{eff} = 142 000 K, $L = 3500\ L_\odot$, $R = 0.09\ R_\odot$) and from a Raymond-Smith type plasma possibly due to colliding winds. The model reproduces the ROSAT count rates when attenuated by interstellar extinction and convolved with the instrumental response matrix (Jordan et al. 1994).

Nova-type thermo-nuclear calculations seem appropriate to model the outbursts. Indeed, some calculations of this kind have been devoted particularly to symbiotic novae, e.g. by Kenyon & Truran (1983), Livio et al. (1989), Sion & Starrfield (1994). We have no space to discuss these calculations, but we want to make two comments for consideration in future work:

i) HR diagram tracks which slowly and horizontally approach the visual maximum do not fit the observations. Such an object would be detected as a symbiotic star with decreasing ionisation. That has never been the case.

ii) According to today's theory the objects should move back to high temperatures on a strictly horizontal path, subject to a core-mass luminosity relation. Not all empirical tracks agree with that law. The luminosity of V1016 Cyg, for instance, increased by one order of magnitude on the way to the left.

3.   Elemental abundances

Observations provide a clue to nuclear processing in the outburst. In the early stages the photospheric composition of the expanded object can be determined. Later, the hot star is difficult to observe directly, but the ejecta are ionized and their emission spectrum can be analyzed. Unfortunately, H I and He I emission lines are not easy to interprete in symbiotic spectra. Only relative C, N, and O abundances are reliably known for most objects. We give the ratios N/(C+N+O) versus O/(C+N+O) in Figure 5.

i) WN PHASE: In the absence of any detailed analysis of a symbiotic nova in the WN stage, the best guess for the chemical composition of its atmosphere is the ratios typical for WN stars, as marked in Figure 5.

ii) EARLY NEBULAR PHASE: The youngest symbiotic novae (SMC 3 and

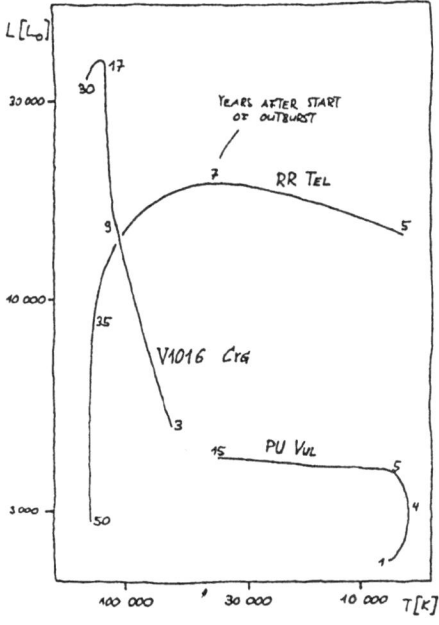

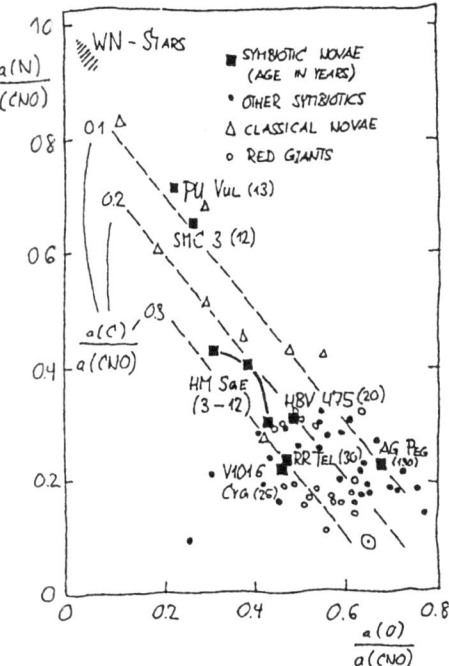

Fig. 4. Evolutionary paths of PU Vul, RR Tel, and V1016 Cyg in the HR diagram (Mürset & Nussbaumer 1994). The figures give the year after the beginning of the outburst.

Fig. 5. CNO abundance ratios in symbiotic novae. log(N/CNO) is plotted versus log(O/CNO). The ratios for the symbiotic stars are from Nussbaumer et al. (1988), Schmid & Schild (1990), Nussbaumer & Vogel (1990), Vogel & Nussbaumer (1992), Schmid & Nussbaumer (1993). Vogel & Morgan (1994). The position of WN stars is adopted from the calculations of Maeder (1990), ratios of classical novae and red giants are compiled in Nussbaumer et al. (1988).

PU Vul) have just entered the nebular phase. The ionized matter shows a significant over-abundance of N. We probably observe, at least in part, emission from the wind of the hot star, which seems to consist of matter processed in the outburst.

iii) LATE NEBULAR PHASE: The composition of the nebulae of AG Peg, RR Tel, HBV 475, and V1016 Cyg are now close to those of red giants. Interpretation: The hot source now ionizes the wind of the cool companion which swept away the ejecta of the outburst. "Non-nova" symbiotics have such CNO abundances as well.

4. Mass-loss

Before outburst, the white dwarf has most probably accreted matter from the wind of the cool giant. At some stages of the outburst, the symbiotic nova seems to lose mass (cf. Section 2). At what time does it start to accrete again, and what total mass is expelled over an outburst? This is important for understanding the outburst evolution as well as for determining the time scales and the net mass gain of a full outburst cycle. Unfortunately, with increasing nebular emission, wind features become hard to detect. However, Nussbaumer et al. (1994) observed AG Peg with HST: 150 years after the start of the outburst N v still exhibits a P Cyg component.

Acknowledgments. We thank H.M. Schmid for valuable comments.

References

Belyakina T.S., Gershberg R.E., Efimov Y.S., Krasnobabtsev V.I., Pavlenko E.P., Petrov P.P., Chuvaev K.K., Shenovrin V.I., 1982, SvA 26, 184

Feast M.W., Whitelock P.A., Catchpole R.M., Roberts G., Carter B.S., 1983, MNRAS 202, 951

Harvey P.L., 1974, ApJ 188, 95

Jordan S., Mürset U., Werner K., 1994, A&A 283, 475

Kenyon S.J., 1986, AJ 91, 563

Kenyon S.J., Fernández-Castro T., 1987, AJ 93, 938

Kenyon S.J., Webbink R.F., 1984, ApJ 279, 252

Kenyon S.J., Truran J.W., 1983, ApJ 273, 280

Kenyon S.J., Mikołajewska J., Mikołajewski M., Polidan R.S., Slovak M.H., 1993, AJ 106, 1573

Livio M., Prialnik D., Regev O., 1989, ApJ 341, 299

Maeder A., 1990, A&AS 84, 139

Morgan D.H., 1992, MNRAS 258, 639

Mürset U., Nussbaumer H., 1994, A&A 282, 586

Mürset U., Nussbaumer H., Schmid H.M., Vogel M., 1991, A&A 248, 458

Nussbaumer H., Vogel M., 1990, A&A 236, 117

Nussbaumer H., Schild H., Schmid H.M., Vogel M., 1988, A&A 198, 179

Nussbaumer H., Schild H., 1981, A&A 101, 118

Nussbaumer H., Schmid H.M., Vogel M., 1989, A&A 211, L27

Nussbaumer H., Schmutz W., Vogel M., 1994, A&A letters, submitted

Schild H., Boyle S.J., Schmid H.M., 1992, MNRAS 258, 95

Schmid H.M., Nussbaumer H., 1993, A&A 268, 159

Schmid H.M., Schild H., 1990, MNRAS 246, 84

Schmid H.M., Schild H., 1995, in preparation

Schmutz W., 1995, in preparation

Sion E.M., Starrfield S.G., 1994, ApJ 421, 261

Thackeray A.D., 1977, Mem. R.A.S. 83, 1

Thackeray A.D., Webster B.L., 1974, MNRAS 168, 101

Vogel M., Morgan D.H., 1994, A&A 288, 842

Vogel M., Nussbaumer N., 1992, A&A 259, 525

EMBARASSING ASPECTS OF OBSERVATIONS OF CNE, RNE AND SYMBIOTICS

M. FRIEDJUNG
Institit d'Astrophysique (CNRS)
98 bis Boulevard Arago, 75014, Paris, France

Abstract. There is a tendency to only take account of recent observations and observations agreeing with theoretical models. This is particularly the case for cataclysmic variables with spectacular outbursts. Some neglected observations are described. It is concluded that inspite of what is said in recent papers, there is good evidence that almost all classical novae have optically thick winds, which collide with slower material ejected in early stages. Many difficulties still exist however in the understanding of these stars, as well as of those of the other clases considered here.

1. Introduction

People working especially in the fields of study of different types of nova, often tend to try to force the observations to fit theoretical models, and in order to achieve this aim, to neglect some of those observations. In addition observing methods have changed, but inspite of great improvements, technical constraints can lead to certain types of observation being no longer made. Old observations tend to be distrusted and in this way important aspects can be "forgotten", especially if they contradict dominant theories. In this way many misconceptions can spread. Some recent papers have upset me and it is now my wish to clarify things.

Astronomical spectroscopy has been fundamentally changed by the advent of electronic detectors. Spectrophotometry is much more reliable than with the old photographic plate. However in general far fewer pixels are available. The result is generally a compromise between spectral resolution and spectral range. One can study a small spectral region at high resolution (and make a wonderful theory based on one spectral line) or a large one at low resolution. Classical novae in relatively early stages after their explo-

65

sions are a very good example of the consequeces of this type of difficulty; they have broad emission components and usualy narrower absorption components. The use of modern detectors will then lead to good photometry of the emission components, which are often overexposed on photographic plates. Many of the narrow absorption components may not even be seen, and the existence of complex line profiles denied! This sort of problem should be resolved in future by the use of echelle spectrographs, but in the meantime dubious models are used to explain recent observations.

2. The Explosions of Classical Novae

Results from photographic spectra were summarized in the classical reviews by Payne-Gaposchkin (1957) and McLaughlin (1943,1960). The line profiles seen soon after optical maximum, are the results of summing P Cygni profiles, with wide superposed emission compoments and several blueshifted absorption components. Such profiles are characteristic of expanding layers with a complex velocity field around a photosphere; if there is spherical symmetry, each layer is expected to produce line emission, which is Doppler broadened around a mean wavelength, and a blueshifted absorption component. The absorption components of different spectral lines with the same (or usually almost the same) Doppler blueshift belong to what is called an "absorption system". The absorption systems mostly seen were classified by McLaughlin. The "pre-maximum" system is seen before optical maximum when its velocity may tend to decrease if the nova is observed early enough according to McLaughlin (1960), but the systen can subsist for some time after. The "principal" system with a higher velocity than that of the pre-maximum system at that time, appears around optical maximum, and seems in the end to contain most of the ejected mass, as its line of sight velocity is similar to that indicated by the emission line width of the expanding nebula, seen in later stages. The later appearing "diffuse enhanced" and "Orion" absorption systems have higher velocities and the usually last appearing one, the "Orion" system, has absorption components of lines belonging to higher ionization stages than the corresponding components of other systems. However it should be noted that more than one absorption system of a given category can be present; the prinipal system can for instance be split into several systems with similar velocities; several diffuse enhanced systems can be seen and nova DQ Her 1934 exceptionally had two Orion systems with very different velocities.

Ultraviolet observations particularly with the IUE satellite, have had a basic influence on ideas about novae. Some novae show the presence of very high velocity ejection in the ultraviolet, not seen in optical spectra. The nean distribution of emitted energy changes with time, shifting in at least one case to longer wavelengths before optical maximum and generally to shorter wavelengths after. When the mean broad band multifrequency energy distribution extending to the infrared is considered, nova FH Ser 1970 had for several weeks after optical maximum a nearly Plamckian energy distribution (Friedjung 1990), suggesting emission of most radiation by a photosphere which was optically thick in the comntinuum. The raduis of this photosphere declined after optical maximum. Such results indicate that the effect of line blanketing in the ultraviolet (often called the "iron curtain") is not as serious as claimed in recent papers, at least for novae like FH Ser.

McLaughlin (1947 in a volune of PASP which is perhaps not in most libraries!) gave strong reasons for believing that the highest velocity material is in the deepest layers, nearest the ejecting object. Such a situation is hard to understand unless one supposes that a high velocity wind, which is optically thick in the continuum at least with respect to electron scattering, is contionuously ejected (see for instance Friedjung 1977). This wind then should produce the already mentioned photosphere. A decrease in the mass ejection rate by the wind will cause the observed photospheric radius decrease. It should also be noted (Friedjung 1990), that the deduced wind velocity of FH Ser appears to have been much larger than the escape velocity at the radius of the photosphere. This as well as as an insufficient radiation density near the phoptosphere of FH Ser for significant acceleration of the wind by radiation pressure (the most convincing mechanism propoposed up to now) according to apprpoximate estimates, suggests its acceleration at a much smaller radius than that of this photosphere.

The explanation of the different absorption systems is not easy, but certain basic features may be simply understandable, if the processes before and after optical maximum are different. Before optical maximum a classical nova could be like present models for supernovae. The premaximum system seems to be ejected by the initial explosion or soon after. An initial shock following this explosion might produce the pre-maximumn system by almost intantanious ejection with velocity increasing outwards, so the highest velocity material is at the outside and that of lowest velocity on the inside (such a situation is called a "Hubble flow"). The optically thick wind seen after optical maximum must then as deduced from observations, have a higher velocity (increasing with time) than the pre-maximum system material; it will collide with the latter and sweep it up to produce a dense layer by a snowplow effect. This dense layer which might be expected to eventually contain most of the ejected mass, should then be that producing the principal system absorption components (Friedjung 1987). The other absorption system properties are probably not explainable if spherical symmetry is assumed; it is possible that instabilities lead to the formation of dense clouds with different radial velocities in front of the photosphere.

Certain recent papers have interpreted novae, in quite different ways. For instance Williams (1992) and Williams et al (1994) divide novae into two classes, one of which does not have optical absorption lines. This seems rather surprising in view of the classical descriptions. Either modern methods have enabled us to observe novae which would not have been detected before, or the stages when optical absorption lines were visible were missed by certain recent observers particularly for novae with a fast development, or the spectral resolution used was too low to see these lines. All this clearly needs to be elucidated. Let it be noted that novae tend to have similar properties when they have faded a cerain number of magnitudes below maximum light; novae in the same stage of fading should be compared. Until this has been done, I shall remain sceptical about the detailed way novae have been divided into two classes.

Recent observations have indicated emission line narrowing (Williams 1992, Shore et al 1993), not so readily seen on photographic plates. The second of these papers interprets nova V1974 Cygni 1992 spectra by supposing most ejection in a "Hubble flow", with the highest velocities at the outside of a thick envelope and the lowest velocities at the inside, this flow not being compressed by a higher velocity wind. In fact emission line narrowing is not in contradiction with the conclusions already drawn here. If the highest velocity premaximum system material is at the outside, only dominating line absorption in very early unobserved stages, it might not be swept up. Its density and emission measure would decrease much more rapidly than those of lower velocity swept up material. In addition as the wind ejection rate decreased with time, the high velocity wings of emission lines due to the wind, would fade with respect the those parts of the profile produced by the swept up material of the principal system. This way of understanding spectroscopic observations is moreover supported by a space resolved image of the material ejected by nova V1974 Cygni 1992, obtained using the Hubble Space Telescope in 1993. Paresce (1994) interpreted the obseved circular ring as a projection on the sky of a thin spherical shell (a later observation showed some ellipticity). This clearly does not suggest a Hubble flow! The asuumption of a Hubble flow by Shore et al (1993) leads moreover to a embarassingly large mass for the ejected envelope.

Interpretations of emission line intensities and profiles have it should be noted, a number of traps. The resonance CIV and NV doublets, each having a common lower level, can have optically thin intensity ratios, even if the optical thickness is much greater than unity, as long as collisional de-excitation is small (Flower et al 1979, Nussbaumer 1994). The profiles of optially thick lines can also be similar to those of optically thin lines, if most emission comes from clumps. Interpretations assuming that lines have beome optically thin at a certain stage, can be therefore sometimes questionable.

Let it be emphasized that much still remains to be done to understand observations of classical novae during their outbursts and that some of them may at first sight appear to contadict the general interpretation of this contribution to the conference. For instance narrow undisplaced spectral lines are occaisionally seen in optical spectra. The fairly narrow emission lines seen in the spectrum of the slowly developing nova HR Del 1967 before optical maximum, are most easily understood as due to the presence of an optically thin wind in this stage (Friedjung 1992). The white dwarf component of this nova probably had an exceptionally small mass, and it could have been a borderline case of a nova outburst. The narrow undisplaced absorption components reported in a few cases are according to McLaughlin (1943, 1960) best interpreted as either circumstellar or as only apparent, due to a minimum in the emission line profile. In view of what has already been said, it is unlikely that they could be produced near the photosphere in layers deeper than that of the wind.

3. Old Classical Novae

Only one possibly embarassing aspect of observations of novae a long time after their explosions will be mentioned. If the distribution of binary periods is considered, the admittedly very small sample shows no clear sign of the presence of the cataclysmic binary period gap. Acording to Baptista et al. (1993) two old novae have periods below the gap and two have periods in the gap. This may indicate a difference in the properties of the nova population compared with that of cataclysmic variables in general. Baptista et al. think this is connected with the role of a strong magnetic field, which may be present in all the white dwarfs of the known short period novae.

4. Recurrent Novae

Recurrent novae do not form a homogeneous population, there being at least two sorts of object. One sort including T CrB and RS Oph has unlike the binaries normally defined as cataclysmic, a giant companion aaociated with a compact object, which according to the present majority opinion is a white dwarf as in the case of classical novae and other cataclysmics. The other sort of recurrent nova appears to ressemble normal cataclysmic binaries, with a cool near main sequence star associated with a white dwarf.

Many phenomena of recurrent novae of the RS Oph category, have been explained by the interaction of the ejected envelope with the strong red giant wind, which can decelerate it. However examination of the strongly blueshifted absorption components may indicate another difference with

classical novae. Their mean blueshift and hence characteristic velocity der-
creases after optical maximum. This ressembles the situation for supernovae, where there is probably a Hubble flow, which starts to become optically thin at the outside. Line absorption near the photosphere, where the
envelope becomes optically thick in the continuum then can dominate; as
the total optical thickness decreases, the radius and hence the velocity of
this photospheric layer should become less. Such an interpretation is supported by the measurements of McLaughlin (1957) for T CrB; he found
that high velocity absorption components appeared and disaapeared before those of lower velocity, with no sign of real physical deceleration. The
meaning of the RS Oph measurements of Dufay et al (1964) is however not
quite so clear cut, two absoprption systems with high and very different
velocities being present according to them, but these measurements look
interpretable in a way similar to those of T CrB. What must be emphasized is that recurrent novae of the RS Oph class appear to show no very
clear sign of the presence of a high velocity wind, as that suggested by
observations of classical novae.

The other category contains the rapidly developing U Sco and the slowly
developing T Pyx and it may not be homogeneous. The presence of both
broad and narrow emission line components may be noted in the spectrum
of U Sco (Barlow et al 1981). The 1966 outburst of T Pyx is described by
Catchpole (1969). It is not completely clear from this description or from
that of Payne-Gaposchkin (1957), to what extent it ressembles a classical
nova. Multiple absorption systems were seen, with some higher velocity
systems being seen later in the 1966 outburst. The highest velocities might
have been asociated with a wind.

5. Symbiotic Stars

Like one of the classes of recurrent nova, these stars are binary with a cool
giant component. There are reasons however which lead many soecialists
to believe that the compact component is not always a white dwarf.

The cool component is usually thought to be a luminosity classs III
giant, but a good determination of the luminosity is not so easy. This is
particularly true for AG Dra, whose cool component is less cool than that
of most other symbiotic stars and for which there would be theoretical
difficulties if it were a class III giant. Huang (1982) presented evidence that
the cool component of AG Dra was much brighter than a cool giant; I did
not want to believe him because of my own prejudices! However a recent
study of very near infrared spetra in which I collaborated (Huang et al
1994), has confirmed this result.

Some symbiotic stars are called symbiotic novae, having only one observed outburst and behaving in many ways like extremely slowly developing novae. Spectroscopic studies of the symbiotic nova RR Tel (summarized by Thackeray 1977) show that the distribution of velocities is not simple. As for a classical nova emission lines from higher and higher stages of ionization appeared during the post outburst development. The mean line width of any of the stages is roughly correlated with the ionization potential, more ionized atoms having wider lines. However the widths of lines of a given ion tended to decrease with time (Friedjung 1966). This suggests a model where the ions are mostly created by photoionization from a hot compact object and where velocities of ejected material (presumably in an optically thin wind) decrease with increasing distance from it, that is that slower material was ejected earlier. The compact object (thought to be a white dwarf after a thermonuclear flash) should have a temperature which increases with time, able to ionize atoms to an increasing extent. As photons able to ionize the ejeted material penetrate regions further away from the centre, lower velocity material should become then more ionized. However the exact "details" of what is happening are not clear; higher velocity material coming from central regions should collide with slower moving material further out. In any case there will be no spherical symmetry, at least because there is a stellar companion which is a cool giant, with its own wind. Much more study than that done up to now of colliding winds is required. Until such effects are taken into account, these observations will also be embarassing.

We can conlude that if the sort of result mentioned in this contribution to the conference is negleted, the various sorts of objects mentioned will not be properly understood. Let it also be noted coming much closer to home, that the solar wind is not that well understood!

References

Baptista, R., Jablonski, F.J., Cieslinski, D. Steiner, J.E. (1993) *ApJ*, **406**, pp L67

Barlow, M.J., Brodie, J.P., Brunt, C.C., Hanes, D.A., Hill, P.W., Mayo, S.K., Pringle, J.E., Ward, M.J., Watson. M.G., Whelan, J.J., Willis, A.J. (1981) *MN*, **195**, pp 61

Catchpole, R.M.,(1969) *MN*, **142**, pp 119

Dufay, J., Bloch, M., Bertaud, C., Dufay, M. (1964) *Ann. d'ap*, **27** pp 555.

Flower, D.R., Nussbauner, H., Schild, H. (1979) *A&A*, **72**, pp L1

Friedung, M. (1966) *MN*, **133**, pp 401

Friedjung, M. (1977) *Novae and Related Stars*, M.Friedjung (ed), pp 61, Reidel, Dordrecht

Friedjung, M. (1987) *A&A*, **180**, pp 155

Friedjung, M. (1990) *Physics of Classical Novae*, A. Cassatella and R.Viotti (eds), pp 244, Springer, Heidelberg

Friedjung, M. (1992) *A&A*, **262**, pp 487

Huang, C.C. (1982) *The Nature of Symbiotic Stars*, M. Friedjung and R. Viotti (eds), pp 185, Reidel, Dordrecht

Huang, C.C., Friedjung, M., Zhou, Z.X. (1994) *A&AS*, **106**, pp 413

McLaughlin, D.B. (1943) *Publ. Obs. Univ. Michigan*, **8**, pp 149

M. FRIEDJUNG

McLaughlin, D.B. (1947) *Publ. ASP*, **59**, pp 244
McLaughlin, D.B. (1957) *AJ*, **62**, pp 94
McLaughlin, D.B. (1960) *Stars and Stellar Systems VI, Stellar Atmospheres*, J.L. Greenstein (ed), pp 585, Univ. of Chicago Press, Chicago
Nussbaumer, H. (1994) personal communication
Paresce, F. (1994) *A&A*, **282**, pp L13
Payne-Gaposchkin, C. (1957) *The Galactic Novae*, North Holland, Amsterdam
Shore, S.N., Sonneborn, G., Starrfield, S., Gonzalez-Riestra, R., Ake, T.B. (1993) *AJ*, **106**, pp 2408
Thackeray, A.D., (1977) *Mem. R. astr. Soc.*, **83**, pp 1
Williams, R.E. (1992) *AJ*, **104**, pp 725
Williams, R.E., Phillips, M.M., Hamuy, M. (1994), *ApJS*, **90**, pp 297

1c. CATACLYSMIC VARIABLES

PERIODIC AND NEAR-PERIODIC DECADAY LIGHTCURVES IN OLD NOVA AND NOVA-LIKE CVS

R.K. HONEYCUTT, J.W. ROBERTSON, G.W. TURNER

Astronomy Department, Indiana University,
Bloomington, IN 47405
Electronic mail: honey, bigjay, turner @algol.astro.indiana.edu

Abstract. Several CVs classified as old nova or nova-like systems have been found to have photometric variations with stable or near-stable periods of tens of days. In RW Tri, DI Lac, and V841 Oph the ~ 0.5 mag variations are sine-like and not always present. In V446 Her the ~ 1.5 mag variations have been present for four years and have the form of outbursts. RZ LMi and PG0943+521 have large variations with a complex but very similar shape to the periodic lightcurves. If these decaday variations have a common cause it is likely to involve a periodic modulation of the mass transfer rate.

1. Introduction

The lightcurves discussed here were obtained with the 41-cm automated CCD Photometric Telescope at Indiana University (Honeycutt et al. 1990; Honeycutt and Turner 1992). Among the approximately 65 old nova and NL systems that are measured typically once each clear night, several show periodicities in the range 19 to 43 days. This behavior is unusual for several reasons. These CVs are generally regarded as high-state systems and are not expected to suffer the kind of disk instabilities (Cannizzo 1993a; Meyer-Hofmeister and Ritter 1993) thought to give rise to dwarf nova eruptions in the decaday period range. Even if some of these periodicities are due to disk oscillations the observed variations are of unexpected stability. Finally, for two systems, the lightcurve shapes are of surprising similarity, complexity, and stability.

75

A Bianchini et al (eds), Cataclysmic Variables, 75-81
© *1995 Kluwer Academic Publishers*

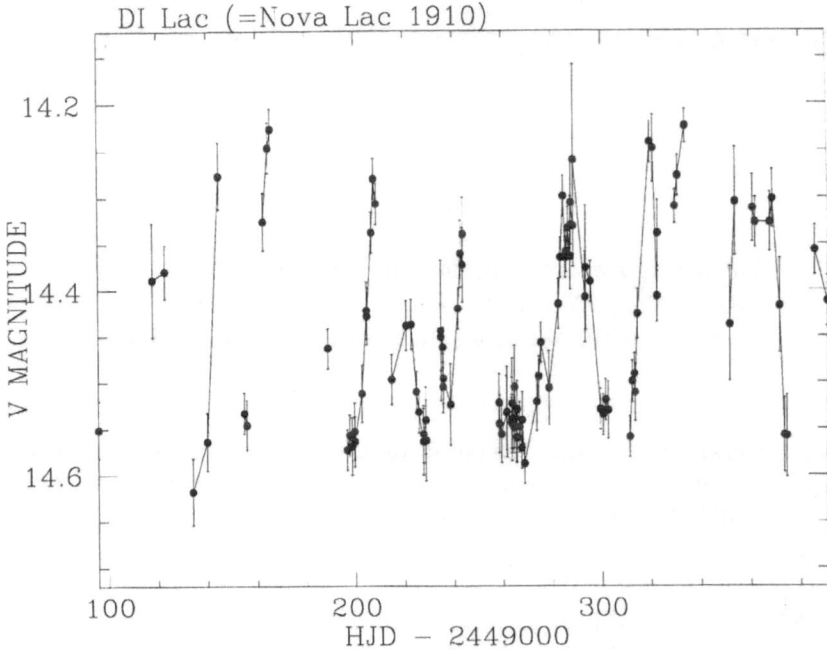

Figure 1.   The 1993 lightcurve of DI Lac. Data points ≤ 6 days apart are connected.

2. Sine-like lightcurves in RW Tri, DI Lac, and V841 Oph

The nova-like RW Tri showed a sinusoidal variation with an amplitude of
0.45 mag and a period of 25 days during 1993 (Honeycutt et al. 1994a).
The oscillation was persistent for about 150 days (or 6 cycles) but was
not apparent for the remainder of the 1990-94 observations. The mean
magnitude was approximately constant for the 4 year interval, including the
oscillation phase. Outside the 1993 oscillation interval RW Tri continued to
show 0.45 mag variations outside eclipse. These latter variations were not
as coherent as in 1993 but nevertheless there was significant excess power
in the periodogram near 25 days for most intervals outside 1993.

DI Lac (=Nova Lac 1910) and V841 Oph (=Nova Oph 1848) have behavior similar to that just described for RW Tri. V841 Oph's lightcurve had
a strong peak in the power spectrum at 35 days in both 1991 and 1992,
but the oscillations were much less coherent in 1993-94. DI Lac oscillated
with a characteristic period of 40 days during 1993 (Fig 1), but the periodic
signal was weaker outside this interval. In both cases the amplitudes are
about 0.5 mag. Like RW Tri both the amplitude of the lightcurve and the
mean brightness in these two stars seems to be independent of whether the
coherent oscillations are present or not.

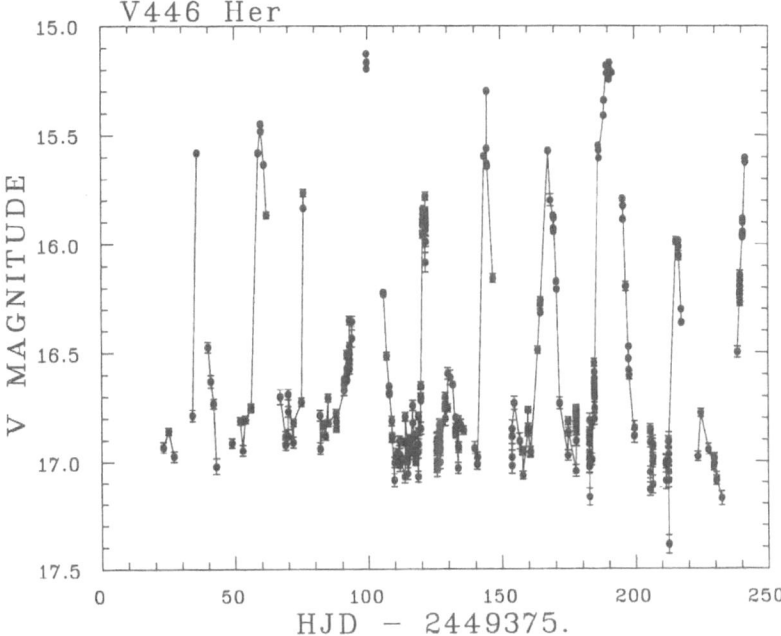

V446 Her

Figure 2. A portion of the 1994 lightcurve of V446 Her. Points spaced closer than 3 days are connected.

Mineshige (1993) summerizes a number of kinds of disk instabilities that may be able to function in a high state disk. These mechanisms are not well studied for the CV regime but some would appear to be interesting candidates for the oscillations described here. Another promising mechanism is a modulation of the mass transfer rate.

3. V446 Herculis Periodic Outbursts

V446 Her (=Nova Her 1960) showed photometric variations before the nova eruption that have sometimes been identified as DN outbursts (Livio 1989); however Robinson (1975) concludes that the flaring is inconsistent with DN behavior. Eight years after the nova Stienon (1971) found no variability exceeding 0.2 mag. Our 1990-94 photometry shows that regularly-spaced outbursts of ~ 1.5 mag are always present at this epoch. An earlier V446 Her lightcurve is found in Honeycutt et al. 1994a and Fig 2 shows some of the later data. A typical outburst is a triangular-shaped event with a full duration that is variable between 5 and 11 days. Altogether 30 flares were detected with good enough sampling to allow timing, giving a mean period of 22.2 days. There is considerable structure in the O-C diagram shown in Fig 5. (Note that because of cycle count ambiguities this linear

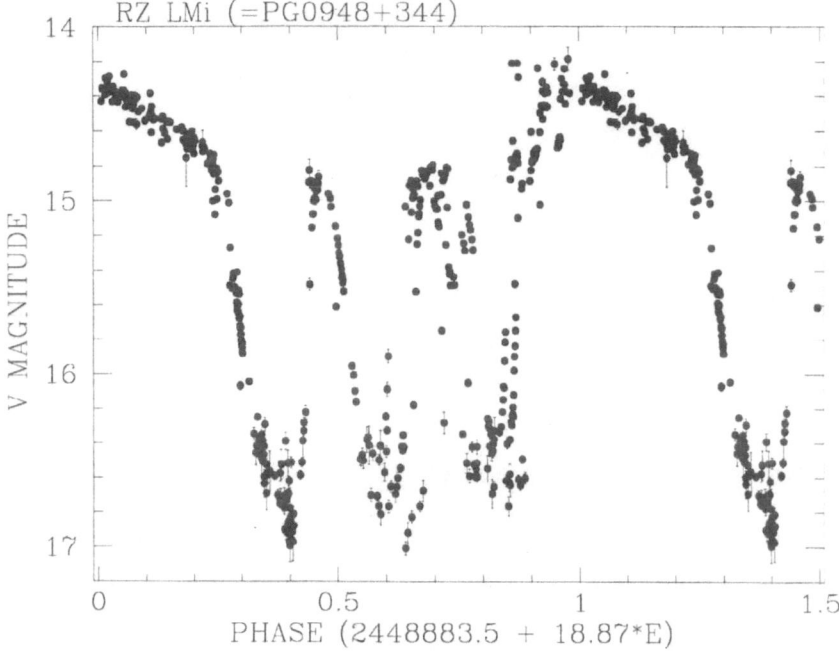

Figure 3.   Folded 1992-94 lightcurve of RZ LMi

ephemeris and O-C diagram is not the only such combination of interest.
Other periods near 22 days will also give smooth O-C curves of differing
character.) The V446 Her outbursts have an amplitude range 0.4-1.8 mag,
weaker than most dwarf nova outbursts. Periodic modulation of the mass
transfer therefore may be responsible for these flares, in a manner perhaps
related to the superoutbursts in SU UMa systems. SU UMa outbursts are
known to be more predictable than dwarf nova outbursts (Vogt 1980), and
the V446 Her outbursts are often predictable to within the observational
error of 2-3 days.

4. RZ LMi (= PG0948+344) and PG0943+521

These two systems have similar lightcurves with periods of 18.9 and 42.9
days respectively. In both cases an interval of high-amplitude oscillation
with a period of $\sim$ 4 days is preceded by a smooth dimming of about
$\sim$ 0.5 mag (see Figs 3 and 4). Robertson et al. (1994) presented an initial
description of the RZ LMi lightcurve, and the data of Shugarov and Pikalova
(1995) show that this particular lightcurve shape was also present 12 years
earlier.

The O-C curve for RZ LMi (Fig 5) shows a relatively smooth change

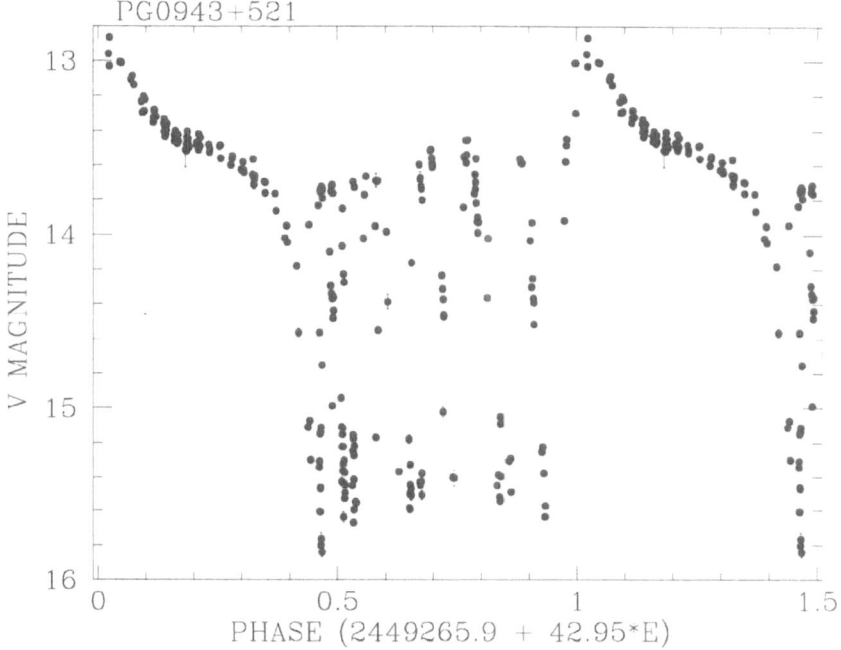

Figure 4. Folded 1993-94 lightcurve of PG0943+521

in period from 18.7 to 19.0 days over 1000 days, while the O-C curve for PG0943+521 is flat with $\dot{P} \leq 10^{-3}$ over 300 days. (The observational error in O-C for these two stars is only a few tenths of day, obtained by interpolating to a specific magnitude on the smooth decay portion of the light curves.) Under certain conditions disk instability models produce lightcurves with alternating wide and narrow maxima (Mineshige 1986; Cannizzo 1993b) with some of the characteristics of Figs 3 and 4. SU UMa-type dwarf nova in particular share a number of lightcurve properties with PG0943+521 and RZ LMi. Richter (1992) shows a number of SU UMa lightcurves that have an oscillating-like behavior during decline from supermaxima and therefore somewhat resemble these lightcurves. Furthermore, models by Osaki (1994) for SU UMa outbursts result in lightcurves that are somewhat similar to Figs 3 and 4. The relationship of these two systems to SU UMa systems and to superoutburst phenomena is very suggestive. However the high stability of the periods, the very short supercycle periods, and the very short oscillation periods are outside the range of currently observed and modeled SU UMa phenomena.

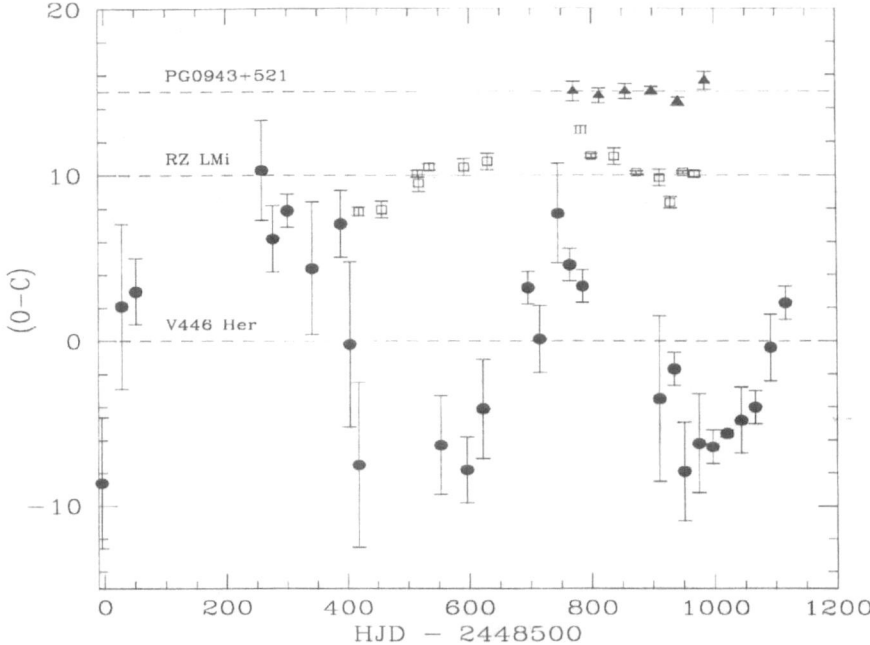

Figure 5.   O-C for V446 Her, PG0943+521 (offset 10 days), and RZ LMi (offset 15 days)

5. Discussion

It is quite possible that several mechanisms are responsible for the various types of decaday lightcurves discussed here. However, the fact that the periods are similar and the stability of the periodicities is often substantially higher than that of dwarf nova phenomena suggests (along with Ockham's Razor) that a single mechanism should first be explored. It may be that an unreasonably steady accretion rate is needed to house a fundamental clock of the required stability in the disk and that a modulation of the mass transfer rate might provide the needed periodicity. (Note however that the stable mean brightness for all these systems supports the idea of a very steady $\dot{M}$).

In a qualitative manner all these decaday variations might be understood as the result of periodic mass transfer bursts with differing mean $\dot{M}$. For sine-like and outburst-like behavior in nova-like and old nova systems the gas can be imagined to fall onto a high-state disk whose high viscosity allows the disk brightness to follow the changes in $\dot{M}$, and $< \dot{M} >$ is large enough to keep the disk in the high state. For RZ LMi and PG0943+521 a short $\dot{M}$ burst might temporarily take the disk to the high state, from which the disk slowly decays as gas drains onto the white dwarf. When the surface density of the disk has fallen to the upper critical surface density for dwarf nova oscillations, a cooling front transforms the disk to the low state and several oscillations ensue before another mass transfer burst transforms the disk back to the high state. Cannizzo's models for V794 Aql (Honeycutt et al. 1994b) show how a disk can undergo such oscillations in the absence of mass transfer, as the disk relaxes from the high state. Four days is shorter than most dwarf nova recurrance times but as the mass transfer rate is increased and the cold-state viscosity is increased the recurrance time can be shortened in the models.

References

Cannizzo, J.K. 1993a, in Accretion Disks in Compact Stellar Systems, ed. J.C. Wheeler (Singapore:World Scientific), 6

Cannizzo, J. 1993b, ApJ, 419, 318

Honeycutt, R.K., Vesper, D., White, J., Turner, G.W., and Adams, B. 1990, in CCDs in Astronomy II: New Methods and Applications of CCD Technology, ed. AGD.Philip, D.J. Hayes, & S.J. Adelman (Schenectady:Davis), 177

Honeycutt, R.K., and Turner, G.W. 1992, in Robotic Telescopes in the 1990s, ed A. Filippenko (San Francisco:ASP), 77

Honeycutt, R.K., Robertson, J.W., Turner, G.W., and Vesper, D.N. 1994a, in Interacting Binary Stars, ed. A.W. Shafter (San Francisco:ASP), 277

Livio, M. 1989, in Physics of Classical Novae, ed. A. Cassatella & R. Viotti (Berlin:Springer), 342

Honeycutt, R.K., Cannizzo, J.K., and Robertson, J.W. 1994b, ApJ, 425, 835

Meyer-Hofmeister, E., and Ritter, H. 1993, in The Realm of Interacting Binary Stars, ed. J. Sahade, G.E. McCluskey, & Y. Kondo (Dordrecht: Kluwer), 127

Mineshige, S. 1986, PASJ, 38, 831

Mineshige, S. 1993, A&SS, 210, 83

Osaki, Y. 1994, in Theory of Accretion Disks-2, ed W.J. Duschl, J. Frank, F. Meyer, E. Meyer-Hofmeister & W.M. Tscharnuter, (Dordrecht:Kluwer), 93

Richter, G.A. 1992, in Vina Del Mar Workshop on Cataclysmic Variable Stars, ed. N. Vogt (San Francisco:ASP), 12

Robertson, J.W., Honeycutt, R.K., and Turner, G.W. 1994, in Interacting Binary Stars, ed. A.W. Shafter (San Francisco:ASP), 298

Robinson, E.L. 1975, AJ, 80, 515

Shugarov, S., and Pikalova, O. 1995 (this volume)

Stienon, F.M. 1971, PASP, 83, 363

Vogt, N. 1980, A&A, 88, 66

MULTI-, QUASI-, A- AND PERIODIC VARIATIONS IN CATACLYSMIC VARIABLES

IVAN L. ANDRONOV
Department of Astronomy, Odessa State University
T.G.Shevchenko Park, Odessa 270014 Ukraine
e-mail: root@astro.odessa.ua

Abstract.
Results of long-term time-resolved photometric studies of the cataclysmic variables AM Her, MR Ser, QQ Vul, MV Lyr, TT Ari, EI UMa, BH Lyn, HQ And, V795 Her, BZ Cam, IW And, and V361 Lyr are summarized. The following topics are reviewed: solar-type activity in secondaries which may modulate the accretion rate; "magnetic valve" and "swinging dipole" models of polars; synchronization of white dwarf rotation with orbital motion; QPOs and shot noise variability in selected systems; and the shape of the power spectra.

Due to lack of the observational time, many interesting systems are "forgotten" after discovery and development of a preliminary model, based on multilateral observations obtained during a short time interval. However, due to instability of the photometric phase curves, regular observations are needed that allow one not only to determine mean curves with higher accuracy, but also to study time scales longer than the primary photometric period. The main results of such monitoring of several CVs are presented below, being collected from numerous articles.

1. AM Her

Regular observations of this prototype star have been carried out since 1978. They were summarized by Andronov (1987a, 1993) with corresponding references, thus we will repeat them only briefly. The phase curve undergoes drastic changes, even from cycle to cycle. The most prominent example is

83

A. Bianchini et al. (eds.), Cataclysmic Variables, 83-91.
© *1995 Kluwer Academic Publishers.*

the change of amplitude from 0.96^m to 0.15^m at JD 2444151 (Andronov et al. 1980). The regular phase curve depends on luminosity, having two maxima at the brightest luminosity state, and usually one in the intermediate states. Transitions between high, intermediate and low luminosity states last from few days to a month, corresponding to modulation of the accretion, possibly caused by the instability of irradiation of the red secondary. The amplitude of the irregular variations increases in the intermediate state. The phase of minimum varies cyclically, with an amplitude of $17°$ and a time scale of ≈ 3 yr, which may be caused by variations of the orientation of the white dwarf's magnetic axis, with respect to the secondary star. At high and intermediate states the phase shift decreases with decreasing luminosity, which is a second observational test of the "swinging dipole" model. Such a correlation is difficult to explain if the magnetic field of the white dwarf captures the accretion stream much closer than the Lagrangian point (e.g. Cropper 1990).

Polarimetric observations (Shakhovskoy et al. 1992) show variations not only of the longitude of the accretion column, but of the latitude ($11°$) as well. In the active state the polarization and flux vary at a time scale 6 minutes. At high frequencies f, the power spectrum obeys a power law, $S(f) \propto f^{-\gamma}$. The values $\gamma = 0$, 1, 2 for "infinite" runs correspond to white noise, flicker noise and random walks, respectively (Terebizh 1992). The autocorrelation function (ACF) does not obey the exponential decay law, which is characteristic for an irregular shot-noise model. This observational fact shows that traditional data reduction methods may give unsatisfactory results, if they do not take into account the run length and trend removal. Recent unsimplified statistical analysis (Andronov 1994) of the properties of the ACF for the detrended runs of finite length shows that the minute-scale variability may be interpreted as a sum of the first-order autoregressive (Markov) process and of uncorrelated noise, rather than by 6-min QPOs. Contribution of the correlated variations decreases with luminosity as well as the circular polarization and the slope of the power spectrum γ, showing the cessation of accretion onto the visible part of the white dwarf. Mechanisms of accretion-induced variability are discussed by Andronov (1987b).

An unprecedented flare was detected on 1992 August 28, simultaneously during photometric ($UBVRI$, $\Delta t = 12$ s) and polarimetric (R, $\Delta t = 4$ s) monitoring in the Crimean Astrophysical observatory (Shakhovskoy et al. 1993a). Contrary to numerous flares caused by the cyclotron emission of accreting blobs, the characteristics of this event shows similarity to the flares of UV Cet stars (e.g. Gershberg 1975). Exponential decay time was ≈ 5 min, and circular polarization was absent within the error estimates. According to the amplitude (7.8^m in U), this event was one of the most energetic ones. Another flare was detected with the 6m telescope (Mason et al. 1994) showing changes of the line emission of H and He II by several times, whereas the continuum changed only by 30%.

2. MR Ser

MR Ser = PG 1550+191 was classified as a polar by Liebert et al. (1982). A detailed history of its study was published by Andronov et al. (1992e). Unfortunately, after discovery of the object, no regular observations were carried out, which could have determined the character of the primary photometric period changes. The phase light curves obtained in 1982 – 1991 showed extreme variations in shape. The amplitude varied from $0.1 - 0.3^m$ on some nights to 1.1^m and even 2.3^m on nearby dates. Such extreme cycle-to-cycle changes resemble those observed in AM Her at JD 2444151 (Andronov et al. 1980). TV observations allowed obtaining an ephemeris for 6 minima in R in 1991 (Andronov et al. 1992e): $Min.\ HJD = 2448446.7284(15) + 0.078795(12)E$.

However, this period differs from the previously published ones by 7 σ, and it is not possible to fit all the times in 1981 – 1991 with any single period. There is a systematic shift between times of minima in B, V, and R ($\Delta t = t_V - t_R = -0.0043$ d) and I ($\Delta t = -0.0009$ d). However, the phase changes have a large amplitude, up to $\Delta \phi = 0.4$. This phenomenon may be explained by the drastic changes in the accretion geometry, although a cycle number miscount cannot be ruled out. If the photometric period changes so rapidly that the phases vary significantly over a decade, this may mean that the photometric period corresponds to that of the white dwarf synchronizing with the orbital motion. Such asynchronism is strange for a magnetic CV below the period gap. Recently Pavlenko et al. (1994) discovered a high-amplutude flare in the system that possibly corresponds to an accretion event or UV Cet-type activity of the secondary. Thus the system is a very interesting target for future *regular* observations.

3. QQ Vul

Photographic study of the object discovered by Nousek et al. (1984) with the Moscow and Sonneberg plate collections, as well as on series of negatives obtained at Odessa and Abastumani, led to detection of inactive states with durations sometimes shorter than 3 weeks. Seasonal photographic phase curves since 1928 showed cyclic phase changes at a time scale of 14 years (Andronov and Fuhrmann 1987). Such variability may be interpreted "swinging dipole" model (Andronov 1987a). The mean amplitude of the longitude variations of the accretion column is $\approx 14°$. This value is close to that observed in AM Her, although the time scale is five times longer.

4. MV Lyr

Variability of the object was discovered by Parenago (1946). Multiple studies were reviewed by Andronov et al. (1988, 1992b). Wenzel and Furmann (1983) reported on some regularity in the excursions from the high ($\approx 13^m$) to intermediate ($\approx 15^m$) luminosity states. The cycle length changed twice between 1928 and 1979 and was equal to 455.5 d, 325.3 d and 357.8 d. In 1979 the luminosity dropped to $\approx 17.5^m$. In 1989 the photometric behavior of the system drastically changed, into the opposite direction. The excursions to the same (!) intermediate states occurred from the low level at JD 2444317, 4720, 5015, 5470, 6352, 6638 (Andronov et al. 1988). One may say, figuratively, that "the VY Scl type of variability was changed into SS Cyg type," with a mean interval of ≈ 400 d. However, these eruptions are *not* similar to those of dwarf novae. Even if one does not accept the lower limit of the time interval 130^d, the mean amplitude is much smaller than the value 4.8^m calculated from the revised "amplitude - cycle length" relationship (Richter 1986).

An outburst up to 14.5^m near JD 2445144 was followed by a prominent sequence of outbursts with height decaying to $\approx 17^m$ over ≈ 90 d. A possible explanation of these outbursts is a trigger mechanism causing irradiation of a secondary slightly underfilling its Roche lobe (Basko and Sunyaev 1973; King and Lasota 1984).

After return of the system to its usual active state in 1989, transient QPOs with frequencies $15 - 30$ and $61 - 92$ cycles/day were detected, as well as variations with near-orbital frequency 8 cycles/day (Andronov et al. 1992b). The mean value of circular polarization is $< q > = -0.03 \pm 0.05\%$, and there is no evidence for the reality of its changes with time scales of dozens of minutes. $UBVRI$ observations show high correlation between variations at different wavelengths, especially in UBV. In RI another additional mechanism of variability with its own time scale is present (Andronov et al. 1992b).

5. TT Ari

TT Ari is one of the brightest cataclysmic variables and remains one of the most interesting objects of this class due to a variety of phenomena observed at different time scales from seconds to months. This argues for active monitoring of the object at different observatories to obtain data with smaller gaps. The first international campaign was organized in 1985, and its results were summarized by Wenzel et al. (1986).

A second campaign covered the autumn months in 1988. Our recent re-analysis of these data (Tremko et al. 1994) has led to several conclusions. Moments of 33 brightness maxima and 35 minima show large scatter of phases within an interval of $\Delta\phi = 0.6$. The ephemeris for maxima from one-frequency fits is $Max. HJD = 2447411.9779(2) + 0.132953(13)E$. Cowley et al. (1975) first mentioned a difference between the values of the spectroscopic P_{orb} and photometric P_1 periods. Udalski (1988) discussed the possible similarity of the difference between the photometric and spectroscopic period to that observed in SU UMa stars. However, the super-hump period P_{sh} is always longer than the orbital period P_{orb}. By using the "$P_{sh} - P_{orb}$" relationship with a corrected zero-point $\log P_{sh}(h) = -0.0043(33) + 1.077(6) \log P_{orb}(h)$ (Andronov 1990), one may calculate the predicted value of $P_{sh} = 0.1493 \pm 0.0007$ d, which corresponds to the spectroscopic value $P_{orb} = 0.13755114$ d (Thorstensen et al. 1985). This value corresponds neither to P_1 (which is even smaller than P_{orb}), nor to P_2. Moreover, there is no evidence for other peaks of height similar to that corresponding to P_1 or P_2.

A secondary oscillation of period $P_2 = 0.1952$ d was found in obser-vations obtained in 1985 (Wenzel et al. 1986) and in 1988 (Andronov et al. 1992d). Analysis of the complete data set showed four secondary peaks in the periodogram in the range $5.4 - 7.3$ h have nearly the same signif-icance. However, three of them are aliases of the fourth one (Tremko et al. 1992d). This secondary wave has no theoretical explanation yet.

Semeniuk et al. (1987) reported on QPOs with a time scale decreasing from 27 min in 1962 to 17 min in 1985. Such behavior may be explained by the "beat-frequency model" (Hollander and van Paradijs 1992). Peri-odograms for detrended nightly runs show significant peaks at different frequencies from 28 to 103 cycles/day, the most prominent of which cor-respond to periods of 23.7, 15.2, 27.5 and 51.6 min (Tremko et al. 1994). This argues for transient quasi-periodic oscillations occurring at few pre-ferred time scales rather than monotone decrease of the cycle length. It seems that in TT Ari we observe contributions of several instability mecha-nisms with similar time scales. At higher frequencies of $90 - 900$ cycles/day, the periodogram obeys a power law with an index $\gamma = 1.600 \pm 0.008$.

Values of $U - B = -1.04, -1.24, -1.35$ and $B - V = 0.19, 0.05, 0.12$, respectively, were determined for mean brightness, fast $(15 - 24$ min) and slow (3 h) variations. Irregular flares with an amplitude up to 1^m (An-dronov et al. 1992d) may occur on a red secondary. However, no apparent consequences on the accretion rate were detected. The next campaigns are planned in 1994 and 1995.

6. BH Lyn = PG 0818+513

This object was classified as a CV by Green et al. (1982, 1986). Spectroscopic study was made by Thorstensen et al. (1991). Andronov et al. (1992c)
collected 929 photographic observations obtained in Abastumani, Kishinev,
Moscow, and Sonneberg. Fifteen minima of this eclipsing CV since 1965 correspond to the ephemeris $Max.\ HJD = 2447180.3364(3)+0.15587490(2)E$.
The brightness out of eclipse varies from 13.7^m to 15.1^m, possibly corresponding to changes of the accretion rate.

7. HQ And

Variability of the object was discovered by Meinunger (1975). Andronov *et
al.* (1992a) detected during one night 3 waves with possible periods 119 ± 9,
40.1 ± 1.1 and 18.4 ± 0.2 min and nearly equal amplitudes $\approx 0.05\pm0.015^m$.
The mean degree of the circular polarization is $<q>= 0.5\pm0.4\%$ for 298
5-s integrations. The periodogram for the brightness variations follows the
power law with $\gamma = 2.29\pm0.12$ for frequencies $90\leq f\leq 550$ cycles/day,
contrary to AM Her, for which γ varies from ≈ 1 at the high state to ≈ 0
at the low state.

8. V795 Her = PG 1711+336

The object was classified as a CV by Green et al. (1982, 1986). Photographic
study (Wenzel et al. 1988) showed slow variations of brightness from 13.2^m
(1936) to 12.6 (early sixties) and then decline to 13.5^m in 1987. Additional
waves with a cycle ≈ 2000 d and an inactive state in 1946 with depth
$\approx 0.45^m$ resemble those of MV Lyr.

9. BZ Cam

Shakhovskoy et al. (1993b) obtained 2 nights of $UBVRI$ observations. Like-
wise TT Ari and MV Lyr, the variations in different bands are highly correlated, with the amplitude increasing with decreasing wavelength. The
periodograms show transient peaks with frequencies slightly depending on
wavelength and centered at $12.8 - 17.6$, $39.3 - 42.2$ and $66 - 70$ cycles/day.
At high frequencies, the slope $\gamma = 1.08\pm0.05$ in U and 0.86 ± 0.04 in R,
i.e., practically equal to unity.

10. EI UMa = PG 0834+488

The UV excess of this object was found by Green et al. (1982, 1986).
Thorstensen (1986) reported a 6.43-h spectroscopic period. Andronov et

al. (1992c) obtained 8 nights of observations. The mean periodogram shows peaks at 9.0 ± 0.4 and 14.5 ± 0.5 cycles/day, but the corresponding peaks occur not at all runs. There is no evidence for a photometric wave corresponding to a spectral period. Brightness variations ranged over $13.41 - 14.87^m$ (B), but no eruptions characteristic of dwarf novae or VY Scl-type "dips" were detected. A recent review on the CV candidates from the Palomar-Green Survey was by Ringwald (1993).

11. IW And

Variability was discovered by Meinunger (1975). Meinunger and Andronov (1987) found exotic photometric behavior: the object mainly spends its time in the intermediate state (72% in the interval $15.1 - 15.3^m$), which sometimes is interrupted by brightenings (18% at $13.7 - 15.0^m$) and weakenings (10% at $15.4 - 17.3^m$) with a duration of few dozen days. Once a brightness variation of 3^m in one day was detected. It seems that bright states gather with weak. In other words, contrary to polars and nova-like variables, the inactive (constant) state corresponds to intermediate luminosity. The object needs detailed study besides photometric monitoring.

12. V361 Lyr

The variability of the object was discovered by Hoffmeister (1963), who classified it as an RR Lyr star. Galkina and Shugarov (1985) pointed out its binary nature. Andronov and Richter (1987) proposed a phenomenological model of this peculiar object. The orbital period of the system $P = 0.3^d$ is close to that characteristic for CVs, although the mass gaining component is probably not a white dwarf. The extremely high depth of the primary *and* secondary eclipse allows the suggestion that the eclipsed object is not a star, but a "hot spot" between the stars, which is brighter than the stars. This model may be indirectly justified by the phase shift of the secondary minimum. For more precise determination of the model parameters, multicolor observations are needed, as well as computations taking into account the reflection effect. If the model could be justified independently, the system may be a prototype of a new class of binaries.

Acknowledgements. The author is grateful to the co-authors of the joint papers cited for fruitful collaboration, and to F. Ringwald for helpful discussions. This research was partially supported by ESO Grant A-04-018.

References

Andronov, I.L.: 1987a, *Astrophys. Space Sci.* **131**, 557.
Andronov, I.L.: 1987b, *Astron. Nachr.* **308**, 229.

Andronov, I.L.: 1990, *Astron. Tsirk.* **1545**, 18.
Andronov, I.L.: 1993, *Odessa Astron. Publ.* **6**, 21.
Andronov, I.L.: 1994, *Astron. Nachr.* **315**, 353.
Andronov, I.L., Borodina, I.G., Kolesnikov, S.V., Pavlenko, E.P., Shakhovskoy, N.M.: 1992a, *Commun. Spec. Astron. Obs.* **69**, 112.
Andronov, I.L., Borodina, I.G., Chinarova, L.L., Fuhrmann, B., Kolesnikov, S.V., Korth, S., Pavlenko, E.P., Pikhun, A.I., Shakhovskoj, N.M., Shugarov, S.Yu., Wenzel, W.: 1992b, *Commun. Spec. Astron. Obs.* **69**, 125.
Andronov, I.L., Fuhrmann, B.: 1987, *Inform. Bull. Var. Stars* **2976**, 4.
Andronov, I.L., Fuhrmann, B., Wenzel, W.: 1988, *Astron. Nachr.* **309**, 39.
Andronov, I.L., Kimeridze, G.N., Richter, G.A., Smykov, V.P.: 1992c, *Commun. Spec. AO Russ. Acad. Sci.* **69**, 102.
Andronov, I.L., Kolosov, D.E., Movchan, A.I., Rudenko, A.N.: 1992d, *Commun. Spec. AO Russ. Acad. Sci.* **69**, 79.
Andronov, I.L., Pavlenko, E.P., Seregina, T.M., Shugarov, S.Yu., Shvechkova, N.A.: 1992e, *Proc. Stellar Magnetism,* St.Petersburg, 160.
Andronov, I.L., Richter, G.A.: 1987, *Astron. Nachr.* **308**, 235.
Andronov, I.L., Vasilieva, S.V., Tsessevich, V.P.: 1980, *Astron.Tsirk.* **1142**, 5.
Basko, M.M., Sunyaev, R.A.: 1973, *Astrophys. Space Sci.* **23**, 117.
Cowley, A.P., Crampton, D., Hutchings, J.B., Marlborough, J.M.: 1975, *Astrophys.J.* **195**, 413.
Cropper, M.: 1990, *Space Sci. Revs.* **54**, 195.
Galkina, M.P., Shugarov, S.Yu.: 1985, *Peremennye Zvezdy* **22**, 225.
Gershberg R.E.: 1975, *Flare Stars of Low Masses* (in Russian), Moscow, Nauka.
Green, R.F., Ferguson, D.H., Liebert, J., Schmidt, M.: 1982, *Publ. Astron. Soc. Pacif.* **94**, 560.
Green, R.F., Schmidt, M., Liebert, J.: 1982, *Astrophys. J. Suppl. Ser.* **61**, 305.
Hoffmeister, C.: 1963, *Astron. Nachr.* **287**, 169.
Hollander, A., van Paradijs, J.: 1992, *Astron. Astrophys.* **265**, 77.
King, A.R., Lasota, J.P.: 1984, *Astron. Astrophys.* **140**, L16.
Liebert, J., Stockman, H.S., Williams, R.E., Tapia, S., Green, R.F., Rautenkranz, D., Fergusson, D.H., Szkody, P: 1982,. *Astrophys. J.* **256**, 594.
Mason, P.A., Andronov, I.L., Borisov, N.V., Chanmugam, G.: 1994, *Astron. Soc. Pacif. Conf. Ser.* **56**, 346.
Meinunger, L.: 1975, *Mitt. Veränderl. Sterne* **7**, 1.
Meinunger, L., Andronov, I.L.: 1987, *Inform. Bull. Var. Stars* **3081**, 3.
Nousek, J.A., Takalo, L.O., Schmidt, G.D., Tapia, S.,Hill, G.J., Bond, H.E., Grauer, A.D., Stern, R.A., Agrawal, P.C.: 1984, *Astrophys. J.* **277**, 682.
Parenago, P.P.: 1946, *Peremennye Zvezdy* **6**, 26.
Pavlenko, E.P., Ketsaris, N.A., Halevin, A.V., Andronov,, I.L.: 1994, *Odessa Astron. Publ.,* **7** (in press)
Richter, G.A.: 1986, *Astron. Nachr.* **307**, 221.
Ringwald, F.A.: 1993, *The Cataclysmic Variables from the Palomar-Green Survey*, Ph.D. Thesis, Dartmouth College, Hanover, New Hampshire.
Semeniuk, I., Schwarzenberg-Czerny, A., Duerbeck, H., Hoffman, M., Smak, J., Stępień, K., Tremko, J.: 1987, *Astrophys. Space Sci.* **130**, 167.
Shakhovskoy, N.M., Alexeev, I.Yu., Andronov, I.L., Kolesnikov, S.V.: 1993a, *Proc. Cataclysmic Variables and Related Physics, Annals Israel Phys. Soc.* **10**, 237.
Shakhovskoy, N.M., Efimov, Yu.S., Andronov, I.L., Kolesnikov, S.V.: 1993b, *IAU Symp. 155 Planetary Nebulae,* Kluwer, 407.
Shakhovskoy, N.M., Kolesnikov, S.V., Andronov, I.L.: 1992, *Proc. Stellar Magnetism,* St.Peterburg, 148.
Terebizh, V.Yu.: 1992, *Time Series Analysis in Astrophysics* (in Russian), Moscow, 392.
Thorstensen, J.R.: 1986, *Astron. J.* **91**, 940.
Thorstensen, J.R., Davis, M.K., Ringwald, F.A., 1991, *Astron. J.* **102**, 683.

Thorstensen, J.R., Smak, J., Hessman, F.V., 1985, *Publ. Astron. Soc. Pacif.* **97**, 437.

Tremko, J., Andronov I.L., Chinarova, L.L., Kumsiashvili, M.I., Luthardt, R., Pajdosz G., Patkos L., Rößiger S., Zoła S.: 1994, *Astron. Astrophys.* (subm. 8.10.1993).

Tremko, J., Andronov, I.L., Luthardt, R., Pajdosz, G., Patkós, L., Rößiger, S., Zoła, S., 1992, *Inform. Bull. Var. Stars* **3763**, 4.

Udalski, A.: 1988, *Acta Astron.* **38**, 315.

Wenzel, W., Banny, M.I., Andronov, I.L.: 1988, *Mitt. veränderl. Sterne* **11**, 141.

Wenzel, W., Bojack, W., Critescu, C., Dumitrescu, A., Fuhrmann, B., Götz, W., Grelczyk, H., Hacke, G, Hudec, R., Huth, H., Kozhevnikov, V.P., Kumsiashvili, M.I., Mrkos, A., Oláh, K., Oprescu, G., Patkós, L., Pereský, R., Pfau, W., Reimann, H.-G., Richter, G., Rößiger, S., Shpychka, I.V., Shult, R., Stecklum, B., Tóth, I., Tremko, J., Valníček, B., Verdenet, M.: 1986, *Preprint Astron. Inst. Czechoslovak Acad Sci.* **38**, 44.

Wenzel, W., Fuhrmann, B.: 1983, *Mitt. Veränderl. Sterne* **9**, 175.

HARD X-RAY OBSERVATIONS OF
MAGNETIC CATACLYSMIC VARIABLES

M. ISHIDA AND R. FUJIMOTO

Institute of Space and Astronautical Science
3-1-1 Yoshinodai, Sagamihara, Kanagawa 229, Japan

1. Introduction

We present the results of the observations of Magnetic Cataclysmic Variables (MCVs) by recent Japanese X-ray satellites, *Ginga* and *ASCA*.

Ginga observed more than a dozen of MCVs through its 4 and a half years life. With the large effective area and high energy sensitivity of the LAC (Large Aare proportional Counter), it is revealed that the hard X-ray emission of MCVs are generally represented by the thermal bremsstrahlung with the temperature of 10-40 keV. *ASCA* is the fourth Japanese X-ray astronomy statellite, launched on Feb. 20, 1993. It is the first satellite which carries the X-ray CCD camera whose energy resolution is 10 times better than that of the LAC onboard *Ginga*. This enables us to carry out a fine spectroscopy of the emission lines, which is completely a new method of diagnosing the postshock hot plasma.

2. *Ginga* Observations

Ginga observed 15 MCVs known at that time. Those include 7 polars (EF Eri, V834 Cen, BL Hyi, ST LMi, AM Her, BY Cam and QQ Vul) and 8 IPs (EX Hya, BG CMi, V1223 Sgr, AO Psc, FO Aqr, TV Col, RE0751+14, and GK Per). In addition, *Ginga* discovered one new intermediate polar (hereafter IP), 1H0253+193 (Kamata *et al* 1991). With the large effective area of the LAC (4000 cm^2 in 1.2-38 keV), all but ST LMi are detected and high quality spectra are obtained. The wide energy range up to ∼40 keV is of essential importance for MCVs, because the standard accretion model (Lamb 1985 and references therein) predicts that the plasma at the shock

A Bianchini et al (eds), Cataclysmic Variables, 93-101
© *1995 Kluwer Academic Publishers*

front is as hot as

$$kT_S = \frac{3}{8}\frac{GM}{R+h}\mu m_{\mathrm{H}}$$

$$= 22 \left(\frac{M}{0.6M_\odot}\right)\left(\frac{R}{5.5 \times 10^8 \mathrm{cm}}\right)^{-1}\left(1 - \frac{h}{R}\right) \quad (\mathrm{keV}), \quad (1)$$

where M and R are the mass and the radius of the white dwarf, h is the shock height, μ is the mean molecular weight (0.615 for matter with solar abundance), and m_{H} is the hydrogen mass.

The characteristics of the spectra obtained by *Ginga* are commom among polars and IPs, and can be summarized as follows (Ishida 1991);

1. The continuum is generally represented by the thermal bremsstrahlung with $kT = 10 - 40$ keV. Non thermal power-law does not work.
2. Low energy part of the spectra can be explained by the absorption by matter with solar abundance (Morrison & McCammon 1983) with at most two different column densities. The absorbing column densities are as much as $10^{24}\mathrm{cm}^{-2}$ in some cases.
3. There is a prominent iron $K\alpha$ emission line with the equivalent width of 200-800 eV.

Figure 1 shows examples of the fits for the rotational phase-averaged spectra of 6 MCVs. The spectral paramaters are summarized at the upper-right corners, and χ_ν^2's are at the lower-left corners. Left panels are the examples of the spectra whose low energy photoelectric absorption can be parameterized by a single columnd density, while the right panels are the examples which require two different column densities (hereafter partial covering model). The spectra are arranged as the hardness increases from top to bottom.

Note that the temperature obtained by *Ginga* may have to be regarded as a kind of an average over the postshock plasma, not the shock temperature in eq.(1); There is probably a temperature distribution in the postshock plasma, because the plasma is expected to be cooled as it descends the accretion column. A spectrum of a multi-temperature, optically thin plasma has a steeper photon index than a single temperature thermal bremsstrahlung in lower energy. This discrepancy is likely to be adjusted by the parameters of the absorption in the *Ginga* spectra. We also note that the absorbing matter may have a continuous distribution in the column density, and discrete two columns may be just a simple parameterization.

Figure 2(a) is a diagram between the orbital period and the temperature of the thermal bremsstrahlung obtained by *Ginga*. Each point corresponds to one object. The open circles indicate polars and the filled circles correspond to IPs. It is remarkable that there are no objects whose temperature

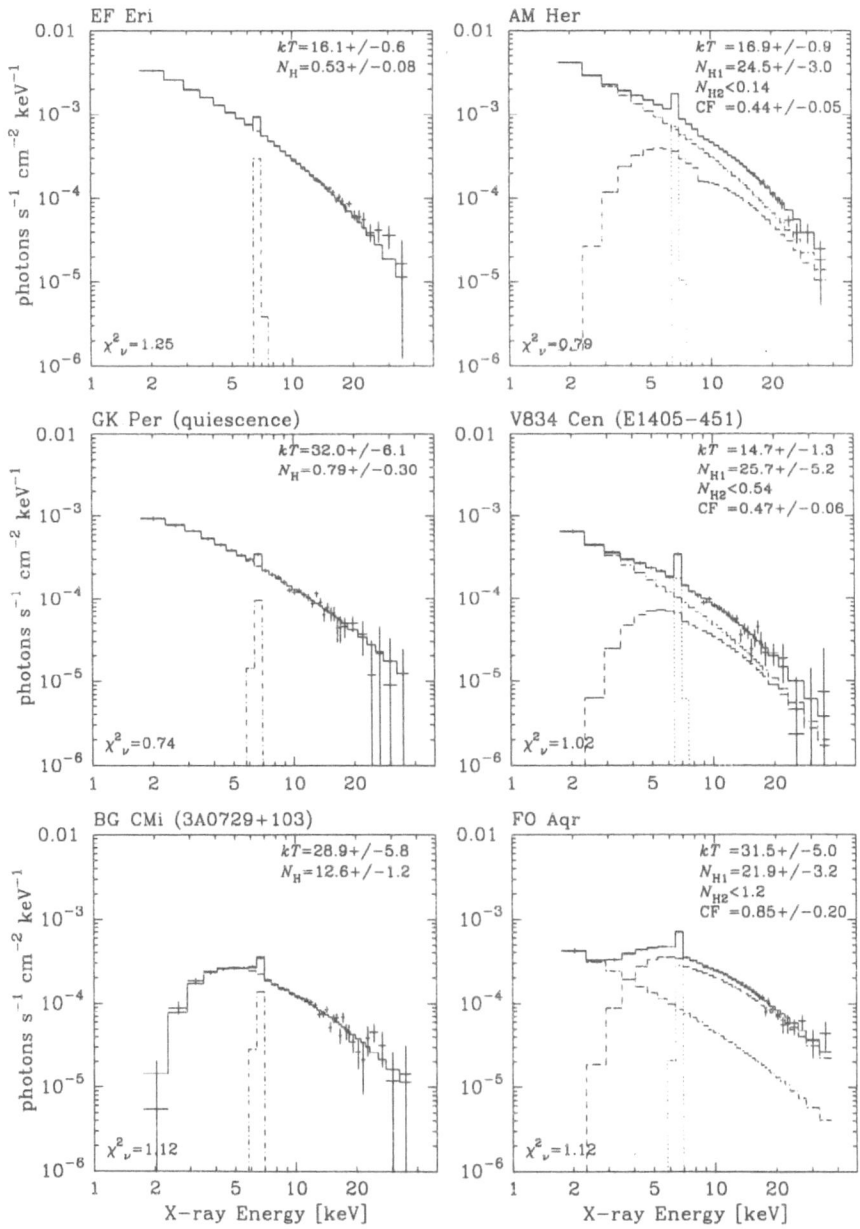

Figure 1. Incident spectra of MCVs observed by *GINGA*. Histogram shows spectral model and crosses are data points. The temperatures and the hydrogen column densities are in the unit of keV and cm^{-2}, respectively. "CF" (Covering Fraction) means the ratio of the continuum which is absorbed more heavily to the entire continuum. See text for more detail.

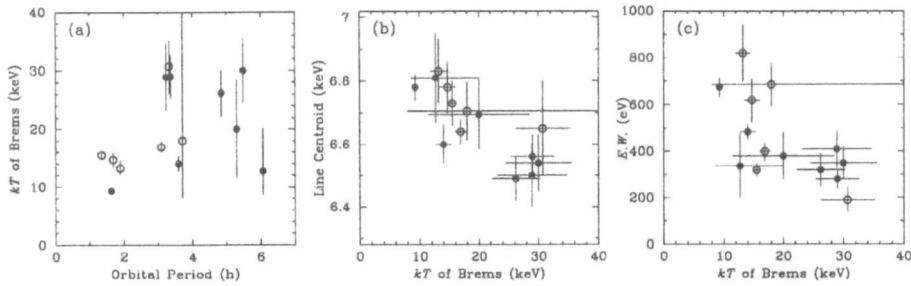

Figure 2. (a) The temperature obtained by fit to the *Ginga* spectra are plotted against the orbital period, (b) iron $K\alpha$ line central energy and (c) its equivalent width are plotted against the temperature used in (a).

exceeds 20 keV below the period gap, whereas the temperature of 50% of the sources above the gap is above 20 keV. Note that the shock height (h in eq.(1)) is very small and at most ∼5% of the white dwarf radius even if only the radiative cooling is taken into account (Aizu 1973). Therefore, assuming the mass-radius relation of the white dwarf ($R \propto M^{-1/3}$), the temperature measured by *Ginga*, although not equal to T_S, can be a relative scale of the white dwarf mass, and it is likely that the white dwarf in the system below the period gap is not as massive as that in the system above the gap on the average.

Figure 2(b) and (c) summarize the iron emission line properties. If the plasma temperature exceeds 20 keV, the plasma cannot produce any line emissions, because the iron in the plasma is completely ionized. The observed line emission is thus solely from cold matter in the preshock accretion column and the surface of the white dwarf via fluorescence. The line central energy is around 6.5 keV, close to the value of the $K\alpha$ emission line from neutral iron (6.40 keV). The line equivalent widths distributes around 300 eV. On the other hand, if the plasma temperature is lower than 20 keV, certain amount of iron populates in He-like and H-like states whose $K\alpha$ emission centroids are at 6.70 keV and 6.97 keV, respectively. As a result, the line central energy becomes an average of the plasma and the fluorescent components and becomes 6.8 keV around $kT = 10$ keV. The line equivalent width also increases by the additional contribution from the plasma component and approaches as much as 700 eV.

3. *ASCA* Observations

ASCA was launched on Feb. 20, 1993, and observed 4 MCVs in the first 6 months. One of the instrument, the SIS, uses X-ray CCD camera for the first time for the satellite, which has extremely high energy resolution

(120 eV at 6 keV). The other instrument, the GIS, has higher sensitivity in high energy, although the energy resolution is not so good as that of the SIS. *ASCA* covers 0.4-10 keV complementarily with these two detectors. In the following sections, we present the first results from *ASCA*.

3.1. FO AQR

Figure 3 is the rotational phase-averaged spectrum of FO Aqr observed by the SIS. As shown in Fig. 1, *Ginga* spectrum of FO Aqr can be pa-

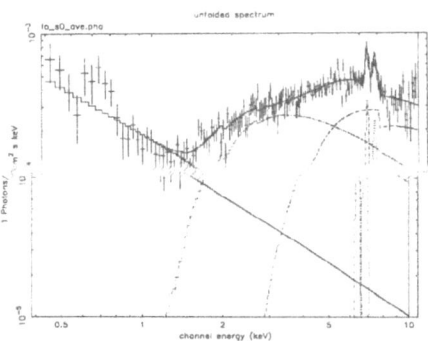

Figure 3. Rotational phase-averaged indicent spectrum of FO Aqr by the SIS.

rameterized by the partial covering abosorber model. *ASCA* observation, however, revealed that another different column is necessary to describe a spectral component which dominates below 1 keV. In Fig. 3, the three components share a common temperature of ~ 30 keV determined by *Ginga*, but with different hydrogen column densities of $< 5 \times 10^{20}$, $\sim 3 \times 10^{22}$ and $\sim 3 \times 10^{23} \mathrm{cm}^{-2}$.

Note that the origin of the softest component first detected by this observation is probably different in its origin from the other two abosrbed components, because only this component shows a strong modulation at a beat frequency, $2\omega - \Omega$, between the rotational and the orbital frequencies (ω and Ω, respectively). This component is thus inferred to be from a structure fixed in the binary frame, such as a hump in the accretion disc, which scatters the emission from the white dwarf. Alternatively, the modulation at the beat period can be regarded as an evidence of the partial accretion disc (Wynn & King 1992). In this case, the softest component is also generated near the surface of the white dwarf. Further analysis is necessary to make it clear the origin of the softest component.

The observed flux in 2-10 keV is $2.8 \times 10^{-11} \mathrm{erg\,s\,cm}^{-2}$ or 1.4 mCrab, which is almost the same as the flux obtained by *Ginga*, of which the softest component contribution is 4.4%. The iron line emission is clearly resolved into the fluorescent ($\sim$6.4 keV) and the plasma ($\sim$6.8 keV) components, although the energy scale calibration is still preliminary. The results of this observation is described in Mukai *et al* (1994) in detail.

3.2. EF ERI

Figure 4 is the average spectrum of EF Eri obtained by the *ASCA* GIS (Gas Imaging Spectrometer). Flux is $5 \times 10^{-11} \mathrm{erg\,cm}^{-2}\,\mathrm{s}^{-1}$ in 2-10 keV. The

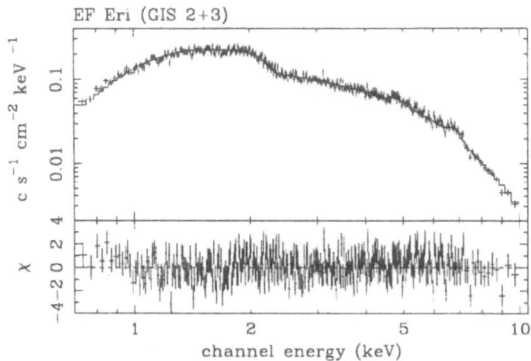

Figure 4. Rotational phase-averaged spectrum of EF Eri obtained by the GIS.

spectrum can be parameterized by the power-law with the photon index of $\sim$ 1.2 plus the iron emission line at $\sim$6.8 keV with the equivalent width of $\sim$150 eV. The temperature is evaluated to be > 50 keV if we use thermal bremsstrahlung. This implies the spectrum is harder than that obtained by *Ginga*. Also the iron line equivalent width is smaller by a factor of 2 than the *Ginga* observation.

The light curve obtained by *ASCA* shows a hump around the linear polarization phase $\phi \sim 0.15$ which is absent in the *Ginga* observation, as well as the main peak around $\phi \sim 0.5$. The overall shape is like that obtained by *Einstein* (White 1981). The detail will be described in Osborne *et al* (1995).

3.3. AM HER

Figure 5(a) is the averaged spectrum of AM Her obtained by the SIS. The spectrum can be parameterized by the partial covering thermal bremsstrahlung

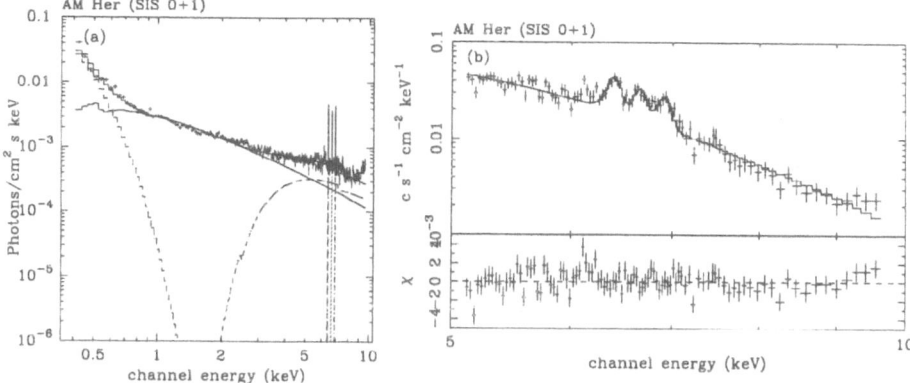

Figure 5. (a) Phase-averaged incident spectrum of AM Her obtained by the SIS, and (b) the spectrum around Fe line energy range. Note that the detector response is not deconvolved in (b).

with the temperature of 17 keV measured by *Ginga*. The flux in 2-10 keV is $4.1 \times 10^{-11} \mathrm{erg\, cm}^{-2}\, \mathrm{s}^{-1}$, which is roughly 40% of that obtained by *Ginga*. In the lower energy band, the wellknown blackbody component is detected whose temperature is 40 ± 8 eV. The flux in 0.4-0.7 keV range, in which the blackbody component dominates the spectrum, is $\sim 4 \times 10^{-12} \mathrm{erg\, cm}^{-2}\, \mathrm{s}^{-1}$.

The iron $K\alpha$ line emission is clearly resolved into three components in Fig. 5(b); a fluorescent component at 6.4 keV, and plasma components coming from the He-like and H-like states (6.70 and 6.97 keV, respectively). The intensity ratio of the latter two lines indicates the ionization temperature averaged over the entire plasma to be 12.7 ± 2.5 keV.

3.4. EX HYA

In Fig. 6(a) shown is the averaged spectrum of EX Hya obtained by the SIS. It is characterized by pairs of emission lines from He-like and H-like states of Mg, Si, S, Ar and Fe. The ionization temperatures calculated by the ratio of each pair distribute from 0.7 keV for Mg to 6 keV for Fe (Ishida *et al* 1994), indicating the continuous distribution of the temperature. This may reflect the plasma cooling below the shock, which is predicted by a number of theoretical work. We thus have attempted to see if the line intensity ratios can be explained by the radiatively cooling plasma. Once the shock temperature and the abundance are assumed, the line intensity of each element in each ionization state is calculated as an integral of the line emissivity weighted by the differential emission measure along the accretion column from the shock front to the surface of the white dwarf. Here we

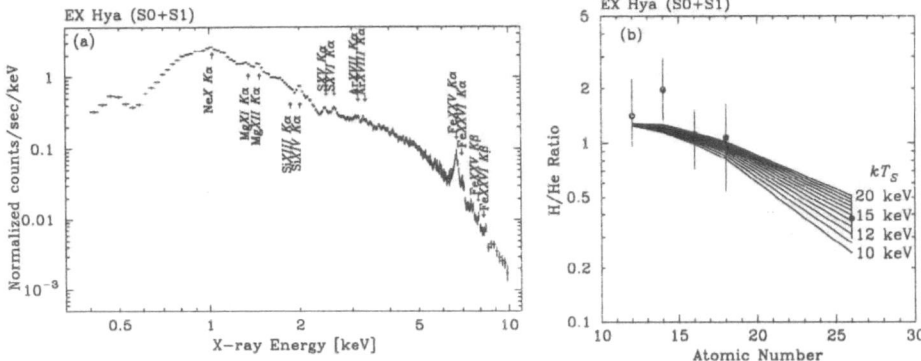

Figure 6. (a) Phase-averaged spectrum of EX Hya obtained by the SIS. The detector response is not deconvolved. (b) Observed line intensity ratio of each element in filled circles compared with those expected from radiatively cooling accretion column model with various shock temperatures.

have adopted the emissivity from Mewe *et al* (1985) and the temperature and the density profile from Aizu (1973), which assumes optically thin thermal plasma emission as a cooling mechanism. Then we have calculated the intensity ratio of the pair of lines for each element to eliminate the uncertainty of abundance. The line intensity ratio thus obtained for each element is compared with that observed in Fig. 6(b). Except for Si, whose ratio is still affected by the calibrational uncertainty of the SIS, the observed line intensity ratios can be explained by the radiatively cooling plasma if the shock temperature is 12-20 keV. Note that the shock is expected to occur near the white dwarf surface. So, assuming the mass-radius relation of the white dwarf, it is possible to evaluate the mass of the white dwarf from the shock temperature. For EX Hya, it becomes $0.5 \pm 0.2 M_\odot$ including the systematic error.

4. Summary

As shown in the previous section, the wide energy range, the high sensitivity and the high energy resolution of *ASCA* develop a new field for the research in MCVs. In particular, we can resolve the emission lines into plasma and fluorescent components. This is particularly useful for estimating the plasma temperature and the mass of the white dwarf, and probably for inferring the geometry around the shocked region.

References

Aizu, K. (1973) X-ray Emission Region of a White Dwarf with Accretion, *Progr. Theoret.*

Phys., **49**, pp. 1184–1194.

Ishida, M. (1991) *X-ray Observations of Accreting Magnetic White Dwarfs*. Ph.D. thesis, University of Tokyo (ISAS RN 505).

Ishida, M., Mukai, K. and Osborne, J. P. (1994) Observation of EX Hydrae with *ASCA*, *PASJ Lett*, **46**, pp. 81–85.

Kamata, Y., Tawara, Y. and Koyama, K. (1991) Discovery of Periodic Eclipses in the X-ray Pulsar 1H 0253+193, *ApJ Lett*, **379**, pp. 65–68.

Lamb, D. Q. (1985) *Cataclysmic Variables and Low-Mass X-ray Binaries*. D. Reidel Publishing Company, Dordrecht.

Mewe, R., Gronenschild, E. H. B. M. and van den Oord, G. H. J. (1985) Calculated X-radiation from Optically Thin Plasmas. V., *A&A Suppl.*, **62**, pp. 197–254.

Morrison, R. and McCammon, D. (1983) Interstellar Photoelectric Absorption Cross Sections, 0.03–10 keV, *ApJ*, **270**, pp. 119–122.

Mukai, K., Ishida, M. and Osborne, J. P. (1994) The *ASCA* PV Phase Observation of FO Aquarii, *PASJ Lett*, **46**, pp. 87–91.

Osborne, J. P. *et al* (1995), *MNRAS, in preparation.*

White, N. E. (1981) An 81 Minute Modulation of the X-ray Flux from 2A 0311-227, *ApJ Lett*, **244**, pp. 85–88.

Wynn, G. A. and King, A. R. (1992) Theoretical X-ray Power Spectra of Intermediate Polars, *MNRAS*, **255**, pp. 83–91.

HIGH GALACTIC LATITUDE CATACLYSMIC VARIABLES

T. AUGUSTEIJN[1] AND R. STEHLE[1,2,3]
[1] *Astronomical Institute, University of Amsterdam, Netherlands*
[2] *Max-Planck Institut for Astrophysik, Garching, Germany*

Recent population synthesis of CVs (Kolb & Ritter 1992, Stehle, Kolb & Ritter 1995) predict the distribution of observable parameters, like the distribution in period for PopI and PopII CVs. Our goal is to test the predictive power of such models by selecting two different samples of CVs, one within the galactic plane (disk–CVs) and one at high galactic latitude (HGL–CVs). The HGL–CVs are expected to be a mixture of PopI and PopII CVs, and their period distribution should be different from the disk–CVs.

To compare our models with observations we need a well–defined sample. We select our sample of HGL–CVs in the following way: **Dwarf novae:** they form a relatively homogeneous group of CVs, and from their outbursts they can be identified out to large distances; **Distance estimates using the magnitude in outburst:** Warner (1987) found that the intrinsic brightness of dwarf novae in outburst is, in contrast to quiescence, very well constrained; **z > 400 pc:** Allen (1976) gives the distance above the galactic plane (z) for PopII stars as starting at 400 pc. For CVs in the disk the scale height is $\simeq$190 pc (Patterson 1984); **b > 20°:** this to avoid including reddened disk sources. Most of the disk dwarf novae which have been studied in any detail have much brighter apparent magnitudes than the HGL–CVs, and the observational selection effects on the two samples may be very different (see, e.g., Ritter & Burkert 1986). We select our sample of disk–CVs as dwarf novae which are at a distance of more than 400 pc and have $z < 190$ pc. In this way we hope to overcome at least part of the complicated selection effects which limits the comparison of observations with the above mentioned population synthesis models. We also believe that our samples are much more homogeneous and better defined than that of Howell & Szkody (1990), who included all types of CVs, and used the magnitude in quiescence as distance indicators for dwarf novae.

---

[3]R.S. is supported by "DAAD–Doktorandenstipendium aus Mitteln des zweiten Hochschulsonderprogramms"

A. Bianchini et al. (eds.), Cataclysmic Variables, 103.
© *1995 Kluwer Academic Publishers.*

A SEARCH FOR CATACLYSMIC VARIABLES IN THE EGRET ALL-SKY SURVEY

P. BARRETT AND E. SCHLEGEL
Universities Space Research Association &
NASA/Goddard SFC, Greenbelt, MD 20771, USA

O.C. DE JAGER
Potchefstroom University for CHE, Potchefstroom, S. Africa

AND

G. CHANMUGAM
Louisiana State University, Baton Rouge, LA 70803, USA

We present results from *Compton*/EGRET observations of the entire class of magnetic Cataclysmic Variables (CV) and many recent novae. The result from this initial survey is negative with no detection greater than 2σ. The average upper limit of the luminosity of a typical (distance $\sim 100pc$) CV is $\approx 7 \times 10^{30} \; ergs \; s^{-1}$ which implies a conversion efficiency of accretion luminosity to γ-ray luminosity of $< 1\%$ for γ-rays above $100MeV$.

This low conversion efficiency places tight constraints on non-thermal models of γ-ray production from accretion-powered, magnetic compact binaries. For diffusive shock acceleration of protons which is the only process possible for the AM Herculis subclass of CVs, we obtain upper limits to the flux above $100MeV$ from VV Puppis, V834 Cen (E1405-451), and AM Herculis of about $10, 5$, and $3 \times 10^{30} \; ergs \; s^{-1}$, respectively. These flux upper limits are more than a factor of ten less than the fluxes claimed by Bhat *et al.* (1989) using COS-B data for VV Puppis and V834 Cen and about 100 less than the TeV flux from AM Her (Bhat *et al.* 1991). These results may mean that the diffusive shock process is not as important in AM Her binaries as is proposed. For the dynamo mechanism of particle acceleration which is the putative process occuring in the DQ Herculis subclass of CVs, the efficiency of converting angular momentum to γ-rays must be less than optimal. This result may be important for the production of γ-rays from neutron star binaries where the dynamo mechanism was first proposed.

A. Bianchini et al. (eds.), Cataclysmic Variables, 104.
© *1995 Kluwer Academic Publishers.*

THE ASIAGO DWARF NOVAE SURVEY

Status and some results

R. U. CLAUDI

Padova Astronomical Observatory
Vicolo dell' Osservatorio,5 I-35122 Padova,Italy

AND

A.BIANCHINI

Padova Departement of Astronomy
Vicolo dell' Osservatorio,5 I-35122 Padova,Italy

1. After two years of observations...

Our survey of faint DNe is carried out using the facilities of the ESO and Asiago Observatories.

The program has two main characteristics:

- a spectroscopic monitoring of selected DNe during their outbursts, mainly to search for orbital periods and other relevant parameters of the binary system;
- a very low resolution spectroscopic survey of the targets at light minimum in the range 3500 Å – 9000 Å to study the continuum energy distribution and the excitation level of their quiescent accretion discs.

In this first stage of the program we have mainly performed a search for orbital periods.

Statistical considerations on the outburst phenomenon (Claudi Bianchini 1994) of the Dwarf Novae suggest that in not excellent observational conditions, we can forecast to observe twelve Dwarf Novae with a ~ 79% of probability to detect at least one object in outburst.

From the end of 1989 to the beginning of 1992, we have observed a subsample of 66 northern and 63 southern DNe detecting light maxima or intermediate states for 8 northern and 6 southern objects. This would indicate a frequence of about 1 outbursting DN every 9 objects which is in good agreement with our previous prediction.

A. Bianchini et al. (eds.), Cataclysmic Variables, 105-106.
© *1995 Kluwer Academic Publishers.*

2. The preliminary Results

CZ Ori

:The star, observed during its maximum and decline phase, shows radial velocity variation with a period of 5.15 hr (Spogli & Claudi 1994). This period determination falls in the range suggested by Szkody and Mattei (1984). This support the validity of their statistical relation.

CC Cnc

: Fourier analysis of radial velocities of both absorption and emission components of Balmer lines gives as the most probable orbital period P=2.26 hr.

BZ UMa

:The spectroscopic observations show a period of 1.62 hr. This value is about half the infrared period (214 minutes) observed by Szkody and Feinswang (1988).

BV Pup

:Fourier analysis of emission Balmer lines radial velocities gives a period of 6.42 hr. This value is slightly different from the IR period of 5.4 hr (Szkody and Feinswong (1988)).

V485 Cen

:This dwarf nova shows a radial velocity modulation with a period of about 7.2 hr. There is a completely different determination of the period by T. Augusteijn (this conference) with a value of 59 min that we have not detected since our time resolution is poor ($\sim 40^m$).

V2101 Oph

:This star, classified in the General Catalogue of Variable Stars as a U Geminorum, was confirmed by Vogt et al. (1982). Three grating spectra taken during its maximum show a late – type spectrum with strong TiO bands.The spectrum obtained the following night do not change and could fitted by M5 II – III type. Thus V2101 Oph is not an U Gem Star, but more likely a semi – regular late type giant.

V598 Sco

: This star was reported by Petit (1960) in his catalogue of dwarf novae where five outbursts of the star are recorded. During our observations we obtained two consecutive spectra of the star ($m_v \sim 14$) at light maximum. As for V2101 Oph we find a late – type spectrum that can be classified as K5 – II. Thus, also in this case we can say that it is not an U Gem Star, but more likely a semi – regular late type giant.

References

Claudi R.U., Bianchini A.,1994, *In preparation*
Petit O., 1960 *J. Obs.*, 43,17
Sekiguchi K. 1992, *Nature*,358,563
Szkody P., Mattei J.A., 1984, *PASP*,96,988
Szkody P., Feinswong L., 1988, *Ap. Journ.*,334,422
Spogli C., Claudi R.U., 1994, *Astron. Astrophys.*,281,808
Vogt et al. ,1982 *Astron. Astrophys.,Suppl.*, 48,383
Warner B. 1987, *Mon. Not.*,227, 23

INFRARED OBSERVATIONS, DISTANCE DETERMINATIONS AND ABSOLUTE MAGNITUDES OF A SAMPLE OF FAINT CATACLYSMIC VARIABLES.

L.N. SPROATS & K.O. MASON
Mullard Space Science Laboratory
Holmbury St. Mary, Dorking, Surrey. RH5 6NT. England

AND

S.B. HOWELL
Planetary Science Insitute
620 N 6th Avenue, Tucson, Arizona, USA.

We have obtained infrared observations of a sample of faint (V>16) CVs which were listed as part of a high latitude CV survey by Howell & Szkody (ApJ, 1990, 356, 623). The motivation behind our work was to investigate whether our CV sample resided at great distances and to determine their absolute magnitudes. Alternatively, if they are nearby, their absolute magnitudes are very low.

The work will be presented in detail in an upcoming MNRAS article (Sproats, Howell & Mason, 1995, submitted), but our main results are summarized below:

- In all cases except one, the observed J-K colour was found to be indicative of an earlier spectral type secondary than is either predicted (by empirical formulations such as Patterson's) or sometimes observed in optical spectra of long period CVs. In one particular case, that of DX And, the J-K colour was found to indicate a "later" spectral type for the secondary than derived from optical spectroscopy.
- We find that the majority of the CVs we observed with periods below the period gap are relatively close by at d<400 pc. The longer period systems reside between ~100-1300 pc.
- For our sample, the *overall* mean M_V was found to be ~11. This is ~3.5 magnitudes fainter than the value presently assigned to typical Dwarf Novae of M_V=7.5 (Warner, 1987, MNRAS, 245, 208).

107

A Bianchini et al. (eds.), Cataclysmic Variables, 107.
© *1995 Kluwer Academic Publishers.*

TWO-HEAD PHOTOMETRY OF THE NOVA-LIKE RW TRI

C. BARTOLINI, G. COSENTINO, A. GUARNIERI, A. PICCIONI AN
L. SOLMI
Dipartimento di Astronomia - Università di Bologna
via Zamboni 33 I-40126 Bologna - Italy

RW Tri is an eclipsing binary classified as a nova-like variable because of its photometric and spectral characteristics similar to those of UX UMa and DQ Her. As in these stars, the O-C curve for the timing of the minima shows significant oscillatory departures from linear ephemeris caused by magnetic cycle in the mass losing star (Warner, 1988). The period variations and their implications on the mass transfer rate have been studied by Robinson et al. (1991), who determined also the value of $\dot{P}$=-(5.6 ± 3.3)x 10^{-12}. New accurate timings can be useful to obtain a better determination of $\dot{P}$.

New observations were carried out by means of an improved version of the double head fast photometer mounted at the Cassegrain focus of the 152 cm telescope of the Bologna observatory, on November 19, 1990 for 3 hours in V light and on January 5, 1992 for 4 hours in B light.

The observations were reduced by means of the program BIC, written by R. Silvotti, that refers the times to the center of mass of the Solar System using the Stumpf (1980) algorithm. The method of Kwee and Van Woerden (1956) was used to determine the times of the primary minima JD baricentric: 2448215.48448 ±.00010 and 2448627.30997± .00010.

These timings confirm the oscillations in the(O-C) curve with the period of 4980 days found by Robinson et al. (1991).

References

Kwee, K.K., van Woerden, H. (1956), A method for computing accurately the epoch of minimum of an eclipsing variable *Bull.Astr.Inst.Neth* **Vol. no. 12**, pp. 327–330.
Robinson, E.L., Shetrone, M.D., Africano, J.L. (1991) New eclipse timings and an upper limit to the rate of mass transfer in the novalike variable RW Tri, *Astron.J.*,**Vol. no. 102**, pp. 1176–1179.
Stumpf, P. (1980), Two self-consistent Fortran subroutines for the computation of the Earth's motion, *Astron. Astrophys. Suppl.*,**Vol. 41no. 41**, pp. 1–8.
Warner, B. (1988), Quasiperiodicity in cataclysmic variable stars caused by solar-type magnetic cycles, *Nature*,**Vol. no. 336**, pp. 129–134.

A. Bianchini et al. (eds.), Cataclysmic Variables, 108.
© *1995 Kluwer Academic Publishers.*

THE CV-LIKE SHORT TIMESCALE LIGHT VARIATIONS OBSERVED IN SYMBIOTIC STARS

D. DOBRZYCKA AND S.J. KENYON
Harvard-Smithsonian Center for Astrophysics
60 Garden Str., Cambridge, MA, 02138, USA

The Symbiotic Stars (SS) form a relatively new and still poorly investigated class of interacting binary stars. Very little is known about the nature of their hot components. There are three types of possible models: disk accretion onto a main sequence star, disk accretion onto a white dwarf, or a very hot stellar source. Rapid light variations were first observed in a few SS several decades ago (CH Cyg, T CrB and RS Oph). Further observations revealed that the amplitude of this "flickering" increases with decreasing wavelength, which implied that it must be connected with the hot components. Flickering is commonly observed in almost all cataclysmic variables (CVs) and has come to be considered an observational proof of accretion onto a white dwarf. By analogy, flickering in SS could also be interpreted in the same way, which would provide us with a confirmation of the model of a hot component as an accretion disk surrounding a white dwarf. So far, there has been only one way of testing the hot component model (Kenyon & Webbink, 1984, ApJ, 279, 252; Mürset *et al.*, 1991, A&A, 248, 458). Unfortunately, it suffers from many difficulties, as it is based on the UV spectra not available for every SS and uncertainties due to interstellar reddening. Observations of the flickering have a chance to become a new method of determining the nature of the hot component.

We present preliminary results of our search for short timescale light variations in a sample of ten SS representing a wide variety of hot components. We found flickering activity in RS Oph and CH Cyg – systems very likely containing a WD as a hot component. We also detected short timescale variability in the MWC 560 light curve. Our data sometime show periodic variations with a characteristic timescale of 23 minutes. Other data reveal more random flickering. We did not find any obvious variations in the light curves of AX Per, Z And, BX Mon, AG Peg and T CrB.

A. Bianchini et al. (eds.), Cataclysmic Variables, 109.
© *1995 Kluwer Academic Publishers.*

ANAYLSIS OF EUVE LIGHTCURVES OF AM HER SYSTEMS

STEVE B. HOWELL
Planetary Science Institute 620 N. 6th Ave.
Tucson, AZ 85705 USA

AND

MARTIN M. SIRK
Center for EUV Astrophysics 2150 Kittredge Street
University of California Berkeley, CA 94720 USA

We present EUV lightcurves for a number of AM Her systems observed either as GO targets or with the Right Angle Program. Our data is either from our own GO pointings or from the EUVE archives. We have formed lightcurves for 10 AM Hers and discuss the similarities and differences that we have found. We draw some conclusions by grouping the systems by inclination, magnetic field strength and other physical parameters. In order to understand the physical structures responsible for these EUV emissions, we have developed a software model to generate synthetic light curves. We find that the EUV accretion regions in the systems UZ For, and VV Pup cannot be fit with a flat spot confined to the white dwarf surface, regardless of its shape or brightness profile. Rather, a small, symmetric raised spot (2-4% R_{wd}) is the only shape consistent with the data. Additionally, we detect a very gradual rise and fall in the EUV flux (<5% of the peak flux) at phases when the spot is completely obscured behind the WD limb. We attribute this to emission from the accretion column at distances up to 15% of the WD radius. Other systems, such as RE1149+28, show two distinct components to their lightcurves, a symmetric hump plus a preceding shoulder. See Howell et al. (1995) and Sirk and Howell (1995) for further details.

References

Howell, S. B., Sirk, M. M., Malina, R. F., Mittaz, J. P. D., and Mason, K. O., 1995 ApJ, 1 Feb issue.
Sirk, M. M., and Howell, S. B., 1995 in preparation.

A Bianchini et al (eds), Cataclysmic Variables, 110
© 1995 Kluwer Academic Publishers

CHAOS, TIDAL CAPTURE AND CATACLYSMIC VARIABLES IN GLOBULAR CLUSTERS

ROSEMARY MARDLING
Department of Mathematics
Monash University
Wellington Rd, Clayton
Victoria 3168
AUSTRALIA

It is somewhat paradoxical that globular clusters appear to contain so few cataclysmic variables (CVs), while containing so many low-mass X-ray binaries (LMXBs). In the galactic disk, CVs outnumber LMXBs by several orders of magnitude (Iben & Livio 1993), while the situation appears to be reversed in globular clusters. In particular, LMXBs are believed to form in the cores of globular clusters via dynamical processes such as tidal capture and exchange, these processes being efficient in such environments. This leads to the paradox: given that white dwarfs outnumber neutron stars by a large factor (and taking into account mass segregation), where are all the CVs?

It may well be that CVs exist in globular clusters in reasonable numbers and are not being observed because of crowding effects. Using ROSAT, Cool et al. (1993) have observed what may be the high energy tail of a larger population of CVs. On the other hand, Shara (see this volume) has pointed out the lack of observations of CVs in outburst, despite much effort using the HST.

Bailyn et al. (1990) have suggested that the tidal capture process only makes CVs containing high mass white dwarfs because other white dwarf binaries formed in this way must go through an unstable mass transfer phase which destroys the binary. This argument is certainly compelling, although it does not take into account the evolution of binaries following capture and thus the final state of such binaries once circularized.

The author has shown (Mardling 1995) that the tidal capture process involves a short initial chaotic phase during which the orbit behaves chaotically and the tides are extremely energetic, followed by a long ($\sim 10^7 - 10^8$

A. Bianchini et al. (eds.), Cataclysmic Variables, 111-112.
© 1995 Kluwer Academic Publishers.

yr?) quiescent phase during which the tides are small and the binary circularizes via normal dissipative processes. It is not clear how a main sequence star which has captured a compact object will respond during the violent chaotic phase; in particular, dissipation may well be dominated by very short timescale processes such as shocks in the extreme outer layers of the star, as low frequency modes of oscillation drive high frequency modes via non-linear interactions. Such processes in turn affect the degree and character of any mass loss and/or transfer which may occur during periastron passage, the ability of the compact object to accrete such matter and ultimately, the state of the binary when it leaves the chaotic phase.

In order to fully understand the problem of CVs in globular cluster cores, it will be necessary to understand in more detail the orbital and stellar evolution of of binaries formed via tidal capture.

References

Bailyn, C. D., Grindlay, J. E. & Garcia, M. R. (1990), ApJ, **357**, L35
Cool, A. M., Grindlay, J. E., Krockenberger, M. & Bailyn, C. D. (1993), **410**, L103
Iben, I. & Livio, M. (1993), PASP, **105**, 1373
Mardling, R. A. (1995) ApJ, submitted

CAN WZ SGE STARS BE DISTINGUISHED FROM ORDINARY SU UMA STARS?

R.E. MENNICKENT

Departamento de Física, Universidad de Concepción, Casilla 4009, Concepción, Chile. e-mail: RMENNICK@buho.dpi.udec.cl

Mean data of WZ Sge and SU UMa stars are given in Table 1. Typical number of stars in each sample was 5 and 15 respectively. The most significant parameter to distinguish between WZ Sge and SU UMa stars is $\Delta T/T_{rec}$, as expected from the observational definition of WZ Sge stars. Significant differences are observed also in the mean P_{orb}, M_2/M_1 and W_α values, whereas UBV colors are similar. In addition, we have compared $\Delta T/T_{rec}$ with the orbital period in a sample of 4 WZ Sge, 6 SU UMa and 10 U Gem subtype dwarf novae. WZ Sge and SU UMa stars show a steep correlation: $\Delta T/T_{rec} = (-0.74 \pm 0.14) + (0.57 \pm 0.09)P_{orb}(h)$, with correlation coefficient 0.91 and $\sigma = 0.07$. In addition, when comparing the general tendency with theoretical models (e.g. Cannizzo et al. 1988, ApJ 333, 227), we have found that WZ Sge stars have very low viscosity disks during quiescence ($\alpha \sim 0.0001$) but disks with normal viscosities during eruptions ($\alpha \sim 0.1 - 1$). Table 1 shows color averages calculed from the Bruch and Engel (1994 A&AS 104, 79) UBV catalogue. Equivalent widths are from references listed in Mennickent (1994, A&A 285, 979) and from Mennickent (unpublished). Orbital periods were obtained from Molnar and Kobulnicky (1992, ApJ 392, 678), Leibowitz et al. (1994, ApJ 421, 771) and from Mennickent (this conference and unpublished). The outburst duration (ΔT) and the recurrence times (T_{rec}) were compiled from several data papers available in the literature.

TABLE 1. Comparison between WZ Sge and SU UMa star properties. The extremly low Hα equivalent width found in CU Vel could be due to a non-quiescence disk and was not included in our calculations

Quantity	WZ Sge	SU UMa
$P_{orb}(hrs.)$	0.062 ± 0.005	0.074 ± 0.003
M_2/M_1	0.14 ± 0.03	0.21 ± 0.03
W_α	122 ± 17	178 ± 29
$U - B$	-0.87 ± 0.07	-0.98 ± 0.05
$B - V$	0.11 ± 0.03	0.13 ± 0.04
$\Delta T/T_{rec}$	$0.005-0.1$	0.24 ± 0.07

A. Bianchini et al. (eds.), Cataclysmic Variables, 113.

THE SPECTRA OF OLD NOVAE

F. A. RINGWALD AND T. NAYLOR
Department of Physics, Keele University
Keele, Staffordshire, ST5 5BG, England

AND

K. MUKAI
Office for Guest Investigator Programs, Code 688
NASA Goddard Space Flight Center
Greenbelt, MD 20771, U. S. A.

The spectra of 54 objects listed by Duerbeck (1987) in A Reference Catalogue and Atlas of Galactic Novae were shown. These novae, which erupted from 1783 to 1986, were observed to study long-term spectroscopic evolution. The spectra were taken in 1990 with Lick Observatory's 3-m Shane telescope (shown at 11-Å/channel) and in 1991 with the 2.5-m Isaac Newton Telescope (binned to 11- and 21-Å/channel).

Over half the sample show obvious Hα emission, but about one-third show featureless or weak-lined spectra with Hα either missing or in absorption. Some objects, especially those in crowded fields, may not even be novae: V1330 Cyg (1970) has two blue candidates with weak lines. Many objects, especially the weak-lined ones, show strong reddening. We redo the analysis of Vogt (1990), using the correlation between Hα equivalent width and orbital inclination (Warner 1987) to correct the outburst amplitudes for inclination effects. We find this correction does narrow the scatter of the amplitudes, although evidence for a secular decline (of about 2 mag/century) is weak in both corrected and uncorrected cases.

Nebular lines are seen to fade away within two decades. There is also surprisingly little evidence for hibernation. Clear red star absorption features appear in only two objects, GK Per (1901) and V2110 Oph (1943). We collect comments of Humason (1938), McLaughlin (1953), and Greenstein (1960), all of whom give descriptions of the spectra consistent with little change over the past 30 – 50 years. We also present spectra of two old novae thought to be hibernating, U Leo (1855) and T Boo (1860).

A. Bianchini et al. (eds.), Cataclysmic Variables, 114.
© 1995 Kluwer Academic Publishers.

A LOW-DISPERSION SPECTROSCOPIC SURVEY

ROBERT CONNON SMITH
Astronomy Centre, Division of Physics and Astronomy,
University of Sussex, Falmer, Brighton BN1 9QH, UK.

MAREK J. SARNA
N. Copernicus Astronomical Center, Polish Academy
of Sciences, ul. Bartycka 18, 00-716 Warsaw, Poland.

AND

D. H. P. JONES
Royal Greenwich Observatory,
Madingley Road, Cambridge CB3 0EZ, UK.

We obtained 146 spectra of 22 cataclysmic variables on the nights of 1991 April 27 to May 1, using the ISIS triple-beam spectrograph at the Cassegrain focus of the 4.2-m William Herschel Telescope at the Roque de los Muchachos Observatory on the island of La Palma. A dichroic was used to split the beam, reflecting the blue light onto the EEV3 CCD chip in the blue arm, and the red light onto the EEV2 CCD chip in the red arm. In both arms a 158 line/mm grating was used giving a resolution of ∼ 5Å, yielding a wavelength range of $\lambda\lambda 3450$–5550Å in the blue and $\lambda\lambda 5810$–8910Å in the red. The reduction procedure followed was standard, using an optimal extraction procedure.

We have obtained spectra of 22 CVs at low resolution. In our sample we have nine U Gem–type CVs, two Z Cam, four SU UMa, two pre-cataclysmic, one AM Her–type, one intermediate polar (DQ Her), one nova-like system and two unknown. Part of our data has been published in two papers: Dhillon et al. (BH Lyn: MNRAS **258** 225 1992) and Catalán et al. (NN Ser: MNRAS **269** 879 1994). In our poster we presented the spectra for five objects (UU Aql, EY Cyg, LL Lyr, DV UMa and RE 1629+78) where the NaI doublet (at $\lambda\lambda$ 8183.3, 8194.8 Å) was detected. The ROSAT system RE 1629+78 shows a clear white dwarf/red dwarf spectrum and the four other systems show evidence for a red dwarf spectrum.

115

A. Bianchini et al. (eds.), Cataclysmic Variables, 115.
© *1995 Kluwer Academic Publishers.*

ON THE STATISTICS OF CATACLYSMIC VARIABLES

KINWAH WU
Research Centre for Theoretical Astrophysics, School of Physics,
The University of Sydney, NSW 2006, Australia

AND

DAYAL T. WICKRAMASINGHE
ANU Astrophysical Theory Centre, School of Mathematical
Science, Australian National University, ACT 0200, Australia

We have conducted a χ^2 test of the magnetic field distribution of AM Hers in comparison with isolated white dwarfs. If we use bins with field ranges of 10-29MG, 30-99MG, 100-299MG and >300MG, we obtain a χ^2 of 10.1 (3 degrees of freedom). This implies a <2.5% chance that the difference is caused by randomness. We have also conducted Kolgomorov tests for the period distribution for different cataclysmic variables subclasses (see Table 1). Our results suggest that the lack of AM Hers with fields > 100 MG is significant, and the period distribution for AM Hers and the other cataclysmic variables are different. For details of the analysis, see Wu, Wickramasinghe & Li (1995, Proc. Astr. Soc. Aus., in press).

TABLE 1. The statistics of the period distribution for different CV subclasses in the Kolmogorov two-sample test (nmCVs = non-magnetic CVs)

sample 1	N_1	sample 2	N_2	D-statistics	significance level
AM Hers	38	nmCVs	160	0.3911	00.02%
IPs	13	nmCVs	160	0.3356	13.33%
DQ Hers	7	nmCVs	160	0.2652	73.32%
IPs + DQ Hers	20	nmCVs	160	0.2812	12.01%
IPs + DQ Hers	20	AM Hers	38	0.6447	<0.01%
IPs	13	AM Hers	38	0.6640	00.04%

116

A. Bianchini et al. (eds.), Cataclysmic Variables, 116.
© *1995 Kluwer Academic Publishers.*

PHOTOMETRIC OBSERVATIONS OF NOVAE AND SYMBIOTIC STARS

G. SOSTERO
Stazione Astronomica di Remanzacco

The Stazione Astronomica di Remanzacco (St.A.R.) is a public Observatory managed by an Italian amateur astronomical society named *Associazione Friulana di Astronomia e Meteorologia* (A.F.A.M.). This Observatory is located in the Noth-East part of Italy, close to the city of Udine. Beside the cultural happenings among the schools, AFAM established a fruitful collaboration with some professional astronomers, mainly concerned on the photometric monitoring of Novae and Symbiotic Stars. To reach this goal the St.A.R. makes available the following equipment: 1) A photoelectric photometer coupled to a 0.45m Newton-Cassegrain telescope. The phototube is the well experienced 1P21, while the filter set is the same UBV combination developed by Johnson. The data handling/storing is provided on line by mean of an A/D converter plus 80486/66MHz PC. In this configuration the normal limiting magnitude is B=12, keeping the RMS lower than 0.02 magn. 2) A couple of Tessar astrographs, 300mm F:4 and 600mm F:6, mainly employed for B, V and R photographic photometry. The limiting magnitude for R band is close to 12 and 15, respectively. 3) A 185/2775mm lens doublet telescope, useful for variable stars field quick inspection and estimation. The Variable Star Section of AFAM is composed by 8 members who have developed by themselves a data aquisition system and a reduction package specifically dedicated.

References

Munari U., Sostero G., Lepardo A. Valentinuzzi T. 1993, IBVS No.3905
Munari U., Tomov T.V., Antov A., Passuello R., Sostero G., Lepardo A. Dalmeri I. 1994, IBVS No.4005
Munari U., Yudin B.F., Kolotilov E.A., Shenavrin V.I., Sostero G. Lepardo A. 1994, A.Ap. 284, L9-L12
Pssuello R., Saccavino S. Munari U. 1994, IAUC No.6065

A. Bianchini et al. (eds.), Cataclysmic Variables, 117.
© *1995 Kluwer Academic Publishers.*

2. INDIVIDUAL OBJECTS

V2 IN 47 TUCANAE: THE FIRST DETECTED DWARF NOVA OUTBURST IN THE CORE OF A GLOBULAR CLUSTER

GUIDO DE MARCHI AND FRANCESCO PARESCE

Space Telescope Science Institute
3700 San Martin Drive, Baltimore MD 21218, USA

The FOC on board the Hubble Space Telescope has been monitoring the cores of some globular clusters (GCs) in the UV since launch, to help the understanding of some peculiarities of the local stellar population. 47 Tucanae has been one of the most observed clusters, for a number of reasons. Chief among these are its still controversial dynamical state, with two red giants being ejected from the core (Meylan 1993), the discovery of 11 millisecond pulsars (Manchester et al. 1991), the presence of the brightest among the low luminosity X-ray sources known (X0021.8-7221, Hertz & Grindlay 1983) plus some dimmer X-ray sources discovered by ROSAT (Verbunt et al. 1993), and, finally, the presence of a centrally concentrated population of blue straggler stars, discovered by the FOC team (Paresce et al. 1991).

We are now progressively convincing ourselves that some, if not all, of these anomalous objects are indeed binaries, and we know and expect binaries to play a crucial role in the dynamical evolution of GCs' cores. It is legitimate, however, as Mike Shara just did in his thorough review, to wonder whether cataclysmic variables (CVs) exist at all in GCs. With their high densities and low velocity dispersions, GC cores are ideal sites for the interactions between stars, and close binary systems are expected to form numerous, through tidal capture or star–binary interaction. Theoretical calculations suggest that 36 CVs should have formed by tidal capture alone in 47 Tuc (Verbunt & Meylan 1988).

CVs, however, have proven incredibly difficult to identify unambiguously in GCs. The dwarf nova V101 in M5 is probably the only such star confirmed so far (Machin et al. 1990), with the possible identification of V4 in M30 as a cluster member. In 47 Tuc, V1, discovered by our team (Paresce, De Marchi & Ferraro 1992), is a good candidate for an asynchronously rotating DQ Her-type star, very likely the UV-optical counter-

121

A. Bianchini et al. (eds.), Cataclysmic Variables, 121-124.
© *1995 Kluwer Academic Publishers.*

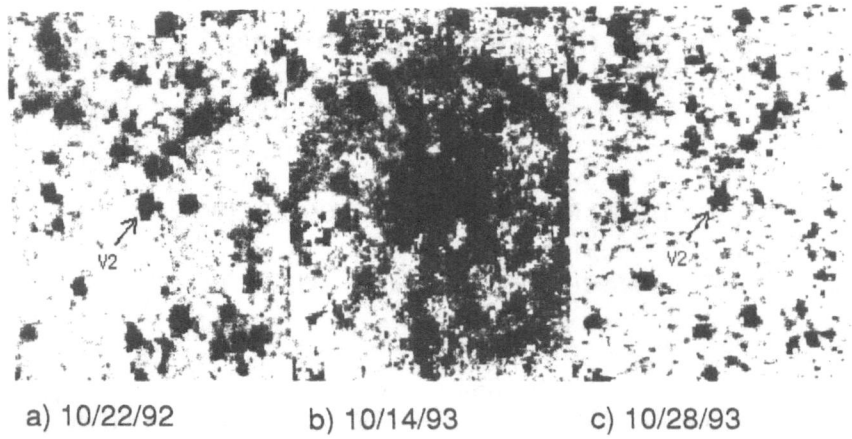

a) 10/22/92   b) 10/14/93   c) 10/28/93

Figure 1. Images of an area of 2."2 × 4."8 around V2 taken at quiescence and at outburst through the F220W filter at different epochs

part to X0021.8-7221. This identification, however, needs to be confirmed with more and better data.

And indeed, we were trying to check further about V1's nature, by taking several exposures through the FOC F/96 F220W filter over a time span of approximately 10 hours, and by adopting also the FOC near UV objective prism, to obtain a low resolution spectrum. An involuntary pointing error, however, made us observe a region a few arcseconds south of the nominal cluster center (our original target), and here is where we discovered a very bright star which we named V2 (Paresce & De Marchi 1994). This star, shown in Figure 1 (central panel) gave us some trouble about pointing at the beginning, since at that position we had never seen such a UV-bright object. We then realized that the star had to have brightened. And indeed, V2 has brightened by at least 4.2 magnitudes at 220 nm with respect to a year before (Fig. 1, left panel), reaching $m_{220} = 15.4$, but had already returned to its low state about two weeks later (right panel). We think that we caught V2 very close to its maximum, since over the 10 hr span covered by our observations we do not detect changes larger than 0.1 mag, as expected for dwarf novae at outburst.

With the canonical distance to 47 Tuc (Hesser et al. 1987) and E(B–V)=0.04 (Webbink 1985) we obtain an absolute U-band magnitude of about 2.6 ± 0.3, which compares very well with the typical absolute U magnitude of dwarf novae at outburst of about 3.2 (Warner 1987). But there is a much more compelling piece of evidence in favour of V2 as a dwarf nova at outburst, namely its spectrum, that we collected with the FOC near-UV objective prism. Figure 2 shows the spectrum of V2 between 2000 and 3500

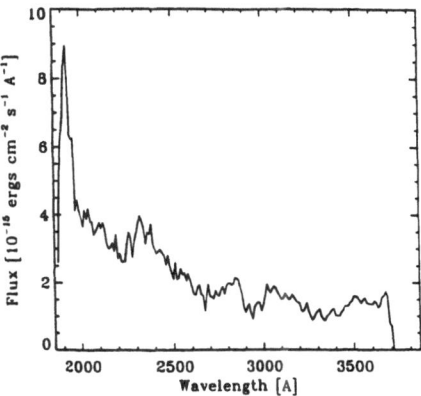

Figure 2. Spectrum of V2 in outburst determined from the FOC NUV Objective Prism

Å. The spectral resolution is only moderate, about 5 Å at 2500 Å. The three "bumps" at ~ 2300, ~ 2800, ~ 3100 Å are due to the contamination by three faint stars along the dispersion direction of V2. The sharp rise below 1950 Å may be due to edge effects at the upper limit of the image, to errors in the prism dispersion curve at the limit of its transmission, or to the strong emission line of C III at 1909 Å. Because of these uncertainties, we limit our investigation to the range 2000 to 3500 Å.

The dereddened continuum of V2 in this spectral range is well fit by a power law distribution, with a slope of 2.3 ± 0.2, which is in excellent agreement with the average slope of ~ 2 found by Verbunt (1987) for dwarf novae at outburst. All these elements taken together ($\Delta m \geq 2.4$, $M_U \simeq 2.6$, $\alpha \simeq 2.3$) strongly support the idea that V2 is a dwarf nova, the first clearly identifiable object of this kind right in the core of a globular cluster.

An intriguing question still open at this point is whether V2 has anything to do with the X-ray sources that inhabit the core of 47 Tuc. X0021.8-7221 (Auriére et al. 1989) is the brightest among the low luminosity X-ray sources. V2 does not fall within the 3σ error ellipse for its position, and indeed there is a much better candidate to its UV-optical counterpart. This is star V1, discovered by our team (Paresce, De Marchi & Ferraro 1992) with the FOC, which has a broad band UV spectrum compatible with a 20,000 K black body, and possibly a 6 hour periodicity. These elements, along with the X-ray to UV luminosity ratio, suggest that V1 is an asynchronously rotating DQ Her-type star.

Verbunt et al. (1993) reported recently the discovery with ROSAT of four other X-ray sources in the core of 47 Tuc. Their positions, however, are only approximately known, so that the identification of a counterpart is still impossible. In addition, these sources seem to be an order of magnitude too bright, in the ROSAT band (0.5–2.4 KeV), to be CVs. This has led

Verbunt et al. to conclude that they may possibly be neutron star binaries. This means that to find the X-ray counterpart to V2 we might need higher sensitivity measurements, like those carried out successfully with ROSAT by Cool et al. (1993) in NGC 6397. They detect a number of X-ray sources with luminosities compatible with that of CVs (of order 10^{32} erg). One of these sources is multiple and sits at the very center of the cluster.

With the FOC we had already found four very blue stars within 6" radius of the cluster center (De Marchi & Paresce 1994). They appear 1 to 2 magnitudes to the left of the TO in our UV color magnitude diagram, with a broad band spectrum compatible with a $15 - 20,000$ K black body. Some of these, but possibly all, show temporal variability, very well evident in the case of DP3, with a clear ~ 0.6 mag modulation over about 4 hours.

In summary, if V1 is confirmed as a DQ Her type star, V2 is the second CV discovered in the core of 47 Tuc. To the list of CVs discovered so far by the FOC we should add the four UV-bright, possibly variable stars found in the core of NGC 6397, which are good candidate to the optical counterpart to CVs detected by ROSAT. On the opposite side, V101 in M5 and perhaps V4 in M30 are the only CVs seen from the ground inside a globular cluster. Then, the discoveries of the HST go a long way towards reconciliation of theory with observation: If a random and sporadic search with the FOC, such as ours, has already turned up several good candidates, we may very well expect that a more systematic and thorough approach should reveal quite a few more. Perhaps as many as predicted by Verbunt & Meylan (1988), or actually less, if tidal capture leads to unstable mass transfer as suggested by Bailyn, Grindlay & Garcia (1990).

References

Auriére, M., Koch-Miramond, L., & Ortolani, S. 1989, A&A, 214, 113
Cool, A., et al. 1993, ApJ, 410, L103
De Marchi, G. & Paresce, F. 1994, A&A, 281, L13
Hertz, P. & Grindlay, J.E. 1983, ApJ, 275, 105
Hesser, J., et al. 1987, PASP, 99, 739
Machin, G., et al. 1990, in Accretion Powered COmpact Binaries, ed. C.W. Mauche (Cambridge: CUP), 163
Manchester, R.N., et al. 1991, Nature, 352, 219
Meylan, G. 1993, in Ergodic Concepts in Stellar Dynamics, ed. D. Pfenninger & V.G. Gurzadyan (Berlin: Springer), in press
Paresce, F. et al. 1991, Nature, 352, 297
Paresce, F. & De Marchi, G. 1994, ApJ, 427, L33
Paresce, F., De Marchi, G., & Ferraro, F. 1992, Nature, 360, 46
Verbunt, F. 1987, A&AS, 71, 339
Verbunt F. & Meylan. G. 1988, A&A, 203, 297
Verbunt, F., et al. 1993, Adv. Sp. Res., 13, 151
Warner, B. 1987, MNRAS, 227, 23
Webbink, R.F. 1985, IAU Symp. 73, 85

SPECTROPOLARIMETRY OF V1315 AQL

An Attempt To Explain The SW Sex Phenomenon

V.S. DHILLON AND R.G.M. RUTTEN
Royal Greenwich Observatory
Apartado 321, Santa Cruz de La Palma, Canary Islands, Spain

1. Introduction

Recent years have seen the emergence of a group of nova-like variables which exhibit remarkably similar behaviour. To date, there exist five such well-studied systems: V1315 Aql (Downes et al. 1986; Dhillon, Marsh and Jones 1991), SW Sex (Penning et al. 1984; Honeycutt, Schlegel and Kaitchuck 1986), DW UMa (Shafter, Hessman and Zhang 1988; Dhillon, Jones and Marsh 1994), BH Lyn (Thorstensen, Davis and Ringwald 1991; Dhillon et al. 1992) and PX And (Thorstensen et al. 1991; Still, Dhillon and Jones 1994). All five objects are eclipsing systems with orbital periods lying in the narrow range of 3.24 to 3.74 hr; all five objects exhibit single-peaked Balmer and HeI emission lines which remain largely unobscured during primary eclipse and which show strong absorption features around the inferior conjunction of the emission-line source; all five objects show single-peaked high-excitation emission lines which are eclipsed by the secondary; and all five objects show radial velocity curves with significant phase shifts relative to photometric conjunction. In addition, a number of systems are believed to be related to these five, exhibiting some, but not all, of the above features: WX Ari (Beuermann et al. 1992), UU Aqr (Diaz and Steiner 1991) and LB1800 (Buckley et al. 1990).

Thorstensen et al. (1991) coined the term 'SW Sex stars' to describe this group of objects. To date, no theoretical model has been able to unambiguously satisfy the observational constraints imposed by these systems (Dhillon et al. 1991). One possible solution to the problem, first proposed by Honeycutt et al. (1986), invoked the existence of a wind region extending above and below the accretion disc, which remains visible during primary eclipse. However, although the abundant IUE observations of ultraviolet lines from the wind lends credence to a wind model, their high ionization implies that the wind can contribute little to the Balmer lines (Hoare 1994).

125

A. Bianchini et al. (eds.), Cataclysmic Variables, 125-128.
© *1995 Kluwer Academic Publishers.*

An alternative possibility is that the Balmer lines originate in the accretion disc and are scattered into the line of sight by the wind. This would explain the uneclipsed single-peaked lines, since we would be in effect looking at the accretion disc at a low inclination (Rutten and Dhillon 1992).

One way of testing this hypothesis is by observing an SW Sex star in polarized light; an asymmetric distribution of electron-scattering material, such as a wind, relative to a source of illuminating radiation, such as the disc, might be expected to produce linear polarization. Szkody et al. (1982) and Cropper (1986) have already shown that the light in some nova-like variables is linearly polarized by typically up to a few tenths of a percent, but as yet it is unclear whether this polarization is intrinsic or interstellar. These studies used broad-band polarization measurements, which do not allow one to separate the contributions from spectral lines and the continuum. This is unfortunate, since any unpolarized continuum might dilute strongly polarized line emission in broad-band polarimetry, making it difficult to determine whether an intrinsic component is present. Spectropolarimetry permits a separation of the line and continuum regions, providing a more accurate way of studying linear polarization and its variability. In this paper we present such a spectropolarimetric study of V1315 Aql.

2. Results

We followed V1315 Aql for approximately one orbital period with the 4.2 m William Herschel Telescope on La Palma on 1993 July 26/27. The upper panel of Figure 1 displays the resulting summed spectrum. No corrections for instrumental response and atmospheric losses have been made and therefore the shape and level of the continuum remains uncalibrated in terms of flux. The spectrum appears almost identical to the one presented by Downes et al. (1986), displaying single-peaked Balmer, HeI, HeII, CIII/NIII and CII emission lines. In addition, the spectrum appears to show absorption features due to CaII and FeII.

The lower panel of Figure 1 displays the resulting linear polarization spectrum of V1315 Aql, consisting of the mean of four individual measurements each of length 40 mins. Two curves are presented in Figure 1: the solid curve represents the polarization spectrum binned to yield errors of 0.10% in each interval. This curve indicates that there are no significant differences between the line and continuum polarizations and there are no gross variations with wavelength. However, since the noise associated with the normalized Stokes parameters always enhances the polarization value, the mean level is biased towards a higher value. To determine the mean level we therefore binned the data to yield a single bin spanning the entire wavelength range (with a polarization error of 0.02% – this is a statistical

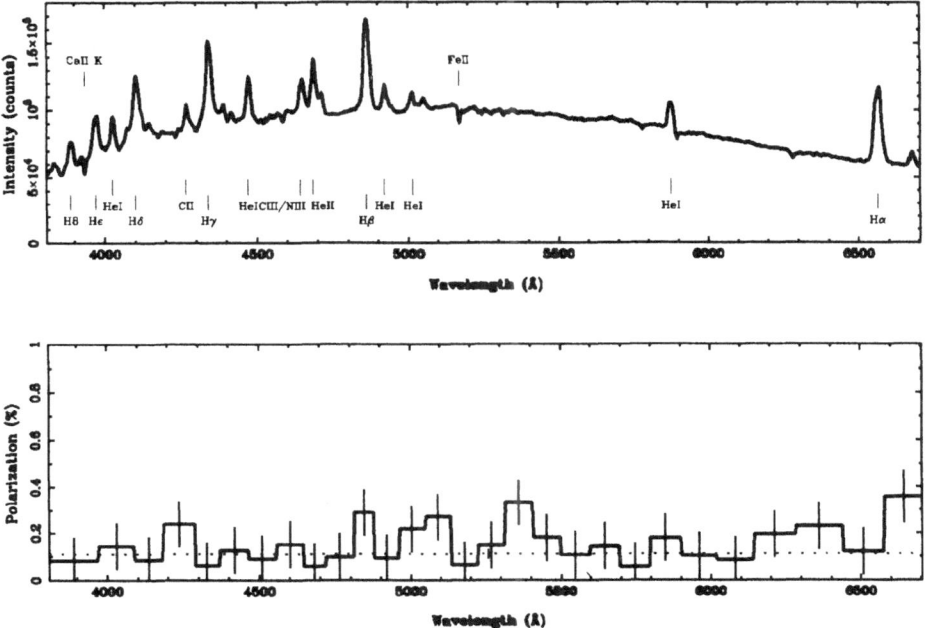

Figure 1. Upper panel: Summed spectrum of V1315 Aql. Lower panel: Percentage linear polarization spectrum of V1315 Aql.

value and does not include systematic errors). The result is shown by the dotted line in Figure 1, which gives a mean polarization for V1315 Aql in this wavelength range of 0.11%. The polarization values determined from the four individual sets of observations over the orbital period are equivalent to within the measurement errors, and hence there is no evidence for any dependence of polarization on binary phase.

3. Discussion

Our observations show that there is evidence for only very low levels of polarization in the spectrum of V1315 Aql. The question naturally arises, is the polarization we measure intrinsic to the system? It is well known that interstellar absorption by dust grains aligned in the galactic magnetic field causes linear polarization. To estimate the fraction of the measured polarization due to interstellar material we can use the result of the polarization survey by Serkowski, Mathewson and Ford (1975), which provides an estimate of the maximum level of interstellar polarization as a function of colour excess: $P_{max} \leq 9.0$ E(B–V). The colour excess of V1315 Aql can be determined from IUE spectra of the interstellar absorption bands around 2200Å, which gives a value of E(B–V)=0.0 (Rutten, van Paradijs and Tinbergen 1992), or from the distance of V1315 Aql (300 pc; Rutten et al. 1992), which gives a (less accurate) mean value of E(B–V)=0.14 (Scheffler 1967). Adopting E(B–V)=0.05, this yields P_{max}=0.45%, which is sufficiently high to explain some, if not all, of the observed polarization. From the dependence of interstellar polarization on wavelength (Serkowski et al. 1975), the polarizations at 3900Å and 6600Å are expected to be approximately 87% and 96% respectively, of P_{max} (at ∼5500Å), which is in broad agreement with the observed data (see Figure 1). In conjunction with the fact that there is no evidence of orbital variability in the polarization signal, the above arguments lead us to conclude that the low level of linear polarization we have measured in V1315 Aql is consistent with the expected behaviour of the insterstellar medium and hence that there is no evidence for any polarized emission intrinsic to V1315 Aql.

This result does not necessarily discount electron scattering of disc emission in accretion disc winds as the cause of the uneclipsed, single-peaked lines observed in V1315 Aql and other nova-likes. One alternative possibility is that the line emission is multiply scattered in the wind. Another option is that the Balmer line radiation is for the larger part produced in a wind with a non-radial outflow (Hoare 1994).

References

Beuermann K., Thorstensen J. R., Schwope A. D., Ringwald F. A., Sahin H., 1992, A&A, 256, 442
Buckley D. A. H., Sullivan D. J., Remillard R. A., Tuohy I. R., Clark M., 1990, ApJ, 355, 617
Cropper M., 1986, MNRAS, 222, 225
Dhillon V. S., Jones D. H. P., Marsh T. R., 1994, MNRAS, 266, 859
Dhillon V. S., Marsh T. R., Jones D. H. P., 1991, MNRAS, 252, 342
Dhillon V. S., Marsh T. R., Jones D. H. P., Smith R. C., 1992, MNRAS, 258, 225
Diaz M. P., Steiner J. E., 1991, AJ, 102, 1417
Downes A. R., Matteo M., Szkody P., Jenner D. C., Margon B., 1986, ApJ, 301, 240
Hoare M. G., 1994, MNRAS, 267, 153
Honeycutt R. K., Schlegel E. M., Kaitchuck R. H., 1986, ApJ, 302, 388
Penning W. R., Ferguson D. H., McGraw J. T., Liebert J., Green R., 1984, ApJ, 276, 233
Rutten R. G. M., Dhillon V. S., 1992, A&A, 253, 139
Rutten R. G. M., van Paradijs J., Tinbergen J., 1992, A&A, 260, 213
Scheffler H., 1967, Z. Astrophys., 65, 60
Serkowski K., Mathewson D. S., Ford V. L., 1975, ApJ, 196, 261
Shafter A. W., Hessman F. V., Zhang E. H., 1988, ApJ, 327, 248
Still M. D., Dhillon V. S., Jones D. H. P., 1994, MNRAS, in press
Szkody P., Michalsky J. J., Stokes G. M., 1982, PASP, 94, 137
Thorstensen J. R., Davis M. K., Ringwald F. A., 1991, AJ, 102, 683
Thorstensen J. R., Ringwald F. A., Wade R. A., Schmidt G. D., Norsworthy J. E., 1991, AJ, 102, 272

V485 CEN: A DWARF NOVA WITH A 59 MIN ORBITAL PERIOD

T. AUGUSTEIJN

thomas@astro.uva.nl
Astronomical Institute, University of Amsterdam, Netherlands

Abstract. From time-resolved spectroscopy of the Hα emission line it is found that the orbital period of the dwarf nova V485 Centauri is 59 min. This is well below the well known cut-off in the orbital period distribution of cataclysmic variables at ∼80 min.

1. Introduction

The variable star V485 Cen is classified as a U Gem dwarf nova type cataclysmic variable (CV). In an earlier article (Augusteijn et al. 1993) we reported the discovery of a 59 min photometric period in this dwarf nova. The average light curve shows a ∼0.2 mag "hump" which extends for approximately half a cycle. Here I present time resolved spectroscopy of V485 Cen to clarify the nature of this photometric period.

2. Spectroscopy

A total of 31 spectra, covering the wavelength range 5200–6950 Å at a resolution of 5 Å, were obtained with an integration time of 9 min on March 23rd 1993, with the 3.6m telescope at the European Southern Observatory in Chile. The slit was orientated such that a nearby star was observed simultaneously. For each exposure the ratio was determined of the spectrum of the second star on the slit, with a flux calibrated spectrum of this "comparison" star (approximate spectral type G2 v). This "ratio" spectrum was used to obtain a flux calibrated spectrum of V485 Cen in the same exposure. In Fig. 1 I show the average flux calibrated spectrum of V485 Cen.

A Bianchini et al (eds) Cataclysmic Variables, 129-132

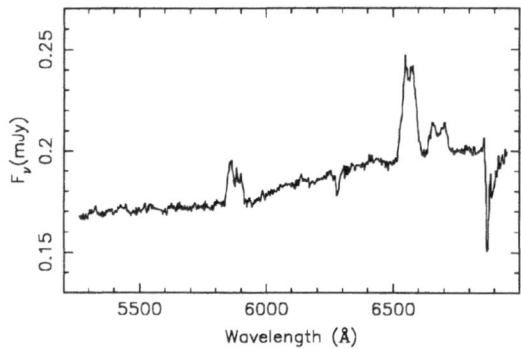

Figure 1. The average flux calibrated spectrum of V485 Cen

To look for radial-velocity variations in the individual spectra we used the double Gaussian convolution technique (see, e.g., Shafter et al. 1986). We looked for periodic variations in the radial velocity measurements for Hα and He I 6678 Å using a range in separations between the two Gaussians of 200–3000 km/s. Significant variations were only found in the Hα line. In the middle and bottom panels of Fig. 2 we present a Fourier spectrum analysis using the Lomb-Scargle method and the CLEAN algorithm of the radial velocities of Hα for separations between the two Gaussians of 600 and 1800 km/s, respectively. In the top two panels of Fig. 2 we present a Fourier spectrum and the CLEANed spectrum of the total flux of each individual spectrum integrated over the wavelength range 5200–6950 Å. It is clear from Fig. 2 that the brightness and the radial velocities vary with the same period within the accuracy of our data, and these periods are equal to within their error with the previously detect 59 min photometric period. We do not detect any significant radial velocity variations with a period other than the $\sim 1^{\text{hr}}$ period. The peak seen at low frequency in the Fourier spectrum of the total intensity shown in the top two panels of Fig. 2 are the result of a small off-set between the intensity scale of the first and second half of the observations as a result of a slight difference in the position of the slit with respect to V485 Cen and the second star on the slit.

In Fig. 3 we present a grey-scale plot of the Hα emission line as a function of phase at the 59 min period, where phase zero corresponds to photometric maximum. From this figure it can clearly be seen that there are two components which vary with this period. The velocity variations of these two components have been determined separately, and the resulting orbital element are presented in Table 1. In this table phase refers to superior conjunction of the line emission region with respect to photometric maximum. If, like in other dwarf novae, maximum light corresponds to viewing the hot spot face-on, the phasing of the two components presented in Table 1 are consistent with the identification of the component which

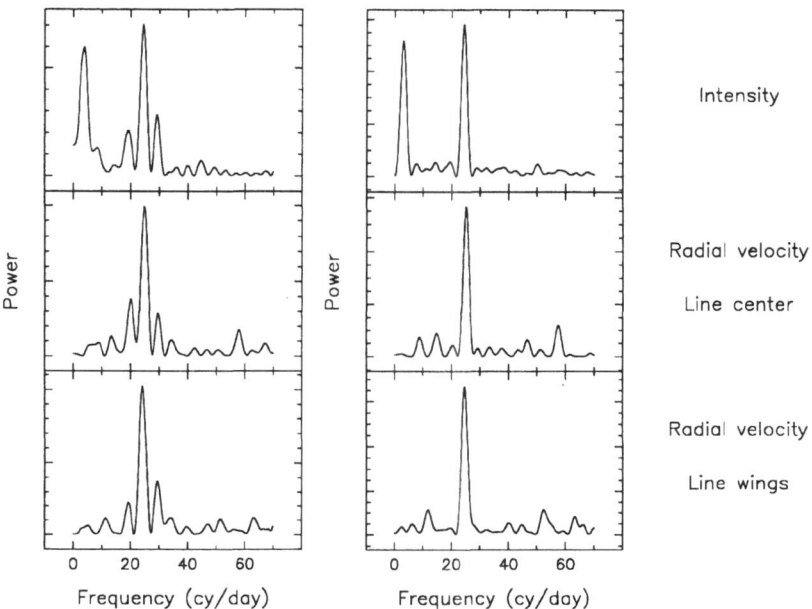

Figure 2.   In the top two panels we show a Fourier spectrum analysis using the Lomb-Scargle method (on the left) and the CLEAN algorithm (on the right) of the integrated flux over the wavelength range 5200–6950 Å. In the middle and bottom panel we present a Fourier spectrum and the the CLEANed spectrum of the radial velocities of Hα for separations between the two Gaussians of 600 and 1800 km/s, respectively

dominates the radial-velocity variations in the line center with emission from the hot spot (the socalled S-wave), and the component which dominates the line wings with emission from the accretion disk center on the white dwarf primary. I, therefore, conclude that the 59 min period can be identified with the orbital period of the system.

3. discussion

The only physical property known accurately for a large number of CVs is the orbital period (Ritter & Kolb 1994). The period distribution of CVs

TABLE 1. Orbital elements

	γ (km/s)	K (km/s)	Phase
line center	19 (28)	389 (38)	0.551(16)
line wings	12 (7)	79 (10)	0.138(20)

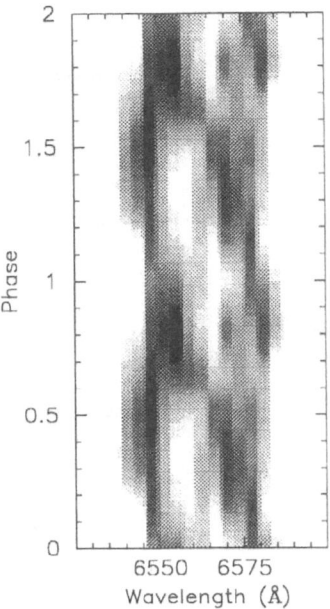

Figure 3. Grey-scale plot of the Hα emission line as a function of phase at the 59 min period. In this figure the data have been smooth in the phase direction for representation purposes only. Phase zero corresponds to photometric maximum. Two periods are shown for clarity

has a cut-off at a minimum period of ∼80 min. There is presently only one group of CVs known with orbital periods less than 80 min: the AM CVn stars. They are characterized by the total absence of hydrogen lines in their spectra, and they show photometric and/or spectroscopic periods of 17.5–46.5^m. These systems are probably CVs containing a (helium degenerate) white-dwarf secondary. However, the detection of Hα in the spectrum of V485 Cen excludes the possibility that the source is an AM CVn type CV.

Theoretical calculations have been made to determine the minimum period of CVs as a function of hydrogen content of the secondary (see, e.g., Sienkiewicz 1984, Nelson et al. 1986), which show a smooth decrease in the minimum period as a function of the hydrogen content, and I conclude that the secondary in V485 Cen probably has a low, but finite, hydrogen content. A more detailed discussion of the observations and the evolutionary status of the system is presented elsewhere (Augusteijn et al. 1995).

References

Augusteijn, T., van Kerkwijk, M.H., van Paradijs, J. 1993, A&A, 267, L55
Augusteijn, T., van der Hooft, F., de Jong, J.A., van Paradijs, J. 1995, A&A, in press
Nelson, L.A., Rappaport, S., Joss, P.C. 1986, ApJ, 304, 240
Ritter, H., Kolb, U. 1994, in *X-ray Binaries*, Eds. W.H.G.Lewin, J. van Paradijs and
    E.P.J. van den Heuvel (Cambridge Univ. Press), in press
Shafter, A.W., Szkody, P., Thorstensen, J.R. 1986, ApJ, 308, 765
Sienkiewicz, R. 1984, Acta Aston., 34, 325

EUVE OBSERVATIONS OF U GEM

KNOX S. LONG
Space Telescope Science Institute
3700 San Martin Dr., Baltimore, MD 21218

CHRISTOPHER W. MAUCHE
Lawrence Livermore National Laboratory
L-401, P.O. Box 808, Livermore, CA 94450

PAULA SZKODY
Dept. of Astronomy, University of Washington
Seattle Washington 98195

AND

JANET A. MATTEI
American Assoc. of Variable Star Obs.
25 Birch St., Cambridge, MA 02138

Abstract. We have observed U Gem during the peak and declining phases of a wide outburst in 1993 December with *EUVE*. At peak, U Gem was one of the brightest EUV sources in the sky. The spectrum of the source is complex. Fitted to a blackbody spectrum, the apparent temperature at peak is $\sim$ 130,000 K, the luminosity is $\sim 6 \times 10^{34}$ ergs s^{-1}, and the minimum size of the emitting region is comparable to that of a white dwarf. If the EUV emission arises primarily from the boundary layer, then the boundary layer luminosity in U Gem is comparable to the disk luminosity. The source counting rates are strongly modulated with the orbital phase.

1. Introduction

U Gem is the prototypical dwarf nova, undergoing quasiperiodic outbursts of up to six magnitudes at optical wavelengths. It is also one of the very few dwarf novae in which the soft X-ray flux is known to increase greatly (20–50 times) during optical outburst (Córdova & Mason 1984; Córdova et

A. Bianchini et al. (eds.), Cataclysmic Variables, 133-138.
© 1995 Kluwer Academic Publishers.

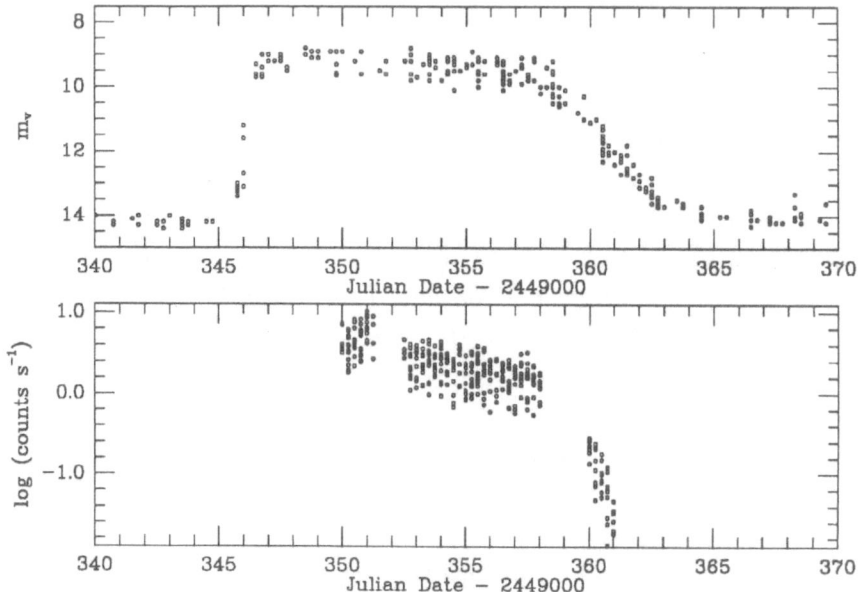

Figure 1. The AAVSO visible light curve and the count rate from U Gem as observed with the short wavelength spectrometer on *EUVE*.

al. 1984), as predicted by simple accretion disk and boundary layer theory. In order to study the soft X-ray spectrum of U Gem in detail, we recently observed U Gem with *EUVE*. Since all earlier observations of the soft X-ray spectrum were made with proportional counters ($\lambda/\delta\lambda < 1$ at 100 Å), these are the first observations with sufficient spectral resolution ($\lambda/\delta\lambda \sim 100$) to characterize accurately the soft X-ray (or extreme ultraviolet) spectrum of U Gem. Here we give a preliminary report of the observations and the results which are being obtained.

2. Observations

The observations began on 1993 December 28 following reports on the evening of 1993 December 24 that an outburst of U Gem had begun. Three sets of observations were carried out extending through 1994 January 8, covering the peak and much of the decline of the optical outburst.

At peak, U Gem was a bright source for *EUVE*. The average counting rate from U Gem in the short wavelength (SW) spectrometer declined from $\sim 5 \, \text{cts s}^{-1}$ to $\sim 2.1 \, \text{cts s}^{-1}$ to $\sim 0.1 \, \text{cts s}^{-1}$ during the observations. The AAVSO light curve and the light curve obtained from the SW spectrometer

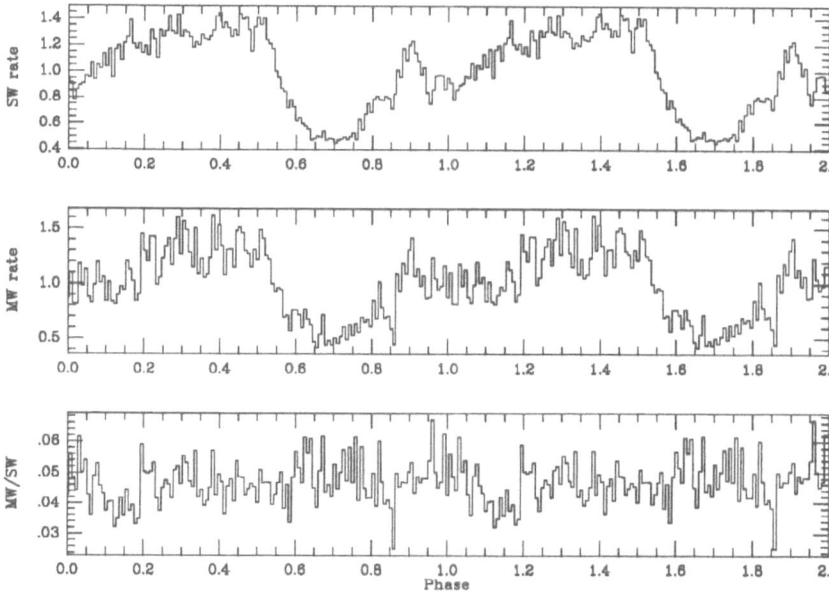

Figure 2.   The normalized count rate observed with *EUVE* during the second observing interval folded on the orbital period. Phase 0 corresponds to secondary conjunction. The upper panel is from the short wavelength spectrometer (75–180 Å); the second panel is from the medium wavelength spectrometer(140–210 Å); and the bottom panel is the ratio of the count rates in the two spectrometers. No major changes in the spectrum are seen during the absorption dips near phase 0.7.

are shown in Fig. 1. Outbursts of U Gem are characterized as narrow or wide, depending on the length of time at maximum. The December outburst is typical of a wide event. During the peak of the optical outburst, the *EUVE* light curve exhibits a slow, relatively continuous decline on long (12 hour) timescales but shows considerable variability on short (100 second) timescales.

Folding the light curve on the orbital period of U Gem revealed that much of the variability is due to orbital variations. This is illustrated in Fig. 2, which is the light curve obtained from interval II. The folded light curves are complex. There are detailed differences in the light curves from intervals I and II as well as from orbit to orbit, but the broad minimum near phase 0.7 and the secondary minimum near phase 0.0 or 0.1 were seen throughout intervals I and II. The count rates were lower during interval III, but it is quite clear that the dip near phase 0.7 is still present.

The fluxed spectrum obtained from the SW spectrometer during interval

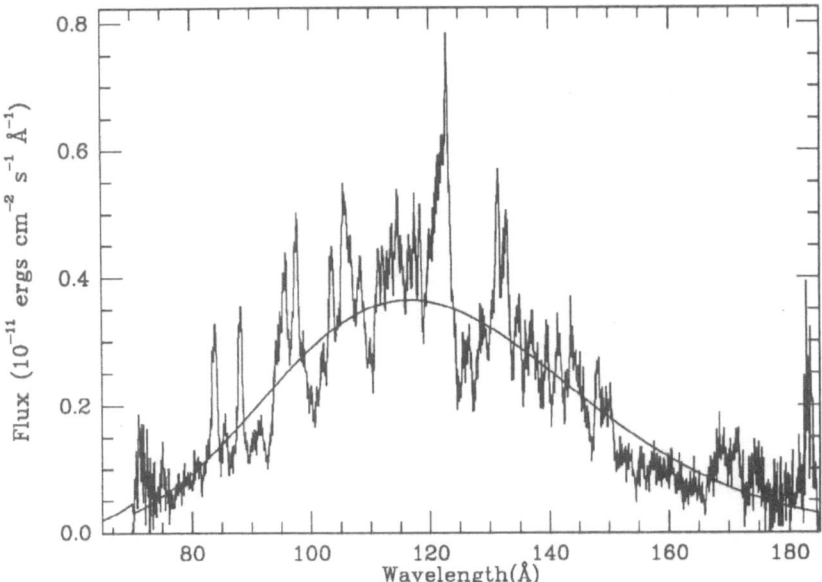

Figure 3.   The fluxed spectrum in the short wavelength spectrometer from the second observing interval fit to a black body spectrum assuming that $N_H = 3 \times 10^{19}$ cm^{-2}. The best fit has $T = 125{,}000$ K.

II is shown in Fig. 3. The spectrum is quite complex, very different from that expected from a hot white dwarf surface or from an optically thin plasma. Individual peaks can be identified with lines expected in a low-density, collisionally ionized plasma with a temperature of order 3×10^5 K. The solid lines give the best fit black body approximations to the spectrum assuming that N_H is 3×10^{19} cm^{-2}, as suggested by Mauche (1991), based on an analysis of interstellar absorption lines observed in high resolution *IUE* spectra of U Gem in outburst. The best fit temperatures are 134,000 K and 125,000 K for the SW spectra from the first and the second observing intervals respectively.

3.   Discussion

In the standard model for dwarf novae in outburst (Lynden-Bell & Pringle 1974), the UV and optical emission from the disk accounts for 1/2 of the accretion luminosity. The rest should be radiated from a boundary layer between the inner edge of the accretion disk and the white dwarf. In dwarf novae in outburst, models predict that the boundary layer will be optically

thick in outburst and will radiate with an effective temperature of 200,000
– 500,000 K (Pringle 1977; Kley 1989).

From a UV/optical spectrum of of a typical outburst of U Gem, Panek
& Holm (1984) estimated the accretion rate during outburst is 5×10^{17}
$(D/90\,\mathrm{pc})^2$ g s^{-1}, assuming a white dwarf mass of 1.2 $M_\odot$. The implied
disk luminosity is $\sim 8 \times 10^{34}\,(D/90\,\mathrm{pc})^2$ ergs s^{-1}. Based on our blackbody
fit to the $EUVE$ spectrum, we estimate a luminosity during the first observ-
ing interval for the boundary layer of $6 \times 10^{34}\,(D/90\,\mathrm{pc})^2$ ergs s^{-1}. Thus
the EUV luminosity is consistent with the standard model. The tempera-
ture of the boundary layer appears to be somewhat lower than standard
model would suggest. If one assumes that the source geometry is that
of an isotropically radiating sphere, the implied radius of the sphere is
$4 \times 10^8\,(D/90\,\mathrm{pc})^2$ cm, roughly that of the white dwarf surface.

The spectrum of U Gem qualitatively resembles the spectrum obtained
of an anomalous outburst of SS Cyg by Mauche, Raymond & Mattei (1994).
However, for SS Cyg, the lines that are observed indicate a temperature
of $\sim 10^{5.8}$ K. Furthermore, if N_H is the value indicated by the strength of
interstellar absorption lines in high resolution IUE spectra, then blackbody
fits to the continuum for SS Cyg suggest a temperature of 430,000 K and a
luminosity of $\sim 4 \times 10^{32}$ ergs s^{-1}. If this is the case, then, in contrast to U
Gem, the boundary layer luminosity in SS Cyg is much less than the disk
luminosity and the area involved is a small fraction (4×10^{-5}) of the white
dwarf surface. Detailed spectral analyses of both sources are needed to
determine how different these two systems really are. As noted by Mauche
et al., if there were additional internal absorption in SS Cyg, then estimates
of the luminosity of the EUV emission would increase, which would reduce
the discrepancy between the boundary layer and disk luminosities in SS
Cyg.

The spectrum of U Gem is quite different from the relatively featureless
continuum spectrum obtained by Vennes et al. (1995) of the polar VV Pup
in the high accretion state and modeled as a white dwarf atmosphere with
a temperature of $\sim 300,000$ K. EUV model spectra for white dwarfs are
much more featureless at temperatures of 300,000 K than at 130,000 so the
qualitative difference between the spectra of VV Pup and U Gem does not,
by itself, argue that the U Gem emission is created very far from the white
dwarf surface.

The existence of dips in the X-ray light curve of U Gem was first noted
by Mason et al. (1988) who observed U Gem with $EXOSAT$ during the declining phase of a very wide (40 day) outburst. The $EXOSAT$ observations
spanned about 5 orbital periods of U Gem (spread out over about 5 days).
The most persistent feature was a dip near phase 0.7, and, as in the $EUVE$
data, there is clear evidence for variations in the light curve from orbit to

orbit. The orbital dips were observed in the LE but not the ME detector, which lead Mason et al. to suggest that the dips were due to absorption by cold, cosmic abundance material with N_H of about 3×10^{20} cm^{-2}. An alternative which seems more likely is that the harder X-ray emission is produced over a more extended region. Our new observations suggest that the dips are a common feature in outbursts of U Gem.

The explanation for the dips is unclear, especially since the inclination for U Gem is about 65°, which implies that the obscuring material must lie far above the disk plane. Mason et al. suggested that there was a similarity between the orbital dips in U Gem and the orbital dips which are seen in some low mass X-ray binaries, such as XB 1916-053 (Smale et al. 1988) and EXO 0748-676 (Parmar et al. 1986). These systems also have dips around phase 0.7 or 0.8, but are much higher inclination systems than U Gem. There also may be a connection between the U Gem phenomena and the phase-dependent line absorption seen in some intermediate polars and SW Sex stars (Hellier, Mason & Cropper 1990; Hellier & Mason 1990). Three-dimensional structure calculations of accretion disks by Hirose, Osaki, & Mineshige (1991) do appear to suggest that there are situations in which the half thickness of the disk can be as large as 20% of the disk radius and that, due to the "bounce" of material hitting the disk, the disk thickness is greatest at phases near 0.8 and 0.2. In the case of U Gem, the material must be at a height which is 42% of the disk radius.

These observations would not have been possible without the dedicated supported of the observers of the AAVSO and of the *EUVE* mission support teams at Berkeley and at GSFC. This work is supported by NASA contract NAS 5-2572 to the Space Telescope Science Institute.

References

Córdova, F. A., Chester, T. J., Mason, K. O., Kahn, S. M., & Garmire, G. P. 1984, ApJ, 278, 739
Córdova, F. A., & Mason, K. O. 1984, MNRAS, 206, 879
Hellier, C., Mason, K. O., & Cropper, M. 1990, MNRAS, 242, 250
Hellier, C., & Robinson, E. L. 1994, ApJ, 431, L107
Hirose, M., Osaki, Y., & Mineshige, S. 1991, PASJ, 43, 809
Kley, W. 1989, A&A, 222, 141
Lynden-Bell, D., & Pringle, J. E. 1974, MNRAS, 168, 603
Mason, K. O., Córdova, F. A., Watson, F. G., & King, A. R. 1988, MNRAS, 232, 779
Mauche, C. W. 1991, private communication
Mauche, C. W., Raymond, J. C., & Mattei, J. A. 1994, ApJ, submitted
Panek, R. J., & Holm, A. V. 1984, ApJ, 277, 700
Parmar, A. N., White, N. E., Giommi, P., & Gottwald, M. 1986, ApJ, 308, 199
Pringle, J. E. 1977, MNRAS, 178, 195
Smale, A. P., Mason, K. O., White, N. E., & Gottwald, M. 1988, MNRAS, 232, 647
Vennes, S., Szkody, P., Sion, E. M., & Long, K. S. 1995, ApJ, in press

RADIO IMAGES AND LIGHT CURVES FOR NOVA V1974 CYGNI 1992

R.M. HJELLMING
National Radio Astronomy Observatory
P.O. Box O, Socorro, NM, USA 87801-0379

1. Introduction

Extensive VLA and MERLIN radio observations of Nova V1974 Cygni 1992 provide the best chance ever to determine the mass and kinematics of a nova shell from radio data. Examination of the image sequences, ranging from optically thick to optically thin states, indicates that there is initially one elliptical shape related to the outer boundary of the shell, while later optically thin images are dominated by a different elliptical inner boundary for the shell.

2. Analysis of Radio Light Curves and Images

2 1 SPHERICALLY SYMMETRIC MODELS WITH LINEAR VELOCITY GRADIENTS

Radio observations of Nova V1974 Cygni 1992 have been made by Hjellming *et al.* 1995 using the VLA at frequencies of 1.49, 4.9, 8.4, 14.9, and 22.5 GHz. Spherically symmetric shell models, with a linear velocity gradient, can be used to fit the radio light curves, as was done for the novae HR Delphini 1967, FH Serpentis 1970, and V1500 Cygni 1975 by Hjellming *et al.* 1979 and Seaquist *et al.* 1977. The essence of these models is assuming that the inner and outer radii are described by $R_1(R,t) = v_1 \cdot (t - t_0)$ and $R_2(R,t) = v_2 \cdot (t - t_0)$, where t is the time, t_0 is an initial starting time, and v_1 and v_2 are fixed velocities for the inner or outer boundary of the shell, which are both assumed to be spherically symmetric. Inside the shell the velocity is given by $v(R,t) = v_2 \cdot R/R_2(R,t)$ (with $v_1 \leq v \leq v_2$) and the density is given by $\rho(R,t) = [M_{shell}/(4\pi R^2)]/[R_2(R,t) - R_1(R,t)]$, where M_{shell} is the mass in the shell. The emission and absorption coefficients

139

A Bianchini et al (eds) Cataclysmic Variables 139-144

are those for thermal bremmstrahlung radiation, so the radiative transfer needed to compute both the surface brightness in the shell and the total flux density at any time is straightforward (Hjellming *et al.* 1979).

The separate determination of distance and absolute inner and outer velocities requires other information. This is best derived from optically thin nebular line profiles for lines emitted by the entire ionized shell. It can be shown that the linear velocity gradient model predicts a simple analytic equation for the velocity profile of shells that obey this model. Expressed in the form of the normalized line profile function, $\phi(v)$, one obtains

$$\phi(v) = \frac{(v_1 \cdot v_2)}{4(v_1 - v_2)}[v^{-1} - v_2^{-1}] \tag{1}$$

for $v_1 \leq |v| \leq v_2$, and $\phi(v) = \phi(v_1)$ for $|v| \leq v_1$. These line profile functions have been used by Hjellming *et al.* 1994 to fit the N IV] 1486 Å, He II 1640 Å, and C III] 1908 Å HST/GHRS nebular spectrum profiles on Sept. 7, 1992 (Shore *et al.* 1993) giving $v_1 = 890$ km/sec and $v_2 = 1940$ km/sec. The general shape of the line profiles matches these data very well, but, of course, there major fluctuations about the mean for the region $|v| \leq v_1$ due to the clumping in the inner portions of the shell.

2.2. MODEL FITS AND SMALL DISCREPANCIES

A sequence of radio images of Nova QU Vul 1984, made with the VLA during the late optically thin stages of this nova, were the first used to independently determine (Taylor *et al.* 1988) the mass and electron temperature in a nova shell. The mass was found to be $M_{\text{shell}} = 3.6 \cdot 10^{-4} M_\odot$ and the electron temperature $T_e = 17,000$K.

A preliminary analysis (Hjellming *et al.* 1993,) of a Dec. 27, 1992, 1.3 cm image of nova V1974 Cygni (cf. third image in Figure 2), which was made with the VLA 313 days after outburst, showed consistency with an optically thin, spherically symmetric shell with inner and outer radii of 0.08″ and 0.175″, with $M_{\text{shell}} = 3.1 \cdot 10^{-4}(d_{\text{kpc}}/2)^{5/2})(T_e/10^4)^{-0.58}M_\odot$, $v_1 = 890(d_{\text{kpc}}/2)$ km/sec, and $v_2 = 1940(d_{\text{kpc}}/2)$ km/sec. Note that the velocities determined from inner and outer boundaries match those determined from nebular line profiles (Hjellming *et al.* 1994) only for a distance of 2 kpc, so a combination of the profile fit to the early nebular spectra, and the inner and outer boundaries in the image model fit, determine $d_{\text{kpc}} = 2$.

Figure 1 shows the observed, radially averaged, brightness temperature profile for the Dec. 27, 1992 image as filled circles, together with the model that fits the data (dashed lines) and the model smeared to the resolution of the VLA image (solid lines). The quantities determined from fits to the data depend on distance (d_{kpc}) and an average electron temperature (T_e). These

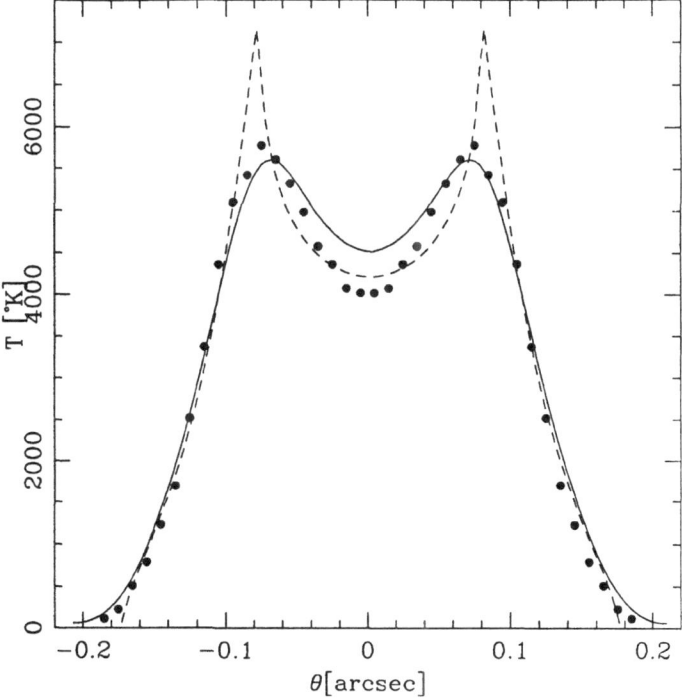

Figure 1. Brightness temperature plotted as a function of angular radius. Filled circles indicate the radially averaged brightness temperatures determined from the Dec. 27, 1992 image of nova V1974 Cygni at 1.3 cm; the dashed line is the model distribution and the solid line is that model smeared to the resolution of the VLA observations.

combinations of parameters are determined solely by the surface brightness in the optically thin image of Dec. 27, 1992. As discussed by Hjellming *et al.* 1979, in additon to the above parameters there is one additional combination, $(v_2/d_{\rm kpc}) \cdot T_{\rm e}^{1/2}$, that is determined by the radio light curves, which in effect determines the electron temperature. As discussed by Hjellming *et al.* 1993 and Hjellming *et al.* 1994, with $T_{\rm e} \approx 10^4$K there is a good fit to the pre-maximum to decay portions of the light curves. However, the early optically thick portions of the light curves are lower than predicted by this model, with a systematic gradient in which the data points are below, but converge to, the model light curves just before maxima. Fortunately Pavelin *et al.* 1993 obtained a series of radio images of V1974 Cygni during the early optically thick phases of the shell, and because they resolved the emission they were able to determine that the brightness temperature, or average electron temperature, was increasing with time. This supports the assumption of a linear increase of electron temperature with time, from 5000 K to 15,000 K between 30 and 150 days after outburst, that Hjellming

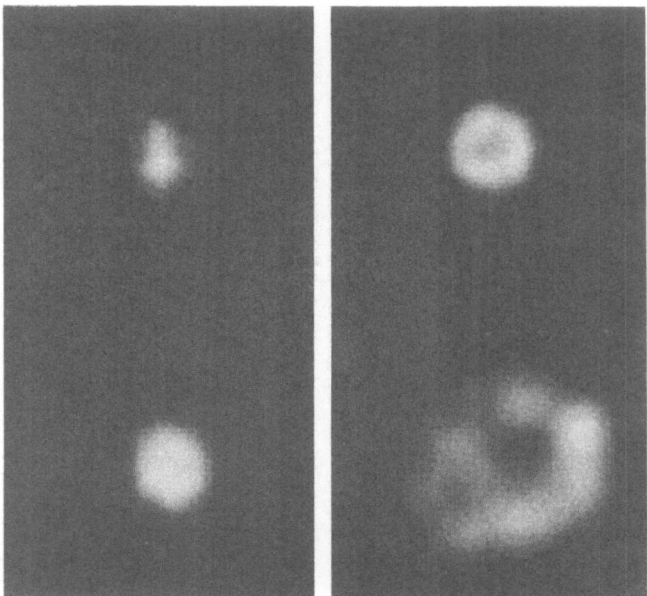

Figure 2. Four 1″ by 1″ images of nova V1974 Cygni. A MERLIN 6 cm image (top left) made 173 days after outburst, two VLA 1.3 cm images made 255 days (bottom left) and 313 days (top right) after outburst, and a VLA 2 cm image made 769 days after outburst (bottom right).

et al. 1994 showed was sufficient to fit the early light light curve data.

Although the main features of the radio light curves were fit by the abovementioned model fit, there are still small discrepancies near maximum and during the transition from maximum to the optically thin decay line. Therefore other effects must be included in the model to more accurately fit the radio light curve data. As we discuss next, the asymmetries seen in both optically thick and optically thin images indicate that it may be just a modification of the geometry of inner and outer boundaries that is needed.

3. Radio Image Sequences

In Figure 2 we show a sequence of images for Nova V1974 Cyg made with MERLIN (Pavelin *et al.* 1993) and the VLA (Hjellming *et al.* 1994). The first image (MERLIN 6 cm), corresponding to the early stage where the source is optically thick, indicates a roughly North-South elongation; the second (VLA 1.3 cm) image has the shape of an ellipse with major axis at a position angle of 25° and a ratio of major to minor axis of about 1.3. The third image (VLA 1.3 cm) shows a roughly spherically symmetric shell

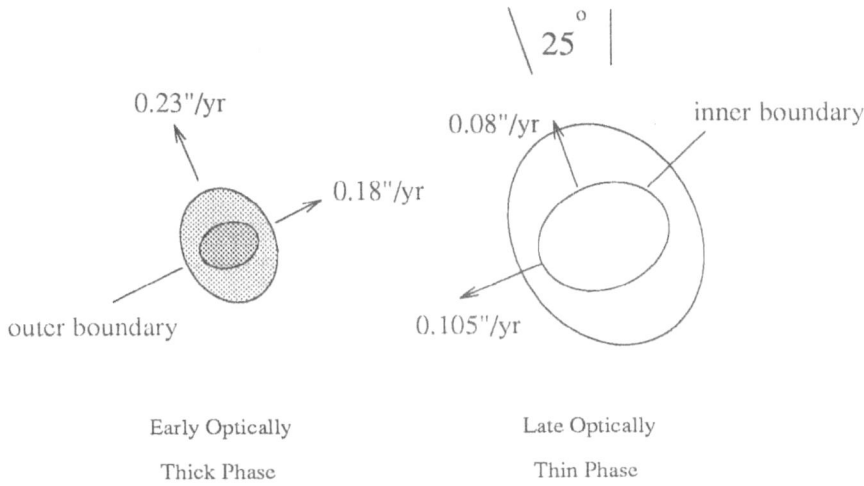

Figure 3. The apparent geometry of elliptical inner and outer boundaries indicate the kinematics of the nova V1974 Cygni shell at optically thick and thin phases.

during a phase when the source is mostly optically thin. The fourth image (VLA 2 cm), made when the shell was very optically thin, indicates an elliptical shape with major axis at an position angle of about 115°, and a ratio of major to minor axis of about 1.3.

A very simple modification of the shell model discussed above is sufficient to explain the different elliptical shapes seen during optically thick and thin states of V1974 Cygni. One need only assume that the outer boundary projected on the plane of the sky is an ellipse with a major axis expanding at a rate of 0.23″/yr along a line with a position angle of 25°, and a minor axis expanding by 0.18″/yr – while the inner boundary projected on the plane of the sky is an ellipse with a major axis expanding at a rate of 0.108″/yr along a line with a position angle of 115°, and a minor axis expanding by 0.08″/yr. Figure 3 shows schematic diagrams of this geometry during early optically thick and late optically thin phases. Since the distance is 2 kpc, these angular rates translate into velocities of 2200 and 1700 km/sec for the major and minor axes of the outer boundary, and 1000 and 770 km/sec for the major an minor axes of the inner boundary.

The projected inner and outer boundaries in Figure 3 can have this appearance with either prolate or oblate spheroids for the true three-dimensional geometry, so this introduces two new optional parameters into the problem.

Since programs doing the radiative transfer with prolate or oblate ellipsoidal boundaries have not yet been developed, we computed the brightness temperature profiles for cross-cuts at position angles of 25° and 115°, using the parameters shown in Figure 3. The results were exactly the profile evolution needed to have the shape and surface brightness variations seen in Figure 2. How the detailed appeerence of the images evolves, and whether the fine tuning of more complicated boundary parameters improves the fits of the light curve data, is still to be determined.

The reasons for the transition from an optically thick radio source, with the projected elliptical shape of the outer boundary, to an optically thin radio source, with the projected elliptical shape of the inner boundary, is obvious. Initially it is an optically thick surface near the outer boundary that contributes most of the observed emission. Later on the R^{-2} density distribution between shell boundaries ensures that, particularly with the shell increasing with time because of the velocity gradients, the densest regions will be at the inner boundary, so the bulk of the contribution to the emission measure is very close to the inner radius (cf. Figure 1).

3.1. CLUMP COMPLEXES AND FOUR MAJOR LINE COMPONENTS

The fourth image in Figure 2 shows that the elliptical shell is clumpy, and higher resolution imaging with the same 2 cm data, and 1.3 cm data, indicates that there are four distinct cloud complexs, one pair on a line with position angle 25° and one on a position angle 115°. Each is associated with with apparent velocities of 1000 and 770 km/sec for major and minor axes, so the receding and approaching portions of these systems can produce the four component nebular spectra found in most novae.

References

Hjellming, R.M., Wade, C.M., Vandenberg, N.R., and Newell, R.T. (1979) Radio Emission from Nova Shells, *Astron. J.*,**84**, pp. 1619–1631.

Hjellming, R.M. (1993) Radio Imaging and Spectrum Variations in Nova Cygni 1992, *Bull.A.A.S.*, **24**, p. 1257.

Hjellming, R.M. (1994) Radio Light Curves and Images of Nova Cygni 1992, *Bull.A.A.S.*, **26**, pp. 946–947.

Hjellming, R.M., Palma, C., Seaquist, E.R., Taylor, A.R., and Gehrz, R.D. (1995) Radio Light Curves and Images for Nova V1974 Cygni 1992, *Astrophys. J.*, in preparation.

Pavelin, P.E., Davis, R.J., Morrison, L.V., Bode, M.F., and Ivison, R.J. (1993) Radio observations of the classical nova Cygni 1992 eighty days after outburst, *Nature*,**363**, pp. 424–426.

Seaquist, E.R. and Palimaka, J. (1977), Thick Inhomogenous Shell Models for the Radio Emission from Nova, Serpentis 1970, *Astrophys. J.*,**217**, pp. 781–787.

Shore, S.N., Sonneborn, G., Starrfield, S., Gonzalez-Riestra, R., and Ake, T.B. (1993), The Early Ultraviolet Spectral Evolution of Nova Cygni 1192, *Astron. J.*,**106**, pp. 2408–2428.

Taylor, A.R., Hjellming, R.M., Seaquist, E.R., and Gehrz, R.D. (1988) Radio Images of the expanding ejecta of nova QU Vulpeculae 1984, *Nature*,**335**, pp. 235–238.

LONG-TIME VARIABILITY OF 1H 1752+081 = V2301 OPH

S.V. ANTIPIN AND S.YU. SHUGAROV
*Sternberg Astronomical Institute, Universitetskij pr., 13,
119899, Moscow V-234, Russia; e-mail: shugarov@sai.msk.su*

The star was investigated on 224 plates taken from JD 2441893 to JD 2448092. The data reduction revealed the existence of regular light variations with a $148\overset{d}{.}3$ period. The light curve is presented in the Fig.1. It is clear that a sine-like modulation with superposed more rapid light variations with amplitude close to $1.^m$ are present.

Note that the mean exposure time is equal to 45 minutes while the duration of the eclipse is about 12 minutes (Silber *et al.*, 1994), and the depth of minima caused by eclipses does not exceed 0.3 magnitudes. Hence these rapid light variations are due to some other reasons, such as, unstable accretion rate.

References

Silber A.D., Remillard R.A., Horne K., Bradt H.V. //*Astroph. J.*, 1994, **424**, 955.

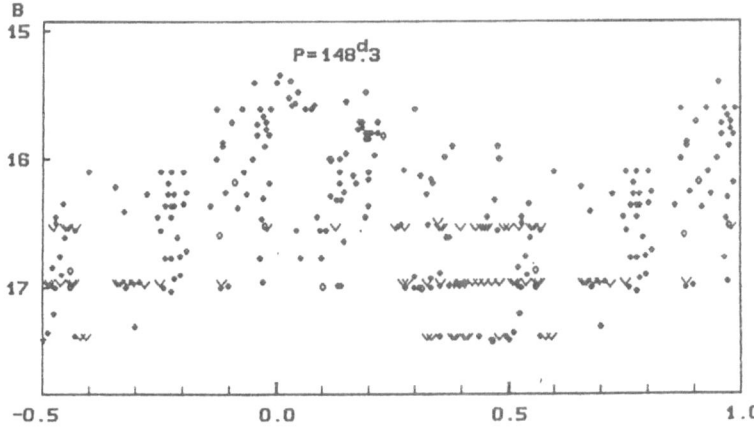

Figure 1. The B light curve of 1H 1752+081. Uncertain data are shown as open circles, upper limits, as "v" symbols.

145

A. Bianchini et al. (eds.), Cataclysmic Variables, 145.
© *1995 Kluwer Academic Publishers.*

IUE OBSERVATIONS OF THE IP CANDIDATE H0857-242 DURING A DECLINE OF AN OUTBURST

R. ALVAREZ[1], M. MOUCHET[2], D. DE MARTINO[3], J. DREW[4], D. BUCKLEY[5]
[1] *Laboratoire d'Astronomie, Université, F-34095 Montpellier*
[2] *DAEC, Observatoire de Meudon, F-92190 Meudon*
[3] *ESA IUE Observatory, E-28080 Madrid*
[4] *Department of Physics, University, Oxford OX1 3RH*
[5] *Southern African Astronomical Observatory, Cape Town, SA*

The X-ray source H0857-242 has been optically identified by Remillard et al. (1994) with a new cataclysmic variable. An orbital period of 114 min has been proposed together with a shorter period of 49 min. Though it has been tentatively classified as an intermediate polar (IP) (hard X-ray emission, weak HeII 4686Å emission line, possible detection of the rotation of white dwarf), it differs from IPs by exhibiting frequent outbursts of 4 magnitudes. H0857-242 was observed with IUE from March 6 to 10 1992 in the context of the Whole Earth Telescope project during an outburst (de Martino et al., 1992). The source has returned in a quiescent state on March 10; the UV spectrum shows the typical emission lines seen in dwarf novae.

During high states, the UV continuum has been fitted with an optically thick accretion disc model. The derived disc temperatures of the best fits for the 7 and 8 March spectra correspond to accretion rates of $6.3 \ 10^{-8}$ and $2.7 \ 10^{-8}$ $M_\odot$/yr respectively. A flux excess is detectable at short UV wavelengths, which could be associated either with the boundary layer or with the white surface heated by the accreting material.

On March 7 and 8, the resonance lines exhibit clear P-Cygni profiles while they are in emission on March 10. Changes in the CIV profiles are also observed within one day and probably have an orbital origin. We assume the existence of an additional emission component which is partly eclipsed at phase 0. The P-Cygni profiles at phase 0 are interpreted in terms of a bipolar wind issuing from the white dwarf, as described by Drew (1987). The accretion rates are fixed to the values derived from the continuum fit. Different ionisation profiles are used for March 7 and March 8. The best solutions are found for an inclination angle of 40°.

References

de Martino, D. et al. (1992), *IAU Circ.* No. 5481
Drew, J. (1987), *Monthly Notices Roy. Astron. Soc.* **224**, 595
Remillard, R.A. et al. (1994), *Astrophys. J.* **428**, 785

A Bianchini et al (eds) Cataclysmic Variables 146

9.6 YEAR PERIODICITY IN OPTICAL LIGHT CURVE OF THE SYMBIOTIC NOVA RT SERPENTIS (1909)

V.V. BOCHKOV AND E.P. PAVLENKO
Crimean astrophysical observatory, Nauchny, Crimea, 334413, Ukraine.

RT Ser is a prototype of symbiotic nova. The brightness data from outburst decline of RT Ser from 1927 up to present days have been analysed with Diming's method. The analysed material contains observations obtained in Harvard observatory [1,2] from 1927 to 1941 (photography, 26 points), in Gissar observatory [3] from 1939 to 1978 (photography, 20 points) and in Crimean astrophysical observatory from 1983 to 1994 (B band of Johnson photometric system, 45 points). The photographic data were reduced to B band of standard system [3]. Linear trend (0.06 mag per year) has been previously subtracted from these data. The frequency spectrum displays maximum amplitude at the frequency, corresponding to the period of 3497 +- 50 days (9.6 year). This analysis point out the fact, that light variations with typical time of about 10 years are possibly periodic process for symbiotic nova RT Ser.

1. References.

1. Campbell, L. (1927-35), American Association of Variable Star Observers - monthly report, Popular Astronomy, Vol. No. 35-43.
2. Campbell, L. (1936-41), American Association of Variable Star Observers - monthly report, Annals of Harvard College observatory, Vol. No. 104-11.
3. Bochkov, V.V, Pavlenko, E.P., Vasiljanovskaja, O.P. (1993), Odessa astronomical publications, Vol. No.6

A. Bianchini et al. (eds.), Cataclysmic Variables, 147.
© *1995 Kluwer Academic Publishers.*

OPTICAL QUIESCENT STATE OF GRO J0422+32:
CONSTRAINTS ON THE COMPACT STAR MASS

J.M. BONNET-BIDAUD[1], M. MOUCHET[2]
[1] *DSN/DAPNIA/SAp, CE-Saclay, F-91191 Gif sur Yvette.*
[2] *DAEC Observatoire de Meudon, F-92190 Meudon.*

GRO J0422+32 is a hard transient X-ray source (X-ray nova) discovered by BATSE on board GRO on August 1992 (Paciesas et al. 1992, IAU Circ. 5580). It has exhibited four successive outbursts within two years. This class of X-ray novae has focussed particular interest since the discovery among them of at least three black hole candidates.

We have observed GRO J0422+32 on February 4 1994, using the multi-object spectrograph MOS, at the 3.6m CFHT telescope. Three spectra of 30 min. each have been obtained at a dispersion of 3.6 Å/pixel.

The averave spectrum is typical of what observed for the other X-ray novae in quiescence confirming that the source has now returned to a state of very low excitation (V $\sim$ 20.5), showing strong broad emission Balmer lines and no HeII line. The H_α profile shows a double-peaked shape with a velocity peak separation of $\sim$ 900 km.s^{-1}. The absence of lines characteristic of a late type secondary indicates that the star has not yet reached the full quiescent state.

A value of $\sim$ 50° for the inclination is derived, assuming that the intrinsic widths of the lines is comparable in the different sources and attributing their observed values as a consequence of the inclination of the disc. A mass of the compact star of $\sim$ 1.5M$_\odot$ is estimated from the peak separation of the H_α line profiles, for an orbital period of 5.1h (Chevalier and Ilovaisky 1993, IAU Circ. 5692; Kato et al. 1993, IAU Circ. 5704); allowing for uncertainties a maximum mass of 2.2M$_\odot$ is thus derived. Our results do not confirm the high mass (lower limit of M$_c \geq 2.9$M$_\odot$) derived on the basis of a superhump period 2.2% larger than the orbital period (Kato et al. 1993, IAU Circ. 5704). Only in the case of a double orbital period of 10.2h (Chevalier and Ilovaisky 1994, IAU Circ. 5928), the maximum mass (M$_c \leq (3.5)$M$_\odot$) may be consistent with a black hole.

The confirmation of the nature of the companion has to wait the detection of the companion absorption lines after a further decline of the disc residual luminosity.

A more detailed study is reported in Bonnet-Bidaud and Mouchet (1994, A&A, in press).

A. Bianchini et al. (eds.), Cataclysmic Variables, 148.
© *1995 Kluwer Academic Publishers.*

THE DETACHED BINARY RE1629+781:

A LOW-MASS MAGNETIC SECONDARY STAR?

M. S. CATALÁN
Astronomy Centre, University of Sussex, U. K.

M. J. SARNA
N. Copernicus Astronomical Center, Poland

C. M. JOMARON
Dept. of Physics and Astronomy, University of St. Andrews, U. K.

AND

ROBERT CONNON SMITH
Astronomy Centre, University of Sussex, U. K.

We present low resolution spectra (3400–8900 Å) of the close detached white dwarf-red dwarf binary RE1629+781, discovered by ROSAT. The spectra clearly show a composite nature and resemble those of pre-cataclys-mic binaries. They show H-Balmer absorption features from the white dwarf and TiO and NaI in absorption from the red dwarf secondary. Narrow Balmer and CaII in emission, most likely originating from the cool star, are also present. Using a pure hydrogen model, fits to the H_β, H_γ and H_δ absorption yield a temperature of $41\,800 \pm 2000$ K and surface gravity of $\log g = 8.0 \pm 0.5$ for the white dwarf. Using the TiO bands to type the secondary star, we estimate the red dwarf to be a dM4. A main-sequence assumption for the secondary results in a mass range 0.16–0.24 $M_\odot$ and a radius 0.2–0.3 $R_\odot$. Earlier observations by Sion, Holberg & Barstow (1994, PASP, submitted) discuss the active nature of the low-mass star in this close binary is confirmed, then not only would it be the first detection of a magnetically active M dwarf in a close binary, but it also suggests the likelihood that magnetic braking could still be an efficient mechanism of angular momentum loss, which has far reaching evolutionary implications on the formation and evolution of cataclysmic binaries.

References

Cooke, B. A. et al. (1992) *Nat.*, **355**, 61
Wade, R. A. and Horne, K., 1988. *ApJ*, **324**, 411
Sion, E. M., Holberg, J. B. and Barstow, M. A., 1994 *PASP*, submitted

A Bianchini et al (eds) Cataclysmic Variables 149

SEASONAL OUTBURST CYCLE LENGTH VARIATIONS IN THE DWARF NOVA EM CYGNI

L. L. CHINAROVA AND I. L. ANDRONOV
Department of Astronomy, Odessa State University
T. G. Shevchenko Park, Odessa 270014 Ukraine
e-mail: root@astro.odessa.ua

The mean brightness of EM Cyg varies with a cycle of ≈ 3000 d and is highly correlated with the amplitude of the sinusoidal variations. Although the time intervals P between subsequent outbursts vary by a few dozen percent, the seasonal values show changes with a few-year scale of two types: with nearly harmonic shape (Bianchini 1988, 1990) and with abrupt switchings (Andronov and Shakun 1990). This phenomenon may be important for understanding the existing theoretical models of the outbursts as the limit cycles of the accretion disk structure variations.

For analysis we have used 4027 (from 4506) observations from the AFOEV database. Corresponding mean and r.m.s. deviations are equal to $< m >= 13.15$, $\sigma_O = 0.532$. For the periodogram analysis a least squares fit $m(t) = a + b \sin 2\pi f t + c \cos 2\pi f t$ was used and a test function $S(f) = \sigma_C^2 / \sigma_O^2$, where "O" and "C" correspond to r.m.s. deviations of the original and values smoothed with a sine fit of their sample means. The expectation for the noise is $\bar{S} = 2/(n-1) = 0.0005$.

The data were subdivided into 24 seasons and periodograms were computed for each. Results are tablulated below: t_1 and t_n are times $(JD - 2440000)$ of start and end of the season, $< t >$ is the sample mean time. Best fit periods P are expressed in days, corresponding amplitudes $r = (b^2 + c^2)^{1/2}$ in units 0.01 mag, as well as r.m.s. deviations σ_O, σ_C of the observed and calculated values from their sample means. L_p is a decimal logarithm of the false alarm probability of the appearance in a periodogram of a peak of given height from pure white noise (Tremko et al. 1994).

One may note significant variations of the seasonal mean cycle length from season to season. As in other dwarf novae, they may be explained by variations of the accretion rate (Andronov and Shakun 1990) due to solar-type activity of the secondary (Bianchini 1990) filling its Roche lobe.

150

A. Bianchini et al. (eds.), Cataclysmic Variables, 150-151.
© *1995 Kluwer Academic Publishers.*

TABLE 1. Characteristics of the seasonal periodograms of EM Cyg. a is expressed as $100(magnitude - 13.00)$.

t_1	t_n	$<t>$	n	$S(f)$	L_p	r	P	a	σ_O	σ_{O-C}
739	952	844	149	.222	6.6	27±4	25.70±.36	10±3	41	36
1007	1303	1174	201	.236	10.1	30±4	21.43±.14	8±3	43	37
1392	1676	1537	173	.350	14.4	42±4	19.83±.09	24±3	49	40
1824	2044	1926	118	.171	3.3	26±5	25.61±.41	14±4	45	41
2189	2364	2272	101	.341	7.6	37±5	32.53±.47	9±4	45	36
2512	2774	2640	131	.352	10.6	37±4	20.64±.14	7±3	43	35
2926	3073	3011	78	.348	5.8	37±6	18.04±.23	3±4	41	34
3347	3482	3404	111	.199	4.1	20±4	12.65±.12	1±3	30	27
3715	3844	3775	83	.261	4.1	30±6	28.44±.65	4±4	36	33
4008	4188	4092	42	.533	5.2	54±8	27.13±.45	20±6	51	35
4433	4602	4504	56	.559	8.2	59±7	21.59±.20	42±5	55	37
4751	4958	4862	109	.247	5.2	45±8	20.54±.21	31±5	60	52
5100	5298	5212	149	.434	16.8	53±5	25.47±.17	27±3	55	42
5472	5745	5599	168	.225	7.7	41±6	20.63±.15	31±4	61	54
5818	6124	5958	164	.381	15.3	52±5	21.31±.12	25±4	61	48
6315	6435	6366	98	.493	12.9	53±6	24.51±.40	28±4	51	36
6553	6802	6684	271	.192	11.0	24±3	18.99±.13	7±2	37	33
6886	7205	7036	250	.206	10.9	36±4	22.65±.17	17±3	53	47
7302	7540	7411	249	.305	18.0	54±5	24.60±.17	34±4	65	54
7645	7913	7782	291	.356	26.1	59±5	32.66±.22	26±3	65	52
7999	8274	8132	284	.387	28.4	53±4	25.51±.14	22±3	59	46
8340	8625	8496	323	.206	14.6	35±4	44.41±.53	17±3	54	48
8719	8992	8862	214	.209	9.3	34±5	17.86±.11	14±3	51	45
9095	9166	9130	60	.606	10.7	76±8	18.90±.27	29±6	65	41

All these parameters vary with time. Variations of a seem to be regular, and have a cycle of $\Pi \approx 3000$ d. This value is close to that observed in other cataclysmic variables (e.g. Bianchini 1988, 1990; Andronov and Shakun 1990). However, this object shows sharper variations of r and P.

Acknowledgements. We thank E. Schweitzer for use of the AFOEV database, and M. Orio and A. Bianchini for helpful discussions. This research was partially supported by ESO Grant A-04-018.

References

Andronov, I.L., Shakun, L.I.: 1990, *Astrophys. Space Sci.*, **169**, 237.
Bianchini, A.: 1988, *Inf. Bull. Var. Stars*, **3136**, 4.
Bianchini, A.: 1990, *A. J.*, **99**, 1941.
Tremko, J., Andronov, I.L., Chinarova, L.L., Kumsiashvili, M.I., Luthardt, R., Pajdosz, G., Patkós, L., Rößiger, S., Zoła, S.: 1994, *As. Ap.*, submitted.

A SPECTROSCOPIC STUDY OF WX CEN

M. P. DIAZ and J. E. STEINER
Instituto Astronômico e Geofísico - Universidade de São Paulo

WX Cen is currently classified as a nova-like (Vogt, 1989 in Classical Novae, edts. Bode & Evans, p225) but its peculiar spectral characteristics have resisted interpretations since its discovery. Time-resolved spectrophotometry was obtained with the 1.5m and 1.0m telescopes of CTIO in 1991. The emission line spectrum of WX Cen is similar to that of V Sge, V617 Sgr and GQ Mus. An E(B-V) = 0.3 was derived from the HeII4686/10124 ratio. The strong HeII4686 line and Hβ show an orbital radial velocity modulation with a period of 0.416994(±42) days, while the continuum at 5135 Å displays a quasi-sinusoidal orbital hump with a semi-amplitude of 0.14^m. The possibility that the system contains an evolved low-mass secondary is unlikely due to the lack of molecular bands up to 1μ. No signatures of a main sequence companion were found. The diagnostic diagram of the Hβ wings presents an unusual situation where the radial velocity semi-amplitude (K) increases continuously with the sampling velocity. The radial velocity curves derived from the wings above 1100 km/s have a non-sinusoidal shape. The Doppler tomogram in HeII4686 shows a symmetric component centered about *the center of mass of the binary* which may be understood as a *circunstellar/wind* emission. In addition, a central dip in the map is probably associated with self-absorption in this transition. Using the HeII4686 and the Hβ maps it is possible to trace the abundance of helium (He++/H+) in the velocity space. A preliminary analysis suggests that He/H varies from ~0.3, in the symmetric component, to 0.15-0.20 near the secondary. A sharp asymmetric component present in the Hβ maps is similar to those seen in systems where the secondary is heated and the stream is clearly mapped (e.g., VV Pup in Diaz & Steiner 1994, A&A, 283, 508). The constraints on K_2 velocity derived from such feature have important dynamical consequences: for an inclination between 30° and 75°, it is straightforward to find that if the primary is a white dwarf then M_2 ≤ 0.35 M⊙. On the other hand, if the secondary follows a ZAMS mass-radius relation then M_1 ≥ 3.5(±0.5) M⊙. At the moment, considering the absence of X-rays detection, the most likely hypothesis about the nature of the primary seems to be a helium main sequence star. This scenario may explain the helium overabundance as well as the companion illumination effects.

A. Bianchini et al. (eds.), Cataclysmic Variables, 152.
© *1995 Kluwer Academic Publishers.*

EXPANDING ENVELOPE OF NOVA CYGNI 1992

D. CHOCHOL[1], R. KOMŽÍK[1], L. HRIC[1] AND J. GRYGAR[2]
1) *Astronomical Institute, Slovak Academy of Sciences*
059 60 Tatranská Lomnica, The Slovak Republic
2) *Institute of Physics, Czech Academy of Sciences*
Na Slovance 2, 180 40 Praha 8, The Czech Republic

Available spectroscopic, HST and radio imaging observations of Nova Cygni 1992 (= V 1974 Cyg) were used to derive the kinematic model of expanding nova envelope which consists of two major components: a fast-moving low-mass outer envelope and a slow high-mass inner envelope. Both envelopes can be well approximated by geometrically thin equatorial rings and axially symetric polar blobs. Outer envelope was detected as the outer edge of the radio shell as well as spectroscopically as double absorptions in P Cygni spectral line profiles. The blobs and rings of the inner envelope was resolved as very stable peaks in emission line profiles during the nebular stage. Inner envelope images were taken also by HST. We have calculated the inclination angle of the ring plane against the cellestial sphere using RV of absorption lines, RV of the peaks of emission line profiles and HST image of the shell taken 467 days after the outburst. The resulting value of $i = 32°4 \pm 0.6$ enabled us to determine the true expansion velocities of the gaseous structures of both envelopes. During the first 50 days after outburst the outer envelope was accelerated by the hot wind from 1500 km s^{-1} to 5300 km s^{-1} reaching its terminal velocity. On the day 91 the envelope encountered the preoutburst circumstellar shell (containing material expelled during previous outbursts)and decelerated. The slow high-mass envelope was accelerated by the wind from 830 km s^{-1} to 895 km s^{-1} (ring) and 1015 km s^{-1} (blob). This envelope reached the circumstellar shell after the day 403 and decelerated, too. Using this model we have calculated the radius of the envelope at the time of HST observation to be 34 bilions km. Corresponding distance of Nova V 1974 Cygni: **1.76 kpc** is in good agreement with the value 1.77 kpc derived by Chochol et al. (1993, A&A 277, 103) using indirect methods, but disagrees with distance 3.2 ± 0.5 kpc determined by Paresce (1994, A&A 282, L13).

153

UBV PHOTOMETRY OF NOVA CASSIOPEIAE 1993

D. CHOCHOL, L. HRIC, R. KOMŽÍK AND Z. URBAN
Astronomical Institute, Slovak Academy of Sciences
059 60 Tatranská Lomnica, The Slovak Republic

Nova Cas 1993 was discovered by K. Kanatsu (1994, IAU Circ. 5902) on Dec. 7, 1993, ten days before it reached its maximum visual brightness $V_0 = 5.68$. Progenitor of the nova was identified by Skiff (1993, IAU Circ. 5904) on the POSS as B $\sim$18 mag star. No systematic trend in the mean brightness was observed in prenova stage during the period 1970–85 (Munari et al., 1994, IBVS 3977). Our UBV photoelectric photometry taken by two 0.6 m telescopes at the Skalnaté Pleso and Stará Lesná Observatories in 13 nights from Dec. 29, 1993 to Feb. 17, 1994 as well as published photometric data (IAUCs; Munari et al., 1994, IBVS 4005) were used to derive the basic parameters of Nova Cas 1993 (see the Table), which can be classified as the slow nova of DQ Her type. Three long-run UBV observations taken on Dec 29/30, 1993 (298 min); Jan 12/13 (322) and Jan 21/22 (300), 1994 were used to serch for the periodicity of the light variations observed in the range 0.14, 0.10, 0.21 mag (ΔU), 0.09, 0.07, 0.19 mag (ΔB) and 0.10, 0.07, 0.19 mag (ΔV). No periodicity was found.

Maximum apparent mag.	$V_0 = +5.68$, $B_0 = +6.29$
Colour index at max. light	$(B - V)_{max} = +0.61$
Apparent mag. 15 days after max.	$V_{15} = 7.05$ $B_{15} = 7.5$
Time t_2 of decline	$t_{2,V} = 32$ d, $t_{2,B} = 42$ d
Time t_3 of decline	$t_{3,V} = 57$ d, $t_{3,B} = ?$
Mean absolute mag. at max.	$M_{0,V} = -7.17$, $M_{0,B} = -6.95$
Intr. colour index at max. light	$(B - V)_{max}^{in} = M_{0,B} - M_{0,V} = +0.22$
Dist. mod. (affected by extinction)	$V_0 - M_{0,V} = 12.85$, $B_0 - M_{0,B} = 13.24$
Adopted distance modulus	13.04 ± 0.14
Colour excess at max. light	$E_{B-V} = +0.38 \pm 0.02$
Visual extinction	$A_V = 3.1 \times E_{B-V} = 1.18 \pm 0.06$
Distance of the nova	$r = 2.36 \pm 0.14$ kpc

A. Bianchini et al. (eds.), Cataclysmic Variables, 154.
© *1995 Kluwer Academic Publishers.*

HIGH RESOLUTION SPECTROSCOPY OF CP PUP.

S.W. DIETERS
Astronomical Institute, University of Amsterdam
Kruislaan 403, 1098 SJ, Amsterdam, Netherlands.

AND

R.D. WATSON, J. GREENHILL. & K. HILL
Physics Dept., University of Tasmainia
PO Box 252c Hobart, Tasmania 7000., Australia.

CP Pup erupted as a bright, energetic nova in 1942. The photometric and spectroscopic (orbital) periods are both very short ($\sim$1.5 Hrs) and slightly different. This is a characteristic of the intermediate polars which contain magnetic white dwarfs or the SU UMa stars. CP Pup is an unusual member of either CV subclass. Even the lowest of previous estimates of the radial velocity amplitude (K) imply a white dwarf mass <0.6 $M_\odot$. This is below the mass limit required by the thermonuclear runaway (TNR) model to produce a nova explosion. We obtained high signal to noise, high resolution (1.2 Å FWHM), blue (3940—4918 Å) spectra using the AAT in an attempt to resolve this discrepancy.

The short spectroscopic period was confirmed. Two additional line components were found. One was an S-wave with K = 310 ± 13.5 km/sec and phase 0.25 ± 0.03 with respect white dwarf's superior conjunction. The second was an out-of-plane component with K = 137 ± 15 km/sec, $\gamma = +360 \pm 9$ Km/sec and phase = 0.08 ± 0.03. The S-wave originates in a hot spot near the outer edge of the disk, while the phasing of out-of-plane component is consistent with a stream running above the disk. To avoid the effects of these components the radial velocities of the line wings were measured using the double gaussian convolution method. The estimated radial velocity amplitude of the white dwarf is about 60 km/sec which allows quite a large region in the mass-inclination diagram above the theoretical limit for a nova explosion.

A. Bianchini et al. (eds.), Cataclysmic Variables, 155.

SIMULTANEOUS MULTIWAVELENGTH OBSERVATIONES OF DWARF NOVAE SU UMA: MINIHUMPS AT MINIOUTBURST?

J. ECHEVARRIA, M. TAPIA J. BOHIGAS, R. COSTERO, J.A. LOPEZ, AND
M. ALVAREZ, L.F.RODRIGUEZ, J. BARRAL, E. DE LARA
OAN IA UNAM, Aptdo.Post. 877, Ensenada, B.C., Mexico

G. TOVMASSIAN, N. ASATRIAN
Byurakan Observatory, 378433 Byurakan, Armenia

M. SHARA
STScI, 3700 San Martin Drive, Baltimore, MD 21218, USA

D.H.P. JONES, R. WALLIS, M. ROTH
RGO, Madingley Road, Cambridge CB3 0EZ, UK

R. STOVER
Lick Observatory, Santa Cruz, CA 95064, USA

C. MARTINEZ, F. GARZON
IAC, Via Lactea, E-38200 La Laguna, Tenerife, Canary Islands

R. GILMOZZI
IUE Observatory, Villa Franca 54065, E-28080 Madrid, Spain

N. VOGT
ESO, La Silla observatory, La Serena, Chile

E. ZSOLDOS
Konkoly Observatory, P.O. Box 67, H-1525 Budapest, Hungary

P. SZKODY
Univ. of Washington, Seattle, WA 98195 USA

J. MATTEI
AAVSO, 25 Birch Street, Cambridge, MA 02138, USA

AND

F. BATTESON
Roy. Astron. Soc. of NZ, P.O.Box 3093, Greeton, New Zealand

A. Bianchini et al. (eds.), Cataclysmic Variables, 156-157.
© 1995 Kluwer Academic Publishers.

SU UMa — the prototype of a subgroup of dwarf novae, which displays superhumps in superoutbursts, has been observed[1] during an international campaign dedicated to the observation of the first day of outburst of dwarf novae during February 1986. After the start of a brightening was reported, the star was monitored by IUE, ground-based photometry and spectrophotometry, IR-photometry and Radio observations by VLA. However, it did not undergo to a normal outburst or a superoutburst.

We observed a short and low-amplitude flare, consisting of two phases. The first phase lasted 22 hours, during which the system brightened in the UV and optical wavelengths by factor of 2 and 3.5 respectively; then declined to initial values at long wavelengths, while in the far UV remained at mid-brightness. In the second phase it brightened again mainly at long-wavelengths and reached half the intensity of the primary brightening. The spectral lines have varied, but remained in emission during the entire flare, therefore no transformation to the optically thick state occured. The variation of flux of the UV emission lines are correlated with changes in the continuum , while the optical lines observed with higher time resolution showed great variability of relative intensities and equivalent widths. The Balmer lines ratios indicate the higher densities and temperatures of the disc after the primary brightening, when drop of luminosity in longer wavelenghts was observed.

Furthermore, multichannel photometry obtained during the first phase of the flare displayed a modulation with near half of the orbital period.

The observation of the flux distribution during the flare are inconsistent with the simple steady-state model of accretion discs, described by single power law. Also, the separate behavior of the UV and optical continuum, reported from previous observations, was outlined.

We conclude that the observed flare is consistent with the mass transfer instability model of outburst of the Dwarf Nova. According to this model the primary brightening could be interpreted as a burst of mass transfer which impact into the accretion disc.. The boldend bright spot gave rise to the light modulation. The consecutive drop and secondary brightening at longwavelenghts could resulted from the disc shrinking due to the low angular momentum interference into the disc and it's succesive restoration..

---

[1] The addresses of authors above refer to the time of observational campaign.

Nova Cygni 1992: MERLIN Observations from 1992 to 1994.

S. P. S. Eyres[1], R. J. Davis[1], M. F. Bode[2] & P. E. Pavelin[3]

1. *University of Manchester, N.R.A.L., Jodrell Bank, Macclesfield, Cheshire, SK11 9DL, England*

2. *School of Chemical and Physical Sciences, Liverpool John Moores University, Byrom Street, Liverpool, L3 3AF, England*

3. *HIROS group, School of Physics and Space Research, University of Birmingham, Birmingham, England*

Nova Cygni 1992 was observed with the *Multi Element Radio Linked Interferometer Network* (MERLIN) between 1992 May 9 and 1994 January 1 as a target of opportunity. Six observations were made at 6 cm, and a further seven at 18 cm. We were able to follow the evolution of the expanding nova, and trace the radio light curves.

The first four maps, at 6 cm, were published by Pavelin *et al.* (1992), and show both expansion and increasing flux. Further observations at 6cm at on 1992 November 15 and 1992 December 30 show that this trend continued. However, the structure varies between epochs. As the expansion continues, the peak brightness drops as the expanding components are resolved. By the time of the final observation in December, the circumstellar gas has separated into several separate peaks arrayed around a central depression.

At the beginning of 1993, MERLIN receivers were switched to 18 cm. The nova was observed at this frequency at seven epochs between April 1993 and January 1994. Variation in structure appears to continue, but expansion slows. Part way through this period, integrated flux began to decline, in keeping with VLA measurements at other frequencies (Hjellming 1994). This indicates that the nova has become partially optically thin to radio emission, as the gas becomes more rarefied.

At both frequencies, we observe expansion of a roughly circular structure, although there is little symmetry, indicating that the gas is not expanding as a homogeneous sphere. These variations may be due in part to optical depth effects during the partially optically thick phase. Brightness temperatures were found to exceed 40 000 K at the end of 1992, a value higher than that expected for radiatively excited hydrogen plasma. As suggested by Pavelin *et al.* (1992), this indicates that some shock heating mechanism may well be operating

MERLIN is a national facility operated by the University of Manchester on behalf of PPARC.

References.

Hjellming, R. M. 1994 these proceedings
Pavelin, P. E. *et al.* 1992 Nature Vol. 363 pp 424-426

158

A. Bianchini et al. (eds.), Cataclysmic Variables, 158.
© *1995 Kluwer Academic Publishers.*

INFRARED SPECTROSCOPY OF BL HYI

LILIA FERRARIO AND DAYAL WICKRAMASINGHE
The Australian National University Astrophysical Theory Centre,
School of Mathematical Sciences, Canberra, ACT 0200, Australia

AND

JEREMY BAILEY
Anglo-Australian Observatory, PO Box 296, Epping, NSW 2121,
Australia

Observations of the AM Her system BL Hyi covering the wavelength range 0.9 to 2.5 μm were made on 1992, Oct 5 using the IRIS instrument on the 3.9m Anglo-Australian Telescope. During these observations, cyclotron emission harmonics were detected near 1.25, 1.60 and 2.20μm. To interpret these data, we have adopted the linearly extended cyclotron emission region models of Ferrario and Wickramasinghe (1990).

We have assumed that the basic model parameters are as deduced by Piirola *et al* (1987), namely the orbital inclination is $i = 70°$ and the dipole inclination is $\theta_d = 150°$. We have found that the emission region is located below the orbital plane between magnetic colatitudes $\theta_1 = 15°$ and $\theta_2 = 17°$ and magnetic longitude $\psi_1 = 175°$ and $\psi_2 = 185°$. The electron temperature of the shock is $T = 20$ keV and the optical depth parameter is $\Lambda = 7.9$. The magnetic field strength at the pole is $B_p = 23$MG.

Wickramasinghe *et al* (1984) found $B_p = 30$MG through a Zeeman study of BL Hyi. This value differs from ours, thus suggesting that the dipolar field is not centred and the pole with the weaker field is located on the hemisphere that faces the companion star and accretes more strongly.

References

Ferrario, L., & Wickramasinghe, D.T. (1990) Arc-shaped cyclotron emission regions in AM Herculis systems, *Astrophysical Journal*, **Vol. no. 357**, pp. 582–590.
Piirola, V., Reiz, A., & Coyne, G.V. (1987) Five colour (UBVRI) polarimetry of H0139-68=BL Hydri, *Astronomy and Astrophysics*,**Vol. no. 185**, pp. 189–195.
Wickramasinghe, D.T., Visvanathan, N. & Tuohy, I.R., (1984) The magnetic field of the AM Herculis object H0139-68, *Astrophysical Journal*, **Vol. no. 286**, pp. 328–331

A. Bianchini et al. (eds.), Cataclysmic Variables, 159.
© *1995 Kluwer Academic Publishers.*

AN INTRIGUING PHASE 'GLITCH' IN THE V/R RATIO SPIN MODULATION OF GK PERSEI

M. A. GARLICK
Astronomy Centre, University of Sussex, U. K.

AND

J. P. D. MITTAZ, S. R. ROSEN AND K. O. MASON
Mullard Space Science Lab, University College London, U. K.

The intermediate polars (IPs) are defined by the presence of a spin-pulsed X-ray and/or optical modulation which in GK Per has a period of 351 s. Hutchings & Cote (1986) reported spin phase-resolved spectroscopy of this system, but their data were not sufficient to resolve the spin modulation properly. However, although limited in scope, their work did establish that radial velocity and equivalent width spin modulations were evident in GK Per, as is seen in other IPs. Later, Reinsch (1994) confirmed the presence of radial velocity modulations at the spin period.

We performed a similar analysis of spin phase-resolved optical spectroscopy of GK Per (full details are published in Garlick et al. 1994). In common with other such systems, we found the spectrum to be modulated at the spin period of the white dwarf. However, the characteristics of the modulation in GK Per are not typical of the class and, more interestingly, our results differ from those reported by Reinsch (1994). We detect a V/R ratio spin modulation which exhibits a 'glitch' in phase in the space of one day; an equivalent width modulation demonstrates no such phenomenon.

We conclude either that the accretion geometry in GK Per is unstable, and/or that the modulation pattern is produced by the interplay of multiple spin-modulated components.

References

Garlick M. A. et al., 1994, MNRAS, 269, 517
Hutchings J. B., Cote J., 1986, PASP, 98, 104
Reinsch K., 1994, MNRAS, 181, 108

A. Bianchini et al (eds.), Cataclysmic Variables, 160.
© 1995 Kluwer Academic Publishers.

THE X-RAY ECLIPSING POLAR RX J1802.1+1804

J. GREINER[1], R. REMILLARD[2], C. MOTCH[3]
[1] *MPI für Extraterrestrische Physik, 85740 Garching, FRG*
[2] *Center for Space Research, MIT, Cambridge, MA 02139 USA.*
[3] *Observatoire de Strasbourg, 67000 Strasbourg, France*

A new AM Her system (RX J1802.1+1804) was discovered during the *ROSAT* All-Sky-Survey as a highly variable and soft X-ray source. Follow-up pointed observations revealed a periodic flux modulation including a sharp and nearly complete eclipse of the X-ray emission. The X-ray period is derived to be 1.885±0.002 hr. The X-ray spectrum averaged over the full period is dominated by soft emission below 0.5 keV. We have fitted a blackbody model plus a thermal bremsstrahlung model with a fixed temperature of 10 keV to the data. The best fit values are $kT_{bb}=30\pm20$ eV, $N_H=(1.9\pm1)\times10^{20}$ cm^{-2}.

RX J1802.1+1804 is optically identified with a V=15 mag object. The optical spectrum is typical for a high-excitation cataclysmic variable with strong emission lines of H, HeI and HeII. Phase-resolved spectroscopy shows a strong velocity wave, but the amplitude and velocity midpoint may vary from night-to-night, suggesting complex structure along the accretion stream. The best fit period is 1.8841±0.0017 hr. Photometric observations have revealed 0.1 mag variations, but no eclipse. The maxima in the optical continuum also vary in amplitude from cycle to cycle, but they consistently appear near or just after the 'zero-crossing' in the radial velocity, from red-shifted to blue-shifted phases.

The evidence of an X-ray eclipse without a corresponding optical eclipse implies that the X-ray emitting region is confined to a small region on the white dwarf surface and that the eclipse is due to the disappearance of this region behind the limb of the rotating white dwarf.

Acknowledgements

This work was supported by the German Bundesministerium für Forschung und Technologie under contract No. 50 OR 9201 (JG).

A. Bianchini et al. (eds.), Cataclysmic Variables, 161.
© *1995 Kluwer Academic Publishers.*

THE X-RAY ECLIPSING CATACLYSMIC VARIABLE 1H1752+081

F.V. HESSMAN, K. BEUERMANN, V. BURWITZ AND K. REINSCH
Universitäts-Sternwarte
Geismarlandstr. 11, D-37083 Göttiingen, Germany

AND

H.-C. THOMAS
Max-Planck-Institut für Astrophysik
Karl-Schwarzschild-Str. 1, D-8xxxx Garching, Germany

The orbital light curve of 1H1752+081 pieced together from the ROSAT All-Sky Survey (RASS) data shows there is considerable variation from orbit to orbit: an orbital hump around phases 0.5-0.9 can be seen. 1H1752+081 was fairly hard (HR1=+0.8) but still showed an unusually high 0.1-3 cts/s in the ROSAT PSPC, making it one of the brightest X-rays sources in the CV sky. We have obtained ROSAT PSPC pointed observations centered on two eclipses on 1993 Sept. 10 and 11. The eclipse of the X-ray "accretion spot" agrees with the optical eclipse and is total as is expected for the estimated inclination of 80°. The hump at 0.7-1.0 keV in the X-ray spectrum can only be reproduced using fairly "warm" gas (0.3-1.4 keV) with lots of X-ray emission lines, and the hard emission must be due to a dominant, optically thin "hot" gas (> 3 keV) component. The X-ray spectrum does not change in shape with orbital phase; particularly, the amount of absorption is constant.

Using the ancient technique of trailed spectroscopy, we have obtained several hours of data in 1994 March using the Calar Alto 3.5m Twin-Spectrograph. The trailed spectrogram shows the classical spectrum of an AM Her: broad components from the stream as well as the narrow emission component from the secondary. Given the effective time resolution of about 3 minutes, the He II λ4686 broad emission line can be differentiated into a high-velocity component whose egress is later (i.e. the stream) and another at lower velocities which – at least in projection – must come from around the white dwarf (i.e. near the accretion shock) and which suffers an abrupt but small(!) velocity shift during the eclipse.

162

A Bianchini et al (eds) Cataclysmic Variables, 162

THE OPTICAL LIGHT CURVE OF V664 CAS

N.A. KETSARIS

*Faculty of Physics, Moscow State University, Moscow V-234,
Russia; e-mail: ketsaris@sai.msk.su*

AND

S.V. ANTIPIN AND S.YU. SHUGAROV

*Sternberg Astronomical Institute, Universitetskij pr., 13,
119899, Moscow V-234, Russia*

V664 Cas is the central star of the planetary nebula HFG 1 discovered in 1982 (Heckathorn *et al.*). The star was observed using *UBV* photometer during JD=24,49245-24,49257. We found the brightness variations V=13.^{m}46-14.^{m}70, B=13.^{m}81-14.^{m}94, U=12.^{m}97-14.^{m}16. We also investigated this star on each of 265 plates obtained by using the 40-cm astrograph and equatorial camera of Sternberg Astronomical Institute in JD=24,14715-24,48627. V664 Cas showed the brightness variations 13.^{m}8-15.^{m}1 (see Fig.1). This data were combined with the photoelectric B-observations. They revealed a more precise value of the orbital period to be $P_{orb} = 0.^d581647 \pm 0.^d000001$. The period seems to be very stable.

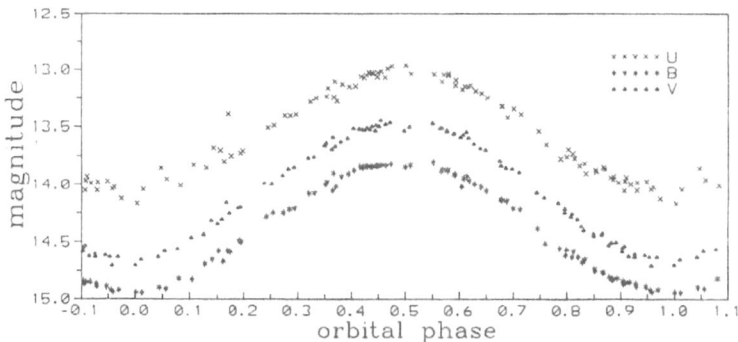

Figure 1. *UBV* light curves of V664 Cas.

163

A. Bianchini et al (eds.), Cataclysmic Variables, 163-164.

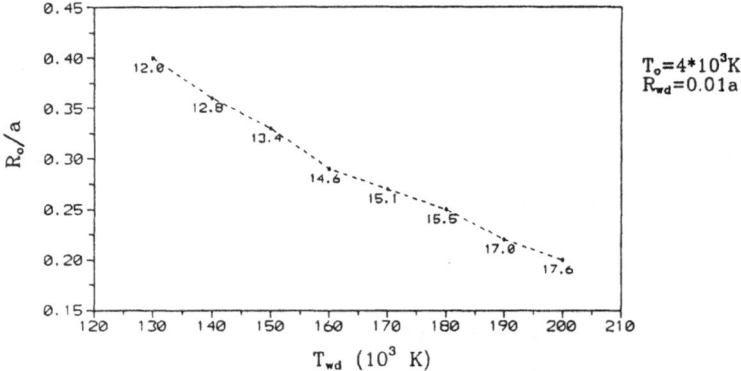

Figure 2. The parameter space at fixed R_{wd} and T_o.

The orbital period stability implies no mass transfer, so it is the detached system. The light curve of V664 Cas was modeled in a way similar to BE UMa (see Raguzova *et al.*). The assumed radius of the PN nucleus R_{wd} was $2.5 \times 10^4 km$ ($\approx 0.01a$), which is typical for such objects, and the temperature of K-M dwarf T_o was assumed to be $4,000K$. The main result of the modeling is the determination of a boundary dependence of the cold component radius R_o on temperature T_{wd} of the PN nucleus at fixed R_{wd} and T_o (see Fig.2). The region of permitted values of parameters lies above the dotted line. The numbers near the points show value of i for some R_o and T_{wd}.

V664 Cas unsignificantly reddens in light minima. These results are possibly due to the system is seen almost head-on. The heated part of the red dwarf is seen all the time and only visible area of the spot varies.

The mass of the hot planetary nebula nucleus is in the range from 0.6 to 1.2 $M_\odot$. The cool component is a K-M dwarf (Graner *et al.*, 1987), so its mass is 0.4-$0.8 M_\odot$ and the binary separation is $2.3 \pm 0.3 \times 10^6 km$ ($3.35 \pm 0.35 R_\odot$).

Although the variability of the orbital period has not been found, we do not exclude that more precise observations can change our conclusion. The systems are detached but because of a very high temperature of the primary component the mass transfer by stellar wind is possible and this process lead to mass exchange and period change.

References

Graner A.D. *et al.*   *BAAS*, 1987, **19**, 643.
Heckathorn J.N., Feson R.A., Crull T.R.   *A&A*, 1982, **114**, 414
Raguzova N.V., Ketsaris N.A., Shugarov S.Yu.   *"The Optical Light Curves of BE UMa"*, this book.

SYMBIOTIC BINARY CH CYGNI IN 1992–1994

L. LEEDJÄRV
Tartu Astrophysical Observatory
Tõravere, EE 2444 Tartumaa, Estonia

AND

M. MIKOŁAJEWSKI
Institute of Astronomy, N. Copernicus University
Chopina 12/18, PL 87 100 Toruń, Poland

CH Cygni is considered as a peculiar symbiotic star (P $= 5700^d$) consisting of a SRb variable M6.5–M7.2III and a white dwarf with a strong magnetic field ($B \sim 10^7$Gs). An oblique magnetic rotator model (Mikołajewski & Mikołajewska, 1994, *MNRAS*, submitted) allows to interpret the behaviour of CH Cyg as transitions between three stages: inactive, low activity (*propeller*), high activity (*accretor*). The most spectacular active period was observed from 1977 to 1986. By 1988 CH Cyg reached the lowest activity of last decades, and in 1990 the star was at the lowest brightness ever observed – V $\approx 9^m.0 - 9^m.2$. Since 1989, however, some episodes of erratic activity have been observed (Tomov *et al.*, 1989, *IBVS*, 3389, Mikołajewski *et al.*, 1992, *IBVS*, 3742).

The UBV observations carried out in 1992–1994 at Tartu, Toruń and Olsztyn Observatories show a continuous tendency of increasing brightness, especially in the U filter bandpass. The U brightness reached U $\approx 7^m.0$ by May 1994 which is by only $\sim 1^m.5$ less than the brightest U magnitude during the accretor state in 1981–1984. Spectroscopic observations of CH Cyg carried out by the Tartu Observatory 1.5 m telescope in 1992–1994 show the spectrum of an M giant superimposed by the hot continuum and by Balmer and other emission lines. The most intriguing elements are weak variable blue- or red-shifted emission components of Balmer lines at radial velocities from -700 to $+700$ kms^{-1}. Those observations are interpreted on the basis of the magnetic rotator model as follows: CH Cyg is in a propeller state now, with the white dwarf's rotating magnetosphere repelling outward the accreted matter and forming a quasi-stable envelope which is becoming optically thick.

A. Bianchini et al. (eds.), Cataclysmic Variables, 165.

SPECTROSCOPIC VARIATIONS AND STELLAR
PARAMETERS OF 6 SU UMA TYPE STARS

R.E. MENNICKENT

Departamento de Física, Universidad de Concepción, Casilla
4009, Concepción, Chile. e-mail: RMENNICK@buho.dpi.udec.cl

We present the main results of time resolved spectroscopy of CU Vel, AQ Eri, EK TrA, TY PsA, TU Men and WX Cet, with 2-4 Å resolution. Spectra were taken at La Silla and Las Campanas in 1991 with the 2.2 and 2.5 meter telescopes. The time resolution was $\sim$ 6 minutes. Radial velocities were obtained with the double gaussian routine kindly provided by A. Shafter and P. Szkody. Typical uncertainties in the K values were 7 km s^{-1}. Orbital periods with $\sim$ 0.5 % accuracy were calculed by using the *pdm* IRAF routine. Mass ratios were obtained by using the Shafter and the Jurcevic et al. (1994, PASP 106, 427) calibrations. Results are given in Table 1. When compared our new $\Phi_{V/R}(máx)$ values with those found by Mennickent (1994, A&A 285, 979), we observe two groups of stars in the $\Phi_{V/R}(max)$ vs. mass ratio diagram: one of them grouped around phase 0.6 (the expected phase when V/R variations are caused by a normal hot spot), and the other group following a tight correlation with the system mass ratio. Detailed studies will be published elsewhere. This research was supported by Fondecyt Grant 1940708 and by a fund of the Facultad de Física de la Pontificia Universidad Católica de Chile.

TABLE 1. Orbital periods, mass ratios and spectroscopic quantities. Phase 0 refers to superior conjunction of the white dwarf. Double minima in the diagnostic curves of the radial velocities gave two possible solutions for TU Men and AQ Eri.

Star	P_{orb} (days)	$K(kms^{-1})$	M_2/M_1	$\Phi_{V/R}(máx)$
CU Vel	0.0785(2)	49 ± 6	≈ 0.11	0.53 ± 0.03
AQ Eri	0.061(2)	91 ± 9 or 53 ± 5	0.26 ± 0.03 or 0.12 ± 0.01	0.28 ± 0.03
TY PsA	0.0852(5)	72 ± 6	≈ 0.21	0.60 ± 0.04
EK TrA	0.0636	65 ± 9	0.19 ± 0.03	0.10 ± 0.05
WX Cet	0.05497(3)	45 ± 5	0.11 ± 0.01	0.12 ± 0.03
TU Men	0.1172(2)	108 ± 10 or 87 ± 2	0.455 ± 0.045 or ≈ 0.33	0.93 ± 0.04

A. Bianchini et al. (eds.), Cataclysmic Variables, 166.

CH CYGNI AS A PROPELLER STAR

M. MIKOŁAJEWSKI
Institute of Astronomy, N. Copernicus University
Chopina 12/18, PL-87100 Toruń, Poland

AND

J. MIKOŁAJEWSKA
Copernicus Astronomical Center
Bartycka 18, PL-00716 Warsaw, Poland

Different variable phenomena have been revealed in symbiotic binary CH Cygni with timescales ranging from hundreds years, down to minutes. The most spectacular phenomenon was, however, the onset of the bipolar jet ejection just after a sudden drop of brightness in July 1984. We propose a model of the activity based on the theory of an oblique rotator, and assuming a wind accreting magnetic white dwarf as the energy source for the active component (Mikołajewski & Mikołajewska, 1994). Two stages of the rotator, *propeller* and *accretor*, are possible. The star spends most of time in the subsonic propeller state, when magnetically driven outflow leads to formation of quasistatic envelope. Simultaneously, the infalling matter gains some angular momentum from the rotator, and slow accumulation of a large dead disk (without any accretion) can take place. Once the disk mass exceeds some critical value, transition to the accretor stage occurs, and the hot component's luminosity reaches maximum. After accreting certain amount of the matter accumulated in the disk the rotator returns to the propeller state. As a result of increased density near the rotator the sound speed is reduced, and the propeller rotates supersonically driving bipolar jets. CH Cyg seems to be a prototype of a new class of interacting binaries – *propellers* or *propeller stars*.

This study was partially supported by Polish KBN research grant No. 2 P304 007 06.

References

Mikołajewski, M., Mikołajewska, J.: 1994, *MNRAS*, submitted

A. Bianchini et al. (eds.), Cataclysmic Variables, 167.

QU VUL: A NEW ECLIPSING NOVA IN THE PERIOD GAP

K. A. MISSELT AND A. W. SHAFTER
San Diego State University
Department of Astronomy, San Diego, CA, 92182, USA

P. SZKODY
University of Washington
Department of Astronomy, Seattle, WA, 98195, USA

AND

M. POLITANO
Arizona State University
Department of Physics and Astronomy, Tempe, AZ, 85287, USA

1. Abstract

We report the preliminary results of a photometric and spectroscopic study of the classical nova, QU Vulpeculae (= Nova Vul 1984 2). QU Vul was discovered by Collins on 23 Dec 1984 and reached a peak of V~5.7. Subsequent work by Gehrz *et al.*, and by Starrfield *et al.* revealed QU Vul to be member of the ONeMg sub-class of the classical novae. This sub-class shows an overabundance of medium weight nuclei, e.g. O, Ne, Na, Mg and Al, in their ejecta, which are believed to be produced when a thermonuclear runaway occurs on the surface of a massive, ONeMg white dwarf.

Here we report the discovery the eclipses in the light curve of QU Vul, which recur with a period of 2.68 hr, and place QU Vul in the well-known "gap" in the period distribution of cataclysmic variables. QU Vul now becomes the second classical nova to be found in the period gap, the other being V Per, with a period of 2.57 hr. The eclipsing nature of QU Vul is significant because this system may now offer the best chance to determine the mass of an ONeMg white dwarf in a classical nova system.

A Bianchini et al (eds), Cataclysmic Variables 168
© *1995 Kluwer Academic Publishers*

GEOMETRY OF THE POLAR AN UMA

M. MOUCHET[1], J.M. BONNET-BIDAUD[2], T. SOMOVA[3], N. SOMOV[3]
[1] *DAEC, Observatoire de Meudon, F-92190 Meudon*
[2] *DSN/DAPNIA/SAp, CE-Saclay, F-91191 Gif sur Yvette*
[3] *Special Astrophysical Observatory, 357147, Russia*

We report on simultaneous photometric and spectroscopic observations of AN UMa obtained in high time resolution mode at the Russian 6m telescope in March 1991 and January 1992. Spectra in the wavelength range [4200-5200Å] are obtained with a spectral resolution of 2Å and with a maximum temporal resolution of 32 ms.

A study of the velocities of the narrow and broad Balmer and HeII emission line components reveals an inaccuracy in the rotational/orbital period (Liebert et al. 1982). A new ephemeris is derived with a refined period of (6890.6436 ± 0.0035)s. With this new period all published photometric X-ray and optical light curves are correctly phased over more than 15 years.

The geometry of the system appears stable. A modelling of the narrow and broad emission line velocities allow to constrain the system parameters. Two possible regions responsible of the narrow component have been considered; the heated hemisphere of the secondary and the curved stream connecting the inner Lagrangian point to the magnetosphere, while the broad line component is assumed to arise from the magnetosphere (Mukai 1988, Mouchet 1993). The best solution is found for a low mass white dwarf $(M_{wd} = 0.4\text{-}0.6M_{\odot})$, an inclination $i = 40 - 60°$, an angle of the magnetic field axis with respect to the perpendicular to the orbital plane of $\theta_d = 25 - 45°$ and the narrow component of the emission lines produced in a stream near the capture radius.

These results are reported in Bonnet-Bidaud et al. (1994).

References

Bonnet-Bidaud J.M., Mouchet M., Somova T.,Somov N. (1994), submitted to *A&A*
Liebert J., Tapia S., Bond H., Grauer A. (1982), *Astrophys. J.* **254**, 232
Mouchet M. (1993), *Annals of the Israel Physical Society*, vol. 10, p. 208
Mukai K. (1988), *Monthly Not. Roy. Astr. Soc.* **232**, 175

A. Bianchini et al. (eds.), Cataclysmic Variables, 169.
© 1995 Kluwer Academic Publishers.

TT ARIETIS OBSERVED AT BUCHAREST OBSERVATORY

GABRIELA OPRESCU AND ALEXANDRU DUMITRESCU
Astronomical Institute of the Romanian Academy
Str. Cutitul de Argint 5 ,75212 Bucharest 28, ROMANIA

The blue peculiar variable TT Arietis (BD + 14°341), which is now thought to be an accreting (magnetic ?) white dwarf in a low-mass binary, was firstly taken as an eclipsing binary by Strohmeier et al. (1957). It was initially considered to be a nova-like variable but lately it is believed to be a possible - intermediate polar. The main argument for this classification is the difference between the photometric and spectroscopic periods of the system. Semeniuk et al. (1987) searched the quasi - periodic variability on time scale of minutes. They found a 20 min. period which appeared in the majority of nights of all observed seasons. More recently Kraicheva et al. (1987) and Roessiger (1987) have observed sudden dips in the light curve. The extraordinary character of this star requires that continous observations be made in different spectral ranges. In 1985 the Bucharest Observatory participed to an international coordinated observational campaign. TT Ari was observed on August 20/21, 21/22 and 23/24 in filters B and V, about 100 days after the return to the active state. A great amount of simultaneous optical photometric and spectroscopic data was obtained by the participating observatories. On September 27, 1992 we observed TT Ari in filter B. DM +14°0336 was used as comparison star. The observations were carried out with the 50cm Cassegrain telescope of the Bucharest Observatory. An EMI 9502B unrefrigerated photomultiplier and B filter (Schott 1mm BG 12 + Schott 1mm GG 13) were used. For each observed point the integration time is about 15sec.Δm_B was estimated taking into account the extinction correction. The individual errors do not exceed 0.01 mag. For phase calculations we used the Wenzel's ephemeris (Wenzel et al. 1986), 'Min HJD = 2437646.6650 + 0.132771 E' One can observe random fluctuations in magnitude proper to this system. The amplitudes of these fluctuations lie in the range of 0.4 mag. A dip in the brightness ($\Delta m_B \simeq 0.5$ mag.) with the duration of about 10 min. was observed at 1h 11min in local time. Using Wenzel's ephemeris and timing the maximum deduced from our light curve we got (O-C) = 0.0187 days = 0.14 P. With Udalski's (1988) ephemeris given by him for 1985 'Min HJD = 2446287.1281 + 0.132765E,' we obtained (O-C) = 0.0117 days = 0.088 P. This value is consistent with those given by Udalski. On the other hand, if we consider the ephemeris proposed by Udalski for 1987/1988, 'Min HJD = 2447031.5811 + 0.132957E,' we get (O-C) = 0.1054 days = 0.792 P, value which is inconsistent with 1992's data.

A. Bianchini et al. (eds.), Cataclysmic Variables, 170.
© *1995 Kluwer Academic Publishers.*

IUE SPECTRA OF THREE Z CAM TYPE DWARF NOVAE: A STUDY OF P CYGNI PROFILES

M. T. Özkan
İstanbul University Observatory Research and Application Center, 34452 University, İstanbul - TURKEY

H. H. Esenoğlu, T. Ak, A. T. Saygaç, S. Güler
İstanbul University Observatory, Astronomy and Space Sciences Department, 34452 University, İstanbul - TURKEY

ABSTRACT: We study the P Cygni profiles in three dwarf novae. In a study of Z Cam by Szkody and Mateo (1986), it was claimed that the mass loss continues for 1 month after the start of a standstill and ceases within 6 months. This results of the work depends on only one standstill spectrum. On the other hand, in another work carried out by Özkan and Esenoğlu (1994) it was found that the case may not be true for Z Cam, because two ultraviolet spectra obtained during the sixth month of another standstill configuration shows the P Cygni profile having a stronger absorption component and a weaker emission component which implies that the mass loss may continue during the standstill. Following these works in the present study, we aimed to investigate this situation for Z Cam and two other Z Cam type dwarf novae, RX And and HL CMa, considering all ultraviolet spectra obtained with IUE and AAVSO observations (Mattei, 1994).

RESULTS

When we examine all P Cygni profiles belonging to Z Cam, RX And and HL CMa and AAVSO light curves in more detail we can arrive at the following results.

1) With the exception of the profile seen in the first standstill spectrum of Z Cam (December 3, 1979), all of them are well defined in all spectra. The strengths of absorption and emission components in P Cygni profiles of Z Cam do not change so much during declines from outburst and during standstills.

2) Another important point is that in the case of Z Cam the equivalent width of the absorption component is always greater than the emission component for all spectra covering outburst, decline, and standstill states. This is also true for two other Z Cam stars, except for one case of RX And and for three cases of HL CMa. This confirms similar results obtained in TZ Per, which is also a Z Cam star, by la Dous et al. (1985) and in Z Cam by Szkody and Mateo (1986). The standstill spectra of Z Cam and RX And lead to the suggestion that mass loss may continue until the end of a standstill.

3) If we take into account the mean equivalent widths in Z Cam, it appears that the C IV absorption of the decline spectrum is little stronger than that of standstill spectrum, however, and the C IV emission little weaker. In RX And both the absorption and the emission components of the C IV feature of the decline spectrum are somewhat stronger than those of the standstill spectrum.

The main aim of our study was to determine whether the behaviour of the P Cygni profiles is related to the time elapsed the start of a standstill. The IUE and the AAVSO data suggest that there is no correlation between the strengths of these features and the time since the last outburst. In other words, although Z Cam stars release at least an order of magnitude less energy flux at standstill than at outburst maximum, the above results and spectral indices describing the continuum flux distributions show no large changes in the accretion disk structure and mass loss rates during decline from outburst and during standstill.

Acknowledgements
In this research we have used data and information from the AAVSO international database. This research was partly supported by TÜBİTAK.

References
la Dous, C., Verbunt, F., Schoembs, R., Argyle, R. W., Jones, D. H. P., Schwarzenberg-Czerny, A., Hassall, B. J. M., Pringle, J. E. and Wade, R.A., 1985, Mon. Not. R. Astr. Soc., 212, 231
Lynden-Bell, D., 1969, Nature, 233, 690
Mattei, J. A., 1994, Observations from the AAVSO International Database, private communication
Özkan, M. T., Esenoğlu, H. H., 1994, in preparetion
Szkody, P., Mateo, M., 1986, Ap. J., 301, 286

A. Bianchini et al. (eds.), Cataclysmic Variables, 171.
© 1995 Kluwer Academic Publishers.

THE PECULIARITY OF *V1500 CYG* SYNCHRONIZATION

E.P. PAVLENKO, V.P.MALANUSHENKO
Crimean Astrophysical Observatory
334413 Ukraine, Crimea, p/o Nauchny

V1500 Cyg is the first recognized magnetic nova where the primary rotational period is 3.5 minutes shorter than the orbital one. The statistical data analysis based on the 11-years cycle of photometric research in Crimean astrophysical observatory was made. It showed that in spite of the mascing effects due to the complex accretion geometry, there exists the photometric beat period.

Using the known values of orbital and beat periods one can obtain the non-observable photometricaly rotational period of the white dwarf from the relation: $\qquad 1/P_{beat} = 1/P_{rot} - 1/P_{orb}$

To investigate detaily the behaviour of P_{rot} begining from the second year after outburst up to 1987 year, we combined every two-year runs of data, shifting on a one year for computing of frequency spectra. All periodograms show a several peaks in the vicinity of the beat period. For every spectrum we selected one from the 5 most sugnificant periods which is closer to the mean value $P_{beat} = 7.69$ days. The current values of the P_{rot} (folded circles) are presented in Fig. The open one is the independent estimate obtained from polarimetry by Schmidt and Stockman. There was the rapid spin down of the primary with rate $\dot{P}_{rot} = 1.8 \times 10^{-6}$ in 1977-1979

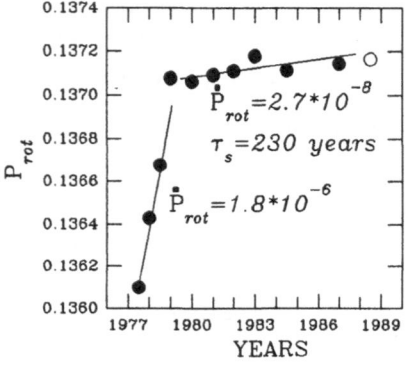

yrs and with $\dot{P}_{rot} = 2.7 \times 10^{-8}$ in 1979 - 1987. The last corresponds to the time-scale of synchronization equal to 230 yrs.

Authors thank Ukrainian ISF for the support of this work and tor funding to attend this conference by grant N 1477-3.

A. Bianchini et al. (eds.), Cataclysmic Variables, 172.
© 1995 Kluwer Academic Publishers.

THE PHOTOMETRIC INVESTIGATION OF RZ LMI.

O.D. PIKALOVA
Faculty of Physics, Moscow State University, Moscow V-234, Russia;

AND

S.YU. SHUGAROV
Sternberg Astronomical Institute, Universitetskij pr., 13, 119899, Moscow V-234, Russia; e-mail: shugarov@sai.msk.su

RZ LMi is the object with a strong UV excess. Depending on the brightness of the variable, emission or absorption hydrogene lines are observed.

We have studied the object by using 300 photographic plates taken in JD 24,45433-24,48777 and UBV photoelectric photometry obtained in JD 24,46581-24,48777. The system was found to vary in a range 14^m-18^m. In contrast to the typical U Gem stars, RZ LMi shows short term brightening (or fading) by 2-3 magnitudes on a time scale of several days, and then returns to its previous state. We found no periodic light variations within all interval; however, in JD 24,46019-24,46208 we do found a probable brightness variation cycle with a period of 21^d167 or 23^d313. The data folded with one of these periods is shown in Fig.1.

We conclude that RZ LMi is likely to be the prototype of a relatively unique CV subclass – VY Scl binary type.

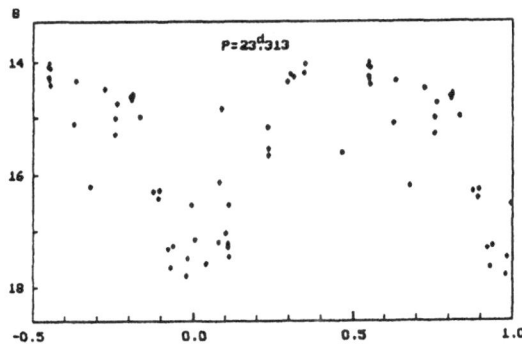

Figure 1. The light curve of RZ LMi.

173

A. Bianchini et al. (eds.), Cataclysmic Variables, 173
© *1995 Kluwer Academic Publishers.*

CAN HYDROGEN EMISSION LINES OF PU VUL BE PRODUCED IN THE OUTER ATMOSPHERE OF THE RED GIANT COMPONENT?

D. PROGA AND S.J. KENYON
Harvard–Smithsonian Center for Astrophysics
60 Garden Street, Cambridge MA 02138, USA

We calculated atmospheric models for a cool giant star illuminated by a hotter companion. This situation occurs in some symbiotic novae – such as PU Vul – shortly after eruption. We used a modified version of the UMA code (Gustafsson et al. 1975, Nordlund & Vaz 1990 and references therein) to calculate radiative atmosphere models for the cool component of the system; one model without illumination and four with illumination by a black body plane-parallel beam for different incident angles. We adopted parameters of the companions from Kenyon (1986). We assumed the orbital separation to be 1AU and the atmosphere is classical plane-parallel, with solar abundance and LTE. We used the atmospheric structure of the illuminated models and the MULTI code (Carlsson 1986) to solve the radiative transfer problem for hydrogen. We have modified the MULTI code to include the illumination effect. In the non–LTE case only the Lyman lines are in emission; the rest are in absorption. For comparison we also calculated line profiles assuming LTE. All LTE hydrogen lines are in emission with exception for Paschen β which is in absorption with central emission.

The upper atmosphere of the red giant is very likely in non–LTE condition due to the low density and the strong radiation field. Thus we do not believe that strong emission lines can be produced in the red giants atmosphere. These lines must form in material ejected by the hot component.

References

Carlsson M. (1986) *A Computer Program for Solving Multi-level Non–LTE Radiative Transfer Problems in Moving or Static Atmosphere*, Uppsala Astronomical Observatory Report No. 33

Gustafsson B., Bell R.A., Eriksson K., Nordlund A. (1975), *A&A*, **42**, 407

Kenyon S.J. (1986) , *AJ*, **91**, 563

Nordlund A., Vaz L.P.R. (1990) , *A&A*, **231**, 231

A Bianchini et al (eds), Cataclysmic Variables, 174
© *1995 Kluwer Academic Publishers*

THE OPTICAL LIGHT CURVE OF BE UMA

N.V. RAGUZOVA AND N.A. KETSARIS
Faculty of Physics, Moscow State University, Moscow V-234, Russia

AND

S.YU. SHUGAROV
Sternberg Astronomical Institute, Universitetskij pr., 13, 119899, Moscow V-234, Russia; e-mail: shugarov@sai.msk.su

BE UMa is a classical example of an eclipsing binary system with a strong reflection effect (Margon & Downes, 1981; Ando *et al.*, 1982; Ferguson *et al.*, 1987). A hot star looks like the central star of a planetary nebula, and illuminates the faced-on surface of the cool component. However, no planetary nebula around BE UMa was discovered. The hot star has a temperature of $1\text{-}2 \times 10^5 K$, so that spectral maximum of its proper radiation falls in the far ultraviolet thus contributing unsignificantly to the optical brightness of the system. On the other hand, a hot spot on the surface of the cooler component produced by the reflection effect has a central temperature of about $1.5 \times 10^4 K$ giving maximal radiation output in the near ultraviolet. Changing of the visible area of the spot leads to sinusoidal brightness variations of the whole system.

We measured the star on 544 plates obtained at Crimean laboratory of Sternberg Astronomical Institute with the 40cm astrograph and the 50cm Maksutov telescope. In addition, we performed 121 *UBV* electrophotometrical observations of BE UMa at the 60cm and 125cm reflectors.

The B-observations were reduced jointly with Harvard photographic observations (Ferguson *et al.*, 1987). As result, a centennial series of observations were obtained to derive an improved ephemeris:

$$HJD_{\min} = 47628.5381 + 2.^{d}2911667\,E$$

The light curve folded with new elements is shown in Fig.1. No significant orbital period changes were noticed. The light curve amplitude is different in different photometric systems: $\Delta U = 5^m$, $\Delta V = 2^m$. This difference

A Bianchini et al (eds), Cataclysmic Variables, 175-176
© *1995 Kluwer Academic Publishers*

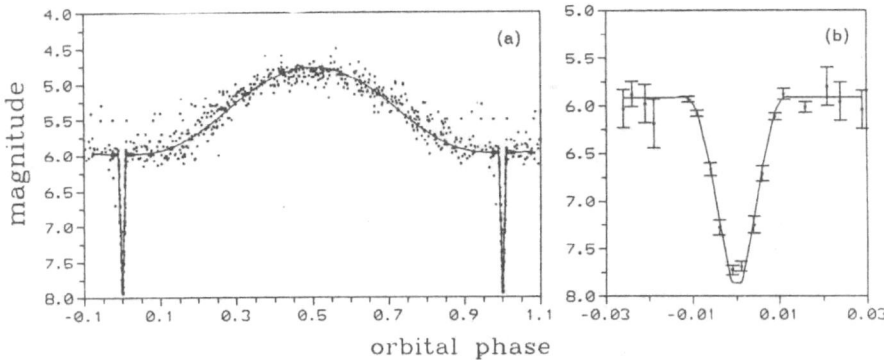

orbital phase

Figure 1. The observed and synthesized *B*-light curves of BE UMa: (a) total; (b) only in the eclipse.

enabled us to determine temperatures of both components of the binary. Similar results were obtained by other authors (e.g. Ando *et al.* (1982)).

Using light curve synthesis method, we found some parameters of BE UMa – orbital inclination i, radius of the primary R_{wd} and of the cool star R_o, and temperatures of both components T_{wd}, T_o:

i	$82.°9 \pm 0.°8$
R_{wd}	$1.18 \pm 0.10 R_\odot$
T_{wd}	$5.40 \pm 0.45 \times 10^3 K$
R_o	$0.090 \pm 0.007 R_\odot$
T_o	$1.25 \pm 0.13 \times 10^5 K$

A few stars are known to show similar characteristics. However, they all have lower temperatures of the degenerate component. For example, temperature of the nucleus of V477 Lyr and of UU Sge is $4 \times 10^4 K$ and $5 \times 10^4 K$, respectively.

A separation between the components of BE UMa was estimated from the third Kepler's law. The mass of the hot planetary nebula nucleus ranges from 0.6 to 1.2 $M_\odot$. If the cool component is a G-K star, then its mass is 0.6-$0.8 M_\odot$ and the components are separated by $6.0 \pm 0.5 \times 10^6 km$ $(8.7 \pm 0.7 R_\odot)$.

References

Ando H., Okazaki A., Nishimura Sh. //*PASP*, 1982, **34**, 141.
Ferguson D.H. *et al.* //*ApJ*, 1987, **316**, 399.
Margon B., Downes R. //*PASP*, 1981, **93**, 548.

S 10932 COMAE – A JUMPING JACK AMONG THE CVS

G.A. RICHTER[1], J. GREINER[2]
[1] *Thüringer Landessternwarte Tautenburg, Außenstelle Sonneberg, 96515 Sonneberg, FRG*
[2] *Max-Planck-Institut für Extraterrestrische Physik, 85740 Garching, FRG, jcg@mpe-garching.mpg.de*

As part of a programme of investigating the optical long-term behaviour of selected X-ray sources, we studied RX J1239.5+2108 in more detail. This source was observed during the *ROSAT* All-Sky-Survey in December 1990 for a total of 470 sec at a mean count rate of 0.05 cts/sec. The hardness ratio suggests a hard spectrum. The Palomar Observatory Sky Survey prints revealed only one object at the X-ray position (with a 2σ error circle of $30''$) within $2'$distance. Spectroscopic observations show this very blue object to have a low exited spectrum of strong Balmer emission lines and fainter ones of HeI and HeII, resembling that of a dwarf nova.

Photometrically, this star displays fireworks of brightness changes combining characteristics of different subclasses of cataclysmic variables (CV): short brightness dips (eclipses?) and humps (both ≈ 1 mag) within one night, dips and humps (≈ 1.5 mag) within several days, high ($m_{pg} \approx 16.5$) and low ($m_{pg} \approx 18$) states of brightness. To carry the matter to an extreme, the object had one eruption (≈ 4.5 mag) which lasted at least several days, and which may be that of a dwarf nova put on record on Sonneberg plates.

So far, there is only one other CV known which shows both eruptions similar to dwarf novae and different brightness levels, namely V426 Oph (Wenzel & Splittgerber 1990, IBVS No. 3532). But in contrast to S 10932 the magnitude difference between the two levels is small in V426 Oph (only about 0.6 mag) and the cycle lengths are much smaller (13^d in the high state, 22^d in the low state) than in the case of S 10932.

Acknowledgements

This work was supported by the German Bundesministerium für Forschung und Technologie under contract Nos. 05 2S0524 (GAR), 50 OR 9201 (JG).

A. Bianchini et al. (eds.), Cataclysmic Variables, 177.

UNDERSTANDING THE BEHAVIOR OF THE ECLIPSING POLAR DP LEO

CRAIG R. ROBINSON AND FRANCE A. CÓRDOVA
The Pennsylvania State University
525 Davey Lab., University Park, PA 16802 U.S.A.

Systematic variations are suspected in both the orbital period and accretion longitude in the DP Leo system. The variations in accretion position on the surface of the white dwarf, relative to the position of the secondary star, are not dependent upon the mass transfer rate. Combining our ROSAT data with all other available data, we find further evidence to support our hypothesis that the accretion longitude variations are produced through orbital period oscillations. The mechanism producing the orbital period changes is likely driven by the growth and decay of the magnetic field of the secondary star during solar-like magnetic activity cycles. This is the same mechanism suspected of producing orbital period oscillations in other close binaries with late-type stars. Evidence for similar variations in other polars exists, but the baseline of observations for the other eclipsing systems is limited. Further details are available in Robinson & Cordova (1994, *ApJ*, in press). An additional paper on this subject is in preparation.

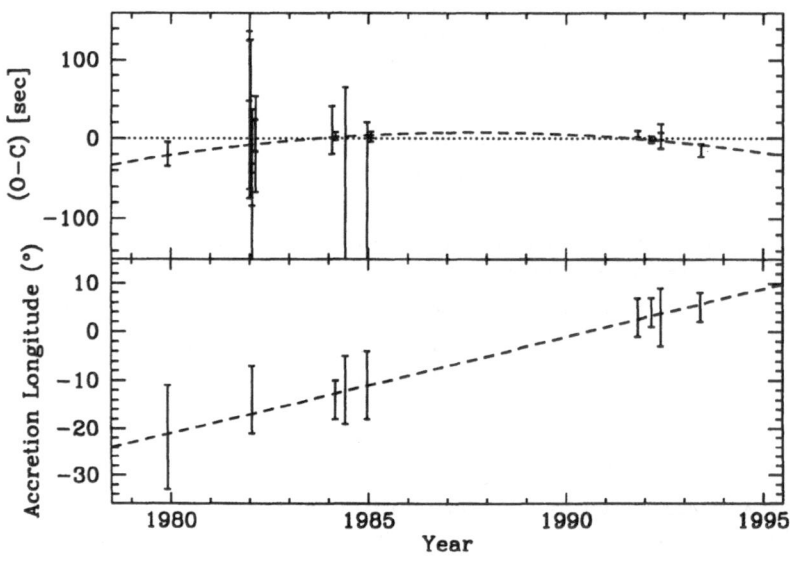

A. Bianchini et al. (eds.), Cataclysmic Variables, 178.

THE VARIABLE NATURE OF TT ARI IN THE ULTRAVIOLET

C.R. ROBINSON, F.A. CÓRDOVA AND G.S. STRINGFELLOW
The Pennsylvania State University
525 Davey Lab., University Park, PA 16802 U.S.A.

The complete IUE database of TT Ari observations has been obtained and analyzed. These data include the period during and following our August 1991 observing campaign. Throughout this time period, the P Cygni-like C IV resonance line exhibited great variability. The variations in the equivalent width of the absorption component of the C IV line are consistent with cyclic behavior at the orbital period of the system (3.3 hr). The data do not show line variability consistent with the photometric period (3.2 hr). The X-ray observations during this time period do not show variability at either period. These data are dominated by a 15.8 keV bremsstrahlung continuum with a line at 0.92 keV (Robinson, Córdova & Ishida 1994, in preparation).

The observations from mid-1979 to mid-1982 show a transition in the relative strengths of the absorption and emission components of the C IV line with the absorption component dominating the first 12 months of this time period, nearly absent in the second 12 months and only slightly stronger in the third 12 month period. During these three time spans, variations consistent with the orbital period again appear in the equivalent widths of the components of the C IV line.

Models of the system during the low, intermediate and high states, and the relative contribution of the UV component to the hard X-ray to near-IR spectrum of TT Ari are the subjects of papers in preparation.

A. Bianchini et al. (eds.), Cataclysmic Variables, 179.
© 1995 Kluwer Academic Publishers.

A NEW PERIOD DETERMINATION FOR EY CYG

MAREK J. SARNA
*N. Copernicus Astronomical Center, Polish Academy
of Sciences, ul. Bartycka 18, 00–716 Warsaw, Poland.*

W. PYCH
*Warsaw University Observatory,
Al. Ujazdowskie 4, 00–478 Warsaw, Poland.*

AND

ROBERT CONNON SMITH
*Astronomy Centre, Division of Physics and Astronomy,
University of Sussex, Falmer, Brighton BN1 9 QH, UK.*

We obtained CCD photometry in V, R, I of EY Cyg during two nights in 1993 (August 17 and October 13), using the 60–cm Cassegrain Telescope located at the Warsaw University Observatory at Ostrowik. We used the standard reduction procedure available in Warsaw, which is based on the IRAF reduction package. EY Cyg is a U Gem–type dwarf nova. The first orbital period determination was made by Hacke & Andronov (MVS **11** 74 1988) from noisy photographic observations. They found a possible period of 0.181228 day. In our R and I observations we detected a very smooth light curve modulation with an amplitude of about 0.05 mag. The formal error of observation is ±0.01 mag. The power spectrum periodogram show two peaks at frequencies: $f_1=3.800\pm0.004$ day^{-1} and $f_2=4.576\pm0.004$ day^{-1}. These give orbital periods: $P_1=0.2630\pm0.0005$ and $P_2=0.2185\pm0.0005$ day, respectively. The second period has a higher statistical weight; further, there are evolutionary and spectroscopic arguments to support this value. From the spectra of EY Cyg we can classify the red dwarf spectral type as dM2 – dM3. From the calibration by Popper we found that the mass of the red dwarf component lies in the range 0.5 – 0.4 $M_\odot$. Next from the Echevarría mass–radius relationship, assuming that the red dwarf is a main-sequence star and fills its Roche lobe, we found a possible range for the orbital period: 0.22 – 0.182 day. Only the second value $P = 0.2185 \pm 0.0005$ day is consistent with this constraint.

180

A Bianchini et al (eds), Cataclysmic Variables, 180
© *1995 Kluwer Academic Publishers*

A LONG ORBITAL PERIOD AND THE NATURE OF V1017 SGR : INTER-RELATION

KAZUHIRO SEKIGUCHI
South African Astronomical Observatory
P.O. Box 9, Observatory 7935 South Africa

V1017 Sgr is an unusual classical nova with a long orbital period of 5.714 day (Sekiguchi 1992). It shares common observational characteristics with two other long orbital period cataclysmic variables (CVs) GK Per (Crampton *et al.* 1986) and BV Cen (Gilliland 1982). These systems are characterized by (i) a long orbital period (P > 0.5 day), (ii) slow dwarf nova type outbursts with a symmetric light curve, (iii) classical nova type outbursts (GK Per and V1017 Sgr) and (iv) a hard X-ray source.

Their relation to other CV subclasses has been obscure. They may be a long orbital period tail of the CV distribution. Their accretion is most likely to be driven by nuclear evolution of the secondary so that a high rate of mass transfer is expected. The differences between these systems and the U Sco subclass of the recurrent novae, whoes orbital periods are also longer than 0.5 day, may be a mass of the white dwarf in the system. The U Sco subclass RNe have a massive (> 1.37 $M_{\odot}$) and a hotter WDs than the V1017 Sgr group.

References

Crampton, D., Cowley, A.P. and Fisher, W.A. (1996), *Ap. J.*, **300**, 788.
Gilliland, R. (1982), *Ap. J.*, **263**, 302.
Sekiguchi, K. (1992), *Nature*, **358**, 563.

A. Bianchini et al. (eds.), Cataclysmic Variables, 181.
© *1995 Kluwer Academic Publishers.*

THE RECURRENT NOVA T PYX: AN OUTLINE OF IUE OBSERVATIONS

P.L. SELVELLI
CNR-GNA-Osservatorio Astronomico di Trieste-Italy

R. GILMOZZI
ESO-Garching-Germany

AND

A. CASSATELLA
CNR-Istituto di Astrofisica Spaziale-Frascati-Italy

1. Introduction Recurrent novae, where we know in advance that an outburst will occur, provide a unique opportunity to study the pre-outburst behavior in the nova phenomenon. T Pyx is particularly well suited for this purpose because the last outburst was recorded in 1966 and the average recurrence time between the previous outbursts was 19 ± 5.3 years.

In 1986 we started a monitoring programme with IUE for the purpose of following its behavior during quiescence and the phases prior to the expected new outburst. Surprisingly, the star has somehow managed to postpone this long-awaited event, and has presently (Dec. 1994) surpassed by 4 years the longest previous interoutburst interval of 24 years: 1920-1944. The IUE spectra obtained during our observations cover the time interval from Sept. 27,1986 to Nov. 23, 1994. In addition, we have retrieved from the IUE archive the spectrum of May 11,1980.

We outline here the main results and some problems raised by these observations.

2. The continuum distribution. After correction for reddening, $E_{B-V} = 0.31$, and the variation with time in the IUE sensitivity, the continuum energy distribution is remarkably constant in slope and intensity over 14 years and can be represented over the entire IUE range by a single power-law spectrum $F_\lambda = A\lambda^{-\alpha}$ with $\alpha = 2.47 \pm 0.05$. The total UV continuum flux is $2.48 \pm 0.08 \times 10^{-10}$ erg cm^{-2} s^{-1}.

3. The emission lines In contrast with the constancy of the continuum, the emission lines show substantial and sometimes dramatic changes in their intensities, so that there are barely two emission-line spectra that look alike. Only the lines of CIV λ 1550 and HeII λ 1640 are present in all spectra. The variations (especially for the emission lines of HeII λ 1640 , and of NV λ 1240) take place on timescales as short as a few hours (this lower limit is set by the length of a proper exposure time, on the order of 200 min.). The variations are not produced either by instrumental effects (e.g. different centering in the IUE large aperture of the 10" nebular shell surrounding the system), or by aspect-dependent and/or orbital effects: the system is viewed at low inclination ,$\leq$ 20 degrees, and the average exposure time is on the order of one orbital period or longer.

182

A. Bianchini et al. (eds.), Cataclysmic Variables, 182-183.
© 1995 Kluwer Academic Publishers.

4. Distance , L_{UV} , $\dot{M}$.

A new estimate of the distance d = 3500 pc has been made using the new values for the reddening, as determined from the λ 2175 feature, and the absolute magnitude at maximum in 1966, $M_v = -6.7$, as determined from recent MMRD relations given by Cohen (1985) and by Capaccioli et al.(1989). This distance provides a UV continuum luminosity $L_{UV} = 90 L_\odot$ and an absolute visual magnitude at minimum $M_v = 1.6$. These values are much higher than those of other nova remnants which have L_{UV} in the range 1-25 $L_\odot$ (Gilmozzi,Selvelli, and Cassatella 1994) and Mv in the range 2.2-4.8 (Warner 1987), with the low-inclination systems being the more luminous ones. If accretion powers the observed UV luminosity, a lower limit of $2.1 \times 10^{-8} M_\odot$ yr^{-1} for the mass accretion rate $\dot{M}$ can be derived. This rather high value, remarkably constant over 14 years, is in good agreement with the expectations of TNR models for RNe (see Livio (1993) for a review).

5. The origin of the emission line variations

In other CVs, intrinsic variations in the emission lines are observed only in systems which are not stable accretors, (e.g. DNe from a low to a high state) where $\dot{M}$ changes take place in an optically thin accretion regime. Apparently this is not the case of T Pyx, where the high UV luminosity indicates stationary optically thick accretion. We think that an explanation for the peculiar behavior of the emission lines must be sought in terms of high chromospheric and flare activity in the main-sequence companion, enhanced by the strong irradiation by the EUV flux of the hot component (see for example Jomaron et al. 1993). Several detached systems are known (e.g. V 471 Tau (Young et al. 1991) , BE UMa (Ferguson et al. 1987)), which show irradiation-induced effects and clear evidence that the atmosphere of the late-type star is highly inhomogeneous, with the presence of active regions and magnetic loops (see also Jeffries et al. 1993, and Vennes and Thorstensen 1994). These structures have sizes comparable to the binary separation and are unstable, having a lifetime on the order of hours/days (Doyle et al.,1994).

We guess that similar phenomena take place in T Pyx also, and can explain the peculiar variations in its emission-line spectrum.

A comprehensive discussion of this and other problems raised by our IUE observations of T Pyx is to appear soon in a forthcoming paper.

References

Capaccioli M.,Della Valle M.,D'Onofrio M.,Rosino L. 1989,*Astron.J.97,1622*
Cohen J.B.,1985,*Astrophys.J.292,90*
Doyle J.G.,et al.,1994,*Astron.Astrophys.283,522*
Ferguson, H.D.,et al.,1987,*Astrophys.J.316,399*
Gilmozzi R.,Selvelli P.L.,Cassatella A. 1994, in *"Evolutionary Links in the Zoo of Inter-acting Binaries"*, F.D'Antona et al. *(eds.)*, *Mem. SAIt.,65,199*
Jeffries R.D.,Elliot K.H.,Kellett B.J.,Bromage G.E.,1993,*MNRAS 265,81*
Jomaron C.M.,et al.,1993,*MNRAS,219,227*
Livio M.,1993, in *"Interacting Binaries"*, H. Nussbaumer (ed.), Springer, in press
Vennes S.,Thorstensen J.R.,1994,*Astrophys.J.433,L29*
Young A.,Rottler L.,Skumanich A.,1991,*Astrophys.J.378,L25*
Warner B.,1987,*MNRAS,227,13*

15 YEARS OF IUE OBSERVATIONS OF THE SYMBIOTIC NOVA RR TEL

P.L. SELVELLI
C.N.R.-G.N.A.-Osservatorio Astronomico di Trieste, Italy

AND

M. HACK, R. ZUCCOLO
Dipartimento di Astronomia-Università di Trieste, Italy

Introduction. The IUE spectrum of RR Tel has been extensively studied by Penston et al (1983) and by Hayes and Nussbaumer (1986), but the limited time coverage in the data which were then available prevented a specific study of the long-time variations in the UV emission line intensities and excitation.

We have inspected all high-resolution spectra obtained with IUE from 1978 to 1993, with the purpose of detecting any trend over time, both long-term and period-like, regarding the changes in the emission line intensities, excitation and profiles. We report here some preliminary results.

The results. In general, the observed decrease in the emission line intensities is of a factor of about two for the medium-high excitation lines (e.g. CIV λ 1550, Si IV λ 1398, etc.) and of a factor larger than three for the low-ionization lines (e.g. MgII λ 2800, SiII, CII, OI, FeII, etc.). Instead, the high-ionization lines (e.g. MgV λ 1324, OV λ 1370, λ 1844, λ 2941, Ne IV λ 2421, etc.), which are all characterized by a broader profile, have maintained their original strength or have suffered only a slight weakening. Two high ionization lines (based on their rather high FWHM, Penston et al., 1983) at λ 2558.0 and λ 2934.4 have appeared in recent spectra. They have so far defied any attempt for a reliable identification: possible candidates are OIII UV 21 λ 2558.06, and CIV UV 18 λ 2935.12 or [NiV] λ 2935.0, respectively, but none is fully convincing.

The OI lines. The changes in the OI emissions are remarkable and deserve special comment. These lines are produced by the well known Bowen mechanism by which the accidental coincidence in wavelength between photons of Ly_β λ 1025 Åand the $2p^4\ ^3P \rightarrow 3d\ ^3D^o$ transition of OI pumps this upper term. In the subsequent cascade the upper term $3s\ ^3S^o$ of the λ 1300 resonant triplet is excited and this results in the observed emission lines. Optical depth effects (and I.S. absorption) are severe in these lines and this results both in the presence of anomalous relative intensities in the triplet, and in the appearence of the semiforbidden λ 1641.3 Å emission, a transition which has the same upper term as the λ 1303 triplet. In practice, the great optical depths in the λ 1303 photons has the effect of converting these "trapped" photons into λ 1641.3 photons that easily escape from the nebula. For details see for example Hack and Selvelli (1982).

In RR Tel the changes in the λ 1303 triplet from 1978 to 1993 were dramatic: the total emission intensity decreased from 3.88 $erg\ cm^{-2}\ s^{-1}$ in 1978 to a very small value with an upper limit of 0.33 $erg\ cm^{-2}\ s^{-1}$ in 1993. , The semiforbidden λ 1641.3 emission, instead, decreased much less, from 0.93 $erg\ cm^{-2}\ s^{-1}$in 1978 to 0.43 $erg\ cm^{-2}\ s^{-1}$ in 1993, and this gives rise to a remarkable situation whereby of two transitions arising from the same upper level it is the one which is less favoured by the atomic probabilities that is actually stronger. We have interpreted

A. Bianchini et al. (eds.), Cataclysmic Variables, 184-185.

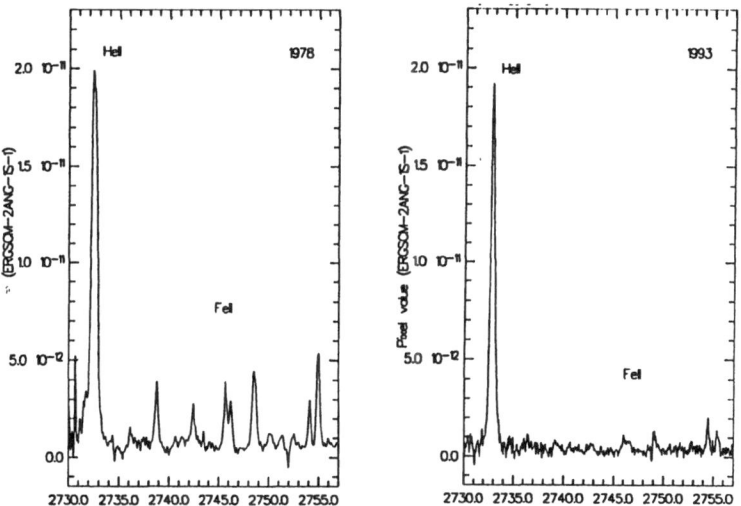

Figure 1.   The variations in the FeII lines (UV 62 and 63) between 1978 and 1993

these variations as evidence of higher optical depth in the OI resonance triplet, an indication of an increase in the column density of OI in the emitting region. However, the fact that the sum of the intensities of the λ 1303 triplet and of the λ 1641.3 line decreased significantly is an indication of a decrease in the fluorescent pumping efficiency. We interpret this as a consequence of the fact that fewer Ly_β λ 1025 photons are available which, in turn, is an effect of the gradual decline in the luminosity of the EUV ionizing source, as reported by Jordan et al.(1994).

The FeII emissions.   The richness in the FeII spectrum of RR Tel was reported by Penston et al.(1983), and by Johansson (1983) who pointed out that in addition to collisional excitation, responsible for most of the flux in the decays from the odd levels near 5 eV, a selective excitation or recombination mechanism was required in order to explain the great strength of several multiplets (e.g. 391,.. 399) which represent decays from even levels near 10 eV.

The comparison between the IUE spectra of 1978 and 1993 shows that the low-lying multiplets (e.g. 62,63) suffered a substantial decline and almost vanished (see Fig.1) while the higher multiplets (e.g. 391...399) kept most of their intensity. This is remarkable because the lower level of multiplet 399 (z^4 D^o) is the upper level of multiplet 63. The lack of photons in multiplet UV 63 could be explained by opacity effects in this multiplet which has a metastable lower level a^4 D from which substantial reabsorption might occur if its population is significantly increased with respect to the first observations of 1978. We suggest tentatively that the missing flux in UV 63 has been converted into flux in VIS 6 with whom UV 63 shares the same upper level. The lower level of this multiplet is high-lying compared with that of UV 63, and therefore it is less affected by reabsorption.

These short considerations show the complexity in the behavior of the OI and FeII emissions in RR Tel, but also the great potential these lines have for diagnosing the conditions (excitation and opacities) in the emitting region.

References

Hack M. and Selvelli P.L., 1982, *Astron.Astrophys.107,200*
Johansson S., 1983, *MNRAS 205,71p*
Jordan S.,Muerset U.,Werner K.,1994,*AStron.Astrophys.,283,475.*
Penston M.V. et al., 1983, *MNRAS 202,833*

THE ECLIPSING, MAGNETIC CV RX J0515.6+0105

A. W. SHAFTER AND K. A. MISSELT
San Diego State University
Department of Astronomy, San Diego, CA, 92182, USA

AND

K. REINSCH AND K. BEUERMANN
Universitäts-Sternwarte
Geismarlandstr. 11, D-37083 Göttingen, Germany

1. Abstract

Multiwavelength observations of the optical counterpart of the *ROSAT* X-ray source RX J0515.6+0105 were presented. The X-ray data show that the system is characterized by a soft thermal spectrum, with $kT_{bb} \simeq 33$ eV. Follow-up optical photometric data spanning ~ 1 yr revealed eclipse features that are deep, total, and variable in profile. The eclipse timings, including one provided by V. Burwitz, establishes an orbital ephemeris of $JD_{\odot} = 2449087.8173(3) + 0.3326120(7)$ E, with considerable phase jitter caused by the variable eclipse profiles. In addition, the optical light curve is characterized by a strong and asymmetric brightness modulation at twice the orbital frequency, by quasi-periodic oscillations with a characteristic time scale of 6 – 15 minutes, and, most significantly, by significant circular polarization (D. A. H. Buckley, private communication). Spectroscopic observations reveal strong He II $\lambda 4686$ emission with large-amplitude radial velocity variations, and suggest the possibility of cyclotron features near 8500 Å and 5900 Å– roughly consistent with the expected wavelengths of the second and third harmonics of cyclotron radiation originating from an ~ 61 MG field. Taken together, these characteristics strongly suggest that RX J0515.6+0105 is a magnetic cataclysmic variable, most likely a polar (AM Her system) in which the accretion column/stream is eclipsed. However, the possibility that the white dwarf rotates asynchronously, and that RX J0515.6+0105 is an intermediate polar (DQ Her system), has not yet been ruled out unequivocally.

A. Bianchini et al. (eds.), Cataclysmic Variables, 186.
© *1995 Kluwer Academic Publishers.*

HIGH-SPEED PHOTOMETRY AND POLARIMETRY OF THE ECLIPSING CATACLYSMIC VARIABLE PX AND = PG 0026+27

N. M. SHAKHOVSKOY
Crimean Astrophysical Observatory Nauchny, 334413 Crimea, Ukraine

AND

S. V. KOLESNIKOV AND I. L. ANDRONOV
Department of Astronomy, Odessa State University
T. G. Shevchenko Park, Odessa 270014 Ukraine

In 1992 we observed the object at 2.6m and 1.25m telescopes of the Crimean Astrophysical Observatory. No circular polarization was found with a typical error of < 0.1 % for a run with 256 integrations. Individual light curves show a variable shape, with eclipses occurring in agreement with the ephemeris of Hellier and Robinson (1994, *Ap.J.*, **431**, L107). However, contrary to previous observations, a wide minimum is seen, which may be even deeper than the narrow eclipse. We list below the magnitudes and corresponding phases obtained at JD 2448916. The data were smoothed by "running parabolae" (Andronov I.L.: 1990, *Kinem. Phys. Celest. Bodies*, **6**, 6, 87) with $\Delta t = 0.01^d$. The UV excess is larger just before and at eclipse, which is puzzling, as at this phase the disk must be partially eclipsed by the red dwarf. Waves with full amplitudes 0.11-0.14 mag and cycles of $24-29^m$ are seen. Periodogram analysis of the residuals shows transient peaks at 213, 720 and 560 cycles/day, none of which are observed at all runs.

	$HJD - JD$	ϕ	U	B	V	R	I
Max 1	.3112	-2203.755	14.46	14.64	14.27	13.71	13.47
Min 1	.3570	-2203.442	14.96	15.13	14.69	14.10	13.74
Max 2	.4026	-2203.137	14.31	14.58	14.20	13.67	13.44
Min 2	.4212	-2203.004	14.66	15.06	14.68	14.10	13.84
Max 3	.4572	-2202.758	14.26	14.40	14.04	13.49	13.34

This research was supported by ESO Grant A-04-018.

187

UV PULSATIONS OF DQ HERCULIS AT THE WHITE-DWARF SPIN PERIOD

A.D. SILBER, B. MARGON, S. ANDERSON
University of Washington, Dept. of Astronomy FM-20
Seattle, WA 98195

AND

R. DOWNES
Space Telescope Science Institue, User Support Branch 3700
San Martin Drive Baltimore, MD 21218

The old nova DQ Herculis was observed with the Hubble Space Telescope (HST) Faint Object Spectrograph [1150-2500 Å] over three orbits of the satellite. These observations were carried out in the "rapid" mode, with a time resolution of 4.08 s, allowing us to search for pulsations of the UV light at the 71-s white-dwarf spin period, analogous to the low amplitude oscillations long known to be present in visible light. The integrated spectrum shows strong line emission in Lyα, NV, SiIV/OIV], CIV and HeII. The flux of all of these lines drops during the eclipse of the white dwarf by the late-type star, indicating that most of the emission is not nebular, geocoronal, or from an extended wind. The SiIV/OIV] complex is marginally resolved, with OIV] perhaps surprisingly contributing the majority of the flux. Furthermore, none of 300 simple photoionization models examined yields SiIV/OIV] as strong as that observed. During the first and second HST orbits, we detect coherent 71-s pulsations in the continuum, with amplitudes of 9from DQ Her. The amplitude is time-variable, as no pulsations are seen during the third orbit, with an amplitude upper limit of 2the second HST orbit, the strong Lyα emission is seen to pulse with a 12and the phase of maximum light is displaced 0.3 later than that of the continuum pulse. No significant pulsations were detected in any other emission line, nor in Lyα during the other two orbits. The line profile also shows no variation at the spin period. The line and continuum pulsations clearly hold great promise as an aid in unraveling the complexities of this well-studied but still poorly understood system.

A. Bianchini et al. (eds.), Cataclysmic Variables, 188.
© *1995 Kluwer Academic Publishers.*

ON THE OPTICAL PULSATIONS IN DQ HER STARS

PHILLIP J. MARTELL
Space Telescope Science Institute
3700 San Martin Drive, Baltimore, MD 21218 USA

AND

WILLIAM F. WELSH
Keele University, Department of Physics
Keele, Staffordshire, ST5 5BG UK

DQ Her stars are cataclysmic binary systems in which matter from a late-type near-main sequence star is accreted onto an asynchronously spinning magnetized white dwarf. X-rays generated at shocks at the magnetic poles of the white dwarf sweep the system with a "lighthouse" beam, producing optical signals modulated at the spin period of the white dwarf and/or at one or more spin–orbit sidebands. The shape of the *pulse amplitude* spectrum yields information about the emission mechanism of the pulsed component of the light. [See de Martino *et al.* in these proceedings.]

We have fitted pulse amplitude spectra derived from published data for AE Aqr, FO Aqr, BG CMi, DQ Her, AO Psc, and V1223 Sgr with power law and blackbody models. In all cases, a power law yields an excellent fit. AO Psc (805 s period) and BG CMi (913 s) each require an infinite blackbody temperature. All other systems yield blackbody temperatures ranging from $\sim$10,000 K (DQ Her) to $\sim$50,000 K (AE Aqr).

Assuming the pulsations are due to a single blackbody modulated in area, our blackbody fits yield a projected radiating area, given a distance. The results for AE Aqr are consistent with pulsed radiation originating in a small area on or near the white dwarf's surface. For FO Aqr (spin), DQ Her (spin), AO Psc (sideband), and V1223 Sgr (sideband), the projected radiating areas suggest reprocessing of X-rays in large structures. Independent evidence implies the DQ Her pulses originate over much of the disk, and in general, sidebands probably originate in X-rays reprocessed on the secondary star.

A Bianchini et al (eds), Cataclysmic Variables, 189

IUE AND OPTICAL OBSERVATIONS OF AM HERCULIS IN ITS LOW STATE

A.D. SILBER
University of Washington, Dept. of Astronomy FM-20
Seattle, WA 98195

J. C. RAYMOND
Center for Astrophysics, 60 Garden St.
Cambridge, MA 02138

N.M. SHAKHOVSKOY
Crimean Astrophysical Observatory, P/O Nauchny
334413 Crimea, Ukraine

P.A. MASON
Case-Western Reserve University, Warner and Swasey Observatory
Dept. of Astronomy, 10900 Euclid Ave., Cleveland, OH 44106

AND

I.L. ANDRONOV
Department of Astronomy, Odessa State University
270014 Ukraine

Ten spectra of AM Herculis were obtained with the SWP instrument of the IUE over 2.5 orbital periods during a faint state in 1992. Significant periodic variability was seen in the UV flux and spectra. On the same day, UBVRI photometry was obtained at the 1.25 meter optical telescope of the Crimean Astrophysical Observatory. The UV spectra and optical photometry at the same orbital phase were fit to spectral models to determine the temperature. When AM Her was faint a single cool (T~20,000 K) model fit the data. At the bright phase, a second, hotter component covering a small portion of the white dwarf surface improved the fit to both the continuum shapes and Lyα profiles of the spectra. This component could be the high-state hot spot cooling slowly after the accretion stops, or it might be heated by continuing accretion at a low rate. The high temperature (T~ 50,000 K) three weeks after the transition to the low state and the modest deviation from a smooth UV light curve lead us to favor the model of continued accretion at a low accretion rate.

190

A. Bianchini et al. (eds.), Cataclysmic Variables, 190.
© 1995 Kluwer Academic Publishers.

ANALYSIS OF THE ECLIPSE LIGHT CURVES OF UX UMA

J. SMAK
N. Copernicus Astronomical Center
Bartycka 18, 00-716 Warsaw, Poland

Light curves of cataclysmic variables with stationary accretion disks are a superposition of a *disk* light curve and a *hot spot* light curve. The contribution from the spot complicates the analysis of the disk eclipse. As a new approach to this problem, it is proposed to perform the light curve analysis in three parts: (1) decomposition of the observed light curve into the *spot* and *disk* light curves, (2) analysis of the *disk* light curve, and (3) analysis of the *spot* light curve, with parts (2) and (3) providing independent constraints on system parameters. (For details see *Acta Astronomica*, **44**, 45, 1994).

Light curves of UX UMa, according to their shapes, can be divided into two types: (1) the *standard* light curves, showing a well defined hump, due to the hot spot, and (2) the *peculiar* light curves, showing either an ill defined hump of a much smaller amplitude, or even a depression in the phase interval, where the hump is normally present.

Standard light curves of UX UMa, available in the literature, were successfully decomposed into their *disk* and *spot* components, and from the analysis of the *disk* and *spot* eclipses the basic system parameters were determined. (For details see *Acta Astronomica*, **44**, 59, 1994).

The analysis of *peculiar* light curves shows that their peculiarities are due to absorption produced by "circumdisk" material. Such light curves can be analyzed by assuming a simple formula for the phase dependence of the absorption produced by "circumdisk" material. (For details see *Acta Astronomica*, **44**, 59 and 257, 1994).

A. Bianchini et al. (eds.), Cataclysmic Variables, 191.
© *1995 Kluwer Academic Publishers.*

RJ051542+0104.7: A NEW LONG PERIOD (8 HR) ECLIPSING MAGNETIC CV!

PAULA SZKODY
University of Washington, Seattle, WA 98195

PETER GARNAVICH
Dominion Astrophysical Observatory, Victoria, BC V8X 4M6

RUSSELL ROBB AND DAVID ZUREK
University of Victoria, Victoria, BC V8W 3P6

AND

D.W. HOARD
University of Washington, Seattle, WA 98195

We present photometric and spectroscopic observations of an X-ray bright cataclysmic variable, RJ051542+0104.7 which was discovered serendipitously by Walter and Zoonematkermani (1994, AIP Conf Proc. 313, 287) in their Orion fields. The overall evidence from their X-ray light curve and spectrum suggests that RJ0515 is a magnetic CV.

Our 9 nights of CCD photometry between December 1993 and January 1994 show that the system undergoes eclipses with a period of 7.9835 ±0.0002 hrs (ephemeris of HJD = 2449391.8232(11) + 0.332647(10)E), which is unusually long for an AM Her system and provides interesting evolutionary constraints. The visual magnitude outside of eclipse shows a double-humped structure ranging between V of 15.5 and 16.2. The eclipse is one magnitude deep and appears to be total.

Spectra obtained on November 20 and 21 with the 2.1m telescope at KPNO using GoldCam show a strong blue continuum with broad asymmetric Balmer emission as well as HeII (4686 and 5411Å), CIII/NIII(4650Å) and HeI lines. During eclipse, the blue continuum and emission lines disappear, revealing the continuum of an M0 star. With the observed eclipse magnitude (17.3), the distance to RJ0515 is about 500 pc.

This research was partially supported by NSF grant AST 9217911 and NASA grant NAGW 3158 to P.S.

A. Bianchini et al. (eds.), Cataclysmic Variables, 192.

TIME-RESOLVED SPECTROPHOTOMETRY OF INTERMEDIATE POLARS AND CANDIDATES.

I. SS Auriga.

GAGHIK H. TOVMASSIAN
Instituto de Astronomía, UNAM,
P.O.Box 439027, San Ysidro, Ca, 92143-9027, USA

NIKOLAY V. BORISOV
Special Astrophysical Observatory,
Nizhny Arkhyz, Zelenchuksij Region, 357147, Russia

AND

IVAN M. KOPYLOV
Pulkovo Observatory,
Pulkovo, 196140 Sankt-Peterburg, Russia

From UBVRI photometry we have discovered (Efimov et al., 1986, Tovmassian 1987) quasiperiodic variations in the light curves of SS Aur which are similar to those of an IP. However the periods of variations changed from night to night in the range of 19–30 minutes. Moreover SS Aur was observed in a low state not usual for a DN (Efimov et al., 1986). The time-resolved spectrophotometry was carried out at 6 m. telescope of SAO (Russia) on 27 Sept. 1987 in order to examine this phenomenon. The spectra were obtained in $\lambda\lambda$ 3400 − 5200 Å range. Spectral resolution was 4 Å. During 2.2 hours 27 scans were obtained with 240 sec exposure each.

Fitting the measurements of V/R ratio of the hydrogen lines by a sine curve with the period equivalent to orbital, showed, that observation covered the half of the period around 0.75 orbital phase. To determine the variability in the continuum, the fluxes corresponding to standard U and B photometric bands were integrated and multiplied by the U & B filter sensitivity. The following features in the spectral behavior of SS Aur were observed.

a. The continuum level at the long wavelength end of the spectra decreased by a factor of 2, while at the short end it remained at the same level even as airmass decreased during the observations.

A. Bianchini et al. (eds.), Cataclysmic Variables, 193-194.
© *1995 Kluwer Academic Publishers.*

b. The amplitude of variation of the fluxes corresponding to the U and B bands exceeds 30%, that is much more than the expected errors, and they are wave shaped.

c. There is a large variability in the Ca II lines compared to the hydrogen lines. These variations are correlated with the continuum variations.

Keeping in mind that previous photometry showed highly modulated radiation from SS Aur, we applied a DFT method for examination of intensity variations in lines and continuum. The power spectrum showed a strong peak at 40 day^{-1} ($\approx 36^m$). This peak is of great interest, because it's close to the periods obtained earlier from photometry. It has no alias in the spectral window, and it disappears after subtraction of a sine curve of corresponding period from the data set.

The time-resolved observations of SS Auriga confirm the presence of high amplitude, probably modulated variations in its radiation. However, we failed to identify SS Aur as a member of the Intermediate Polars, as they are described by adopted models. The origin of the modulated radiation from SS Aur is probably due to accretion disk, rather than a WD, based on the wide range of observed periods and apparent correlation with Ca II line emission, rather than any high excitation line.

References

Efimov Yu.S., Tovmassian G.H., Shakhovskoy N.M., 1986, *Astrofizika*, **24**, 227
Tovmassian G.H., 1987, *Astrofizika*, **27**, 231

THE 1985-86 EPISODES OF AG DRACONIS

R. VIOTTI[1], P. GIOMMI[2], M. FRIEDJUNG[3], A. ALTAMORE[4]
[1] *Istituto di Astrofisica Spaziale, Via Fermi 21, Frascati, Italy*
[2] *ESIS, Information System Div. of ESA, ESRIN, Frascati*
[3] *Institut d'Astrophysique, 98bis Bvd. Arago, 750014 Paris*
[4] *Dip. Fisica, Terza Univ., Via E. Segre 2, 00146 Roma*

EXOSAT observations of the peculiar *halo* symbiotic star AG Dra during its recent light maxima of February 1985 and January 1986 have disclosed a large X–ray fading with respect to quiescence. Simultaneous IUE observations have on the contrary shown an increase of the continuum and emission line flux, especially of the high temperature Nv 124 nm and Heii 164 nm lines, at the time of the light maxima. Two distinct components might exist, in the far–UV spectrum of AG Dra's companion, associated with a hot subdwarf and an accretion disk. The behaviour in 1985–86 might be due to a temperature drop of a non–black body hot component, or to a continuous absorption of the X–rays, shortwards the N^{+3} ionization limit.

This work is based on archive data of the IUE Observatory collected from ULDA, and of the EXOSAT satellite obtained using the ESIS system.

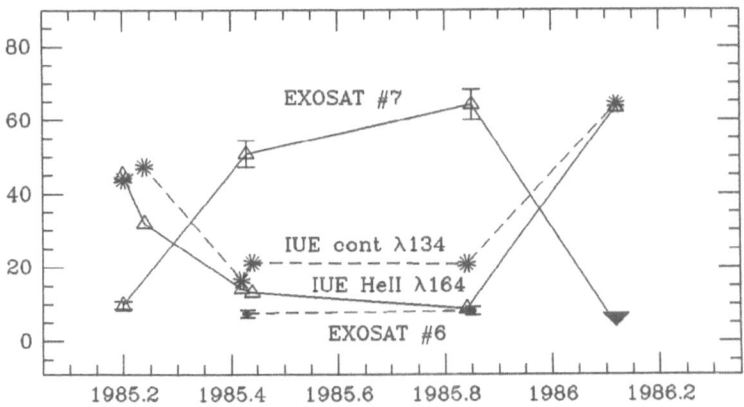

Figure 1. EXOSAT X-ray and IUE ultraviolet variation of AG Dra during 1985–86.

195

A. Bianchini et al. (eds.), Cataclysmic Variables, 195.

THE ECLIPSE MINIMUM IN DWARF NOVA SS CYGNI: THE NEW DATA

IRINA VOLOSHINA AND VICTOR LYUTY

Sternberg Astronomical Institute
Universitetskij prospect 13,
119899 Moscow, Russia

SS Cyg is the brightest and therefore well-studied classical dwarf nova with the orbital. period $\approx 6^h.6$. The photometric investigation of SS Cygni in quiescence was fulfilled in frames of large observational program "The investigation of close binary system on late evolutionary stages" that is performing during several years in Sternberg Astronomical Institute. On the base of numerous observations during period 1982-1990 the narrow eclipse minimum with $\Delta V = 0^m.05$ $\Delta B = 0^m.07$ and $\Delta U = 0^m.09$ at the $\varphi = 0.54$ was detected on the light curve of this star at the moment of the white dwarf superior conjunction (Voloshina, 1986).

The new night–long observations of SS Cyg have been carried out in UBV photometric system using the technic and method discribed earlier (Voloshina, 1992) during 7 nights in August–September 1992. Measurements were done between two consequent outbursts when SS Cyg was in quiet state. The whole sample consists of 400 UBV–measurements. As an example one of the individual light curves obtained during one orbital period is shown on the Fig.1. A narrow dip (marked by arrow) with amplitude more than $0^m.25$ is clearly seen on this curves.

The analysis of our data shows that the shape of the light curve of SS Cyg do not depend on the particular observing season, but strongly depends of the outburst cycle phase. The eclipse minimum near $\varphi = 0.54$ is observed on the light curves for two dates just after outburst: 25–26 August 1992 and 27–28 August 1992. but is absent on the others curves obtained in the middle of quiet state. After reducing all individual curves to the common level we constructed the mean light curves of SS Cyg for each band in the same manner as discribed early using our published data and the elements from (Voloshina and Lyuty, 1993):

$Min \ I = JD \ 2444841.9378 + 0^d.2751302 \cdot E$

These curves are plotted in Fig.2.

From consideration of all our data the following conclusions can be drawn:

1. The existance of eclipse minimum at $\varphi = 0.54$ on the light curve of SS Cyg is confirmed.

2. This eclipse minimum is absent on the light curves corresponding to the middle of quiescence.

3. The color characteristics and the duration of this minimum indicate on

196

A. Bianchini et al. (eds.), Cataclysmic Variables, 196-197.
© 1995 *Kluwer Academic Publishers.*

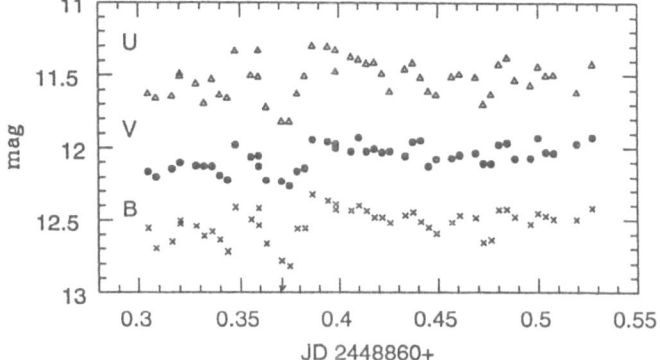

Fig. 1.   The light curves of SS Cyg obtained on 25–26 August 1992.

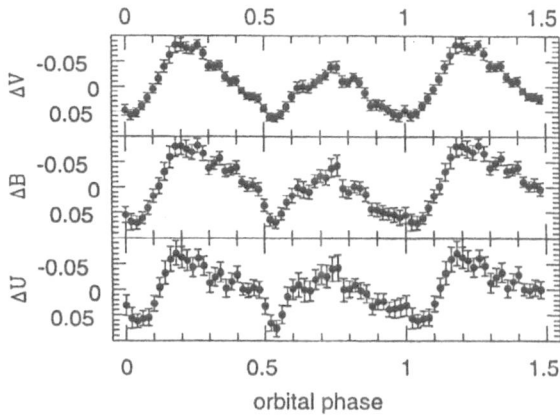

Fig. 2.   The mean light curves of SS Cyg for all data (old and new)

the eclipse of the hot spot, arising at the place where the matter, flowing from the secondary component, meets the accretion disc, definetely show that the dimensions of accretion disc increase before the outburst and became minimal in the middle of the quiescence stage.

References

Voloshina, I.B.: 1986, *Pis'ma Astron. Zh.* **12**, 219

Voloshina, I.B.: 1992, '' in Vina del Mar Workshop on Cataclysmic Variable Star, ed(s)., *N. Vogt*, A.S.P. Conf. Ser., 29, 343

Voloshina, I.B., Lyuty, V.M.: 1993, *Astronomicheskij Zh.* **70**, 61

X-RAY OBSERVATIONS OF HT CAS AND UX UMA

J.H. WOOD ET AL.
Department of Physics, Keele University
Keele, Staffordshire, ST5 5BG, England

X-ray and optical observations of the dwarf nova HT Cas were obtained (Wood et al. 1994a). The X-ray source is eclipsed and comes from the near vicinity of the white dwarf. It must therefore be a boundary layer. The X-ray flux (0.1-1.2 keV) is $\sim 5 \times 10^{-13}$ ergs s^{-1}. The spectrum is consistent with an absorbed thermal bremsstrahling model with temperature, $kT \sim 2.4$ keV and hydrogen column density, $n_H \sim 1.8 \times 10^{20}$ cm^{-2}. The low luminosity in the boundary layer, $L \sim 5 \times 10^{30}$ ergs s^{-1}, may be the result of having a low mass transfer rate onto the white dwarf, of $\sim 1 \times 10^{-12} M_\odot yr^{-1}$. The optical observations showed that HT Cas has low states where the accretion rate in the disc is dramatically reduced from the usual quiescent level. The white dwarf fluxes in the low state show that the white dwarf has a temperature of ~ 13200 K, cooler than the temperature determined from the normal state of 18700 K. The cooling of the white dwarf must occur in less than two weeks. The ROSAT observations were obtained during a low state.

ROSAT observations of the eclipsing novalike variable UX UMa (Wood et al. 1994b) show that the X-ray spectrum is extremely soft, with a hardness ratio of ~ 0.3. The X-ray luminosity, $L_x \sim 10^{30}$ ergs s^{-1}, is much lower than the luminosity expected from a boundary layer. Unlike HT Cas, phase folded light curves show no evidence for the presence of an X-ray eclipse, indeed we can rule out the presence of a total eclipse of a source of X-rays close to the white dwarf at the 2.9σ level. A model where the X-rays from the boundary layer are obscured from view at all orbital phases can explain the low luminosity of the observed X-rays and the lack of an eclipse. The observed X-rays may come from a corona or wind above the accretion disc or the secondary star, probably a combination of these.

References

Wood J.H., Naylor T., Hassall B.J.M., Ramseyer T.F. 1994a, MNRAS, in press.
Wood J.H., Naylor T., Marsh T.R. 1994b, MNRAS, in press.

A. Bianchini et al. (eds.), Cataclysmic Variables, 198.
© 1995 Kluwer Academic Publishers.

3. THERMONUCLEAR RUNAWAYS

MIXING BEFORE AND DURING OUTBURST
IN CLASSICAL AND SYMBIOTIC NOVAE

Icko Iben, Jr.
Departments of Astronomy and Physics, University of Illinois,
1002 West Green Street, Urbana, IL 61801, USA
and
Alexander V. Tutukov
Institute of Astronomy, Russian Academy of Science
48 Pyatnitskaya Street, Moscow 109017, Russia

ABSTRACT
An outine is given of the physical processes which are involved in the development of the abundance distributions found in classical and symbiotic novae. A helium buffer layer at the surface of the accreting white dwarf in the precursor binary system prevents the mixing between white dwarf matter and hydrogen-rich accreted matter until the mass of the buffer layer becomes large enough for the ignition of helium. The time required for this to occur is $\sim$ 2-6 $\times$ 10^6 years, longer than the lifetime of a symbiotic star, and this accounts for the absence of heavy element enhancements in symbiotic novae. Mixing between buffer-layer helium and accreted matter does, however, lead to large enhancements in the abundance of helium during the symbiotic nova phase. The cataclysmic variable precursors of classical novae live long enough for a helium shell flash to be ignited, and, if the flash is violent enough to remove the buffer layer entirely, large overabundances of heavy elements characterize subsequent nova outbursts. Detailed comparisons with observed abundances in nova ejecta suggest that mixing by particle diffusion and by pre-outburst convection must be supplemented by rotation-induced mixing.

1. ABUNDANCES IN CLASSICAL AND SYMBIOTIC NOVAE
The abundance distributions of the elements in the ejecta of classical novae (CNe) cover a broad range that presents a daunting challenge to the theory of mixing in accreting white dwarfs. Table 1 (from Livio & Truran 1994) shows estimates of the abundances by mass of hydrogen (X), helium (Y), and heavy elements (Z) in eighteen novae. Entries with quoted errors are the straight mean of two or more independent estimates which differ by the total error bar. Despite the obvious uncertainties, it is clear that enhancements of heavy elements in the majority of cases are considerably in excess of solar and, in half of these, Y is less than solar. In four out of the five cases in which $Z < 0.1$, Y is substantially larger than solar. Because of uncertainties in abundance estimates, there is the possibility that, in one or more of the four cases with large Y and small Z, the heavy element abundance is actually similar to the solar value.

Table 1. Abundance Distributions in Novae

Star	year	X	Y	Z
T Aur	1891	0.47	0.40	0.13
RR Pic	1925	0.53	0.43	0.04
DQ Her	1934	0.31 ± 0.04	0.13 ∓ 0.04	0.56

A Bianchini et al (eds) Cataclysmic Variables 201 209

HR Del	1967	0.45	0.48	0.08
V1500 Cyg	1975	0.53 ± 0.03	0.24 ∓ 0.03	0.23 ∓ 0.07
V1688 Cyg	1978	0.45	0.23	0.32
V693 CrA	1981	0.23 ± 0.07	0.25 ∓ 0.07	0.52 ∓ 0.14
V1370 Aql	1982	0.049 ± 0.005	0.094 ∓ 0.006	0.86
GQ Mus	1983	0.27	0.32	0.41
PW Vul	1984	0.57 ± 0.12	0.25 ∓ 0.03	0.18 ∓ 0.12
QU Vul	1984	0.32 ± 0.02	0.43 ∓ 0.17	0.25 ∓ 0.15
V842 Cen	1986	0.41	0.23	0.36
V827 Her	1987	0.36	0.29	0.35
QV Vul	1987	0.68	0.27	0.05
V2214 Oph	1988	0.34	0.26	0.40
V977 Sco	1989	0.51	0.39	0.10
443 Sct	1989	0.49	0.45	0.06
LMC	1990A	0.37	0.59	0.04

Seven symbiotic novae (SyNe) are known (see, e.g., Mürset & Nussbaumer 1994), and there is no evidence in their spectra for an enhancement of heavy elements, although there is evidence for a reshuffling among CNO nuclei of the sort expected due to reprocessing through CN-cycle hydrogen burning. This difference with respect to the majority of CNe is, at first sight, surprising. The range of accretion rates in symbiotic star (SyS) precursors of SyNe overlaps the range of accretion rates in the cataclysmic variable (CV) precursors of CNe and one might reasonably expect similar mixing processes to be occurring in both types of system.

2. INITIAL COMPOSITION AND TEMPERATURE OF AN ACCRETING WHITE DWARF

In typical CVs, the donor is a low mass main-sequence star and mass transfer is by Roche-lobe overflow driven by either a magnetic stellar wind (MSW) or by gravitational wave radiation (GWR). The majority of novae occur in systems with periods in the range 3-10 hours, and MSW-driven accretion occurs at a rate in the range $\sim 10^{-9}$-10^{-8} $M_\odot$ yr^{-1}. The precursor of a CV is a close binary in which the primary is of intermediate mass and the secondary is of low mass. When the primary fills its Roche lobe, a common envelope (CE) is formed, the primary loses almost its entire hydrogen-rich envelope, and the orbit shrinks dramatically. The remnant of the primary is a CO white dwarf (if the initial mass of the primary is less than ~ 8 $M_\odot$) or an ONe white dwarf (if the initial mass of the primary is between ~ 8 $M_\odot$ and ~ 10.5 $M_\odot$). Above the white dwarf there is a layer of nearly pure helium which has a mass M_{He}^{init} given approximately by (Iben & Tutukov 1995)

$$\log M_{He}^{init} = -1.835 + 1.73 M_{WD} - 2.67 M_{WD}^2, \qquad (1)$$

where M_{WD}, the mass of the white dwarf, and M_{He}^{init} are in solar units. Equation (1) describes the mass in the helium shell of a single asymptotic giant branch (AGB) star at the termination of helium burning in a thermal pulse cycle (Iben 1977), and it is not surprising that it also describes the mass of the helium layer left after helium burning is completed in the compact remnant of a star which has experienced a CE event (Iben &

Tutukov 1985, 1993). If $M_{CO} \stackrel{<}{\sim} 5\ M_{\odot}$, there is also a thin layer of hydrogen-rich material above the helium layer.

In SySs which can experience nova outbursts, the donor is a red giant or a supergiant which does not fill its Roche lobe, but which blows an intense wind from which the white dwarf accretes at a rate in the range $\dot{M} \sim 10^{-9}$-$10^{-7}\ M_{\odot}\ \mathrm{yr}^{-1}$. The precursor of an SyS is a wide binary in which the primary is of intermediate mass and the secondary is massive enough to evolve into an AGB star in less than a Hubble time. If the primary leaves the AGB during the quiescent hydrogen-burning phase, the mass of the helium layer in the white dwarf which it becomes is given by eq. (1).

In both types of binary, the white dwarf component is expected to have cooled down considerably prior to the onset of mass transfer. Interior temperatures in a white dwarf drop to $\sim 20 \times 10^7$ K in only $\sim 10^8$ years (e.g., Iben & Tutukov 1984a). In a CV, this much time may be needed for the MSW to decrease the semimajor axis of the system which emerges from the CE stage to the extent that the secondary comes into contact with its Roche lobe. In an SyS, the secondary is probably still a main-sequence star when the primary evolves into a white dwarf. Thus, when mass transfer begins in both CVs and SySs, the white dwarf has a cool interior and a surface helium layer of mass given by eq. (1).

3. CRITICAL MASSES FOR SHELL HYDROGEN BURNING AND SHELL HELIUM BURNING

An understanding of the surface element abundances in CNe and SyNe requires an exploration of the consequences of mixing between hydrogen-rich accreted matter and matter which is near the surface of the white dwarf following the nuclear burning phase of each outburst cycle. In the absence of rotation, mechanisms which can cause mixing are particle diffusion and convection *prior* to the outburst. The study of the effectiveness of particle diffusion has been pioneered by Prialnik & Kovetz (1984), Kovetz & Prialnik (1985), Prialnik (1986), Prialnik & Shara (1986), and Kovetz & Prialnik (1990). Similar results have been obtained by Iben, Fujimoto, & MacDonald (1991, 1992b). A very recent calculation which explores extensively the dependence of these processes on white dwarf mass and interior temperature is that of Schwartzsman, Kovets, & Prialnik (1994).

The results of several illustrative experiments are given in Tables 2-4 for a CO white dwarf accretor of mass 1 $M_{\odot}$ (Iben et al. 1992b). Accretion rates are $\dot{M} = 10^{-10}, 10^{-9}$, and $10^{-8}\ M_{\odot}\ \mathrm{yr}^{-1}$, and the various column headings have the meaning: Cold(1) — the initial white dwarf is nearly isothermal and cold; SS(2) — the initial white dwarf is in a thermal steady state configuration, and pre-outburst convection occurs; SS(3) — the initial white dwarf is in thermal steady state but pre-outburst convection is suppressed; and BL(4) — the initial white dwarf is cold but is capped by a helium "buffer" layer of mass 0.001 $M_{\odot}$. In Table 2, the entries give the mass M_{crit} of the hydrogen-rich layer when a hydrogen-burning thermonuclear runaway occurs. Entries in Table 3 give the distance in mass M_{dg} to which mixing extends below the initial surface of the white dwarf. When an outburst occurs, a transitory convective shell extends to the surface and has a mass $M_{CS} = M_{crit} + M_{dg}$. Entries in Table 4 give the surface enhancements of CNO elements in each case.

In the Cold(1) case, the mean interior temperature is about 10^7 K. After approximately

10^6 years of accretion, the interior temperature structure between outbursts approaches a "thermal steady state" equilibrium wherein the rate at which energy is released locally equals the rate at which energy leaves the local region (Neo et al. 1977; Fujimoto & Sugimoto 1979, 1982; Kato 1980). The condition for a time independent solution is that, at every Lagrangian mass point M, the rate at which "gravitational energy" is released is

$$\epsilon_g = -(TdS/dt)_M = +T(\partial S/\partial M)(M/M_*)(dM_*/dt), \qquad (2)$$

where S is the entropy at the mass point M, and M_* is the total mass of the star). A convective zone forms in the thermal steady state SS(2) models because of an opacity discontinuity between carbon-rich and hydrogen-rich matter. Suppression of this convection in the SS(3) models is approximately equivalent to assuming that a helium buffer layer exists above CO matter.

Table 2. M_{crit} ($10^{-5} M_\odot$) for a 1 $M_\odot$ White Dwarf

$\dot{M}$ ($M_\odot$ yr^{-1})	Cold (1)	SS (2)	SS (3)	BL (4)
10^{-8}	2.26	0.60	1.01	
10^{-9}	5.57	1.33	1.51	
10^{-10}	10.2	14.5	14.5	20.9

Table 3. M_{dg} ($10^{-5} M_\odot$) for a 1 $M_\odot$ White Dwarf

$\dot{M}$ ($M_\odot$ yr^{-1})	Cold (1)	SS (2)	SS (3)	BL (4)
10^{-8}	0.25	0.25	0.10	
10^{-9}	1.3	0.41	0.32	
10^{-10}	2.4	3.5	3.5	2.9

Table 4. Surface Enhancements [CNO/H] for a 1 $M_\odot$ White Dwarf

$\dot{M}$ ($M_\odot$ yr^{-1})	Cold (1)	SS (2)	SS (3)	BL (4)
10^{-8}	1.00	1.34	0.82	
10^{-9}	1.15	1.31	1.21	
10^{-10}	1.28	1.19	1.19	-0.06

Under appropriate circumstances, the initial helium layer on the white dwarf will grow in mass with each outburst cycle (see §4). This growth will continue until the helium layer is compressed and heated sufficiently for helium to be ignited. If the white dwarf is in thermal steady state with regard to the accumulation of helium, the mass M_{crit}^{He} of the helium layer when helium is ignited is (Iben & Tutukov 1989)

$$M_{crit}^{He} = 10^{6.65} R_{WD}^{3.75} M_{WD}^{-0.3} \dot{M}_{-8}^{-0.57}, \qquad (3)$$

where R_{WD} is the radius of the white dwarf in solar units and $\dot{M}_{-8}$ is the accumulation rate of the helium layer in units of $10^{-8} M_\odot$ yr^{-1}. If the accumulation rate is taken to be the rate at which helium is formed during the quiescent hydrogen-burning phase of a nova outburst, a good approximation is (Iben & Tutukov 1995)

$$M_{\text{crit}}^{\text{He}} \sim 0.17 \times (1 - 0.59 M_{\text{WD}})^{3.75} M_{\text{WD}}^{-0.3} (M_{\text{WD}} - 0.26)^{-0.57}. \tag{4}$$

Equations (3) and (4) are actually lower limits. When $M_{\text{WD}} = 1\ M_\odot$, $M_{\text{crit}}^{\text{He}} \sim 0.01\ M_\odot$. For a very cold white dwarf, the critical mass can become as large as $0.14\ M_\odot$ (see, e.g., Iben & Tutukov 1991).

4. ABUNDANCE STEADY STATE MODELS AND FORMATION OF A CRITICAL HELIUM-LAYER MASS

Assuming a knowledge of M_{dg}, M_{crit}, and three additional quantities which can be calculated, it is possible to estimate the element enhancements achieved after enough outbursts have occurred for an "abundance steady state" to have been achieved (Fujimoto & Iben 1992, see also Iben 1992). The three additional quantities are (1) the mass $M_{\text{rm}}(\text{H})$ of the hydrogen-rich layer which remains at the surface of the white dwarf after hydrogen burning has ceased following an outburst, (2) the mass $M_{\text{rm}}(\text{He})$ of the layer containing fresh helium produced during outburst, and (3) the amount of hydrogen ΔX_{nuc} which must be burned during the convective shell phase of an outburst in order to expand the hydrogen-rich envelope to beyond the Roche lobe. When $M_{\text{CO}} = 1\ M_\odot$ and the Roche lobe radius is $R_\odot$, one has (Fujimoto & Iben 1992)

$$M_{\text{rm}}(\text{He}) \sim 1.12 \times 10^{-5} \left(\frac{0.01}{Z} \right)^{0.25}, \tag{5}$$

$$M_{\text{H}} \sim \frac{1}{3} M_{\text{He}}, \tag{6}$$

and

$$\Delta X_{\text{nuc}} \simeq (G M_{\text{WD}} / R_{\text{WD}}) / E_{\text{H}} \sim 0.04. \tag{7}$$

Here G is the gravitational constant, $G M_{\text{WD}} / R_{\text{WD}}$ is a measure of the energy per unit mass required to expand the envelope to a radius much larger than the white-dwarf radius, and $E_{\text{H}} = 6 \times 10^{18}$ erg g^{-1} is the energy released per gram by the complete burning of hydrogen.

In each cycle, the mass of the helium layer increases by

$$\Delta M_{\text{He}} = M_{\text{rm}}(\text{He}) - M_{\text{dg}}. \tag{8}$$

Thus, if $M_{\text{dg}} < M_{\text{rm}}(\text{He})$, the total mass of the helium layer above the CO or ONe white dwarf grows with time until it reaches a critical value somewhere between a minimum value given by eq (3) or eq. (4) and a maximum value $\sim 0.14\ M_\odot$. For $M_{\text{WD}} = 1\ M_\odot$, the minimum critical value is $\sim 0.01\ M_\odot$.

Let X, Y, and Z be the steady state abundances by mass of hydrogen, helium, and elements heavier than helium at the surface of the model at the end of the convective shell phase of the outburst, and let X_0, Y_0, and Z_0 be the abundances by mass of these same entities in the accreted material. In the presence of a helium buffer layer (within which $Y \sim 1$) and for $M_{dg} < M_{He}$,

$$M_{CS}Y = (M_{crit} + M_{dg})Y = M_{acc}Y_0 + M_{rm}(H)Y + M_{dg} + M_{CS}\Delta X_{nuc}, \qquad (9)$$

where

$$M_{acc} = M_{crit} - M_{rm}(H). \qquad (10)$$

If the Cold(1) case with $\dot{M} = 10^{-8}$ $M_\odot$ yr^{-1} is applicable, $M_{crit} = 2.26 \times 10^{-5}$ $M_\odot$ and $M_{dg} = 2.5 \times 10^{-6}$ $M_\odot$. If $Y_0 = 0.27$ and $Z_0 = 0.02$, the abundance of helium in the nova ejecta is $Y = 0.37$. Adopting, instead, the SS(3) case with $\dot{M} = 10^{-8}$ $M_\odot$ yr^{-1}, one has $M_{crit} = 1.01 \times 10^{-5}$ $M_\odot$, $M_{dg} = 1.0 \times 10^{-6}$ $M_\odot$, and Y = 0.43. Comparison of the abundance steady state models with observed nova abundances suggests that rotation-induced mixing must be operating (Fujimoto & Iben 1992). In order to understand qualitatively the effect of including rotation-induced mixing, one may double the values of M_{dg} given by the numerical experiments without rotation-induced mixing. Then, the two examples give $Y = 0.49$ (Cold[1] with $\dot{M} = 10^{-8}$ $M_\odot$ yr^{-1}) and $Y = 0.57$ (SS[3] with $\dot{M} = 10^{-8}$ $M_\odot$ yr^{-1}).

The next task is to estimate how long it takes to build up the critical mass of helium for a helium-burning thermonuclear runaway to occur. In the Cold(1) case with $\dot{M} = 10^{-8}$ $M_\odot$ yr^{-1}, $M_{dg} = 2.5 \times 10^{-6}$ $M_\odot$, $\Delta M_{He} = 0.87 \times 10^{-6}$ $M_\odot$, and it requires ~ 1149 cycles to increase the mass of the helium layer to 0.01 $M_\odot$. The duration of an outburst cycle is $T_{cycle} = (M_{crit} - M_{rm}(H))/\dot{M} = (2.26 - 0.37) \times 10^{-5}/10^{-8} = 1.89 \times 10^3$ years. Hence, it requires at least $1149 \times 1.89 \times 10^6 \sim 2.2 \times 10^6$ years to ignite helium. In case SS(3) with $\dot{M} = 10^{-8}$ $M_\odot$ yr^{-1}, $M_{dg} = 1.0 \times 10^{-6}$ $M_\odot$, $\Delta M_{He} = 1.02 \times 10^{-6}$ $M_\odot$, and it requires ~ 1000 cycles or $1000 \times (1.01 - 0.37) \times 10^{-5}/10^{-8} \sim 6 \times 10^6$ years to increase the mass of the helium layer to 0.01 $M_\odot$. These times are longer than those needed to establish thermal steady state conditions for hydrogen-burning outbursts, but perhaps not long enough for the center of the helium layer to be warm enough for eqs. (3) and (4) to be applicable.

In any case, the time to build the helium layer up to a critical value is large compared to the time which a star can live on the AGB ($\sim 10^5$-10^6 years). This, then, may account for the absence of heavy element overabundances in SyNe (Iben & Tutukov 1995).

5. ON THE DEVELOPMENT OF OVERABUNDANCES OF HEAVY ELEMENTS AFTER REMOVAL OF THE HELIUM BUFFER LAYER

In CVs, with typical lifetimes of $\sim 10^8$ years, there is ample time to build up a critical mass of helium, and, once helium is ignited, the explosion may be violent enough to remove the entire helium layer. Once the helium buffer layer has been removed, the potential for the mixing of hydrogen-rich material with CO or ONe white dwarf material in successive outburst cycles is improved. The key to this improvement is the fact that the mass $M_{rm}(He)$ decreases with increasing Z (see eq. [5]).

Assuming that $M_{dg} > M_{rm}(He)$, the abundance steady state equation to solve for Z is

$$(M_{crit} + M_{dg})(Z - \Delta Z_{nuc}) = M_{acc}Z_0 + (M_{rm}(H) + M_{rm}(He))Z + (M_{dg} - M_{rm}(He)). \qquad (11)$$

To find X, one solves

$$(M_{\mathrm{crit}} + M_{\mathrm{dg}})X = (M_{\mathrm{crit}} - M_{\mathrm{rm}}(\mathrm{H}))X_0 + M_{\mathrm{rm}}(\mathrm{H})X - (M_{\mathrm{crit}} + M_{\mathrm{dg}})\Delta X_{\mathrm{nuc}}. \qquad (12)$$

When Z is large and the white dwarf is of the CO variety, most of the hydrogen burned during the convective shell phase converts $^{12}\mathrm{C}$ nuclei into $^{14}\mathrm{N}$ nuclei, so that $\Delta Z_{\mathrm{nuc}} \sim \Delta X_{\mathrm{nuc}}$ (~ 0.04 when the mass of the CO white dwarf is $M_{\mathrm{CO}} = 1\ M_\odot$).

For example, when $M_{\mathrm{WD}} = 1\ M_\odot$ and $\dot{M} = 10^{-9}\ M_\odot\ \mathrm{yr}^{-1}$, the thermal steady state SS(2) solution gives $M_{\mathrm{crit}} = 1.33 \times 10^{-5}\ M_\odot$ and $M_{\mathrm{dg}} = 4.1 \times 10^{-6}\ M_\odot$. Doubling M_{dg} to $\sim 8 \times 10^{-6}\ M_\odot$ in order to approximate the effect of rotation-induced mixing, a solution of eqs. (5), (11), and (12) gives $Z \sim 0.28$, $X \sim 0.38$, and $Y \sim 0$ not unlike the case of V827 Her in Table 1.

6. COMPARISON WITH THE OBSERVATIONS: EVIDENCE FOR ROTATION-INDUCED MIXING

Much work remains to be done to understand the details of mixing in nova precursors and to pin down more precisely the critical mass for shell helium burning in realistic situations. It is particularly important to explore quantitatively the effect of rotation-induced mixing. The fact that such mixing occurs can be demonstrated by using abundance distributions in the ejecta of observed novae in conjunction with abundance steady state models to determine the extent of mixing in the nova precursors. The extent of mixing that is found is larger than given by thermal steady state models in which this mixing has not been taken into account (Fujimoto & Iben 1992). Results in columns 3-5 of Table 5 are for a CO white dwarf of mass 1 $M_\odot$ which is accreting matter of solar composition.

Table 5. Estimated Parameters of Shell Flashes for Nova Explosions.
$(M_{\mathrm{WD}} = 1 M_\odot (\Delta X_{\mathrm{nuc}} = 0.04)$ and $X_0 = 0.72)$

Star	Speed Class	M_{acc} $(M_\odot)$	M_{CS} $(M_\odot)$	$\zeta =$ $\zeta\,(1.0)$	$M_{\mathrm{dg}}/$ $\zeta\,(0.8)$	$M_{\mathrm{rm}}(\mathrm{He})$ $\zeta\,(0.6)$
V1370	VF	2.9E-06	1.2E-05	1.90	1.75	1.67
DQ Her	M	6.0E-06	1.6E-05	1.90	1.78	1.70
V693 CrA	VF	9.7E-06	1.9E-05	1.57	1.51	1.43
GQ Mus	M	1.1E-05	2.0E-05	1.31	1.30	1.25
T Aur	M	1.7E-05	2.8E-05	1.57	1.50	1.45
RR Pic	S	2.6E-05	3.7E-05	1.13	1.14	1.11
HR Del	S	2.2E-05	3.3E-05	1.23	1.17	1.15
CP PuP	VF	4.2E-05	1.0E-04	14	11	9.6
V1668 Cyg	F	1.0E-04	1.5E-04	9.5	5.2	5.0
V1500 Cyg	VF	4.6E-04	6.4E-04	39	8.5	6.6
PW Vul	M	8.5E-05	1.1E-04	3.9	3.2	2.6

The abundance ratios that have been used to derive these results come from Williams (1985) and from Truran & Livio (1986) and the speed class designation is from Bode et al. (1989). The extent of the mixing required by the observed abundances can be described by the ratio $\zeta = M_{dg}/M_{rm}(\text{He})$. Values of this ratio are given in columns 5-7 of Table 5 for three different choices of the mass of the white dwarf: 1.0, 0.8, and 0.6 $M_\odot$.

The novae fall into two groups: one for which [CNO/H] (or [ONeMg]) $\sim 1.7 \pm 0.3$ and [He/H] $\sim 0.12 \pm 0.03$ (case A); and another for which [CNO/H] (or [ONeMg]) $\sim 1.7 \pm 0.8$ and [He/H] $\sim 0.5 \pm 0.1$ (case B). The last four novae listed in Table 5 belong to group A and the first seven belong to group B. For case A systems, the average value of M_{acc} is $\sim 1.7 \times 10^{-4}$ $M_\odot$, and, for case B systems, the average value of M_{acc} is $\sim 1.4 \times 10^{-5}$ $M_\odot$. Results are not dramatically sensitive to the choice of white dwarf mass and similar values of M_{acc} and M_{CS} are obtained on the assumption that the white dwarf is an ONe white dwarf of mass 1.25 $M_\odot$.

Comparing values of M_{acc} in Table 5 with entries in Table 2, it is clear that the accretion rate in case A systems is on average considerably less than in case B systems. In case B systems for which estimated values of M_{acc} indicate accretion rates in the range 10^{-9}–10^{-10} $M_\odot$ yr^{-1}, not only are the observed enhancements of elements heavier than helium far larger than predicted by non-rotating models with accretion rates in this range, but the enhancements of helium also are larger than the predicted ones (Fujimoto & Iben 1992). Thus, for case B stars, some mode of mixing in addition to particle diffusion must be operating between outbursts and one might guess at some form of rotation-induced mixing.

Several mechanisms for rotation-induced mixing have been suggested (Kippenhahn & Thomas 1978; MacDonald 1983; Kutter & Sparks 1987; Sparks and Kutter 1987; Livio & Truran 1987; Fujimoto 1988). The baroclinic instability explored by Fujimoto (1988, 1993) appears promising and preliminary results support this promise (Fujimoto & Iben 1994).

Acknowledgements

Preparation of this manuscript has been supported by the NSF grants AST91-13662 and AST94-17156.

REFERENCES

Bode, M. F., Duerbeck, H. W., & Evans, A. 1989, in Classical Novae, ed. M. F. Bode & A. Evans (John Wiley & Sons LTD; Chichester), ch. 13, p. 249
Fujimoto, M. Y. 1988, A&A, 198, 163
——1993, ApJ, 419, 768
Fujimoto, M. Y., & Sugimoto, D. 1979, PASJ, 31, 1
——1982, ApJ, 257, 291
Fujimoto, M. Y., & Iben, I. Jr. 1992, ApJ, 399, 646
——1994, in progress.
Iben, I., Jr. 1977, ApJ, 217, 788
——1992, in Variable Stars and Galaxies, ed. B. Warner (San Francisco: ASP), 307
Iben, I. Jr., Fujimoto, M. Y., & MacDonald, J. 1991, ApJL, 375, L27
——1992a, ApJ, 384, 580
——1992b, ApJ, 388, 521
Iben, I., & Tutukov, A. V. 1984a, Ap.J., 282, 615
——1984b, ApJ, 284, 719
——1985, ApJS, 58, 661

——1989, ApJ, 342, 430

——1991, ApJ, 370, 615

——1993, ApJ, 418, 343

——1995, ApJ, submitted

Kato, M. 1980, Prog. Theoret. Phys., 64, 847

Kippenhahn, R., & Thomas, H.-C. 1978, A&A, 63, 265

Kovetz, A., & Prialnik, D. 1985, ApJ, 291, 812

——1990, in Lecture Notes in Physics, No. 369: Physics of Classical
    Novae, ed. A. Cassatella & R. Viotti (Berlin: Springer), p. 394.

Kutter, G. S., & Sparks, W. M. 1987, ApJ, 321, 386

Livio, M., & Truran, J. W. 1987, ApJ, 318, 316

——1994, ApJ, 425, 797

MacDonald, J. 1983, ApJ, 273, 289

Mürset, U. & Nussbaumer, H. 1994, A&A, 282, 586

Neo, S., Miyaji, S., Nomoto, K., & Sugimoto, D. 1977, PASJ, 29, 249

Prialnik, D. 1986, ApJ, 310, 222

Prialnik, D., & Kovetz, A. 1984, ApJ, 281, 367

Prialnik, D., & Shara, M. M. 1986, ApJ, 311, 172

Schwartzman, E., Kovetz, A., & Prialnik, D. 1994, MNRAS, 269, 323

Skumanich, A. 1972, ApJ, 171, 565

Sparks, W. M., & Kutter, G. S. 1987, ApJ., 321, 394

Truran, J. W. 1985, in Proceedings of the ESO Workshop on Production and
    Destruction of C, N, O Elements, eds. J. Danzinger, F. Matteucci, & K. Kät, p. 221

——1994, private communication

——1990, in Physics of Classical Novae, Proceedings of IAU Colloq.
    No. 122, eds. A.Cassatella & R. Viotti, (Springer), p. 373

Truran, J. W., & Livio, M. 1986, ApJ, 308, 721

Williams, R. E. 1985, in Production and Distribution of CNO Elements, ed. I. J. Danziger
    (Garching: ESO), p. 225

A NEW METHOD OF DETERMINING THE EJECTED MASS OF NOVAE

WARREN M. SPARKS
Los Alamos National Laboratory
MS F663
Los Alamos,NM 87544
USA

Abstract.
    A new method of determining the ejected mass of novae based on simple, reasonable assumptions is presented. This method assumes that the remnant mass on the white dwarf is the same as that from the previous nova outburst. The hydrogen, helium, and metal abundances of the accreted material from the secondary must also be known or assumed. The white dwarf's mass has a small effect because the amount of hydrogen consumed during the thermonuclear runaway only depends weakly upon this mass. If the composition of the ejecta and the time of the remnant shell burnout are determined from observations, then the ejected and remnant masses can be deduced. At present only a sharp decrease in the X-rays observed by ROSAT has been attributed to this remnant burnout and only for two novae: GQ Mus and V1974 Cyg. The ejected and remnant masses for these two novae are calculated. If other indicators of nova remnant burnout, such as a rapid decrease in high-ionization lines, can be identified, then this method could be applied to additional novae.

1. Introduction

The ejected mass is one of the most basic quantities in nova studies, but it has been very difficult to determine observationally. Most ejected mass determinations involve relating an electron temperature from the ratio of a forbidden line pair, the emission measurement from the flux of a hydrogen

211

A. Bianchini et al. (eds.), Cataclysmic Variables, 211-216.

line, and an estimate of the ejecta's volume. The nonuniformity of the ejecta causes a large uncertainity in these mass determinations.

Nova ejecta consists of material from three sources: accreted material from the secondary, remnant material from the previous outburst, and dredged-up white dwarf material. These three sources are shown schematically as layers on the white dwarf in the first panel of the drawing in Figure 1. The next panels picture the mixing due to the thermonuclear runaway (TNR), the ejection and remnant hydrogen shell burning, and finally, the burned-out remnant left on the white dwarf. When more material is accreted from the secondary, the cycle starts over again. This simple drawing illustrates two of the assumptions used. First, materials from the three sources are thoroughly mixed by the TNR, and the mass of the remnant is the same as that from the previous outburst. These assumptions are supported by numerical simulations. Starrfield et al. (1972) found that the entire envelope above the TNR region is strongly convective and Shara, Prialnik, and Kovetz (1993) have shown that the remnant masses are very similar after the first few outbursts.

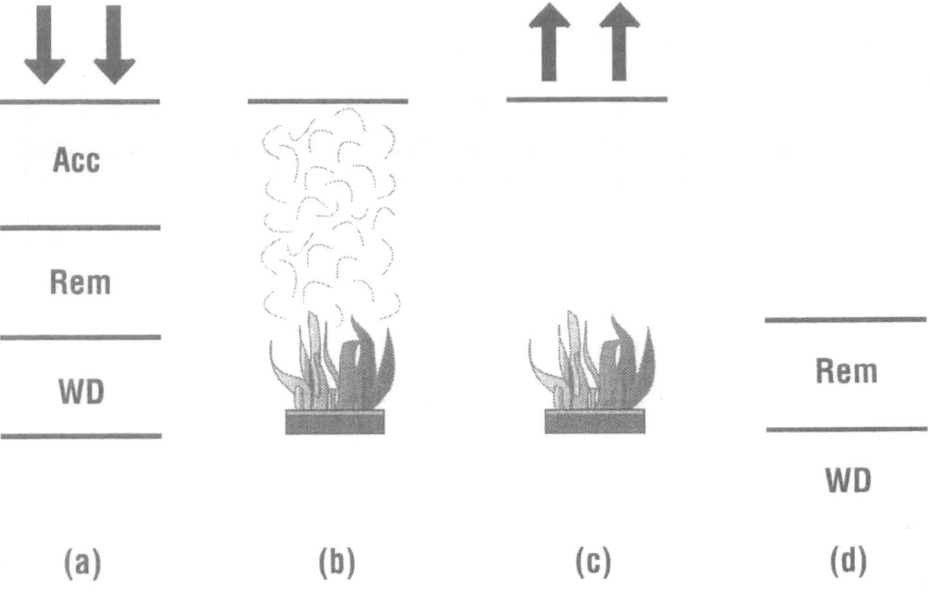

Figure 1. The nova cycle showing (a) the accretion of material onto the white dwarf, (b) the mixing due to the TNR, (c) the ejection and remnant burning, and (d) the burned-out remnant.

2. Formulation of the Problem

With the above assumptions, equating the mass abundances of hydrogen, helium, and the metals before and after the TNR produces the following equations:

$$M_{acc}X_0 = M_{tot}(X + \Delta X_{run}) \tag{1}$$
$$M_{acc}Y_0 + M_{rem}(X + Y) = M_{tot}(Y - \Delta Y_{run}) \tag{2}$$
$$M_{dg} + M_{rem}Z + M_{acc}Z_0 = M_{tot}(Z - \Delta Z_{run}) \tag{3}$$
$$M_{tot} = M_{acc} + M_{rem} + M_{dg}, \tag{4}$$

where the abundances with a zero subscript and no subscript are for the secondary, and the mixed materials, respectively; M_{acc}, M_{rem}, and M_{dg} are the accreted, remnant, and dredged-up white dwarf masses; and ΔX_{run}, ΔY_{run}, and ΔZ_{run} are the changes in the composition due to the TNR. The composition of the white dwarf is assumed to be all metals. This formulation is not new; variations of these equations have been described by Sparks et al. (1988), Prialnik (1990), and Fujimoto and Iben (1992). Note that any mixing mechanism before the TNR does not affect to this formulation.

Solving the above equations gives only a ratio of the individual mass components to the total mass. Another condition for one of these masses must be known in order to determine the individual mass components. The time for hydrogen burnout of the remnant mass provides such a condition. Equating the energy radiated from the remnant to the nuclear energy available from hydrogen burning gives

$$\int L \, dt = E_H \, X \, M_{rem}, \tag{5}$$

where L is the luminosity and E_H is the nuclear energy available from burning hydrogen to helium. We will assume that the luminosity is constant in time and equal to the equilibrium hydrogen shell burning plateau luminosity. Therefore,

$$L_{eq} \, \Delta t = E_H \, X \, M_{rem}. \tag{6}$$

3. Results

Let us now consider GQ Mus, a nova that erupted in 1983. Hassall et al. (1990) derived abundances of X = 0.27, Y = 0.32, and Z = 0.41 for its ejecta. It is assumed that the secondary has solar abundances ($X_0 = 0.71$,

$Y_0 = 0.27$, and $Z_0 = 0.02$). We will further assume that the primary is a 1.0 $M_\odot$ CO white dwarf. The heavy element composition of the ejecta (Hassall et al., 1990) strongly suggests that the TNR occurred on a CO white dwarf, and 1.0 $M_\odot$ is a reasonable mass for such a white dwarf. Fortunately, the results depend only weakly on this chosen mass. For a 1.0 $M_\odot$ white dwarf, the amount of hydrogen burned during the TNR to produce enough energy to lift material to the white dwarf's Roche lobe radius (which for most nova binary systems is ~ 1.0 $R_\odot$) is 0.04 (see Fujimoto and Iben, 1992 for a more detailed discussion). For a TNR on a CO white dwarf, hydrogen is combined with C^{12} to produce mostly C^{13} and N^{14} (Starrfield, Truran, and Sparks, 1975). Thus, we will use $\Delta X_{run} = \Delta Z_{run} = 0.04$ and $\Delta Y_{run} = 0.0$.

Substituting these values into eqs.(1-4) gives $M_{rem}/M_{tot} = 0.34$, $M_{acc}/M_{tot} = 0.44$, and $M_{dg}/M_{tot} = 0.22$. Since ΔX_{run} and ΔZ_{run} are small compared to the observed $X(= 0.27)$ and $Z(= 0.41)$ values, the assumed value of the white dwarf's mass has only a small effect on the final results. Recently, the X-rays from GQ Mus observed by ROSAT (Shanley et al., 1995) have shown a sharp decrease 11 years after outburst. They interpreted this decrease as the turn off of the remnant hydrogen shell source. Using the plateau luminosity of the appropiate hydrogen shell burning model of Fujimoto and Iben (1992) in eq.(6), we calculate a remnant mass of 1.7 $\times$ 10^{-5} $M_\odot$ for GQ Mus. This mass combined with the previous results yields an ejected mass of 3.3 $\times$ 10^{-5} $M_\odot$ for GQ Mus.

One caveat to this simplified picture is that, according to Fujimoto and Iben (1992), the radius of this 1.7 $\times$ 10^{-5} $M_\odot$ remnant will be much larger than any reasonable Roche lobe. In other words, there will be a common envelope around the binary, and the composition of the remnant may be a mixture from the ejecta and the secondary. As an extreme case, we can assume that the remnant material comes entirely from the secondary. Thus, the X in eq.(6) and the X, Y, and Z on the left hand side of eqs.(2) and (3) will have a zero subscript signifying material from the secondary. Under these conditions the remnant mass is 5.0 $\times$ 10^{-6} $M_\odot$, and the ejected mass is 2.0 $\times$ 10^{-5} $M_\odot$. A white dwarf remnant formed from a common envelope will lie somewhere between these two extreme cases, and thus the ejected mass lies between 2.0 and 3.3 $\times$ 10^{-5} $M_\odot$.

V1974 Cyg is the only other nova observed by ROSAT to have a sharp decrease in X-rays. Its X-ray flux turned off 18 months after outburst (Krautter et al., 1995). Recent photoionization models of the emission lines by Austin et al. (1995) constrained the helium abundance from 0.22 to 0.46 and the metals from 0.36 to 0.44. Since the emission lines indicate a large abundance of O, Ne, and Mg, we will assume that the TNR occurred on a 1.25 $M_\odot$ ONeMg white dwarf. The amount of nuclear burning required to lift material off of a 1.25 $M_\odot$ white dwarf indicates that $\Delta X_{run} = \Delta Y_{run}$

= 0.08 and ΔZ_{run} = 0.0 (see Fujimoto and Iben, 1992 and Politano et al., 1995). Here the assumed white dwarf's mass is more critical to the final results than in the case for GQ Mus. Using the largest values for both helium and the metals, we calculate a remnant mass of 8.7×10^{-6} and 8.7×10^{-7} $M_\odot$ and an ejected mass of 7.1×10^{-6} and 1.8×10^{-6} $M_\odot$ when the composition of the remnant mass is like that of the ejected mass and the secondary, respectively. For the smallest deduced values of helium and metal abundances, the remnant mass is 2.1×10^{-6} to 8.7×10^{-7} $M_\odot$ for the same two sources of the remnant mass composition. However, since the lowest limit of the helium abundance can be produced during the TNR, no upper limit constraints can be placed on the ejected mass. Thus, we have found only a lower limit of 1.8×10^{-6} $M_\odot$ for the ejected mass of V1974 Cyg.

4. Conclusions

This method uses the observed abundances from the ejecta and the turn-off time of the remnant burning to determine the ejected and remnant masses of a nova. At present only the rapid decrease in X-rays observed by ROSAT has been attributed to the remnant burnout and only for two novae: GQ Mus and V1974 Cyg. For GQ Mus, we find an ejected mass between 2.0 and 3.3×10^{-5} $M_\odot$, whereas Hassall et al. (1990) estimated it to be 8.4×10^{-5} $M_\odot$. Stringfellow and Bowyer (1994) detected V1974 Cyg with the Extreme Ultraviolet Explorer ∼285 days after its outburst. They fit their data best with a hot white dwarf atmosphere of $4.0 \pm 0.5 \times 10^5$ K and a column hydrogen density near 3×10^{21} cm^{-2}. These high temperatures require that the radius of the remnant on the white dwarf be small. For our assumed white dwarf mass of 1.25 $M_\odot$ and their temperatures, the unburnt remnant mass is estimated to be from 5.0 to 5.6×10^{-7} $M_\odot$, using Fujimoto and Iben's (1992) graphs of equilibrium models. Since at the time of their observation, approximately one half of the remnant mass has burned out, this range of remnant masses is entirely consistent with our calculated range of 8.7×10^{-7} to 2.1×10^{-6} $M_\odot$. Krautter et al. (1995) calculated an upper limit of 2×10^{-5} $M_\odot$ for the remnant mass of V1974 Cyg by equating its potential energy to the emitted energy from the latter part of the X-ray light curve. Although these various deduced masses for the remnant of V1974 Cyg are consistent, there is a large discrepancy in estimations of its ejected mass. We derive a lower limit of 1.8×10^{-6} $M_\odot$ for the ejected mass in V1974 Cyg, whereas Shore et al. (1993) find a lower limit of 10^{-4} $M_\odot$. Future abundance or white dwarf mass determinations of V1974 Cyg can provide upper limits on its ejected mass with this new method. However, it is unlikely that the calculated ejected mass will be as large as 10^{-4} M$_\odot$. These ejected mass discrepancies must be resolved. If other criteria for the burnout of the remnant shell, for example, a rapid decrease of a high-ionization line, could be established, then this method could be applied to other novae.

References

Austin, S., Wagner, M., Starrfield, S., Shore, S., Sonneborn, G., and Burtran, R. 1995, these proceedings

Fujimoto, M. Y., and Iben, I. 1992, *ApJ*, **399**, 646

Hassall, B. J. M., et al. 1990, in Physics of Classical Novae: *IAU* Colloquium No. 122 eds. A. Cassatella and R. Viotti (Berlin: Springer-Verlag), 202

Krautter, J., Ogelman, H., Starrfield, S., Wickmann, R., and Trumpter, J. 1995, in preparation

Politano, M., Starrfield, S., Truran, J. W., Weiss, A., and Sparks, W. M. 1995, to be published in *ApJ*

Prialnik, D. 1990, in Physics of Classical Novae: *IAU* Colloquium No. 122 eds. A. Cassatella and R. Viotti (Berlin: Springer-Verlag), 351

Shanley, L., Ogelman, H., Gallagher, J. S., Orio, M., and Krautter, J. 1995, submitted to *ApJ Letters*

Shara, M. M., Prialnik, D., and Kovetz, A. 1993, *ApJ*, **406**, 220

Shore, S. N., Sonneborn, G., Starrfield, S., Gonzalez-Riestra, R., and Ake, T. B. 1993, *AJ*, **106**, 2408

Sparks, W. M., Starrfield, S. G., Truran, J. W., and Kutter, G. S. 1988, in Atmospheric Diagnostics of Stellar Evolution: Chemical Peculiarity, Mass Loss and Explosion: *IAU* Colloquium No. 108 ed. K. Nomoto (Berlin: Springer-Verlag)

Starrfield, S., Truran, J. W., and Sparks, W. M. 1975, *ApJ*, **198**, L113

Starrfield, S., Truran, J. W., Sparks, W. M., and Kutter, G. S. 1972, *ApJ*, **176**, 169

Stringfellow, G. S., and Bowyer, S. 1994, in Interacting Binary Stars ed. A. W. Shafter (San Francisco: ASP), 315

THE THEORY OF NOVA OUTBURSTS
IN THE LIGHT OF OBSERVATIONS

D. PRIALNIK

Department of Geophysics and Planetary Sciences
Tel Aviv University
Ramat Aviv 69978, Israel

1. Introduction

Nova theory postulates that an outburst results from transfer of hydrogen-rich material onto a white dwarf (WD), leading to a thermonuclear runaway (TNR). The large variation in the observed features — the peak luminosity, the amount, composition and average expansion velocity of the ejected matter, or the duration of the high-luminosity phase — is attributed by the nova theory to differences in the values assumed for three basic parameters: the accreting WD's mass M_{WD} , its temperature T_{WD} (or, equivalently, its luminosity or age), and the mass transfer rate $\dot{M}$ (Fujimoto 1982a,b; MacDonald 1983; Kovetz & Prialnik 1985, Prialnik & Kovetz 1995). These three parameters are, unfortunately, not directly observable and indirect estimates are scarce and uncertain. Hence the theory cannot be straightforwardly tested. One has to resort to roundabout tests in order to confront the theory with the observations and reveal its weak points.

For this purpose an extended set of consistent nova evolution calculations — covering the entire parameter space and following the complete nova cycle — is required. Such a comprehensive study, involving the calculation of a series of full, multi-cycle evolutionary sequences, spanning a sufficiently dense grid of the three basic parameters M_{WD} , T_{WD} and $\dot{M}$, has been recently performed by Prialnik and Kovetz (1995) — referred to hereafter as PK95 — using an updated hydrodynamical stellar evolution code (Kovetz, unpublished). These calculations are sufficiently detailed to enable, for the first time, a thorough comparison with observations.

A Bianchini et al (eds), Cataclysmic Variables 217-230
© *1995 Kluwer Academic Publishers*

D. PRIALNIK

2. Confrontation of theory with observations

2.1. TEST NO. 1: COMPLETENESS

Novae of all classes, including slow, moderate and fast classical novae (CN), recurrent novae (RN) and symbiotic novae (SymN), exhibit an overwhelming variety of properties, which may differ by orders of magnitude between one type of nova to another, and even between individual novae within the same class. Is the entire range of phenomena reproduced by numerical calculations, i.e., by varying the values of the basic parameters?

For each combination of parameters M_{WD} (0.65, 1, 1.25 and 1.4M$_\odot$), T_{WD} (10^7 K - cold, 3×10^7 K - moderate, and 5×10^7 K - hot) and $\dot{M}$ (10^{-6}, 10^{-7}, 10^{-8}, 10^{-9}, 10^{-10} and, for hot models, also 10^{-11} M$_\odot$ yr^{-1}), the PK95 evolutionary calculations provide the general characteristics of a typical nova cycle, such as the accreted mass m_{acc}, the ejected mass m_{ej}, the helium and the heavy element mass fractions of the ejecta, Y_{ej} and Z_{ej}, respectively, the maximal expansion velocity v_{max} and its average over the whole mass loss phase v_{av}, the maximal luminosity attained L_{max}, the outburst amplitude A, and typical time scales: the duration of the mass loss phase t_{m-l}, the time of decline of the bolometric luminosity by, say, 3 mag t_{3bol} and the recurrence period of the outbursts P_{rec}. The overall ranges of observed, or observationally derived, properties are compared with those obtained from PK95's calculations in Table 1.

TABLE 1. Ranges of nova characteristics

| Property | Observations | | Calculations | |
	Range	Exceptions	Range	Special cases
M_{max} mag	-6 to -9	-10 (V1500 Cyg)	-5.6 to -8.0	
A mag	7 to 16	19.3 (V1500 Cyg)	6.5 to 16.8	
t_3 days	4 - 300	SymN: more	0.7 - 680	SymN: $\leq$ 3 yr
m_{ej} 10^{-5}M$_\odot$	1 - 30	RN: less	0.02 - 28	RN: $\leq$ 0.9
Z_{ej}	0.04 - 0.41	0.86 (V1370 Aql)	0.027 - 0.47	
Y_{ej}	0.21 - 0.48	0.1 (V1370 Aql)	0.245 - 0.495	
v_{exp} km/s	350 - 2500	SymN: $\sim$ 100	160 - 2650	SymN: $\sim$ 100
P_{rec} yr	average: 10^4	RN: 10 -100	1.9 10^2 - 3.9 10^6	RN: 16 - 90

Some caution is required when comparing the calculated values with their observed counterparts. The exact decline time in the visual, which is available for most observed novae, is difficult to determine from nova outburst calculations, in which the ejected material that forms an opaque shell, is not included. The emitted radiation shifts toward the blue soon after mass loss ceases and the star contracts rapidly. The central star is seen, however, through the expanding shell, and hence the *observed* shift to the blue is delayed until the shell becomes transparent. Nevertheless, since the ejected mass is relatively small and thus becomes quickly diluted, the duration of mass loss t_{m-l} should be considered a reasonable lower limit to the decline time by a few magnitudes, such as t_3. The calculated outburst amplitude A, which is essentially the ratio between the maximal luminosity and the accretion luminosity, should be regarded as an upper limit to the observed one, which is usually measured from the minimum that occurs a relatively short time after outburst, when the mass transfer rate is still slowly declining (Vogt 1990, Duerbeck 1992).

The conclusion to be drawn from Table 1 is that the theoretical results cover the entire observed ranges, with very few exceptions. In addition, all nova types are accounted for. Accretion onto low-mass WDs leads to outbursts typical of SymN and very slow CN. Observationally, there are systems which have been classified as both, e.g. RR Tel or V407 Cyg (Allen 1980, Kenyon 1986). For higher WD masses, faster outbursts occur — slow, moderate and fast novae — and at still higher masses, very fast novae are obtained. For the highest WD mass, near the Chandrasekhar limit, *only* very fast nova outbursts occur.

Since the recurrence time of nova outbursts is inversely proportional both to $\dot{M}$ and to some power of M_{WD}, it has been assumed that RN must be obtained by accretion at a high rate onto massive WDs (Starrfield et al. 1985, Starrfield et al. 1988, Truran et al. 1988, Kato 1990, 1991, Kovetz & Prialnik 1994). It turns out, however, that while RN *do* require $\dot{M} \geq 10^{-8}$ $M_\odot$ yr^{-1}, the constraint on M_{WD} is less severe. Short recurrence times are obtained for M_{WD} down to $\sim 1 M_\odot$. Consequently, the decline times of RN models are not necessarily very short, as would be the case if their precursors were exclusively very massive WDs. Indeed, the decline times of observed RN span the entire CN range, from 280 d for V616 Mon, to 113 d for T Pyx and down to 6 d for U Sco. In addition, the velocity range of model RN is wide and ejected shells may be as massive as 10^{-5} $M_\odot$.

Another conclusion that emerges from calculations spanning a large domain of the parameter space is that not only do novae represent a three parameter family of events, but all parameters affect all properties to a larger or lesser extent (as discussed in more detail by Prialnik 1993 and by Schwartzman et al. 1994). This is illustrated here in Figure 1 for the PK95 models, which are divided into three groups, according to the initial T_{WD} adopted. The trend of change of t_{3bol}, Z_{ej} and Y_{ej} with the other two parameters is shown by constant value contours in the $(\dot{M}, M_{WD})$ plane, for each value of the temperature.

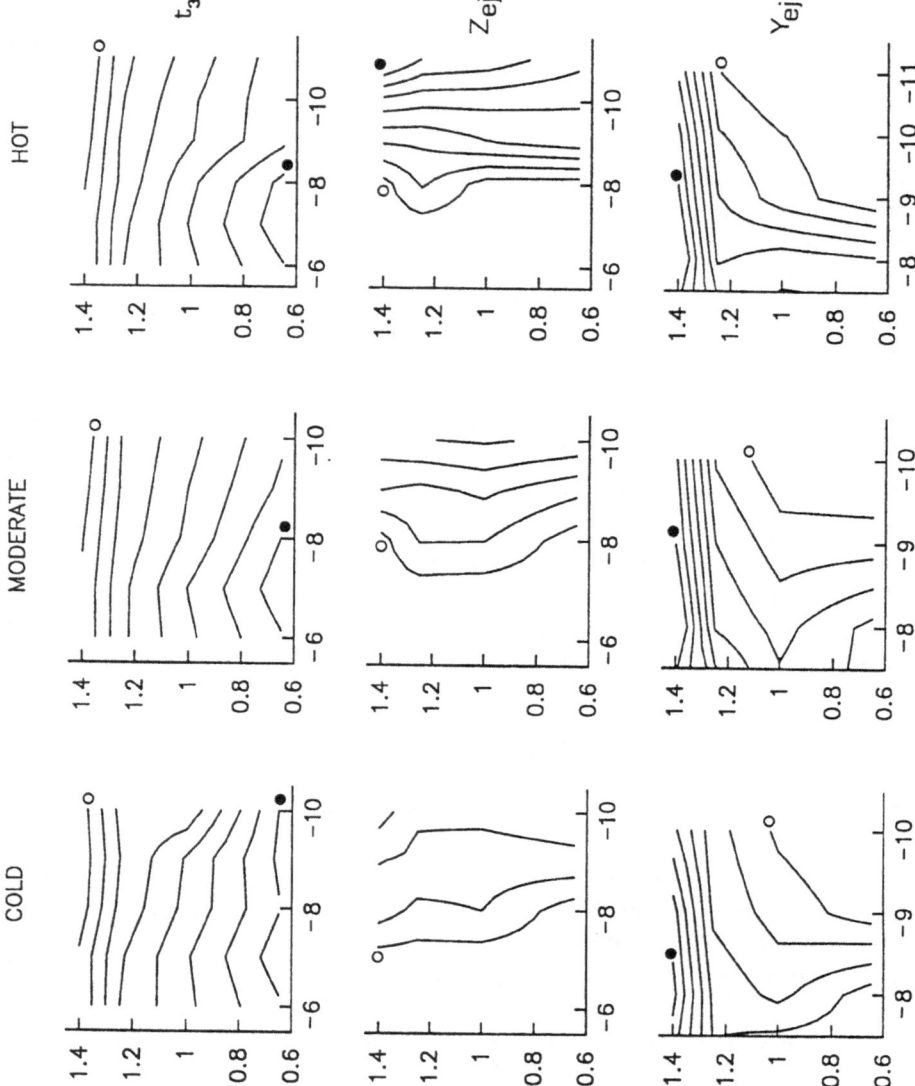

Figure 1. (*rotated*) Contours of times of decline of the bolometric luminosity, $\log(t_{3bol})$ — from 1.0 to 4.5, at 0.5 intervals (upper panels), heavy element mass fractions, Z_{ej} — from 0.05 to 0.40, at 0.05 intervals (middle panels), and helium mass fractions, Y_{ej} — from 0.28 to 0.49, at 0.03 intervals (lower panels), for three different WD temperatures (10^7 — COLD, 3×10^7 — MODERATE, 5×10^7 — HOT). The contour corresponding to the minimum value is marked by an open circle, and that corresponding to the maximum value, by a filled circle.

2.2. TEST NO. 2: CORRELATIONS BETWEEN NOVA PROPERTIES

The most striking feature that emerges from the observation of novae is the correlation between t_3 (or t_2) and M_B. It seems to be so well established, that novae are used as standard candles for distance determinations (e.g. Cohen 1985). Nevertheless, the correlation is not tight and is represented by a wide strip, rather than a line. Two questions arise: a. Do the calculations reproduce the observed correlation? b. Could the correlation be improved by better observational data, or is the spread intrinsic? The results of the PK95 calculations (for both t_{m-l} and t_{3bol}) are shown in Figure 2, together with the linear fit of Pfau (1976) to the observational data (Duerbeck 1987). The slope of the calculated points is less steep than the observational one, especially toward short decline times. The discrepancy may be due to two factors: at short decline times t_{m-l} becomes a poor approximation to t_3 (and the longer t_{3bol} might provide a better approximation), and, more importantly, the calculated L_{max} — chosen as the maximum luminosity obtained during the mass ejection phase, which prevails for a sufficiently long period of time to be observed — may be in error (underestimated). The absolute maximum may be higher (even twice as high), but only for very short periods of time (such behavior has also been encountered by Truran et al., private communication).

Regardless of this discrepancy, however, it is quite obvious that the spread of points occurs naturally, as the result of the effect of T_{WD} and $\dot{M}$ on both L_{max} and t_3, in addition to the more dominant effect of M_{WD}. Thus novae should not be expected to improve as distance indicators.

A less marked, but still obvious correlation may be shown to exist between t_3 and the average expansion velocities deduced from nova spectra (McLaughlin 1960), although systems of very different velocities are often observed in the same nova shell. The results obtained by PK95 are shown in Figure 3 together with velocities derived from observations. There is an excellent agreement between theory and observations, bearing in mind the uncertainties in both. We note that (in all plots) the clustering of computational points may be misleading. It has nothing to do with the frequency of occurrence of corresponding objects; it only means that many parameter combinations (which may rarely occur in reality) lead to similar results. By the same token, a cluster of observational points needs only a single nearby calculated point (parameter combination) to account for them.

Calculations reveal a new correlation, so far undetected observationally, between t_{m-l} and t_{3bol}, as illustrated in Figure 4: t_{3bol} is, on the average, about an order of magnitude longer than t_{m-l} (the ratio t_{3bol}/t_{m-l} varies between 3 and 40). Recent evidence for a correlation between the UV flux decline time and t_3 has, in fact, been deduced from *IUE* observations

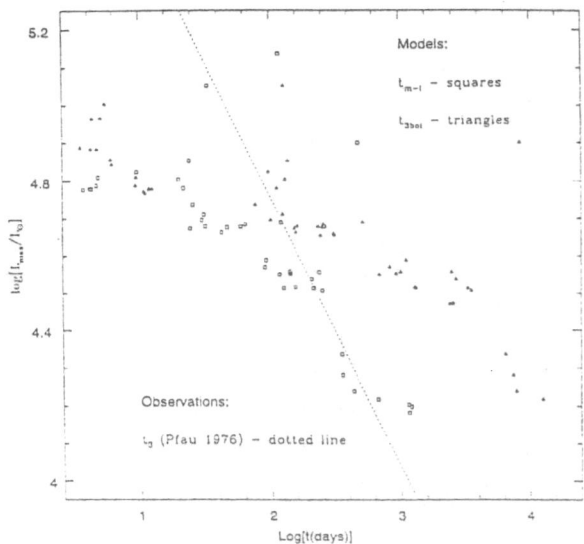

Figure 2. Maximum bolometric luminosity obtained at outburst (see text) vs. duration of mass loss, t_{m-l} (a close lower limit to the observed t_3) — squares, and vs. time of decline of the bolometric luminosity by 3 mag, t_{3bol} — triangles. The dotted line corresponds to Pfau's (1976) fit to observations of M_B and t_3.

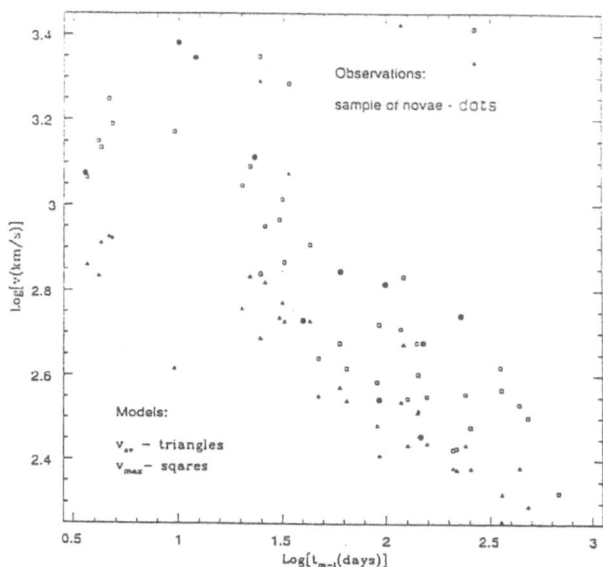

Figure 3. Expansion velocities vs. duration of mass loss t_{m-l}: average velocities — filled triangles, maximal velocities — open squares. Expansion velocities vs. t_3 for observed novae are are shown by filled circles.

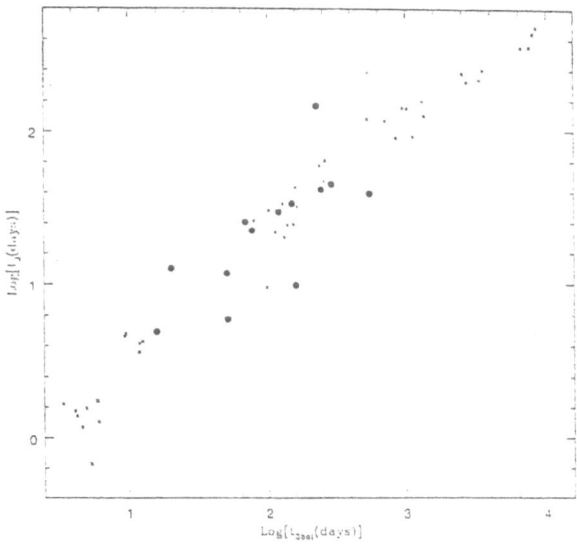

Figure 4. Duration of mass loss, t_{m-l}, vs. time of decline of the bolometric luminosity by 3 mag, t_{3bol}. Results of observations (decline times of UV fluxes, see text) are shown by filled circles.

(Gonzalez-Riestra, private communication). Since the effective temperature of the nova rises to $\sim 10^5$ K after the mass loss phase, the UV decline time is well approximated by t_{3bol}. Observations of X-rays with the *EXOSAT* satellite from three novae indicate that the WD remained luminous — after contracting at the end of the mass loss phase — for at least one year for Nova PW Vul 1984 ($> 2.5t_3$) and Nova QU Vul 1984 ($> 9t_3$) and for almost 3 years for Nova GQ Mus 1983 ($\sim 27t_3$). An even longer constant luminosity phase (~ 5.5 yr$\approx 50t_3$) is inferred for GQ Mus from a later study of the optical emission-line spectrum (Krautter & Williams 1989).

2.3. TEST NO. 3: UNCORRELATED CHARACTERISTICS

One of the puzzling traits of CN is the lack of correlation between t_3 and Z_{ej}, which appears to contradict the straightforward theory: a strong outburst is obtained when the envelope has a high CNO mass fraction and such an outburst should lead to a rapid development, i.e. a rapid decline. The same lack of correlation results from model calculations. Results of observations for a sample of 12 well-observed novae (T Aur 1891, RR Pic 1925, DQ Her 1934, CP Lac 1936, HR Del 1967, V1500 Cyg 1975, V1668 Cyg 1978, V693 CrA 1981, GQ Mus 1983, PW Vul 1984, QU Vul 1984, QV Vul 1987), for which composition, average expansion velocities, and ejected masses are determined (not including nova Aql 1370, which has an extremely unusual

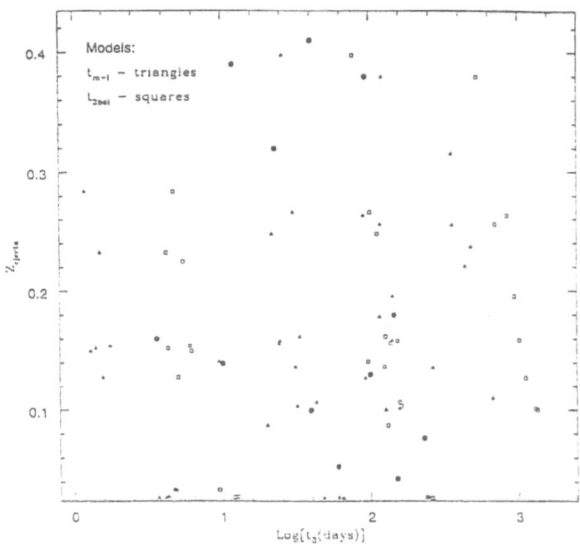

Figure 5. Heavy element mass fraction of the ejecta vs. time of decline. Observations: Z_{ej} vs. t_3 — filled circles. Model calculations: Z_{ej} vs. t_{m-l} — triangles, and Z_{ej} vs. t_{3bol} — squares.

and uncertain composition, Snijders et al. 1987), are plotted in Figure 5 together with results of model calculations (PK95). The agreement (overlap) between theory and observations is quite remarkable. Thus, enhanced CNO is a sufficient, but not a necessary condition for strong outbursts, and strong outbursts are not necessarily short-lived, for example, in cases where the ejection of large shells ensues.

A third independent characteristic of nova outbursts appears to be Y_{ej}. It is not directly correlated to either t_3 or Z_{ej}, as shown in Figure 6a,b. Moreover, a possible dependence of Y_{ej} on a combination of t_3 and Z_{ej} is also ruled out, since no correlation is found between Y_{ej} and Z_{ej} even when sub-samples of the observational data are considered, for which t_3 is more or less constant (see Fig. 6b). Again, this trait is verified by the models, as shown in Fig. 6b, where the results of calculations are also given. The three apparently stray points correspond to models of extreme mass (1.4 M$_\odot$) and extreme (lowest) accretion rate, a combination which should render such events very rare. It should, perhaps, be mentioned that, according to the nova theory, Z is determined by pre-outburst evolution (the outcome of diffusion and convection); t_3 is characteristic of the outburst itself (the expansion and consequent mass loss); and Y is the result of post-outburst evolution (burning of the remnant left over after mass loss). Hence, it is to be expected that these characteristics should be independent of each other.

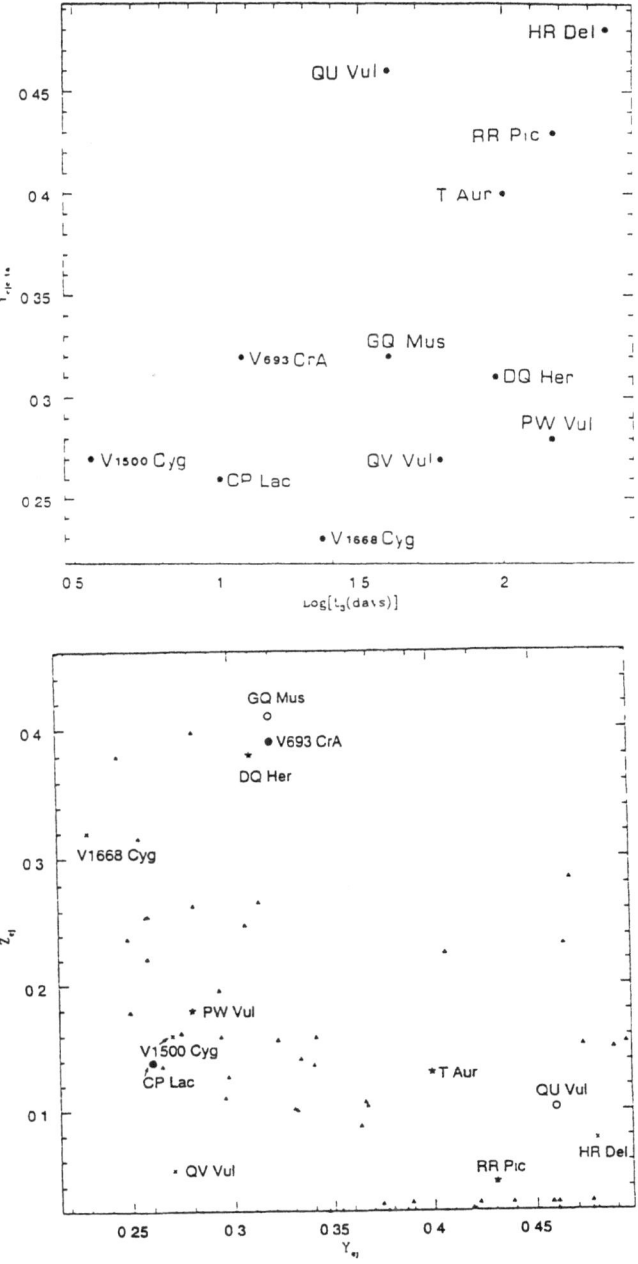

Figure 6. (a) Helium mass fraction of the ejecta, Y_{ej}, vs. time of decline by 3 mag, t_3, for observed novae. (b) Heavy element mass fraction of the ejecta, Z_{ej}, vs. the corresponding helium mass fraction, Y_{ej}, for observed novae — open and filled circles and stars, and for calculated models — triangles. Observed novae marked by the same symbol have similar decay times t_3.

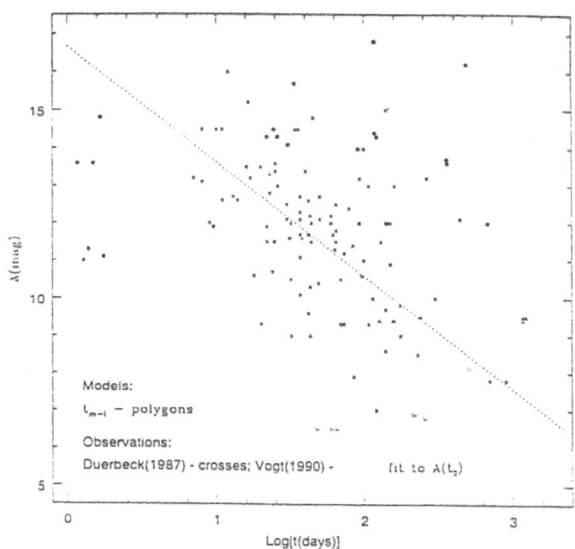

Figure 7. Outburst amplitude A (see text) vs. time of decline. Crosses represent observations (Duerbeck 1987) and Vogt's (1990) linear fit to the observed values. Polygons represent results of calculations, grouped according to the accretion rate: 10^{-8} $M_\odot$ yr^{-1} or more — triangles; 10^{-9} $M_\odot$ yr^{-1} — squares; 10^{-10} $M_\odot$ yr^{-1} — pentagons; and 10^{-11} $M_\odot$ yr^{-1} — hexagons.

The two characteristics of novae that can be straightforwardly determined by observations are t_3, and the amplitude A (which is distance independent). The data provided by Duerbeck's (1987) catalogue was analyzed by Vogt (1990), who found a correlation between A and t_3, with a slope similar (within errors) to that of the (M_B, t_3) correlation. This implies a restricted range of absolute minimum magnitudes (and hence accretion rates) during times of the order of decades prior to or following an outburst — the times of observation of the low-state magnitude. This is in agreement with Patterson's (1984) results and conclusions regarding the high $\dot{M}$ of CN near outburst and with the results obtained by Warner (1987). In Figure 7 we plot the results of PK95 in the (A, t_{m-l}) plane. There is no apparent correlation, although statistical factors (leading to selection effects) have not been considered. This apparent discrepancy with observations provides indirect support to the 'hibernation hypothesis' (Shara et al. 1986, Prialnik & Shara 1986). Calculations predict (deep) quiescence luminosities — mainly determined by $\dot{M}$ — that vary considerably from system to system. Since they are uncorrelated with t_3, the resulting amplitude (the ratio of maximal to minimal luminosity) becomes independent of t_3, in spite of the dependence of the maximal luminosity on t_3.

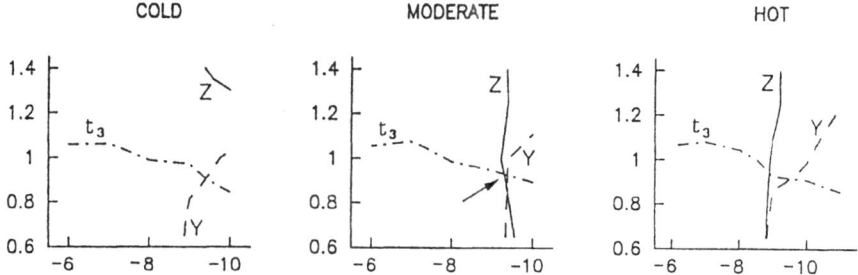

Figure 8. An example of the derivation of nova parameters from observations, for Nova PW Vul 84. The solid line is the $Z = 0.18$ contour in parameter space (for each of the three WD temperatures); the dashed line is the $Y_{ej} = 0.28$ contour and the dash-dotted line is the $t_{3bol} = 1500$d contour. An intersection is obtained in the middle panel, indicating a WD mass of $0.9M_\odot$, a WD temperature of $\sim 3 \times 10^7$ K, and an accretion rate of $\sim 5 \times 10^{-10}$ $M_\odot$ yr^{-1}.

2.4. TEST NO. 4: INVERSE RELATIONS APPLIED TO INDIVIDUAL CASES

Evolutionary calculations yield the values of the independent properties over the parameter grid: $t_{3bol}(M_{WD}, T_{WD}, \dot{M})$, $Y_{ej}(M_{WD}, T_{WD}, \dot{M})$ and $Z_{ej}(M_{WD}, T_{WD}, \dot{M})$. For a well-observed nova, with determined t_3, Y_{ej} and Z_{ej}, one may invert these relations to obtain $M_{WD}(t_3, Y_{ej}, Z_{ej})$, $T_{WD}(t_3, Y_{ej}, Z_{ej})$ and $\dot{M}(t_3, Y_{ej}, Z_{ej})$. If a solution exists and if it is unique, the mass of the nova progenitor, its temperature, and an estimate for the average mass transfer rate can thus be derived. In the few cases in which the white dwarf mass has been determined by observations, the inverse relation $M_{WD}(t_3, Y_{ej}, Z_{ej})$ can of course be directly tested. Other observable properties, such as the expansion velocity or the ejected mass, may also be derived and compared with those actually observed. Unfortunately, this procedure may be as yet premature. From the observational point of view, abundance determinations for nova shells are extremely uncertain (Livio & Truran 1994). From the theoretical point of view, the exact decline time in the visual, or in the blue, is difficult to determine. But, in principle, this method *is valid*, even if, for the time being, it should be applied and regarded with some caution. Since it is based on the simultaneous match of *three* parameters (and, for a given WD composition, there can be only three in a semi-detached binary system, when age is also a factor) it is a far more accurate method than those based on a single parameter match. As an illustrative example, we choose Nova PW Vul 1984 with $t_3 = 147\,d$, $Y_{ej} = 0.28$ and $Z_{ej} = 0.18$ and the grid of models of PK95. Assuming $t_{3bol} \approx 10t_3 \approx 1500$ d (see §2.2), contours corresponding to these values are superposed in Figure 8 (similar to Fig.1), in the $(\dot{M}, M_{WD})$

plane for each of the three temperatures. The three lines intersect for *the moderate temperature*, at a point corresponding to $M_{WD} \sim 0.9$ $M_{\odot}$ and $\dot{M} \sim 5 \times 10^{-10}$ $M_{\odot}$ yr^{-1}. Interpolation between calculated models for these initial parameters, now yields $v_{av} \approx 290$ km s^{-1}, as compared to the observed 285 km s^{-1}, and $m_{ej} \approx 4 \times 10^{-5}$ $M_{\odot}$, significantly lower than the observationally derived value ($\sim 30 \times 10^{-5}$ $M_{\odot}$).

3. Conclusions

The only rigorous way of testing the theory of nova outbursts is by performing calculations of complete cycles that cover systematically and consistently the three dimensional parameter space. Formidable as such a task may seem, it may prove rewarding in the end. Correlations between properties and trends of change of properties in parameter space are far more reliable than specific values obtained for them, either observationally, or by model calculations. Attempts at simulating the behavior of individual systems by ad-hoc assumptions will always leave doubts, whether the assumptions would be generally applicable, and whether a different choice of initial parameters might not have lead to better results. The general theory — with its inherent simplifying assumptions — is not meant to explain all the details of a given system, but rather the typical nova characteristics. These have only been revealed by analyzing the observations of *many* nova systems. The same holds true for model calculations. The main conclusions that emerge from the comparison of the PK95 set of calculations with observations are summarized below.

3.1. DISCREPANCIES BETWEEN THEORY AND OBSERVATIONS

Two main discrepancies are found between calculations and observations:

1. The maximum luminosities obtained, although well above the Eddington luminosities (calculated assuming electron scattering opacities), are not as high as those sometimes inferred from observations, which can surpass the Eddington limit by factors of 10 or more (e.g. V1500 Cyg, Lance et al. 1988).

2. The computed masses of the ejecta — for most parameter combinations — are lower than those derived from observations (e.g., Williams 1994), although the entire observed range is reproduced. There is, however, a large uncertainty in the observational estimates of nova shell masses, especially in view of the clumpiness of these shells (Barger et al. 1993, Saizar & Ferland 1994), which is best shown by recent HST observations.

3.2. AGREEMENT BETWEEN THEORY AND OBSERVATIONS

In all other respects, the agreement between nova theory and observations is outstanding:

1. The TNR-triggered outburst mechanism can explain many different types of novae: CN of all speed classes, including extremely rapid and extremely slow ones, SymN and RN, by varying the values of M_{WD}, T_{WD} and $\dot{M}$.

2. The entire range of observed nova characteristics (with very few exceptions, such as $Z=0.86$ or $A=19$ magnitudes) is covered by the grid of models. Unusual combinations of properties, such as a high Z slow nova (e.g., DQ Her), *can* be obtained for suitable parameter combinations.

3. RN can occur within any CN speed class: from slow ($t_3 \sim 100$ d) to very fast ($t_3 < 10$ d), in agreement with observations.

4. Correlations are obtained between the peak luminosity and time of decline, and between velocity and time of decline, compatible with those derived from observations. These correlations cannot be tight, however, as would be the case if novae were a one parameter family of events. The implication is that novae cannot be considered accurate distance indicators.

5. A distinct correlation is found between the duration of the mass loss phase and the time of decline of the bolometric luminosity (which may be much longer than the observed time of decline in the visual): t_{3bol} is roughly an order of magnitude higher than t_{m-l}. Such a correlation appears to be supported by recent observations of IUE and $EXOSAT$.

6. According to observations, and supported by theory, three nova properties are independent (uncorrelated): the time of decline, and the helium and the heavy element content of the ejecta. In principle, a transformation is possible from the three basic independent parameters to the three observable uncorrelated properties: M_{WD} (t_3, Y_{ej}, Z_{ej}), $\dot{M}$(t_3, Y_{ej}, Z_{ej}) and T_{WD} (t_3, Y_{ej}, Z_{ej}). This should supply the mass, the temperature and the average rate of accretion of a nova progenitor, provided the composition of the nova shell is well determined from observations, and the time of decline is accurately estimated by the theory. We can then test the agreement between theory and observations regarding other characteristics — such as the expansion velocity and the ejected mass. This conclusion should encourage further studies in both directions: observational, as well as theoretical.

4. References

Allen, D. A. 1980, MNRAS, 192, 521

Barger, A. J., Gallagher, J. S., Bjorkman, K. S., Johansen, K. A. & Nordsieck, K. H. 1993, ApJ, 419, L85

Cohen, J. G. 1985, ApJ, 292, 90

Duerbeck, H. W. 1987, Space Sci. Rev., 45, 1

Duerbeck, H. W. 1992, MNRAS, 258, 629

Fujimoto, M. Y. 1982a, ApJ, 257, 752

Fujimoto, M. Y. 1982b, ApJ, 257, 767

Iben, I. Jr., Fujimoto, M. Y. & MacDonald, J. 1992, ApJ, 388, 521

Kato, M. 1990, ApJ, 355, 277

Kato, M. 1991, ApJ, 366, 471

Kenyon, S. J. 1986, The Symbiotic Stars, (Cambridge: Cambridge University Press),

Kovetz, A. & Prialnik, D. 1985, ApJ, 291, 812

Kovetz, A. & Prialnik, D. 1994, ApJ, 424, 319

Krautter, J. & Williams, R. E. 1989, ApJ, 341, 968

Lance, C. M., McCall, M. L. & Uomoto, A. K. 1988, ApJS, 66, 151

Livio, M. & Truran, J. W. 1994, ApJ, 425, 797

MacDonald, J. 1983, ApJ, 267, 732

McLaughlin, D. B. 1960, in Stellar Atmospheres, ed. J. L. Greenstein (Chicago: University of Chicago Press), p585

Ögelman, H., Krautter, J. & Beuermann, K. 1987, A& A, 177, 110

Patterson, J. 1984, ApJS, 54, 443

Pfau, W. 1976, A& A, 50, 113

Prialnik, D. 1993, in Cataclysmic Variables and Related Physics, eds. O. Regev & G. Shaviv (Bristol: Institute of Physics Publishing), p351

Prialnik, D. & Kovetz, A. 1992, ApJ, 385, 665

Prialnik, D. & Kovetz, A. 1995, ApJ, , in press

Prialnik, D. & Shara, M. M. 1986, ApJ, 311, 172

Saizar, P. & Ferland, G. J. 1994, ApJ, 425, 755

Schwartzman, E., Kovetz, A. & Prialnik, D. 1994, MNRAS, 269, 323

Shara, M. M., Livio, M., Moffat, A. F. R. & Orio, M. 1986, ApJ, 311, 163

Snijders, M. A. J., Batt, T. J., Roche, P. F., Seaton, M. J., Morton, D. C., Spoelstra, T. A. T., & Blades, J.C. 1987, MNRAS, 228, 329

Starrfield, S., Sparks, W. M. & Shaviv, G. 1988, ApJL, 325, L35

Starrfield, S., Sparks, W. M. & Truran, J. W. 1985, ApJ, 291, 136

Truran, J. W., Livio, M., Hayes, J., Starrfield, S. & Sparks, W. M. 1988, ApJ, 324, 345

Truran, J. W., Hayes, J., Shankar, A. & Livio, M. , , , (unpublished)

Vogt, N. 1990, ApJ, 356, 609

Warner, B. 1987, MNRAS, 227, 23

Williams, R. E. 1994, ApJ, 426, 279

Williams, R. E., Sparks, W. M., Gallagher, J. S., Ney, E. P., Starrfield, S., & Truran, J. W. 1981, ApJ, 251, 221

THE DYNAMICS OF 2D FINITE LOCAL PERTURBATIONS IN ENVELOPES OF ROTATING WHITE DWARFS

P. GODON AND G. SHAVIV
Dept. of Physics
and
Asher Space Research Institute
Israel Institute of Technology
Haifa, Israel 32000

Abstract

As a first step in the understanding of LTNR on the surface of WD we investigate the behavior of 2D perturbations in the envelopes of newly accreted and hence rotating WDs. For this purpose we developed a time dependent Chebyshev method of collocation and solve with it the full 2D Navier-Stokes system of equations.

Our most important result is that a local perturbation develops in two basic phases: In phase A, which lasts about one dynamic time scale, the perturbation expands radially. In phase B the perturbation expands all over the star by means of pressure waves. Thus, a local perturbation becomes a global one after about one dynamic time scale.

The present calculations were carried out with a Reynolds number of 10^3 and 10^5 and we find that for these Reynolds numbers the viscosity has a minor effect on the development of the local perturbation into a global one.

Introduction

Our goal in this work is to investigate the behavior of finite non spherically symmetric perturbations in rotating envelopes of White Dwarfs. This problem is the first step in an effort to analyze local and finite perturbations in newly accreted envelopes of WDs.

The envelope of an accreting WD is formed via the accretion from an accretion disk. We do not solve the problem of how the matter is accreted onto the surface of the WD and assume that it can retain some of it angular momentum. In principle, the rotating layer can also spin-up the outer layers of the WD. The model assumed in this paper is a static or rotating WD with an envelope that may rotate with different rotational velocities.

Since the angular momentum distribution in the envelop is essentially unknown we start the exploration of this problem by assuming that the initial state is in uniform rotation, $\Omega = const$.

The accreted material is hydrogen rich and when enough matter is accreted, it undergoes a thermonuclear runaway (TNR). All detailed calculations of TNRs were carried out in spherical symmetry. However, the stabil-

231

A. Bianchini et al. (eds.), Cataclysmic Variables, 231-241.

ity of the envelope to perturbations will determine how a local perturbation may develop, to what extent does a local TNR develop into a global one, as well as other properties of the system. The fate of infinitesimal as well as local and finite temperature perturbations was estimated by Shara (1982) and investigated in 1D by Orio and Shaviv (1989). However, a 2D calculation of a TNR was not carried out. We feel that the first step in 2D calculations is to investigate the propagation of perturbations in 2D. The purpose of this paper is to do just this.

A crucial assumption is the axial symmetry and the 2D calculation because it ignores the Coriolis force and its effect in inhibiting energy transfer to smaller scales (Dubrule and Valdettaro (1992). A complete investigation of the Coriolis force effect requires a 3D calculation which is not attempted here. However, the meridional velocities found in this work are rather small and hence we expect the Coriolis force to have a minor effect.

The rotational velocities are of the order of the Keplerian velocities. Laboratory experiments show that shear flow with $Re \geq 10^3$ is turbulent. So our expression for the viscosity is the one appropriate for turbulent shear viscosity. As for the model of the viscosity, we adopt the Navier Stokes one.

Consider the case of differential rotation in the envelop (and the star) and assume a temperature (or density) perturbation around the equator. Clearly, the gas will try to expand towards the poles. However, the centrifugal barrier will try to prevent this trend. The easiest way for the perturbation to decay is to expand outward but then it reduces the pressure along the equator and matter from the pole will try to move towards the equator and enhance, or maintain, the local perturbation. The outcome is therefore, not clear a priori. Furthermore, the question is whether matter moves or the perturbation propagates via a wave. The wave always propagates with the speed of sound while the matter can have small or high velocities.

In the next section we describe the basic assumptions under which we solved the equations. Shortage of space prevents us from a description of the numerical method and we refer the reader to the relevant paper (Godon and Shaviv 1993). in Sec.3 we present the astrophysical model, in Sec.4 we discuss the results and the conclusions are drawn in Sec.5.

2. Governing Equations

Consider a compressible viscous flow, with turbulent viscosity, around a rigid central body. Assume that the total mass of gas in the domain of calculation (an extended atmosphere) is negligible compared to the mass of the star, hence only the gravity of the central object is taken into account. (The effects of gravity will be introduced in a subsequent work when the complete star will be discussed). The equations are written in spherical coordinates. We follow the evolution of a perturbation over several dynamic time scales and assume that the flow is adiabatic. Radiative transfer is not taken into account: the energy equation includes only a source term - the dissipation function. Since we consider here the evolution over only few (up to few tens) dynamic time scales the neglect of the radiative energy transfer is justified since the thermal time is so much longer than the dynamic one. The main energy carrier is the advective term. The radiative term will be included in the equations when we discuss the long term evolution where the velocities are significantly smaller and the radiative transfer is important.

The governing equations are: the continuity equation

$$\frac{\partial \rho}{\partial t} + \nabla \cdot (\rho \mathbf{u}) = 0 \tag{1}$$

the momentum equation

$$\rho \frac{\partial \mathbf{u}}{\partial t} + \rho \mathbf{u} \cdot \nabla \mathbf{u} = -\nabla P - \rho \mathbf{g} + \vec{f_\nu}, \tag{2}$$

the energy equation

$$\frac{\partial \rho \epsilon}{\partial t} + \nabla \cdot (\rho \epsilon \mathbf{u}) = -P \nabla \cdot \mathbf{u} + \Phi. \tag{3}$$

The differential system is supplemented with an equation of state

$$P = P(\rho, T). \tag{4}$$

The equation of state is ideal gas plus radiation pressure. In the above system, $\mathbf{g}$ is the gravitational acceleration, $\mathbf{f}_\nu$ the viscous force, Φ the dissipation function, ρ the density, P the pressure and $\mathbf{u}$ the velocity of the flow.

We assume axial symmetry around the z-axis, and suppose the flow to be symmetric above and below the $z = 0$ plane. In cylindrical coordinates, the dependent variables are ρ, P, v_r, v_θ and v_ϕ. Only the upper right quadrant of the (r, θ) plane is considered.

The domain of computation is:

$$R_* \leq r \leq R_0 \quad 0 \leq \theta \leq \pi/2,$$

where R_* is the radius of the central object and R_0 is the outer radius of the computational domain: $R_0 = 2R_*$.

Our independent variables are the momenta:

$$U = \rho v_r, \quad V = \rho v_\theta, \quad W = \rho v_\phi, \tag{5}$$

and we rewrite the equations in the conservation form. Assuming axial symmetry ($\partial/\partial\phi = 0$), the equations become: The continuity equation

$$\frac{\partial \rho}{\partial t} = -[\frac{1}{r^2}\frac{\partial}{\partial r}(r^2 U) + \frac{1}{r sin\theta}\frac{\partial}{\partial \theta}(V sin\theta)], \tag{6}$$

the radial momentum equation

$$\frac{\partial U}{\partial t} = -\left\{\frac{1}{r^2}\frac{\partial}{\partial r}\left[r^2\left(\frac{U^2}{\rho} - R_{rr}\right)\right] + \frac{1}{r sin\theta}\frac{\partial}{\partial \theta}\left[\left(\frac{UV}{\rho} - R_{r\theta}\right)sin\theta\right]\right\}$$

$$+\frac{1}{r}(\frac{V^2}{\rho} - R_{\theta\theta} + \frac{W^2}{\rho} - R_{\phi\phi}) - (\frac{\partial P}{\partial r} + \frac{GMr^2}{\rho}), \tag{7}$$

the azimuthal momentum equation

$$\frac{\partial V}{\partial t} = -\left\{\frac{1}{r^2}\frac{\partial}{\partial r}[r^2(\frac{UV}{\rho} - R_{\theta r})] + \frac{1}{r sin\theta}\frac{\partial}{\partial \theta}[(\frac{V^2}{\rho} - R_{\theta\theta})sin\theta]\right\}$$

$$+\frac{1}{r}[-\frac{UV}{\rho} + R_{r\theta} + (\frac{W^2}{\rho} - R_{\phi\phi})cot\theta - \frac{\partial P}{\partial \theta}], \tag{8}$$

the angular momentum equation

$$\frac{\partial W}{\partial t} = -\left\{\frac{1}{r^2}\frac{\partial}{\partial r}[r^2(\frac{UW}{\rho} - R_{r\phi})] + \frac{1}{r sin\theta}\frac{\partial}{\partial \theta}[(\frac{VW}{\rho} - R_{\theta\phi})sin\theta]\right\}$$

$$+\frac{1}{r}[-\frac{UW}{\rho} + R_{r\phi} - (\frac{WV}{\rho} - R_{\theta\phi})cot\theta], \tag{9}$$

and the energy equation

$$\frac{\partial P}{\partial t} = (\gamma - 1)\Phi - \frac{U}{\rho}\frac{\partial P}{\partial r} - \frac{V}{\rho r}\frac{\partial P}{\partial \theta}$$

$$-\gamma P[\frac{1}{r^2}\frac{\partial}{\partial r}(r^2\frac{U}{\rho}) + \frac{1}{r sin\theta}\frac{\partial}{\partial \theta}(\frac{V}{\rho})sin\theta]. \tag{10}$$

The dissipation function for the Navier Stokes equation is given by

$$\Phi = 2\nu\rho[\ D_{rr}^2 + D_{\theta\theta}^2 + D_{\phi\phi}^2 + 2D_{r\theta}^2 + 2D_{\theta\phi}^2 + 2D_{r\phi}^2\ ], \tag{11}$$

where D_{ij} is the deformation tensor:

$$D_{rr} = \frac{\partial}{\partial r}(\frac{U}{\rho}),$$

$$D_{\theta\theta} = \frac{1}{r}\frac{\partial}{\partial \theta}(\frac{V}{\rho}) + \frac{1}{r}\frac{U}{\rho},$$

$$D_{\phi\phi} = \frac{1}{r}\frac{U}{\rho} + \frac{1}{r}\frac{V}{\rho}cot\theta,$$

$$D_{r\theta} = \frac{1}{2}[\frac{1}{r}\frac{\partial}{\partial\theta}(\frac{U}{\rho}) + r\frac{\partial}{\partial r}(\frac{1}{r}\frac{V}{\rho})],$$

$$D_{\phi r} = \frac{1}{2}[r\frac{\partial}{\partial r}(\frac{W}{r\rho})], \tag{12}$$

$$D_{\theta\phi} = \frac{1}{2}[\frac{sin\theta}{r}\frac{\partial}{\partial\theta}(\frac{W}{\rho sin\theta})].$$

ν is the coefficient of turbulent viscosity, γ is the adiabatic exponent , G is the gravitational constant, M is the mass of the central object and the Reynolds stress tensor is given by:

$$R_{ij} = 2\nu\rho D_{ij}, \quad i = r, \theta, \phi; \quad j = r, \theta, \phi. \tag{13}$$

The prescription for the turbulent viscosity is based on the Boussinesq approximation.

The Numerical Method The method was discussed at length in Godon and Shaviv (1993) and Godon (1993).

3. The Astrophysical model

3.1 The Initial Conditions

The rotation law of the envelop is not known and hence different degrees of angular velocity in a rigid body rotation law are assumed. We have how-ever, no problem in assuming any other initial rotation law. A perturbation is imposed at $t = 0$ in an otherwise hydrostatic envelope and the evolution of the perturbation is followed. The central star is a White Dwarf with ra-dius $R_* = 0.144 \times 10^{-1} R_\odot$, mass $M = 1 M_\odot$, with an isothermal atmosphere in hydrostatic equilibrium (ideal gas of pure ionized Hydrogen). Three tem-peratures of the isothermal atmosphere were assumed: $T = 0.4 T_v, 0.6 T_v$ and $0.8 T_v$, where the virial temperature T_v is defined by:

$$T_v = \frac{GM}{\frac{\mathcal{R}}{\mu} R_*}. \tag{1}$$

Since the initial atmosphere is assumed to be isothermal, the initial unper-turbed steady state density profile is exponential, namely:

$$\rho_0 = \rho_B exp\left\{-\frac{GM}{\frac{\mathcal{R}}{\mu}T}\left(\frac{1}{R_*} - \frac{1}{r}\right) + \frac{1}{2}\frac{\Omega_0^2 r^2 sin^2\theta}{\frac{\mathcal{R}}{\mu}T}\right\}. \tag{2}$$

The assumption of isothermal atmosphere is not crucial since the most important feature is the exponential decline of the density. The initial un-perturbed steady state pressure P_0 is obtained from the state equation. The initial radial and meridional velocities vanish. We add a *local* perturbation to the temperature and we reevaluate the pressure through the equation of state.

The assumed perturbation is a Gaussian like function of the form:

$$T_{pert} = \alpha_T exp\left\{-A_r(r - R_*)^2 - A_\theta(\theta - \pi/2)^2\right\}, \qquad (3)$$

where α_T, A_r and A_θ are parameters. We use here an ideal gas plus radiation equation of state.

4.2 The Boundary Conditions

We assume in this work that the star behaves as a rigid sphere with no slip boundary conditions on its surface. The outer radius of the computational domain is placed at $r = 2R_*$ and it is a free boundary where no a priori conditions can be prescribed. The total change in density throughout the calculation domain is 4 orders on magnitudes. The equatorial plane $\theta = \pi/2$ is a plane of symmetry and the z-axis ($\theta = 0$) is an axis of rotational symmetry. The physical boundary conditions that can be supplied are therefore: - on the z-axis:

$$v_\phi = 0, \quad v_\theta = 0, \quad \partial P/\partial\theta = 0, \quad \partial\rho/\partial\theta = 0 \quad and \quad \partial v_r/\partial\theta = 0.$$

- on the equatorial plane:

$$v_\theta = 0, \quad \partial v_r/\partial\theta = 0, \quad \partial v_\phi/\partial\theta = 0, \quad \partial P/\partial\theta = 0 \quad and \quad \partial\rho/\partial\theta = 0.$$

- on the stellar surface:

$$v_r = 0, \quad v_\theta = 0, \quad v_\phi = v_*(\theta) \quad and \quad \partial T/\partial r = 0.$$

Where $v_*(\theta)$ is the rotation velocity on the surface of the star. The outer boundary condition calls for a special treatment. In principle, the atmosphere extends to infinity but for practical reasons we have to define a boundary at a finite distance in space. It is possible that the flow would start in the atmosphere and continue beyond the artificial boundary. However, if the velocity is not sufficiently high, it will change its velocity and return eventually to the star, albeit at a point beyond the border of the calculation. Also, the point where the mass returns is not necessarily the one where it left the domain of computation. Thus, we have to find a condition that will allow us to restrict the domain of computation, without hopefully constraining the physics. Clearly, a possible way is to repeat the calculation with the boundary placed at another point and check the consistency of the results. However, this is not the entire picture. Sound waves that are generated in the compressible fluid calculation, simple noise in the calcula-

tion, propagate in the computational domain. These waves hit the artificial
boundary condition and may be reflected back into the computational domain. Hence, it is necessary to guarantee that outmoving or inmoving flows
are properly treated.

The full and detailed mathematical treatment of the open boundary
condition is discussed in a separate paper (Godon and Shaviv , 1993). Here
we only state that the basic idea is to propagate properly the *flow invariants*
through the boundary.

4. Results

The dynamic time scale, namely, the time scale τ_{dyn} for the propagation
of acoustic perturbations is $\tau_{dyn} = (\rho_* G)^{-1/2}$, where $\rho_* = M/(4\pi R_*^3/3)$ is
the mean stellar density. In our case $\tau_{dyn} = 5.6$ sec. Velocities should be
compared with the escape velocity (for non rotating star) which is given by

$$v_{escape} = \left(\frac{2GM}{R_*}\right)^{1/2} = \left(\frac{8\pi}{3}\right)\frac{R_*}{\tau_{dyn}}.$$

For the parameters of our problem $v_{escape} = 14.2 \times 10^3 km/sec.$

The calculation is carried out in the rotating system and J_ϕ is therefore,
the angular momentum in the *rotating system*.

The calculations were carried for the following Reynolds numbers: $Re =$
10^3 and 10^5. The models runs had the following angular velocities: $\Omega =$
$0, 0.5 \& 0.8 \Omega_{Kepler}$. In all models reported here we assumed $A_r = 20\ A_\theta = 30$
at $t = 0$. The value of α_T varied from 10^{-3} to 10^{-1}.

The basic results are shown in the video movie where we show the evolution of the model $\Omega = 0.8\Omega_{Kepler}$ and $Re = 10^3$. We show here only 8
snapshots from the time evolution in figure 1 for a non rotating star and
in figure 2 for a star rotating at $\Omega = 0.8\Omega_{Kepler}$. The figures give the amplitude of the perturbation and not the absolute value, thus continous line
imply positive values and broken lines imply negative values. All times are
given in dynamic time scales. Figure 1a shows the initial state when the
perturbation in the temperature appears. Subsequent times show the evolution of the temperature, density and momentum. At t=0.374 the density
wave reached the surface and the tempretaure perturbaton seems to vanish. However, this is only a transient in the wave as can be seen in figure
1u and v where the temperature wave reached the poles. In figure 2 we
see similar phases but now with angular momentum. As the wave starts
to propagate outward the angular momentum is negative relative to the
unperturbed state. However, at a later phase we see positive and negative
parts in the angular momentum. The lower density region created by the
outward moving density wave forces extra angular momentum flow from
the poles towards the equator. The crucial point from our point of view is

that the perturbation which was restricted to the equatorial belt region has propagated via pressure waves all over the surface within hardly 3 dynamic timescales. While the results shown are for a non rotating WD the results for a rotating star are essentially unchanged.

Conclusions

A local perturbation in the envelope of a non rotating and rotating star develops in two fundamental phases:

Phase A: fast local radial expansion with very little azimuthal expansion. The duration of this phase is about one dynamic time scale.

Phase B: Pressure waves set by the local perturbation expand along the meridian and carry the perturbation to the poles. Due to the present symmetry of the problem the wave that propagates to the pole is reflected back to the equator and vice versa. In a 3D treatment the wave will propagate over the entire star leading to fast pressure and temperature equilibration.

The variations in Ω_0, α_T, the Reynolds number (provided it is sufficiently high), or the initial temperature of the isothermal atmosphere do not change the above picture. Also, the extent of the initial perturbation does not affect the overall fundamental behavior, namely the outcome of a local perturbation in T and ρ is a pressure wave that carries the perturbation along the meridian.

The present results *suggest* that any LTNR induced by a local perturbation in the pressure develops quickly into a global thermonuclear runaway. The process of communication is a series of pressure waves.

References

Godon, P. 1993, Investigations of the Boundary Layer Problem by the Spectral Method, Ph.D. thesis, Submitted to the Graduate School, Technion, Israel Institute of Technology

Godon, P., and Shaviv, G. 1993, in press Comp.Meth.Appl.Mech.Eng.
Dubrule, B. and Valdettaro, L. 1992, *Astron. Astrophys.*, **263**,387.
Orio, M. and Shaviv, G. 1992,Astrophys. & Space Sci., 202,273. 1989
Shara, M. M. Ap.J. 261,649,1982.

Figure Caption

Figure 1. Snapshots from the evolution of a particular model. Each circle is divided into 4 quadrants, the upper right gives iso-temperature lines, the upper left shows the iso-density lines, the bottom right gives the linear momentum and the bottom left gives the angular momentum (zero in this example). The time, measured in dynamic times, is given below each figure. momentum

Figure 2. As figure 1 but for a rotating star.

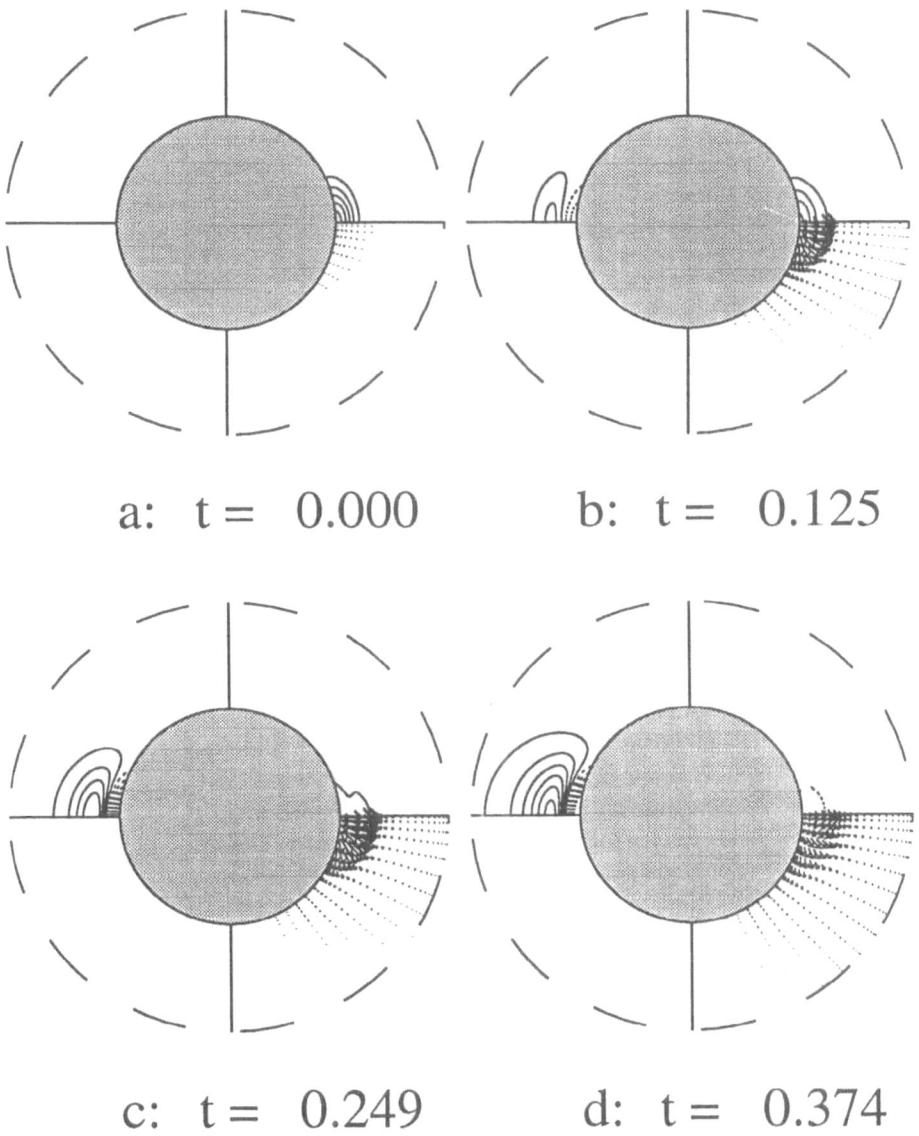

a: t = 0.000   b: t = 0.125

c: t = 0.249   d: t = 0.374

Fig.1   Model A

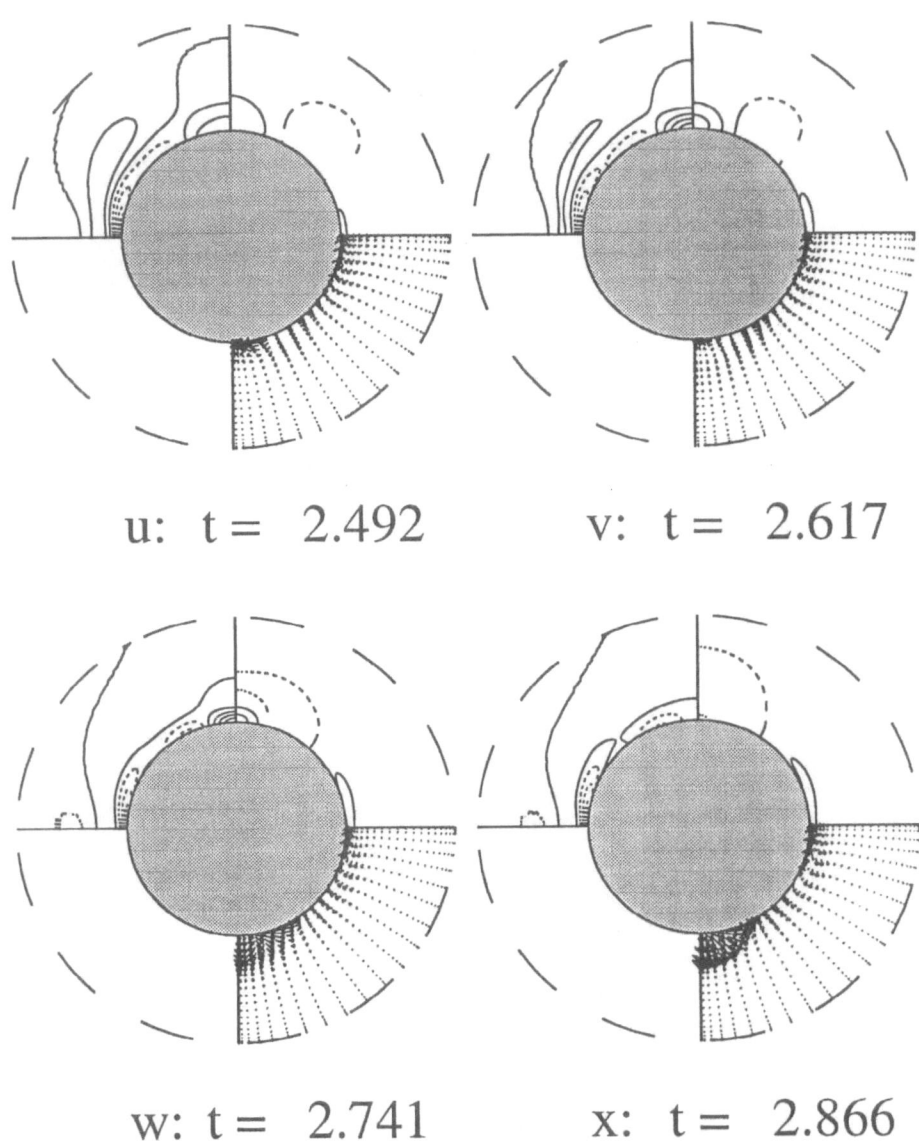

u: t = 2.492   v: t = 2.617

w: t = 2.741   x: t = 2.866

Fig.1   Model A

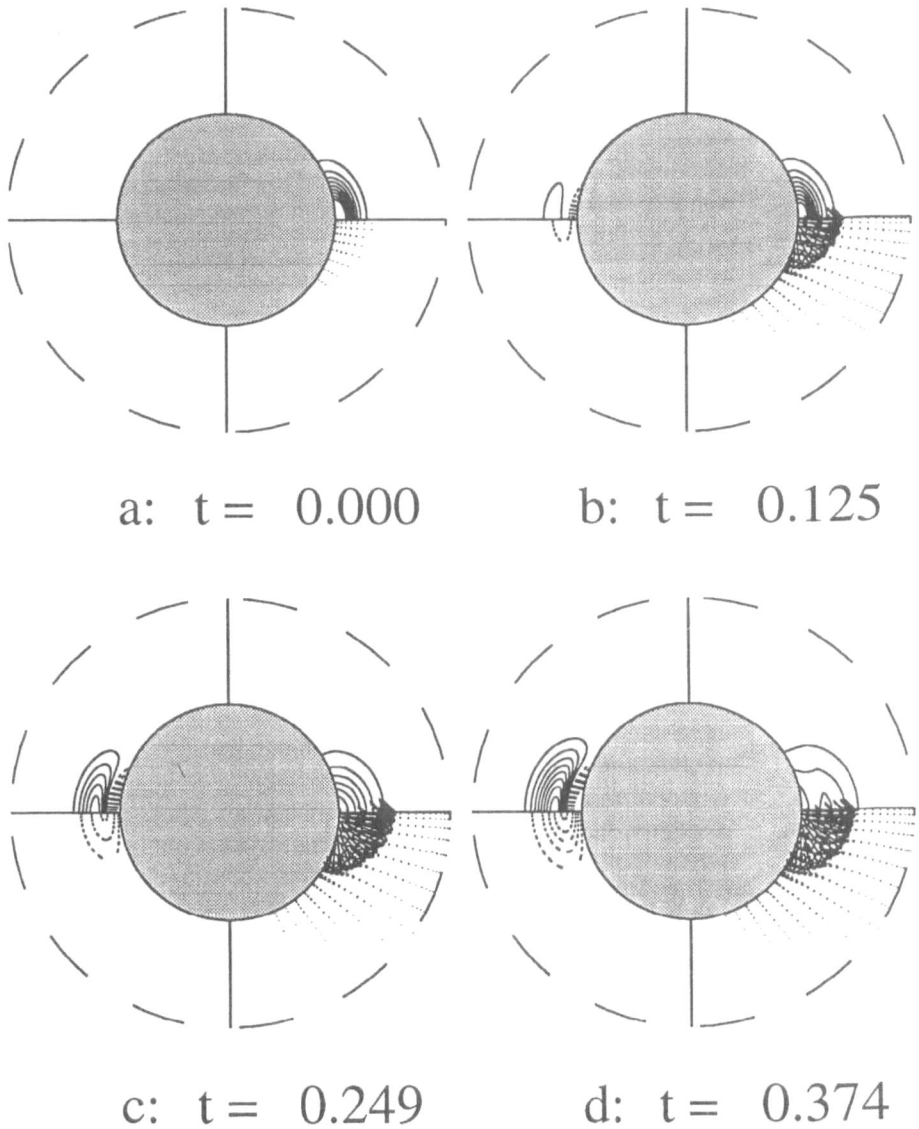

a: t = 0.000   b: t = 0.125

c: t = 0.249   d: t = 0.374

Fig.2   Model C

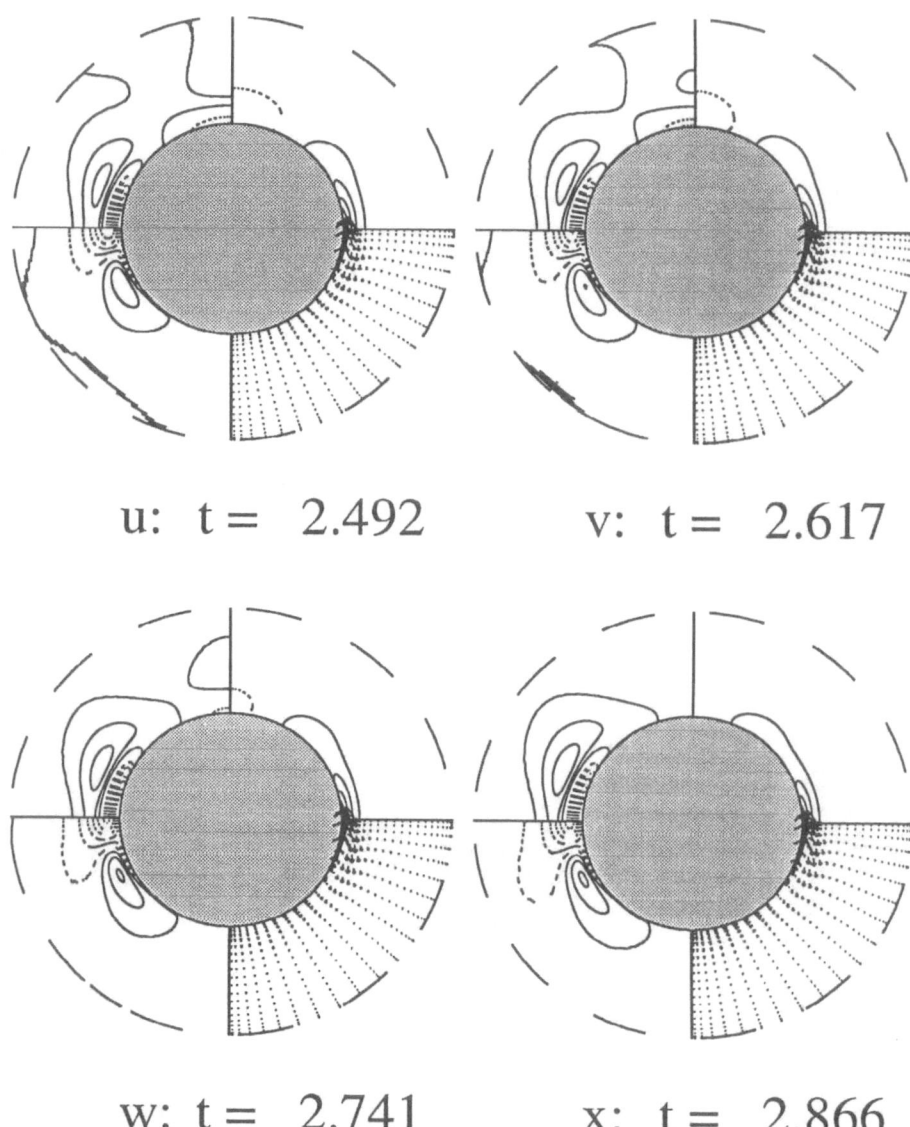

u: t =   2.492   v: t =   2.617

w: t =   2.741   x: t =   2.866

Fig.2   Model C

LIGHT CURVE ANALYSIS OF NOVAE

MARIKO KATO
Dept of Astronomy, Keio University
Hiyoshi, Kouhoku-ku, Yokohama, 223 Japan

ABSTRACT. Light curve fittings for the decay phase of 7 novae are shown. Theoretical light curves are obtained using the optically thick wind theory with the new opacity. Good quality light curves enable us to determine the white dwarf mass, chemical composition and the distance to the star. Multiwavelength observation is very helpful in obtaining these parameters. The theoretical light curve for the helium nova is also given.

1. Introduction

Novae have been widely accepted to be a thermonuclear runaway event on a white dwarf. During the outburst, an envelope on a white dwarf expands to a large size and a significant part of the envelope will be ejected from the system. Such extended stage of the outburst is followed by the optically thick wind theory that is so far the only method in reproducing nova light curves.

Optically thick wind is a continuum-radiation driven wind in which the acceleration occurs deep inside to the photosphere. The structure of the envelope is obtained by solving the equations of motion, continuity, radiative transfer in diffusion approximation, and energy conservation. The steady-state and the spherical symmetry is assumed. The nova decay phase is followed by a sequence of the steady wind solutions. The equations and the numerical method are summarized in detail by Kato & Hachisu (1994).

2. Light curve fitting

The first example of the light curve fitting is Nova Cygni 1978. Figure 1 shows the theoretical light curves in optical and UV regions as well as the observed data. Massive white dwarfs show a rapid evolution because of the small envelope mass. Both of the optical and UV data show a good agreement with the model of the white dwarf mass 1.0 $M_\odot$. The distance to the star is also estimated from the comparison between observed apparent magnitude and the theoretical absolute magnitude to be 2.9 kpc (UV) and 3 kpc (optical). The expansion velocity at the photosphere is also consistent with the IUE data. Details are published by Kato (1994) and Kato & Hachisu (1994).

A. Bianchini et al. (eds.), Cataclysmic Variables, 243-248.

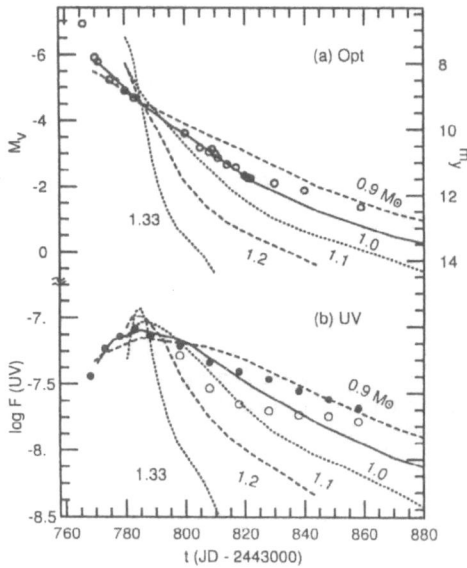

Figure 1. Light curve fitting for classical nova, nova Cygni 1978. Theoretical curves are denoted by thick curves. White dwarf mass is attached to each curve. (a) Optical light curves. The left ordinate shows the absolute visual magnitude for the theoretical curves and the right ordinate the apparent y-magnitude for observed data. Data are taken from Gallagher et al. (1980). (b)UV light curves. Open circles denote the UV flux (1140 Å-3290 Å) and filled circles do the summation of the UV and IR fluxes (> 12000 Å) in the unit of $erg/cm^2 sec^{-1}$ (Stickland et al. 1981). In theoretical UV flux, $F=L_{UV}/4 \pi D^2$, the distance to the star D is assumed to be 2.88 kpc.

Another example of classical nova fitting is nova Muscae 1983. The observational data summarized by Hassall et al. (1990) are denoted by dots which are well fitted by models of white dwarf mass of 0.5 or 0.6 $M_\odot$.

In this way, the light curve fitting gives a new estimate of the white dwarf mass and the distance to the star. Multiwavelength observation is very useful in getting accurate fitting. Especially simultaneous observation of visual and UV region is useful because different decline rates of these two curves limit model parameters into a narrow region.

The light curve fitting of slow nova, RR Pic is shown in figure 3. The observational data shows a good agreement with the curves of models with mass 0.8 or 0.9 $M_\odot$.

Recurrent novae show a very rapid decline rate. Such a rapid evolution can be obtained only in massive white dwarfs. Figure 4 shows theoretical light curves with different set of composition and the white dwarf mass as well as observational data of U Sco and T CrB. These figures give a estimate of the white dwarf mass to be about 1.38 $M_\odot$ and the abundance of heavy element to be Z=0.01-0.02.

The distance determination, however, has a large ambiguity in these objects. If the later part of the light curve is due to the contribution from the accretion disk or the irradiated-secondary, we must shift the observed data upward in these figures.

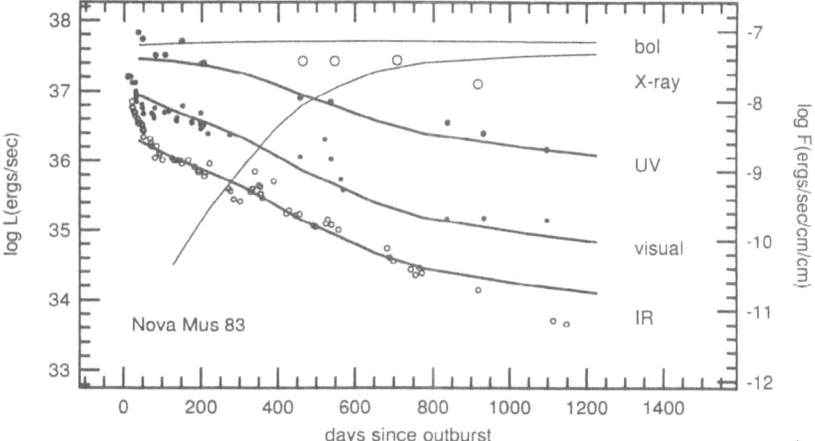

Figure 2. Light curve fitting for classical nova, nova Muscae 1983. The observational data are taken from Hassall et al. (1990). Theoretical curves are for the model of 0.5 $M_\odot$ with chemical composition of X=0.35, Y=0.33, C=0.1,and O=0.2. In obtaining the theoretical magnitude, a rectangular response function is assumed with wavelength region of < 400 Å for X-ray, 1200 - 3300 Å for UV, and 6000-14000 Å for IR region.

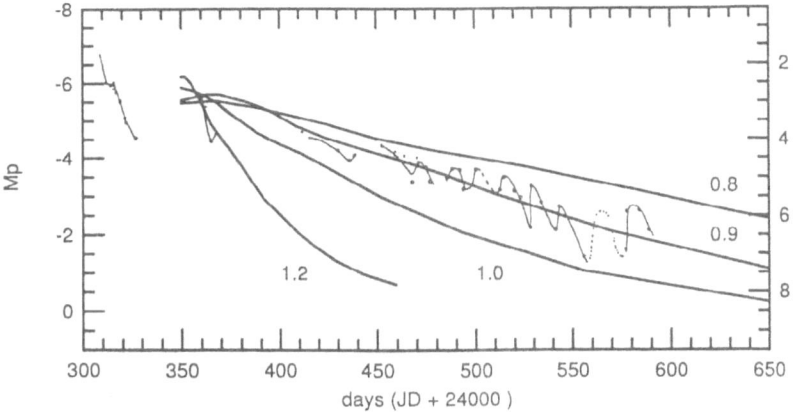

Figure 3. Light curve fitting for slow nova, RR Pic. The data of continuum are taken from Payne-Gaposchkin (1953). The chemical composition of the nova envelope is assumed to be X=0.7, Y=0.28, and Z=0.02.

This correction gives a large distance while the white dwarf mass is unchanged. If UV flux data are available as in figures 1 and 2, we can get much accurate distance determination.

Very similar light curves to U Sco are also shown by V394 CrA and V745 Sco as in figure 5. The light curve fitting gives very massive white dwarfs in these objects, close to the Chandrasekhar limit. The light curve of V745 Sco fits well to both of

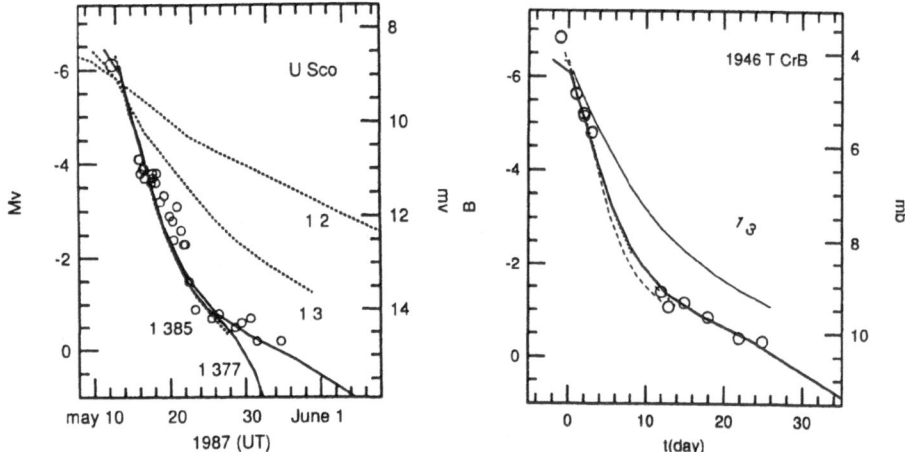

Figure 4. Light Curve fitting for recurrent nova U Sco (left) and T CrB (right). The white dwarf mass is attached to each curve. (left) The chemical composition is assumed to be (X,Z)=(0.5, 0.02) for 1.2 and 1.3 $M_\odot$, (0.7, 0.02) and (0.1,0.02) for 1.377 $M_\odot$ (upper and lower thick curve, respectively), and (0.1,0.01) for 1.385 $M_\odot$. Data are taken from Sekiguchi et al. (1988). The large open circle denotes the peak position estimated from previous outburst data. (right) Observational B-band data are taken from Gordon & Kron (1979). Theoretical curves are for 1.377 $M_\odot$ white dwarf except one denoted by 1.3 $M_\odot$. The chemical composition is assumed to be (X,Z)=(0.5, 0.004) for thin solid, (0.7, 0.02) for thick solid, (0 1,0.02) for dotted, and (0.1, 0.03) for dashed curves.

the models of 1.35 $M_\odot$ with Z=0.02, and 1.377$M_\odot$ with Z=0.01. This suggest that the recurrent novae could belong to the old population. This is consistent with another observational aspect from the space distribution in the Galaxy (Duerbeck 1987).

The success of reproducing theoretical light curves of novae shows that the primary acceleration mechanism is optically thick wind, i.e., the continuum-driven acceleration, and the other mechanism such as drug luminosity in common envelope phase, or line driven wind are the secondary. Detailed discussion can be found in Kato & Hachisu (1994).

3. He nova

When the mass accreting white dwarf has a helium layer of which mass increases after each hydrogen shell flash, unstable He shell flash ultimately occurs that develops a nova-like object. Theoretical light curves of helium novae are shown in figure 6. It will be observed as an extremely hydrogen poor object with very slow development in light curve. Observational identification of such object is important as an evidence of mass-increasing white dwarfs, that suggest a close relation between recurrent novae and Type Ia supernovae.

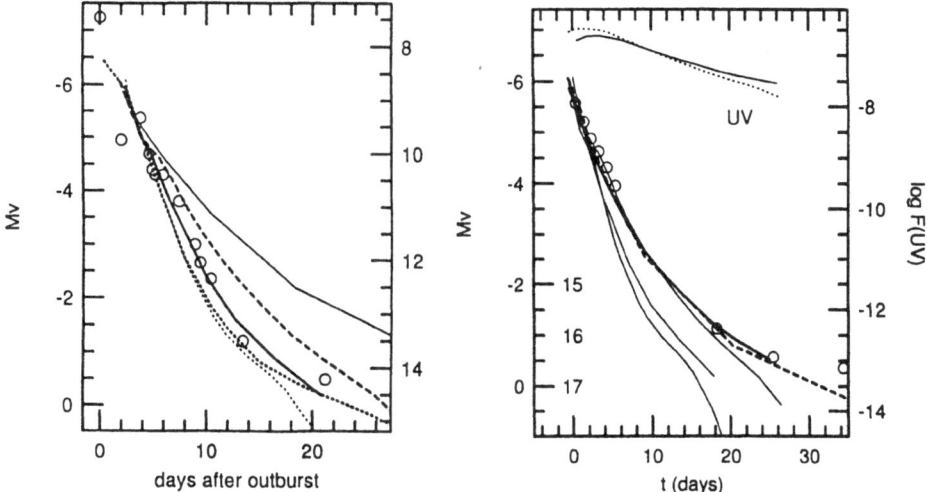

Figure 5. (left) Light Curve analysis of recurrent nova V394 CrA. Theoretical curves are for models of 1.377 $M_\odot$ white dwarf with different chemical composition; From upper to lower, $(X,Z)=(0.5,0.002)$, $(0.1,0.004)$, $(0.1,0.01)$, $(0.7,0.02)$, and $(0.1,0.02)$ are assumed. The observational data by Duerbeck (1988) are also plotted. (right) Light Curve Analysis for Recurrent Nova, V745 Sco. The thick curves are for 1.377 $M_\odot$ model with $(X, Z)=(0.5, 0.004)$ (solid curve) and for 1.35 $M_\odot$ with $(0.5,0.02)$ (dashed curve). Another thin solid curves denote models of 1.377 $M_\odot$ with $(0.1\ 0.004)$, $(0.1,0.01)$, and $(0.1, 0.02)$. Upper two UV light curves are for 1.377 $M_\odot$ with $(0.1,0.004)$ and $(0.5,0.004)$ shown by dotted and solid curves, respectively. The observational data by Sekiguchi et al. (1990) are plotted by open circles with magnitude shown in the inner ordinate.

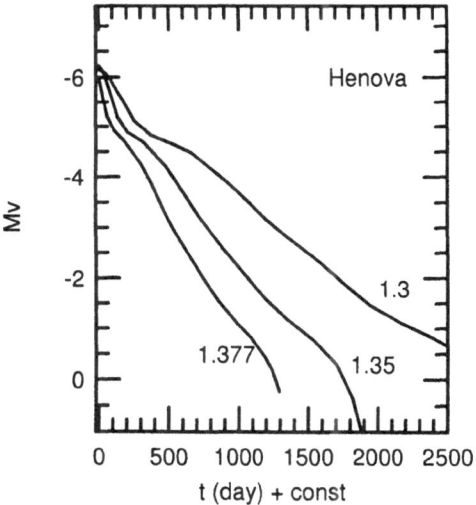

Figure 6. Theoretical light curve of He nova.

References

Duerbeck, H., 1987, *RS Ophiuchi(1985) and the Recurrent Nova Phenomenon*, Bode, M.F. Ed. VNU Science Press, Utrecht, p.99

Duerbeck, H., 1988, A&A, 197,148

Gallagher, J.S., Kaler,J.B., Olson,E.C., Hartkopf,W.I. and Hunter,D.A. 1980. PASP, 92,46

Gordon,K.C. & Kron,G.E., 1979 in IAU Comission 27, Information Bulletin of Variable Stars No 1610

Hassall, B.J.M., Snijders,M.A.J., Harris, A.W., Cassatella, A., Dennefeld, M.,Friedjung, M., Bode,M., Whittet,D., Whitelock,P., Menzies,J., Lloyd Evans,T. and Bath, G.T., 1990, *Physics of Classical Novae*, Springer-Verlag, Berlin.

Kato, M. 1994, A&A, 281, L49

Kato, M. & Hachisu, I. 1994 ApJ, in press (Dec 20)

Payne-Gaposchkin,C., 1957, *The Galactic Novae*, North-Holland, Amsterdam.

Rogers, F. J., & Iglesias, C. A., 1992, ApJS, 79, 507

Sekiguchi, K., Feast,M.W.,Whitelock, P.A., Overbeek, M.D., Wargau,W. and Spencer Jones,J., 1988, MNRAS, 234,281

Sekiguchi, K., Whitelock, P.A., Feast,M.W., Barrett, P.E., Caldwell, J.A.R., Carter,B.S., Catchpole, R.M., Laing, J.D., Laney, C.D. Marang, F. and van Wyk.F, 1990, MNRAS, 246,78

Stickland, D. J., Penn, C. J., Seaton, M. J., Snijders, M. A. J. & Storey, P. J. 1981, MNRAS, 197, 107

THERMONUCLEAR RUNAWAYS EVENTS
IN SYMBIOTIC BINARIES

JOANNA MIKOŁAJEWSKA

Copernicus Astronomical Center
Bartycka 18, PL-00716 Warsaw, Poland

Abstract.

Most symbiotic binaries appear to contain low mass white dwarfs accreting hydrogen rich material from their M giant companions. Depending on the mass accretion rate, the symbiotic white dwarfs may undergo hydrogen shell flashes or burn the hydrogen immediately as it is accreted in a quasi stable shell.

1. Introduction

Accretion of matter onto a white dwarf leads to three possible configurations, delimited by two critical accretion rates (*e.g.* Paczński & Rudak, 1980; Iben, 1982; Kenyon & Truran, 1983). For a small range of $\dot{M}$, hydrogen burns completely as it is accreted. High accretion rates, $\dot{M} \gtrsim \dot{M}_{max}$, lead to accumulation of incoming material above the burning shell and formation of a red giant, while at low rates, $\dot{M} \lesssim \dot{M}_{min}$, hydrogen shell flashes similar to those responsible for nova outbursts occur. A white dwarf stably burning incoming hydrogen should be hot and luminous, with $L_h \sim 1000 - 50,000 L_\odot$, and $T_h \sim 1 - 2 \times 10^5$ K. The shell flash is very strong, and the white dwarf evolves into an A–F supergiant – as classical novae do – when $\dot{M}$ is very low, and the accreted envelope is degenerated. For higher $\dot{M}$'s the envelope is non-degenerated, and the flash is milder; the white dwarf remains at high effective temperatures.

249

A. Bianchini et al. (eds.), Cataclysmic Variables, 249-253.

2. Symbiotic stars

The hot components in many symbiotic systems show intrinsic variability. Based on this activity we distinguish between ordinary or *classical* symbiotic stars which show occasionally 1–3 mag optical eruptions with time scales of a few years, and *symbiotic novae* that have undergone a single outburst of several magnitudes lasting for dozens of years.

The outburst behavior of seven well covered symbiotic novae has been recently discussed by Mürset & Nussbaumer (1994). Four of them – AG Peg, RT Ser, RR Tel and PU Vul – developed A–F supergiant spectra at visual maximum, and are likely candidates for the degenerate hydrogen shell flashes. In the rest – V1016 Cyg, V1329 Cyg, and HM Sge – the outburst led strait to a nebular spectrum, where the hot component remained at high effective temperature, $T_h \approx 40,000$ K. They probably underwent non-degenerated shell flashes.

In all symbiotic novae the outburst develops very slowly; the rise to maximum brightness takes months, and the decline to pre–outburst stage lasts dozens of years. AG Peg is the record–holder: its eruption started in 1850's, and the eruptive component maintained a constant bolometric luminosity, $L_h \sim 3000 L_\odot$ from 1870 to 1985 (Kenyon *et al.*, 1993b). The relatively low – as compared to the classical novae – bolometric luminosity is reasonable for the hot component's mass, $M_h \sim 0.6 \pm 0.1 M_\odot$ implied by spectroscopic orbit, and the hydrogen flashes on accreting low mass white dwarfs calculated by Shara *et al.* (1993) mimic pretty well the evolution of AG Peg on the H–R diagram (Fig. 12 of Kenyon *et al.*, 1993b). The constant luminosity phase in AG Peg was also characterized by mass loss. Kenyon *et al.* estimate the hot component in AG Peg has burnt $\sim 5 \times 10^{-6} M_\odot$, while $\sim 1 - 2 \times 10^{-4} M_\odot$ has been ejected throughout the eruption, and the sum of these two masses compares well with the amount needed to ignite a symbiotic nova eruption (Iben, 1982; Shara *et al.*, 1993).

Recent studies of ~ 20 nearby ordinary symbiotics observed with the *IUE* suggest their hot components cluster around temperatures $T_h \sim 10^5$ K and luminosities $L_h \sim 1000$ L$_\odot$ (Mürset *et al.*, 1991). Similar temperatures and luminosities have been estimated by Morgan (1992) for 10 symbiotics in the Magellanic Clouds. Since most other symbiotics have optical spectra closely resembling those analyzed by Mürset *et al.* and Morgan, luminosities of $\sim 1000 L_\odot$ are probably representative for the entire population. It is hard to believe that these relatively high luminosities are powered solely by accretion. Symbiotic red giants have mass loss rates of $\sim 10^{-7} M_\odot$ yr^{-1}. This represents the maximum accretion rate, and it can be achieved only if the red giant fills its tidal lobe. However, most symbiotic stars interact by wind-driven mass loss instead of Roche lobe overflow, so the expected accretion rate is of order of $10^{-8}M_\odot$ yr^{-1} or less, and the corresponding accretion luminosity is of order of 10–100 L$_\odot$.

The situation changes radically if the symbiotic white dwarfs burn hydrogen–rich material as they accrete it. The minimum accretion rate at which it happens is (Iben, 1982):

$$\dot{M}_{min} \approx 1.32 \times 10^{-7} M_{wd}^{3.57} M_\odot \text{ yr}^{-1},$$

while the maximum steady burning rate is set by the core mass–luminosity relation (e.g. Paczyński, 1970). Although very few symbiotics have good orbital solutions, the radial velocity curves of nearby well studied systems (e.g. Kenyon 1994) suggest $M_h \sim 0.4 - 0.6M_\odot$, which implies $\dot{M}_{min} \lesssim 2 \times 10^{-8}M_\odot$ yr^{-1}. Iben also estimated that the mass accretion rate and the hydrogen–burning luminosity are then related by

$$\dot{M} \approx 1.3 \times 10^{-11}(L/L_\odot)M_\odot \text{ yr}^{-1}.$$

Thus an accretion rate of $\sim 10^{-8}M_\odot$ yr^{-1} is sufficient to power the typical hot component in symbiotic binary, and we can expect that many systems are actually in – or close to – the steady burning configuration.

Unfortunately, it is hard to find a simple interpretation within the steady burning model for the occasional 1–3 mag optical eruptions with time scales from months to a few years, observed in some of these symbiotics.

In BF Cyg and Z And for instance, the hot components appear to maintain roughly constant luminosities even if their effective temperatures changes by a factor of 10 on time scales of years (Mikołajewska & Kenyon, 1992). The outburst thus appears to occur when the hot component expands in radius at roughly constant luminosity. The short, of order of ~ 1 yr, time scales for both systems are more typical of thermal time scales in white dwarf envelops than of nuclear time scales. Thus, in principle, they can be caused by small variations in $\dot{M}$ about the steady burning rate. However the model calculations (e.g. Iben, 1982) suggest such short thermal time scales occur only for massive white dwarfs, with $M_h \gtrsim 1M_\odot$, while both Z And and BF Cyg appear to contain low mass, $M_h \sim 0.6M_\odot$, white dwarfs (Mikołajewska & Kenyon, 1992).

The outburst behavior of AG Dra is even more confusing (Mikołajewska et al., 1995); the hot component's brightness increases by a factor of ~ 10, on time scale of months, while its effective temperature increases during the early stages of each recent outburst and then decreased as the UV and the optical brightness continued to increase. The temperature never changes so much as in Z And and BF Cyg; in particular, it never falls below $T_h \sim 80,000 \div 90,000$ K. The bolometric luminosity of the hot component may have remained constant as T_h decreased; however it clearly increased during the early rise to optical and UV maximum.

Finally, there are many symbiotic systems in which the hot components maintain roughly constant both luminosity and effective temperature – $L_h \sim 1000L_\odot$, $T_h \sim 10^5$ K – over many decades. These non–eruptive symbiotics – V443 Her, RW Hya and SY Mus – may be either in the steady state configuration or may undergo a very slow hydrogen shell flash on low mass white dwarfs. Actually, model calculations by Shara et al. (1993) demonstrated the low mass – $M_h \sim 0.4 - 0.6 M_\odot$ – white dwarfs accreting hydrogen rich gas at a rate of $\sim 10^{-9} M_\odot\, yr^{-1}$ undergo a very slow nova eruption, and the high–luminosity plateau phase can last for ~ 200 yr and more. Simultaneously, the spectroscopic orbit solutions for V443 Her and RW Hya indicate a presence of low mass white dwarfs in both systems (Dobrzycka et al., 1993; Kenyon & Mikołajewska, 1995).

An interesting feature of the systems with luminous white dwarfs is the apparent correlation of hot component luminosities and orbital periods (Mikołajewska & Kenyon, 1992). Again, we can understand it only if the white dwarfs are powered by the thermonuclear runaways. The luminosity–orbital period correlation then implies a relation between white dwarf mass and orbital period if the hot components lie on the plateau portion of white dwarf cooling curve and follow some core mass–luminosity relation. In fact, a correlation between white dwarf mass and orbital period is predicted by some evolutionary scenarios for short–period ($P_{orb} \lesssim 2 - 3$ yr) symbiotic binaries (Kenyon, 1994; and references therein).

3. Concluding remarks

In most symbiotic systems, the hot companion of the M giant is a luminous white dwarf – with $L_h \gtrsim 1000\ L_\odot$ – which seems to be powered by thermonuclear runaways. Almost all symbiotic stars are detached binaries, so mass transfer occurs via a wind instead of tidal outflow. The large orbital periods, $P \gtrsim 2$–3 yr, also guarantee that even when the white dwarf photosphere has expanded in eruption, it will not engulf its M giant companion. The evolution of the thermonuclear runaways thus proceeds, in a first approach, as in the case of an isolated star, and possible effects attributable to its binary nature do not significantly affect the picture. This fact may account for a very slow decline of symbiotic novae, with a time scale more comparable to the nuclear time scale, as opposed to the classical novae. The presence of a red giant primary in symbiotics on the other hand results in generally higher – with respect to classical novae – mass accretion rates. As a result, we can expect that many symbiotic binaries have evolved into a steady burning configuration, where the white dwarf maintains a high luminosity by immediate burning of accreted matter. The symbiotic stars thus provide a very attractive laboratory in which to study thermonuclear runaways in different physical conditions, from degenerate and nondegenerate flashes to steady burning.

Results of the thermonuclear runaways model calculation reproduce at least qualitatively the observed behavior or symbiotic novae. The situation is more complicated with the classical symbiotics. The available observations confirm the main prediction of this model – a long–duration phase of constant luminosity – but some low mass systems appear to evolve on faster time scales than predicted by published calculations.

All symbiotic novae lose most of their accreted mass during an eruption. Moreover, the white dwarfs that burn accreted mass in a quasi steady state also appear to lose mass. On the theoretical side, almost entire envelope is ejected in calculated nova eruptions on low mass white dwarfs (Shara *et al.*, 1993). It is thus very unlikely that the symbiotic white dwarfs can produce SN Ia's by exceeding the Chandrasekhar limit (Kenyon *et al.*, 1993a).

Finally, the steady burning scenario has been recently proposed also for the supersoft X–ray sources (*e.g.* Kahabka 1995, this volume). Both AG Peg and AG Dra have very soft X–ray spectra (Kenyon *et al.*, 1993b; Mikołajewska *et al.*, 1995), while two supersoft X–ray sources, RR Tel and SMC 3 are symbiotic stars. It seems, that these objects are close relatives of symbiotic binaries with hot and luminous white dwarf companions, although their generally higher luminosities indicate higher white dwarf masses and consequently higher accretion rates in the supersoft X–ray sources.

This study was supported by KBN Research Grant No. 2 P304 007.

References

Dobrzycka, D., Kenyon, S.J., Mikołajewska, J.: 1993, *AJ*, **106**, 284
Iben, I.: 1982, *ApJ*, **259**, 244
Kenyon, S.J: 1994, *Mem. S.A.It.*, **65**, 135
Kenyon, S.J., Livio, M., Mikołajewska, J., Tout, C.A.: 1993a, *ApJL*, **407**, L81
Kenyon, S.J., Mikołajewska, J.: 1995, *AJ*, submitted
Kenyon, S.J., Mikołajewska, J., Mikołajewski, M., Polidan, R.S., Slovak, M.H.: 1993b, *AJ*, **106**, 1573
Kenyon, S.J., Truran, J.W.: 1983, *ApJ*, **273**, 280
Mikołajewska, J., Kenyon, S.J.: 1992, *MNRAS*, **256**, 177
Mikołajewska, J., Kenyon, S.J., Mikołajewski, M.: 1989, *AJ*, **98**, 1427
Mikołajewska, J., Kenyon, S.J., Mikołajewski, M., Garcia, M.R., Polidan, R.S.: 1995, *AJ*, in press
Morgan, D.H.: 1992, *MNRAS*, **258**, 639
Mürset, U., Nussbaumer, H.: 1994, *A&A*, **282**, 586
Mürset, U., Nussbaumer, H., Schmid, H.M., Vogel, M.: 1991, *A&A*, **248**, 458
Paczyński, B.: 1970, *Acta Astr.*, **21**, 417
Paczyński, B., Rudak, B.: 1980, *A&A*, **82**, 349
Shara, M.M., Prialnik, D., Kovetz, A.: 1993, *ApJ*, **406**, 220

4. SPECTRAL EVOLUTION OF NOVAE AND PROBLEMS IN NEBULAR PHYSICS

THE PHYSICS OF EARLY NOVA SPECTRA

PETER H. HAUSCHILDT AND SUMNER STARRFIELD

Dept. of Physics and Astronomy, Arizona State University, Box 871504, Tempe, AZ 85287-1504
E-Mail: yeti@sara.la.asu.edu and starrfie@hydro.la.asu.edu

STEVEN N. SHORE

Dept. of Physics and Astronomy, Indiana University South Bend, 1700 Mishawaka Ave, South Bend, IN 46634-7111
E-Mail: sshore@paladin.iusb.indiana.edu

FRANCE ALLARD

Dept. of Geophysics and Astronomy, University of British Columbia, Vancouver, B.C., V6T 1Z4 Canada
E-Mail: allard@astro.ubc.ca

AND

EDWARD BARON

Dept. of Physics and Astronomy, University of Oklahoma, 440 W. Brooks, Rm 131, Norman, OK 73019-0225
E-Mail: baron@phyast.nhn.uoknor.edu

Abstract. We discuss the physical effects that are important for the formation of the early spectra of novae. Nova atmospheres are optically thick, fast expanding shells with flat density profiles, leading to geometrically very extended atmospheres. We show that the properties of early nova spectra can be understood in terms of this basic model and discuss some important effects that influence the structure and the emitted spectrum of nova atmospheres, e.g., line blanketing, NLTE effects, and the velocity field. The proper modeling of nova atmospheres is discussed and we give some computational details.

A. Bianchini et al. (eds.), Cataclysmic Variables, 257-265.

1. Introduction

The modeling and analysis of early nova spectra has made significant progress during the last 2-3 years by the construction of detailed model atmospheres and synthetic spectra for novae by Hauschildt et al.([Hauschildt, *et al.*, *1992; Hauschildt,* et al., 1994a; Hauschildt, *et al., 1994c*]). In the early stages of the nova outburst, the spectrum is formed in an optically *very* thick, in both lines and continua, shell with a flat density profile, leading to very extended continuum and line forming regions (hereafter, CFR and LFR, respectively). Because of the large variation of the physical conditions inside the spectrum forming region, the classical term "photosphere" is not appropriate for novae. The large geometrical extension leads to a very large electron temperature gradient within the CFR and LFR, allowing for the observed simultaneous presence of several ionization stages of many elements. Typically, the relative geometrical extension R_{out}/R_{in} of a nova atmosphere is $\approx 100\ldots1000$, which is much larger than the geometrical extensions of hydrostatic stellar atmospheres (even in giants R_{out}/R_{in} is typically less than 2) or supernovae (SNe). Nova atmospheres are also very different from SN atmospheres with respect to their energy balance. Whereas SNe spectra show constantly decreasing color temperatures and decreasing bolometric luminosities, the color temperatures of nova atmospheres generally increase with time and their bolometric luminosity is constant. This is caused by the presence of a central energy source in novae (the hot white dwarf) which is missing in SNe (where the only sources of energy are the radioactive decays Ni and Co nuclei).

For very low effective temperatures, molecules are present in the nova atmosphere. Therefore, the equation of state (EOS) in nova atmospheres is very complicated and special techniques must be employed so that nova model atmospheres can be constructed. The electron temperatures and gas pressures typically found in nova photospheres lead to the presence of a large numbers of spectral lines, predominantly Fe-group elements, in the LFR and a corresponding influence of line blanketing on the emergent spectrum. The situation is complicated significantly by the velocity field of the expanding shell which leads to an enhancement of the overlapping of the individual lines. This in turns makes simplified approximate treatments of the radiative transfer by, e.g., the Sobolev-approximation, very inaccurate and more sophisticated radiative transfer methods, which treat the overlapping lines and continua simultaneously, must be used in order to obtain reliable models. The line blanketing also leads to strong wavelength redistribution of the radiative energy, therefore, the temperature structure of the shell must be calculated including the effects of the line blanketing.

The situation is further complicated by the fact that the (electron) den-

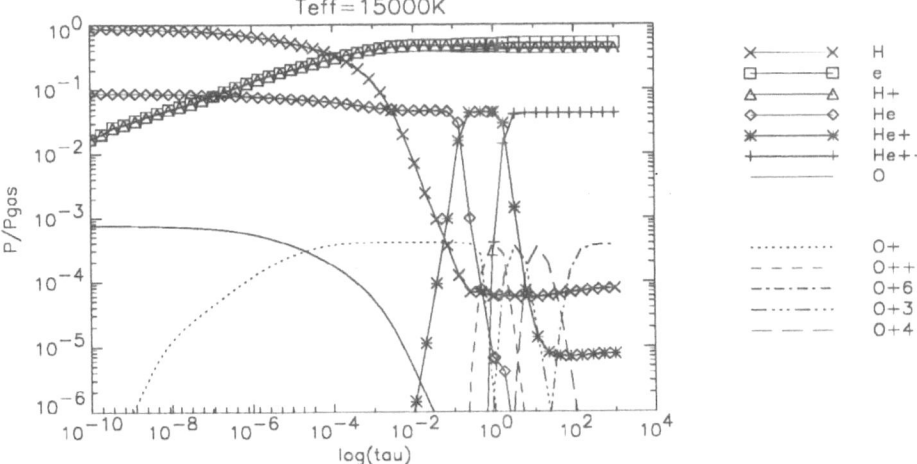

Figure 1. Relative concentration of atoms and ions in nova atmosphere models with $T_{\text{eff}} = 15000$ K, $N = 3$, and solar abundances. The plot gives the partial pressures of the indicated species relative to the gas pressure as functions of standard optical depth τ_{std}.

sities of the CFR and LFR in a nova atmosphere are very low compared to classical stellar atmospheres. This leads to an overwhelming dominance of the radiative rates over the collisional rates. In addition, the radiation field is very non-grey. These two effects lead to large departures from local thermodynamic equilibrium (LTE) in the CFR and LFR. Therefore, NLTE effects must be included self-consistently in the model construction, in particular in the calculation of the temperature structure and the synthetic spectra. As discussed in the previous paragraph, the effects of the line blanketing on the radiative rates requires a careful treatment of the radiative transfer in the NLTE calculations and simple approximations can lead to wrong results.

2. Results

In order to extract detailed quantitative information on velocities, densities, temperatures, and compositions, and constrain theoretical explosion models, it is necessary to analyze the nova spectrum in detail, via synthetic spectral modeling. We use our computer code **PHOENIX** (Version 4.9), to compute our model atmospheres and synthetic spectra for novae. This is an updated version of the code used for the analyses of the early spectra of Nova Cygni 1992 ([Hauschildt, *et al.*, *1994a*]) and SN 1993J ([Baron, *et al.*, *1994*]), a more detailed description can be found in [Hauschildt, *et al.*,

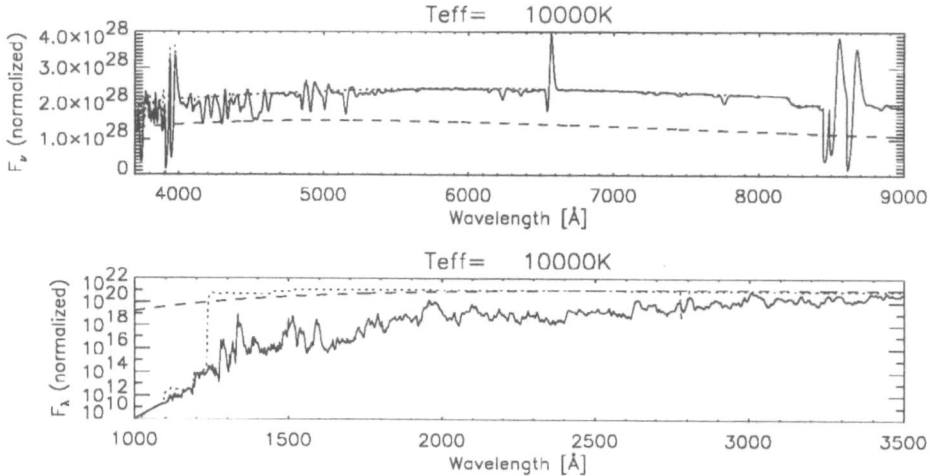

Figure 2. Effect of line blanketing on synthetic nova spectra. The plots show synthetic spectra computed with full line blanketing (full curves), only NLTE lines (dotted curves) and the blackbody energy distribution for a model with $T_{eff} = 10000$ K and solar abundances. Note that the upper part of each plot gives F_ν in the optical on a *linear* scale, whereas the lower part gives F_λ in the UV on a *logarithmic* scale.

1994b].

The low densities and complicated radiation fields in the atmosphere require that the most important species are treated using multi-level NLTE. At present, we can treat H I (15 levels), He I (11 levels), He II (15 levels), Na I (3 levels), Mg II (3 levels), Ca II (5 levels), and Ne I (26 levels) in NLTE. We are currently working on including Fe II in NLTE using a 472 level model atom with ≈ 4500 permitted transitions that are self-consistently included in the radiative transfer and statistical equilibrium calculations and will report the results of these calculations elsewhere.

Our NLTE model atmospheres show that the early evolution of a nova shell, the "optically thick" phase, can be divided into at least 3 very different epochs. The first and very short-lived stage is the "fireball" stage, first detected in the infrared by Gehrz (see [Gehrz, , 1988]) and analyzed in the UV by Hauschildt et al.(1994). In this stage, the density gradient in the nova atmosphere is high, $N \approx 15$ and the effective temperatures are dropping from ≈ 15000 to < 10000 K (lower T_{eff} are probable but have not yet been observed). In this stage, the nova spectrum resembles that of a SN with low velocities ($v_{max} \approx 4000$ km s^{-1} for Nova V1974 Cygni 1992).

As the density and temperatures of the expanding fireball drop, the material becomes optically thin and deeper layers become visible. In this stage, the "optically thick wind phase," the atmosphere evolves to a very

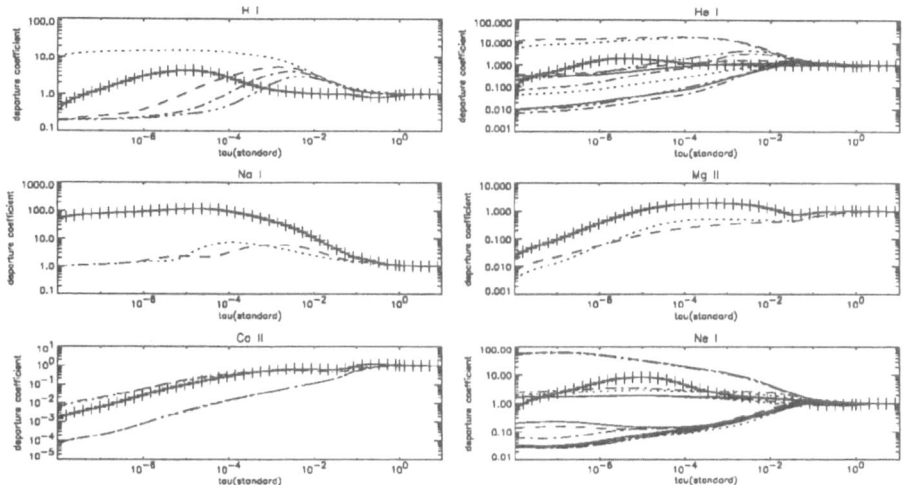

Figure 3   Departure coefficients b_i as functions of standard optical depth τ_{std} for a nova atmosphere model with $T_{eff} = 25000\,K$ In each plot the departure coefficient of the ground state of the species is annotated by the "+" symbols

flat density profile, $N \approx 3$, and the LFR, as well as the CFR, has a very large geometrical extension, values of $\Delta R/R \approx 100$ in the LFR are common. The large geometrical extension causes a very large temperature gradient in the LFR (cf. Fig. 4), typically the electron temperatures range between $1,100\,K$ and $50,000\,K$ for a model with $T_{eff} = 6,500\,K$ and from $3,500\,K$ to $150,000\,K$ for a model with $T_{eff} = 8,000\,K$ (we emphasize that all of the regions can be visible simultaneously in the emitted spectrum). This explains the observed fact that nova spectra can show, simultaneously, lines from different ions of the same element (Fig. 1). The structure of the atmospheres and the calculated spectra are very sensitive to changes of the abundances of iron, carbon, nitrogen, and oxygen, which will make abundance determinations of these elements possible. However, for reliable abundance determinations, NLTE effects need to be included self-consistently for the elements under consideration. Consistent with the results of Pistinner et al.(1994) we also find that the synthetic spectra are insensitive to large changes in the luminosity. However, the spectra are *very* sensitive to changes in the *form* of the velocity profile inside the atmosphere (Fig. 6).

In nova models with low effective temperatures, $T_{eff} \leq 10000\,K$, we find that molecular lines and bands are important. This situation could be realized in the phase between the "fireball" and the "wind" stages. Some novae show CO emission lines due to wide manifold of transitions in their early phases (Evans et al., 1993), such as those reproduced by our models

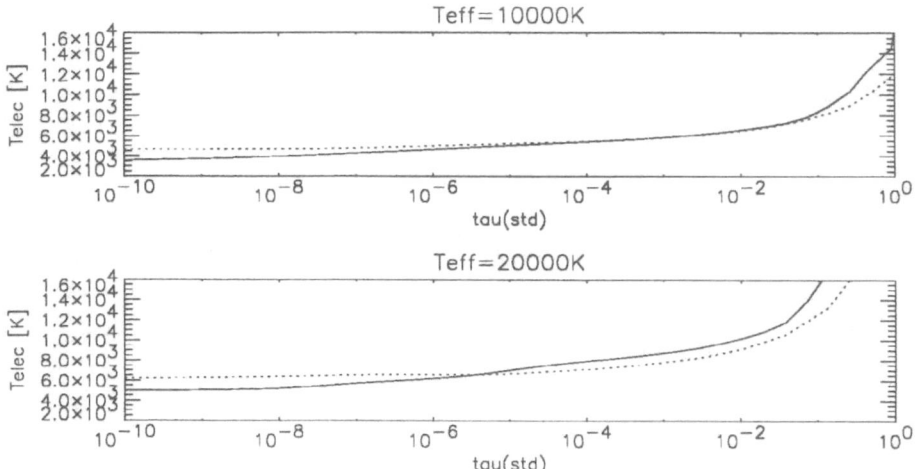

Figure 4   Effects of line blanketing on the temperature structure of nova atmospheres
The plots show the temperature structure computed including full line blanketing (full
curves) or neglecting lines (dotted curves) for models with T_{eff} = 10000 K (top panel)
and T_{eff} = 20000 K (bottom panel) with $N = 3$, and solar abundances

for the early observations of Cas93 ([Hauschildt, *et al.*, *1994c*]. In terms of
the classical scheme of nova spectrum classification ([McLaughlin, , 1960]),
the increase in the effective temperature during the wind stage corresponds
to the transition from to the "premaximum spectrum" to the "diffuse enhanced spectrum."

The UV spectra of novae are totally dominated by the line blanketing of
many thousands of overlapping spectral lines, mostly due to Fe II. The line
blanketing changes the structure of the atmosphere, and the emitted spectrum, so radically that it must be included *self-consistently* in the models
in order to be able to reliably compare the results to observed spectra. In
Fig. 2 we display synthetic spectra for a model with T_{eff} = 10000 K. The
plots also show the spectra obtained by neglecting all but the NLTE lines
(dotted curve) and the blackbody energy distribution for the effective temperature. Line blanketing has an enormous impact on the spectrum emitted
by the model with T_{eff} = 10000 K, in particular in the UV. Practically all
lines in the UV are *absorption* lines, only a few regions of relative transparency in the "line haze" exist, e.g., the broad features at about 1350Å,
1500Å, 1600Å, 2000Å, and 2640Å. These appear as "emission" features, but
are merely holes in the "iron curtain." In the optical spectral range the line
blanketing is more localized in overlapping absorption and P Cygni lines,
but the effect is by far not as strong as in the UV. Note that some fea-

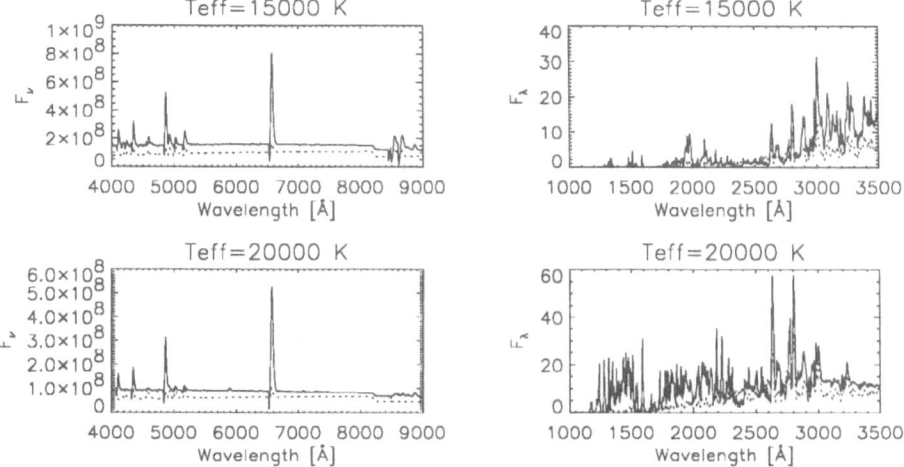

Figure 5   Effect of the temperature structure and NLTE effects on the spectra of nova atmospheres The plots show the synthetic spectra calculated using a fully self-consistent temperature structure including full line blanketing and NLTE effects (full curves) and computed using a temperature structure constructed neglecting line blanketing and NLTE effects (dotted curves) for models with $T_{\text{eff}} = 15000$ K and $T_{\text{eff}} = 20000$ K with $N = 3$, and solar abundances

tures change their character when the effective temperature increases. For example, the 2640Å feature is for $T_{\text{eff}} \approx 15,000$ K formed by a gap between strong Fe II lines, but at $T_{\text{eff}} \approx 20,000$ K it is formed by the overlapping emission parts of Fe II lines just shortward of 2640Å.

The strong coupling between continua and lines, as well as the strong overlap of lines, requires a unified treatment of the radiative transfer to (at least) full first order in v/c (i.e., including advection and aberration terms, our calculations are done using a fully relativistic radiative transfer). Methods that separate continua and lines (e.g., the Sobolev approximation) give unreliable results. The "density" of lines in the UV is so high that practically the complete range is blocked by lines. In fact, most of the observed "emission lines" are merely gaps between "clusters" of lines, thus they are regions of transparency which have *less* opacity than the surrounding wavelength regions. In addition, the radiation fields inside the nova continuum and line forming regions nowhere resemble blackbody or grey distributions. This condition, combined with the low densities, causes very large departures from LTE, cf. Fig 3. The NLTE radiative transfer and rate equations *must* be solved self-consistently including the effects of line blanketing in the UV and optical spectral regions. The effects of neglecting the line blanketing on the departures coefficients are enormous, as shown by Hauschildt

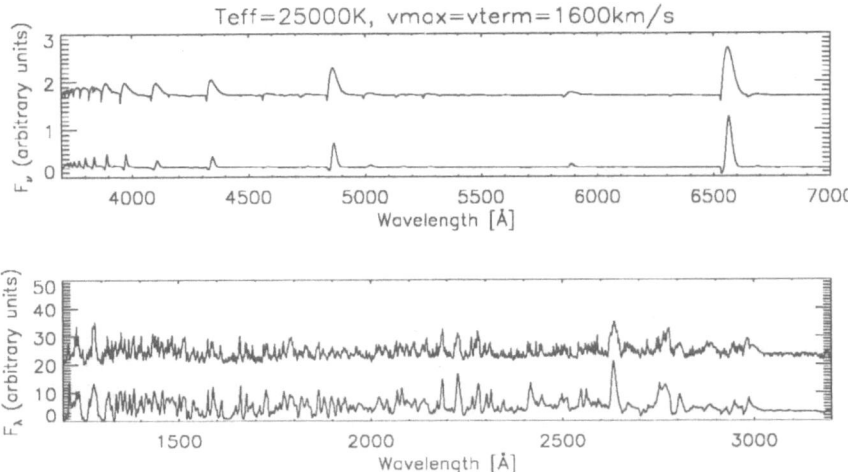

Figure 6. Effect of two different velocity laws on the spectrum for a model atmosphere with $T_{eff} = 25000$ K, and $v_{max} = v_\infty = 1600$ km s^{-1}. In each plot, the top curve is the spectrum calculated using a "wind" velocity field of the general form $v(r) \propto v_\infty (1 - a/r)^b$ whereas the bottom curve represents the results for the ballistic (linear) velocity law. Both models have the common parameter $N = 3$ and twice the solar metal abundances and are self-consistently calculated with their respective velocity fields.

& Ensman (1994) for the case of SN model atmospheres.

The large deviation from a blackbody energy distribution and the extreme non-gray spectrum cause a large effect of the lines on the temperature structure of nova atmospheres. This is illustrated in Fig. 4 in which we compare the temperature structure of two models with $T_{eff} = 10000$ K and 20000 K computed either with line blanketing (full curves) or by neglecting line blanketing (dotted curves). The large effect of line cooling (in the optically thin regions) and back-warming (in the inner regions) is evident. The temperatures changes introduced by line blanketing can amount to more than 2000 K in the CFR and LFR, thus changing the synthetic spectra significantly. This demonstrates that a nova model atmosphere must include the line blanketing self-consistently in order to derive parameters, in particular elemental abundances, otherwise the results are unreliable. We show this in Fig. 5 by comparing the synthetic spectrum calculated using a self-consistent NLTE temperature structure (including line blanketing) to a spectrum calculated using the temperature structure of an LTE continuum model. The differences between the two spectra are dramatic and without the fully self-consistent calculation we would arrive at wrong conclusions about the parameters and, particularly the abundances of the nova atmosphere.

3. Summary

The agreement between our synthetic spectra and observed nova spectra is very good from the UV to the near IR. We are now able to reproduce the UV and optical spectra of very different types of classical novae. The basic physics and the modeling of early nova atmospheres is now well understood and we are currently working on further improvements of the models (more NLTE species) and on the analysis of those novae which have good spectral coverage in the UV and the optical. Another important step will be to investigate the effects of density inversions (clumps) on the emitted spectrum and a more systematic investigation of the effects of different velocity fields on nova spectra is also required. Furthermore, we are currently working on a detailed treatment of the "pre-nebula" phase, i.e., the phase where allowed, semi-forbidden and sometimes forbidden lines are simultaneously present in the observed nova spectra. This will not only require very detailed model atoms for the species showing these lines but will also extend the model atmospheres into later stages of the outburst.

Acknowledgments: It is a pleasure to thank H. Störzer, J. Krautter, G. Shaviv and S. Pistinner for stimulating discussions. This work was supported in part by a NASA LTSA grant to Arizona State University, by NASA grant NAGW-2999; as well as grants to G. F. Fahlman and H. B. Richer from NSERC (Canada). Some of the calculations presented in this paper were performed at the San Diego Supercomputer Center (SDSC), supported by the NSF, and at the NERSC, supported by the U.S. DoE, we thank them for a generous allocation of computer time.

References

Baron, E., Hauschildt, P. H., and Branch, D. 1994, ApJ, 426, 334
Evans, A., et al.. 1993, IAU Circular No. 5916
Gehrz, R. D. 1988, Ann. Rev. Astr. Ap., 26, 377
Hauschildt, P. H. and Ensman, L. M. 1994, ApJ, 424, 905
Hauschildt, P. H., Starrfield, S., Austin, S. J., Wagner, R. M., Shore, S. N., and Sonneborn, G. 1994a, ApJ, 422, 831
Hauschildt, P. H., Starrfield, S., Shore, S. N., Allard, F., and Baron, E. 1994b, ApJ, submitted
Hauschildt, P. H., Starrfield, S., Shore, S. N., Gonzales-Riestra, R., Sonneborn, G., and Allard, F. 1994c, AJ, 108, 1008
Hauschildt, P. H., Wehrse, R., Starrfield, S., and Shaviv, G. 1992, ApJ, 393, 307
McLaughlin, D. B. 1960, in Stellar Atmospheres, edited by Greenstein, J. L., number VI in Stars and Stellar Systems, page 585, University of Chicago Press
Pistinner, S., Shaviv, G., Hauschildt, P. H., and Starrfield, S. 1994, ApJ,, submitted

THE OUTBURST OF NOVA CASSIOPEIAE 1993

SUMNER STARRFIELD AND PETER H. HAUSCHILDT
Dept. of Physics and Astronomy, Arizona State University,
Box 871504, Tempe, AZ 85287-1504
E-Mail: starrfie@hydro.la.asu.edu and yeti@sara.la.asu.edu

STEVEN N. SHORE
Dept. of Physics and Astronomy, Indiana University South
Bend, 1700 Mishawaka Ave, South Bend, IN 46634-7111
E-Mail: sshore@paladin.iusb.indiana.edu

ROSARIO GONZALEZ-RIESTRA
Astrophysics Division, Space Science Department, ESTEC, and
IUE Observatory, ESA, Villafranca, Spain
E-Mail: ch@vilspa.esa.es

GEORGE SONNEBORN
Laboratory for Astronomy and Solar Physics, Code 681, Goddard
Space Flight Center, Greenbelt, MD 20771
E-Mail: sonneborn@fornax.gsfc.nasa.gov

AND

FRANCE ALLARD
Dept. of Geophysics and Astronomy, University of British Columbia,
Vancouver, B.C., V6T 1Z4 Canada
E-Mail: allard@astro.ubc.ca

Abstract.
   In this paper we describe the evolution of Nova Cas 1993 over the first few months of its outburst. We discuss the ultraviolet light curve that covers the period from announcement to just after dust began forming in the ejecta (1994 Feb.15) and IUE spacecraft constraints forced us to halt our observations. The dust formed very quickly despite the initially high temperature of the ejecta. Our results suggest that high energy photons are absorbed efficiently by the innermost material in the ejecta, lowering the temperature while it is still dense. This allows molecules to form and then condense into dust. We used IUE data to obtain an extinction curve as a

A Bianchini et al (eds) Cataclysmic Variables 267-275
© *1995 Kluwer Academic Publishers*

function of time during the dust formation phase. This curve shows that the dust was large - much larger than typical dust in the ISM.

We also used a spherical, expanding, NLTE stellar atmosphere code to compute synthetic spectra and have compared the results to combined ultraviolet (low-resolution 1200Å - 3400Å and high-resolution 2400Å - 3300Å spectra), optical, and infrared spectra. Our fits show that the effective temperature of the ejecta increased from $\sim$ 8000K to $\sim$ 16000K between 1993 Dec. 12 and 1993 Dec. 26. The temperature then increased more slowly to $\sim$ 24000 on 1994 Jan. 28. A preliminary abundance analysis shows evidence for hydrogen depletion, as we also found for Nova V1974 Cygni. However, we found a larger enhancement of carbon, nitrogen, and oxygen. We show that the principal mechanism for mass ejection in this nova is a radiation pressure driven wind and that mechanical driving is not necessary.

1. Introduction

Nova Cas 1993 (hereafter, Cas93), was discovered on 1993 Dec. 7 UT at a photographic magnitude of about 6.5 and announced on Dec. 11, 1993 when it was found to still be at $V = 6.5$. Skiff (1993) measured an accurate position and reported the probable identification of the progenitor. Using his data we derived an outburst amplitude of about 13 mags which is reasonable for a classical nova in outburst (Hauschildt et al. 1994b; Shore et al. 1994a,b).

The optical light curve for the first three months of observations was given in Hauschildt et al. (1994b) and showed that the nova remained at about $V = 6.5$ with only small variations until about 1993 Dec. 18 when it underwent a "flare" to 5.3 mag and a rapid decline back to about 6.5 mag. It underwent smaller "flares" after this time until about Feb. 15, when at $V \approx 9$ mag it began a precipitous decline. If we ignore the flare, then $t_2 = 65$ days. We could not determine t_3 since it occurred after dust formation in the expanding material. An extrapolation of the pre-decline part of the light curve yields $t_3 \approx 100$ days.

The resemblance of the light curve of Cas93 to those designated as the "DQ Her" class of slow novae was remarkable and its light curve most closely resembled DQ Her 1934 itself. Both novae exhibited a pre-maximum halt, a fast rise to maximum, a several magnitude "flare", and a slow decline of two to three magnitudes over a period of two to three months. The slow decline was followed by a sharp drop in the optical and, in a few recent novae, a concomitant rise in the infrared attributed to the formation of dust. The analogy continued to hold, and by about mid-May, the nova had recovered from the deep transition phase. In November 1994 it was still at $V \sim 12.5$.

2. The Dust Formation Phase

Early optical spectra of Cas93 showed the characteristic lines of an optically thick Fe II nova (Williams et al. 1991), thus enhancing the morphological connection between this nova and DQ Her. Woodward and Greenhouse (1993) reported that the infrared Brγ line showed a full width half maximum (FWHM) of 1680 km s^{-1}. This value is consistent with the earliest observations of slow novae. Scott et al. (1994a) detected ^{12}CO infrared emission on 1993 Dec. 13 with possible evidence for ^{13}CO. Early optical spectra of DQ Her also showed CN absorption (Sneden and Lambert 1975), so molecules formed early in both novae. Scott et al (1994b) then reported that observations obtained on 1994 Jan. 10 showed persistent ^{12}CO emission plus 3.38μ emission attributed to polycyclic aromatic hydrocarbons (PAH). They pointed out that the IR evolution of this nova resembled V842 Cen 1986, which exhibited a deep transition and formed dust.

We obtained a set of IUE spectra on Feb. 11.5 UT. The 1200Å - 2000Å spectrum, SWP 49990, is shown in Shore et al. (1994c: Fig. 2a). Our next observation (SWP 50048) was obtained on Feb 17.2 UT, by which time the nova had declined in brightness to $V \approx 10.5$ (Shore et al. 1994c: Fig. 2b). The entire spectrum was depressed by an almost uniform factor of 5. There was no evidence for *any* UV rise toward shorter wavelengths, as is observed for normal interstellar extinction. The 2200Å - 3400Å spectrum was also depressed by a factor of 5 when compared with a long wavelength spectrum obtained on Feb. 10.

The extinction curve (Shore et al. 1994c: Fig. 3) was obtained by taking a ratio of the Feb. 21 spectra to those obtained on Feb. 11. We found that the extinction curve from the dust formed by Nova Cas 1993 was very different from that expected from interstellar dust. It was nearly grey shortward of 2400Å which can best be explained by a population of very large grains with an optical depth of order 4 or 5.

In order to produce flat absorption spanning a factor of three in wavelength in the UV, the grains must have sizes in excess of $2\pi a/\lambda > 4$ (Savage and Mathis 1979), where a is the grain radius and λ is the wavelength, so that the grains must be bigger than about 0.2μ. An optical depth of unity implies that the total mass of dust is about $10^{-7}Q_{abs}^{-1}\rho_{dust}M_\odot$, where Q_{abs} is the absorption efficiency (of order unity for large grains: Savage and Mathis 1979) and ρ_{dust} is the mass density of the dust (in this case assumed to be of order 3 g cm^{-3} for silicates). This dust mass is comparable with values obtained from infrared studies of other novae, especially NQ

Vul (Gehrz 1988), by a completely independent method. These results significantly strengthen the conclusions drawn from the infrared observations that large, warm dust forms quickly in the ejecta of DQ Her-type novae (Gehrz 1988).

The flux removed from the ultraviolet should heat any circumstellar grain population, and, in fact, the initial stages of the extinction event did produce a rapid rise in the infrared (Garnavich 1994). The foreground interstellar extinction appears to be of order E(B-V) = 0.5. This is inferred from the model atmosphere fitting (see below) and also from the requirement that the bolometric luminosity remain constant during at least the optically thick phase of the outburst. We have previously used this procedure on several novae, notably V1974 Cyg (Shore et al. 1993), and it appears to hold for infrared spectra as well (Gehrz 1988). We have used this extinction to derive the bolometric flux at the last stage before the deep minimum. The maximum integrated infrared flux was predicted to be $\sim 8 \times 10^{-8}$ erg cm^{-2} s^{-1}. We have assumed that the grains completely re-radiated the absorbed UV flux. This value agreed well with the observations (Garnavich 1994). The observed dust temperature was about 1200 K on Feb. 15 (Kidger et al. 1994).

3. Ultraviolet Evolution and Atmospheric Modeling

The rapid response capability of IUE made it possible for us to obtain our first UV spectra of Cas93 within a few hours of the discovery announcement. We obtained several exposures with the LWP (2000Å - 3400Å) and SWP (1200Å - 2000Å) cameras. We chose our initial exposure times based on our experiences with previous novae and found that Cas93 was much fainter in the UV than expected. Although there was a possibility that the nova was highly reddened, the extracted spectra showed that, in fact, it was quite cool. Additional data obtained over the next few days showed that we had caught Cas93 at UV minimum.

As discussed in detail in our studies of V1974 Cyg (Shore et al. 1993, 1994a; Hauschildt et al. 1994a), this is the stage when a large number of overlapping absorption lines from the iron group elements, mainly Fe II, block most of the UV light and virtually all of the emitted radiation escapes in the optical. UV spectra showing this phase of the evolution of a nova can be found in Shore et al. (1993) and Hauschildt et al. (1994a).

We then used the computer code PHOENIX (Version 4.9), to compute model atmospheres and synthetic spectra in order to analyze the UV spectra of Cas93. This is an updated version of the code used for the analyses of the early spectra of V1974 Cygni (Hauschildt et al. 1994a) and SN 1993J (Baron et al. 1994) and is described in some detail in these papers. However, in order to calculate fully self-consistent atmospheres with $T_{eff} \leq 10000K$ (required for the first spectra of Cas93), we now include the water vapor line list of Miller et al. (1994). For even lower effective temperatures, molecules are present in nova atmospheres. Therefore, the equation of state (EOS) is very complicated and special techniques must be employed to construct these atmospheres. In addition, the electron temperatures and gas pressures typically found in nova photospheres lead to the presence of a large numbers of spectral lines, predominantly Fe-group elements, in the line forming regions and the corresponding influence of line blanketing on the emergent spectrum. The situation is complicated by the velocity field of the expanding shell which leads to an enhancement of the overlapping of the individual lines. Line blanketing also leads to strong wavelength redistribution of the radiative energy. Therefore, the temperature structure of the shell must be calculated with line blanketing included.

The modeling and analysis of early nova spectra has made significant progress during the last 2-3 years by the construction of detailed model atmospheres and synthetic spectra for novae by Hauschildt et al.(1992, 1994a,b). We have found that in the early stages of the nova outburst, the spectrum is formed in an optically *very* thick, in both lines and continua, shell with a flat density profile, leading to very extended continuum and line forming regions (hereafter, CFR and LFR, respectively). Because of the large variation of the physical conditions inside the spectrum forming region, the classical term "photosphere" is not appropriate for novae. The large geometrical extension leads to a very large electron temperature gradient within the CFR and LFR, allowing for the simultaneous presence of several ionization stages of many elements as is observed. Other parameters necessary to understand the spectra of novae are given in Hauschildt et al. (1995, these proceedings).

In order to analyze the IUE spectra of Cas93, we have computed a grid of co-moving frame NLTE nova model atmospheres with the following set of parameters: $L = 7 \times 10^4\, L_\odot$, $N = 3$, $\xi = 50\,km\,s^{-1}$, $5000 \leq T_{eff} \leq 60000K$, $v_0 = 2000\,km\,s^{-1}$ and solar abundances. We then computed Eulerian frame synthetic spectra for each of the model atmospheres in the grid and used these synthetic spectra to estimate the effective temperature corresponding to the observed IUE spectra. The fit to the observations was iteratively improved by changing the parameters of the model atmospheres, in particular v_0 and the abundances.

Our analysis showed that the metal abundances needed to be increased by about a factor of two over solar abundances in order to obtain good agreement with the UV and optical spectra of Cas 93. The accord is substantially improved in the SWP wavelength range below 1600Å if we use about 100 times the solar abundances of carbon and oxygen. In the follow-

ing discussion, however, we will describe only the results from the models
with twice solar abundances for the metals and defer a discussion of the
more detailed models to a later paper.

4. Results

A plot showing the effective temperature, derived from the fits to the models, as a function of time can be found in Hauschildt et al. (1994b, Fig. 4). The effective temperature of the nova rises rather slowly with time as compared to V1974 Cyg. We also found that the earliest spectrum can be best fit by a model with an effective temperature of $\approx$ 8000K. Nova atmospheres with such low effective temperatures can form molecules in their outer layers. In fact, our model for the earliest epoch of Cas93 predicts emission from the infrared CO first overtone lines, as was observed by Scott et al. (1994a,b). As the nova evolved, the effective temperatures of the models required to fit the spectra gradually increased. As the temperatures increased, the *temperature range* in the LFR also increased and reached a point where lines of very different ionization stages were predicted to be simultaneously present in the spectra. This was observed in Cas93.

IR observations of diffuse band emission at about 3.4μ is also expected if PAH or related molecules have formed in the outermost layers. This emission is pumped by absorption of UV photons and their subsequent redistribution through the many available vibrational modes of a variety of species of intermediate mass molecules of the polycyclic aromatic hydrocarbons (see Desert et al. 1986, Puget and Leger 1989). In order for infrared emission to occur, however, *there must be a sufficiently strong UV radiation source*, and our UV observations show that this did not occur until after the recovery from UV minimum in early 1994 Jan. This is just when such emission was detected. Therefore, we used the results of our model atmospheres to predict that dust would form in the expanding gas (Shore and Starrfield 1994) and this prediction was confirmed in February 1994.

The mechanism responsible for mass ejection is another important result of this study. Mass ejection from a nova may not, and probably does not, proceed in a single event. Unlike a normal stellar wind, the timescale for luminosity evolution during the initial stages is very fast and it may not be possible to establish a steady state wind. Thus, the very high luminosity and temperature that typifies the fireball may initiate a fast wind that quickly shuts off, leading to a secondary phase of lower terminal velocity. The fastest material reaches low temperatures first, and it is this part of the outflow that is responsible for the early CO overtone emission that was observed in Cas93. Subsequently, during the constant bolometric luminosity stage, radiative acceleration may drive a steady state outflow.

In order to demonstrate this, in Hauschildt et al. (1994b) we plotted the ratio of the outward acceleration due to radiative forces (both lines and continua), g_{rad}, to the inward gravitational acceleration, g, due to an assumed 1.25 $M_\odot$ central star. For the coolest model, $T_{\text{eff}} = 8000$K (1993 Dec 13), there is a region where $g_{rad} < g$. However, this zone is rather small and above the region where most of the material is accelerated. In most of the inner regions of the model g_{rad}, is larger than g by a factor of 10. In the hotter models, $T_{\text{eff}} = 16000$K (26 Dec 1993) and $T_{\text{eff}} = 24000$K (28 Jan 1994), g_{rad} always exceeds g. The gravitational acceleration is much larger in the hotter models than in the cooler ones because we have assumed the same (constant) bolometric luminosity of 7.0×10^4 $L_\odot$ for all models. The steep changes in the curves of radiation pressure as a function of depth result from the presence of ionization zones (or regions of molecule formation in the cooler models). These regions produce large changes in the opacity, χ, and corresponding changes in $g_{\text{rad}} \propto \int \chi H_\lambda \, d\lambda$, although the flux, H_λ, itself is varying much more slowly.

We note that the region of radiation pressure driving is smallest in the coolest model and it is possible that at this time radiative driving is insufficient to produce a wind. We emphasize, however, that the nova has cooled from higher temperatures in which radiative driving extended throughout the entire outer layers. Therefore, just before discovery there was sufficient radiative driving, and a few days after discovery there was again sufficient driving, to produce a wind.

The rate of mass loss is a derived quantity for our atmospheres, $\dot{M} = 4\pi r^2 \rho \, v$. Note that the rate of mass loss scales directly with v, so reducing the velocity reduces the rate of mass loss. We find for the cool model that $\dot{M} = 2.5 \times 10^{-3} M_\odot \text{yr}^{-1}$. $\dot{M}$ decreases as the effective temperature increases as so that by the time $T_{\text{eff}} = 16,000$K, the rate of mass loss has decreased to 2.5×10^{-4} $M_\odot \text{yr}^{-1}$ and at $T_{\text{eff}} = 23,000$K, $\dot{M} = 6.8 \times 10^{-5}$ $M_\odot \text{yr}^{-1}$. We note that the nova spent a few days at the lowest effective temperatures when it should have ejected a significant fraction of its accreted envelope.

5. Summary

In this paper we have discussed the early evolution of nova Cas 1993. The light curve and the time evolution of it's UV spectra indicated that Cas93 was a slow nova of the DQ Her type. Based on this similarity we predicted that Cas93 would form dust (Shore and Starrfield 1994) and this prediction was confirmed by observations of the nova in mid-February 1994. Our NLTE, spherical, expanding, atmosphere models of the nova shell indicated that hydrogen was depleted by about a factor of two in the ejecta as compared with solar material, and that carbon and oxygen are possibly enhanced by factors as large as 100.

Our fits to the spectral energy distribution showed that the effective temperature of the shell increased slowly from 8000K (1993 Dec. 13) to about 24000K (1994 Jan. 28) and that the ejecta exhibited the shallow density profile of a nova in the optically thin "wind" stage. In contrast to V1974 Cyg, no fireball stage was observed in Cas93 and it must have occurred prior to the announcement of discovery. The expansion velocities of the shell decreased from $2000\,\mathrm{km\,s^{-1}}$ to about $1300\,\mathrm{km\,s^{-1}}$ and the models are consistent with either a "wind" velocity law of the form $v(r) \propto v_\infty \left(1 - a/r\right)^\beta$ or a homologous (linear) velocity law.

The models indicate that, during most of the outburst, the acceleration of the material by the radiation pressure gradient was much larger than the gravitational deceleration caused by the central star. The radiative acceleration exceeded the gravitational deceleration in the hotter models by factors of order 1000. Therefore, the material could reach the high velocities observed in novae by radiation pressure alone. Only the very coolest model showed a small layer where the radiative acceleration was insufficient to drive mass loss. However, the ejected material must have already reached a very high velocity by this time. It thus appears that radiation pressure is sufficient to drive mass loss in the DQ Her-type slow novae, and that mechanical driving is *not* a prerequisite for mass ejection, although it likely adds to the forces producing mass loss.

Acknowledgments: It is a pleasure to thank R. D. Gehrz, J. Krautter, S. Pistinner, G. Shaviv, and H. Störzer, for stimulating discussions. This work was supported in part by NSF and NASA (LTSA) grants to Arizona State University; as well as grants to G. F. Fahlman and H. B. Richer from NSERC (Canada). Some of the calculations presented in this paper were performed at the San Diego Supercomputer Center (SDSC), supported by the NSF, and at the NERSC, supported by the U.S. DoE, we thank them for a generous allocation of computer time.

References

Baron, E., Hauschildt, P. H., and Branch, D. (1994), *ApJ,*, **426**, 334.
Desert, X. F., Boulanger, F., and Shore, S. N. (1986), *A&A*, **160**, 295.
Garnavich, P. (1994), *IAU Circular*, No. 5941.
Gehrz, R. D. (1988), *ARAA*, **26**, 377.
Hauschildt, P. H., Starrfield, S., Austin, S. J., Wagner, R. M., Shore, S. N., and Sonneborn, G. (1994a), *ApJ*, **422**, 831.
Hauschildt, P. H., Starrfield, S., Shore, S. N., Gonzales-Riestra, R., Sonneborn, G., and Allard, F. (1994b), *AJ*, **108**, 1008.
Hauschildt, P. H., Wehrse, R., Starrfield, S., and Shaviv, G. (1992), *ApJ*, **393**, 307.
Kidger, M., Devaney, N., Sahu, K., Lopez, S. (1994), *IAU Circular*, No. 5936.
Miller, S., Tennyson, J., Jones, H. R. A., and Longmore, A. J. (1994), in *Molecular Opacities in the Stellar Environment*, edited by Thejll, P. and Jorgensen, U., I.A.U.

Colloquium 146, Niels Bohr Institute and Nordita press, Copenhagen, in press.

Puget, J., and Leger, A. (1989), *ARAA*, **27**, 161.

Savage, B. D. and Mathis, J. (1979), *ARAA*, 17, 73.

Scott, A. D., et al. (1994a), *IAU Circular*, No. 5916.

Scott, A. D, Evans, A. and Geballe, T. (1994b), *IAU Circular*, No. 5922.

Shore, S. N., Sonneborn, G., Starrfield, S., Gonzalez-Riestra, R., and Polidan, R. (1994a), *ApJ*, **421**, 344 .

Shore, S. N. and Starrfield, S. (1994), *Sky and Telescope* 4, 42.

Shore, S. N., Starrfield, S.,Gonzalez-Riestra, R., Hauschildt, P. H., and Sonneborn, G., (1994c), *NATURE*, **369**, 539.

Shore, S. N., Starrfield, S., Hauschildt, P. H., Sonneborn, G., and Gonzalez-Riestra, R. (1994b), *IAU Circular*, No. 5925.

Shore, S. N., Starrfield, S., Sonneborn, G., Gonzalez-Riestra, R., and Ake, T. B. (1993), *AJ*, **106**, 2408.

Skiff, B. (1993), *IAU Circular*, No.5904.

Sneden, C., and Lambert, D. L. (1975), *MNRAS*, **170**, 533.dv

Starrfield, S. and Shore, S. N. (1994), *Sky and Telescope* **2**, 20.

Trammell, S. R. and Benjamin, R. A. (1993), *IAU Circular* No. 5910.

Williams, R. E., Hamuy, M., Heathcote, S. R., Wells, L., and Navarette, M. (1991), *ApJ*, **376**, 721.

Woodward, C. and Greenhouse, M. (1993), *IAU Circular*, No. 5910.

SHOCKS IN NOVAE AND IN SYMBIOTIC STARS

M. CONTINI
School of Physics and Astronomy, Tel-Aviv University,
Tel-Aviv, 69978 Israel

Abstract.
Composite models which consistently account for both shocks and a photoionizing radiation flux are successfully used for the calculations of the spectra emitted from RS Ophiuci and HM Sagittae. The composition determined by the model indicate that RS Ophiuci is characterized by a massive white dwarf. Calculated spectra of HM Sagittae show that evolution depends on the shock velocity and on the star temperature.

1. Introduction

Shocks can play an important role in some nova shells, e. g. RS Ophiuci (Contini, Orio, and Prialnik 1993) and symbiotic stars, e. g. HM Sagittae (Formiggini, Contini, and Leibowitz 1994).

Shocks arize in the interaction of high velocity ejecta (2000 - 4000 km s^{-1}) from the hot star with a low velocity stellar wind (20 - 30 km s^{-1}) from the cool component. They are indicated by gas at high temperature, whose signature is recognized as 1) X-ray emission, 2) strong spectral lines from high ionization levels, 3) strong low-ionization and neutral lines, 4) the shape of the spectral energy distribution of the continuum, and - if dust is present - 5) the shape of the infrared bump.

We investigate shocks by fitting the observed spectra by model calculations using the SUMA code (Viegas and Contini 1994, and references therein). SUMA simulates the physical conditions of an emitting gaseus cloud under the coupled effect of photoionization and shocks.

A. Bianchini et al. (eds.), Cataclysmic Variables, 277-280.
© *1995 Kluwer Academic Publishers.*

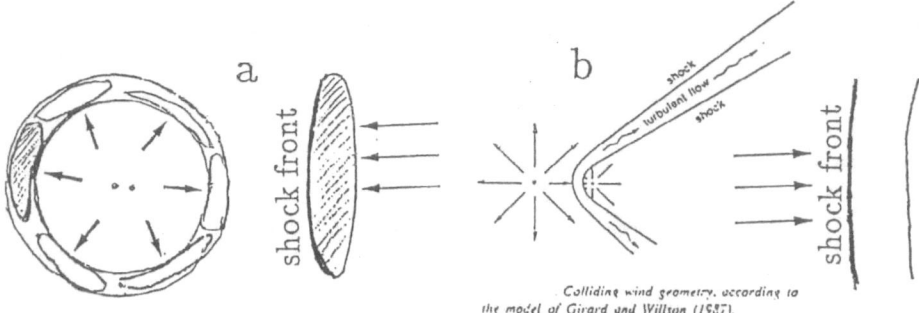

Figure 1.   a: the shocked clump in novae; b: the shocked clump in symbiotics

2. The Models

In the nova case the two stars are very close one to the other, therefore, the shock propagates outwards the stars system (fig. 1a). In symbiotics the two stars are far enough and the shocked nebula resides between them (fig. 1b). Consequently, in nova models the shock front corresponds to the outer edge of the clumps, opposite to that reached by the photoionizing radiation flux. In symbiotics the shock front corresponds to the edge reached by the flux. Steady state plane - parallel models are adopted.

3. RS Ophiuci, a recurrent nova

The observational corrected line spectrum of RS Oph at day 201 (Bohigas et al. 1989) is well fitted (Contini et al. 1993, table 1) by shock models with V_s (shock velocity) = 200 - 300 km s^{-1}, n_0 (preshock density) = 2 − 3×10^4 cm^{-3}, and B_0 (magnetic field) = 10^{-3} gauss. Adopting a composite model (shock + photoionization) a good agreement is obtained with T_s (temperature of the star) = 4×10^4 K and U (ionization parameter) = 10^{-5}. U is rather low, indicating a large distance, R, between the clumps and the star. Main results are the following: 1) A "matter bound" case is indicated by fitting the He II λ 4686 / Hβ line ratio, and the geometrical thickness of the clump agrees with that predicted by theories claiming that the matter was ejected in the first 10 days after outburst with velocity of 3000-4000 km s^{-1}. 2) R > 5×10^{15} cm, derived from the values of T_s, U, and n_0, provides a good fit of calculated integrated fluxes in all ranges (X-ray, UV, optical, IR, and radio) to observational data. 3) the relative abundances of the elements to H, are derived phenomenologically and show low O/H

(8×10^{-5}) and high He/H (0.2 - 0.4) in agreement with TNR theories of accretion onto a massive white dwarf.

4. HM Sagittae, a symbiotic nova

HM Sge has not yet faded after maximum, and displays activity in all spectral ranges, from radio to X-rays. The observations (Schmidt and Shield 1990) provide a very rich line spectrum, in number of lines, ionization levels, elements, and wavelengths. The best fit of calculation results to observational data are presented in Table 1 and the input parameters are given in Table 2. They confirm the physical picture predicted by the observational evidence (cf. fig. 1b) and suggest that dust is present through depletion of Si, S, and Fe. The evolution of the spectrum is characterized by the strengthening of He II lines and of lines from the upper levels (IV, V) together with a decreasing tendency of the UV continuum and of lower levels lines (II, III) (Nussbaumer and Vogel 1990). Calculations by SUMA show that evolution can be explained not only by the increase of star temperature, but, particularly, by decrease of shock velocity. Consequently, early time observations of X-rays are explained by bremsstrahlung from gas heated at temperatures suiting a relatively high velocity shock $(V_s > 200 \, \mathrm{km \, s^{-1}})$.

References

Bohigas, J.,Echevarria, J.,Diego, F., and Sarmiento, J.A. (1989) *MNRAS*, **238**, 1395
Contini, M., Orio, M., and Prialnik, D. (1993) submitted
Formiggini, L., Contini, M., and Leibowitz, E. M. (1994) in preparation
Nussbaumer, H., and Vogel, M. (1990) *A&A*, **236**, 117
Schmid, H. M., and Schild, H. (1990) *MNRAS*, **246**, 84
Viegas, S.M., and Contini, M. (1994) *ApJ*, **428**, 113

M. CONTINI

TABLE 1. The best fit of calculation results to observational data

Line	Ion	obs.	calc.	Line	Ion	obs.	calc.
1240	N V	4.53	4.47	3760	[Fe VII]	0.34b	0.24
1335	C II	< .2	.076	3869	[Ne III]	2.13	2.50
1371	O V	0.18:	.018	4069	[S II]	0.19	0.13
1397	Si IV	0.45	0.62	4363	[O III]	0.85	0.97
1486	N IV]	2.53	2.50	4471	He I	.039	.017
1550	C IV	9.06	10.4	4686	He II	0.68	0.74
1575	[Ne V]	0.27:	.089	4714	[Ne IV]	0.14	0.18
1601	[Ne IV]	0.50	0.70	4740	[Ar IV]	.033	.045
1640	He II	4.27	5.57	5000	[O III]	4.85	5.85
1664	O III]	1.03*	1.38	5146	[Fe VI]	.033	.046
1750	N III]	1.11	1.19	5158	[Fe VII]	.039	.033
1815	Si II*	0.14	.044	5176	[Fe VI]	.064	.081
1892	Si III]	0.73	0.58	5335	[Fe VI]	.011	.015
1909	C III]	4.55	4.20	5755	[N II]	.192	.110
2424	[Ne IV]	.24:	0.26	5876	He I	0.11	0.06
2783	[Mg V]	0.55	0.40	6300	[O I]	.149	.128
2973	[Ne V]	.081*	.032	6312	[S III]	.108	0.11
3203	He II	0.28	0.33	6548	[N II]	0.29	0.22
3426	[Ne V]	4.43	4.50	6731	[S II]	.009	.006+
3663	[Fe VI]	.035	.029	7006	[Ar V]	.082	.070
3727	[O II]	.152b	.013	7136	[Ar III]	0.20	0.36
	+[S III]	-	0.07	6375	[Fe X]	-	3.(-5)

TABLE 2. The input parameters of Table 1

Input Parameters :	Relative Abundances :	
$V_s = 180 \, \text{km s}^{-1}$	He / H = 0.11	Mg / H = 2.6(-5)
$n_0 = 7 \times 10^4 \, \text{cm}^{-3}$	C / H = 1.3(-4)	Si / H = 1.3(-5)
$B_0 = 10^{-3} \, \text{gauss}$	N / H = 1.5(-4)	S / H = 1.6(-5)
$T_s = 2.3 \times 10^5 \, \text{K}$	O / H = 4.6(-4)	Ar / H = 6.3(-6)
U = 0.4	Ne / H = 1.22(-4)	Fe / H = 1.08(-5)

THE MULTIWAVELENGTH CONTINUUM OF NEON NOVAE

PEDRO SAIZAR

Department of Physics and Astronomy
University of Pennsylvania, Philadelphia, PA 19104, U.S.A.

1. Introduction

In this paper we discuss the late evolution of oxygen–neon–magnesium novae, with particular emphasis in the development of the multiwavelength continuum. These studies are important to better understand the physical conditions in the ejected shell and its relationship to the underlying system.

Here we present preliminary results concerning Nova Puppis 1991, a nova that has been intensively observed since discovery (Dec. 27, 1991, IAUC # 5422). Because the only previous observation is from Dec. 11, it is doubtful when the exact maximum actually occured. At discovery time, the apparent magnitude was 6.4. Setting limits on the light curve, and using the t_2–M_V relationship (Warner 1989), Saizar *et al.* (1995, hereafter S95), find the distance to be 2.7±0.7 kpc. The same paper derives a reddening correction E(B–V)=0.72 from various ratios of recombination lines.

The nova was observed by both the IUE and from Cerro Tololo Observatory (Williams 1991). For the continuum analysis presented here, we have matched one optical spectrum taken 370 days after discovery, with one IUE LWP and one SWP spectrum taken on day 359. The smooth decline of the light curves, both in the optical and the UV, suggests that conditions in the shell did not appreciably change in this short period of time. Other spectra were taken at approximately day 136 and 452, and the analysis will be presented elsewhere.

2. Results from the Emission Lines

As part of the analysis of the continuum, we need to have an independent estimate of the physical conditions in the nebular gas. This is presented by S95 in considerable detail. By using collisionally excited and recombination

A. Bianchini et al. (eds.), Cataclysmic Variables, 281-284.
© *1995 Kluwer Academic Publishers.*

lines, they derive for day 370 an electron temperature of 10,000 K, and an electron density of 9.3×10^7 cm^{-3}.

Also, by using the technique of ionization correction factors, they derive the following abundances by number in terms of the solar abundance: He/H=2.1, C/H=2.1, N/H=220, O/H=52, Ne/H=150. Si seems absent or very weak, so it is possible that it is depleted. These are needed to compute the continuum emissivity from metals.

3. The Multiwavelength Continuum

The top panel of Figure 1 shows the continuum in the plane ν vs. $f(\nu)/f(H\beta)$. Features worth noting are: the continuum remains strong throughout the optical/UV region, the Balmer junp appears strong, and there is a considerable excess towards the red, peaking at about 1 micron. In what follows I discuss all possible sources of continuous emission, and how they help interpret the data.

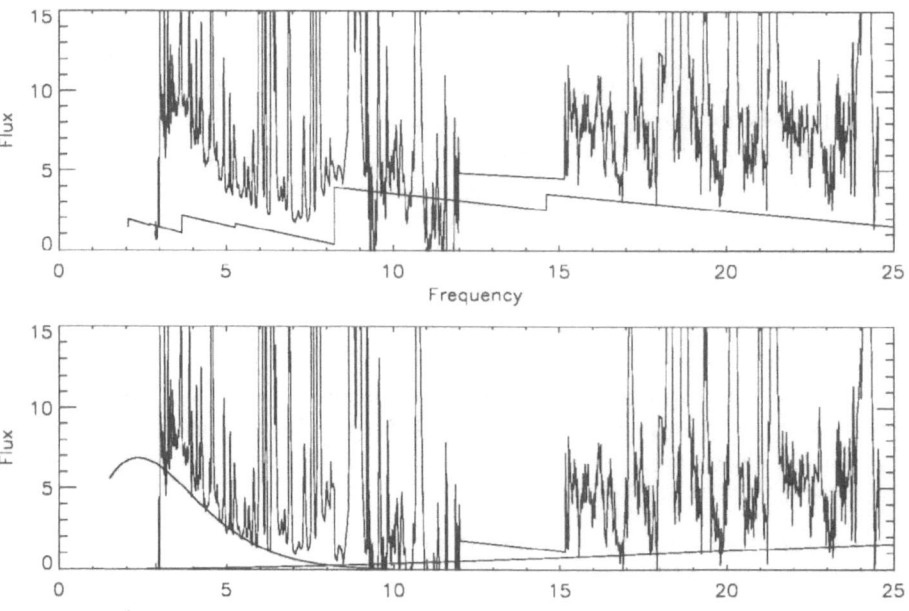

Figure 1. Multiwavelength Continuum at day 370. *Top panel:* observed spectrum and predicted continuum from a nebular gas at T=10^4 K. *Bottom panel:* substracted spectrum showing fits of blackbodys at 3000 K (left), and 10^5 K. Frequency in units of 10^{14} Hz, Flux relative to Hβ flux in units of 10^{-15} Hz^{-1}.

Reprocessed Radiation. The presence of low and intermediate ionization species, at a temperature of about 10,000 K, suggests that the shell is being photoionized by the central source. Under this conditions the gas emits a free–free and bound–free spectrum.

The continuous flux relative to Hβ for a certain ion is given by the product of the relative emission measures times the relative emission coefficients. The total relative flux is then obtained by adding all the observed ionization stages for all the ions. The emission coefficients for H and He are given by Ferland (1980). Because heavier ions are so abundant, they contribute a small but significant fraction to the continuum. Thus, following Ferland's methods, we added the contribution to the continuum from the heavy elements for which we had previously obtained ion abundances (S95). The total emission is shown in the top panel of Fig. 1. It is clear that the observed Balmer jump is well fitted, as it is much of the middle spectral range. The red and the SWP regions still show a considerable excess.

Signature from the Secondary Star? The excess on the red region can be fit with a 4000 K blackbody (Fig. 1, bottom panel). One possibility is emission from hot dust grains. The smooth decline of the optical light curve suggests that this nova did not develop an appreciable amount of dust.

Another source of radiation is the secondary star. Considering that a K–type main–sequence star at nearly 3 kpc is not detectable with an spectrograph (apparent magnitude 20), there are two possibilities: (a) the companion is a giant (which would imply an apparent magnitude of about 14), or (b) the companion is being rendered brighter by irradiation from the white dwarf. More observations focusing on the secondary star are being planned.

Signature of the White Dwarf. An excess still persists on the SWP Camera. Because the remaining flux is essentially zero on the LWP camera, it is reasonable to assume that a power law may be more adequate to represent the far UV. A likely candidate for this radiation source is the underlying white dwarf. A reasonable fit is obtained with a blackbody at 10^5 K (Fig. 1, bottom panel).

4. The Mass of the Ejected Shell

In previous work (Saizar *et al.* 1992, Saizar and Ferland 1994), we found that QU Vul, a very slow oxygen–neon magnesium nova, produced a very strong continuum throughout the optical/UV spectrum, and also a very weak or non–existent Balmer jump. Both were taken as the signature of a very hot, tenuous gas, in addition to the nebular (colder and denser), gas producing the low ionization emission lines.

Given that a hot gas emits a faint bremss (Ferland 1980), it the strength of the continuum implies that the hot shell is significantly massive. Results showed that the total mass ejected by QU Vul was of the order of 10^{-3} solar masses. This result was in agreement with other estimates (Greenhouse *et al.* 1988).

We pointed out (Saizar *et al.* 1992), that a conflict between theory and observation seemed to emerge. The theories of stellar evolution predict that ONeMg white dwarfs should be massive stars, that is, approaching the Chandrasekhar limit (see, e.g., Livio and Truran 1994). The predictions of thermonuclear runaway theories yield very small ejected masses for novae occurring on these massive ONeMg white dwarfs (see, e.g., Starrfield 1989).

The present analysis of Nova Pup 1991 presents a more consistent picture with Starrfield's theoretical predictions. A hot phase is not present or is not significant, and the mass of the nebular phase is likely not higher than 10^{-5} solar masses, based on the strength of Hβ. Thus, we predict that Nova Pup 1991 did occur on a high–mass white dwarf.

Seen in the light of this study, however, the case of QU Vul remains a mysterious one. No satisfactory solution has yet been presented to explain either the high mass of the ejecta or the low mass of the underlying white dwarf.

I would like to thank my colleagues and observers that made this work possible, among them: R. Williams, S. Starrfield, S. Shore, G. Sonneborn, I. Pachoulakis, and E. Rothschild. This work has been funded by a NASA grant administered by the American Astronomical Society. Thanks also to the AAS and the Local Organizing Committee for travel support.

References

Ferland, G. J. (1980), *P.A.S.P.*, **92**, 596.

Greehouse, M. A., Grasdalen, G. L., Hayward, T. L., Gehrz, R. D., and Jones, T. J. (1988) *A. J.*, **95**, 172.

Livio, M. and Truran, J. W. (1994) *Ap. J.*, **425**, 797L.

Saizar, P., and Ferland G. J. (1994), *Ap. J.*, **425**, 755.

Saizar, P., *et al.* (1992), *Ap. J.*, **398**, 651.

Saizar, P., *et al* (1995), preprint (S95).

Warner, B. (1989), *Classical Novae*, ed. N. Evans and M. Bode. Wiley, New York, p. 1.

Williams, R. E., Hamuy, M., Phillips, M. M., Heathcotte, S. R , Wells, L., and Navarrete, M. (1991), *Ap. J.*, **376**, 721.

SPECTRA OF CLASSICAL NOVAE DURING OUTBURST

J. S. GALLAGHER

Dept. Astronomy, University of Wisconsin
5534 Sterling, 475 N. Charter St., Madison, WI 53706 USA

1. Introduction

Classical novae are well known for their post-outburst spectroscopic antics which result from the complex interplay between ejected material and the evolving central binary star system. Thus while we wish to measure fundamental physical properties of the nova, such as ejected mass M_{ej}, chemical abundances, and other parameters, the observations may tell us more about the structure of the nova than we prefer to know!

Some of these difficulties are being overcome through the availibility of sophisticated numerical models for classical novae during all phases of an outburst. Examples include the model atmospheres by Hauschild et al. (1992), application of Ferland's nebular modelling code CLOUDY to nova shells (e.g., Saizar et al. 1991), and the continued advance of models for the basic outburst process (e.g., Starrfield and Sparks 1987; Prialnik and Kovetz 1992). In this paper I emphasize the role of spectroscopic observations as a way to gain insight into the importance of condensations and other asymmetries in shaping the spectra of novae.

That novae are basically a regular phenomenon has been established for more than 50 years (McLaughlin 1960; Payne-Gaposchkin 1957). Classical novae display fundamental regularities in their behavior, exemplified by relationships such as the decline rate-luminosity correlation or the reasonably ordered development of post-outburst spectroscopic stages. Furthermore, these patterns apply not only to novae in the Milky Way, but also to those in nearby galaxies. Perhaps even more notable is the failure to find any large populations of objects which resemble but are not true classical novae in local galaxies (e.g., Tomaney and Shafter 1992, 1993; Williams et al. 1994). The few interesting objects which have turned up in recent years such as the M31 "luminous red-variable" (Mould et al. 1990) or supersoft

A. Bianchini et al. (eds.), Cataclysmic Variables, 285-294.

X-ray sources (Greiner et al. 1991; Orio and Ögelman 1993), should not be ignored but also do not suggest the need for major revisions in our picture of classical novae as a major manifestation of nuclear eruptions on massive white dwarfs (e.g., Iben and Tutukov 1992).

I begin by considering the evolution of a classical nova in terms of a simple spherical model. Spherically symmetric novae have the advantage of being well-represented by one-dimensional numerical models and the possible disadvantage of scarcity in nature. Nova Cygni 1992 (V1974 Cyg) provides a useful recent example of the kinds of effects which occur within active novae.

2. The Basic Spherical Model

Three key parameters control the emergent spectrum of the ideal spherical nova, $L_{bol}(t)$ provided by the nuclear burning white dwarf, the mass loss rate $\dot{M}(t)$, and the velocity of the mass outflow $v(t)$. For this discussion we assume that L_{bol} has three levels, a super-Eddington value for a few days around visual maximum, a plateau during which a nuclear burning shell exists on the white dwarf, and a decline when the shell shuts off (Gallagher and Starrfield 1977). The basic phases of a nova can then be crudely understood in terms of changing redistributions of radiation from the central star.

Optically Thick 'Fireball'. The material ejected by a nova at the start of an outburst is initially highly optically thick. During this phase the radius of the pseudo-photosphere (i.e. defined by the optical depth one radius) R_{ps} thus expands with the ejecta; therefore the effective temperature declines as R_{ps} increases and the luminosity drops from its initial spike. The behavior of many novae just after maximum visual light resembles the cooling, expanding remnant of an explosion, hence the identification with a 'fireball stage'. This simple model does not properly describe the pre-visual maximum light curves, where effects associated with the nuclear runaway are important. However it does explain how a nova will brighten in the visual as its effective temperature and therefore bolometric correction decrease during the rise to visual maximum light.

As the ejecta expand densities and temperatures decrease. In a simple model this leads to a dramatic decline in opacity once the temperature in the pseudo-photosphere is below about 6000 K. As a result the expansion of the pseudo-photosphere stalls and visual maximum takes place. This opacity wall suggests the color temperatures of novae should obey $T_{col} \geq$ 6000 K, which is consistent with the observations. Of course even early spectra show complex emission and a wide range of excitation levels; the underlying assumption of some approximation to thermal equilibrium is not valid. Further quantitative progress depends on fully self-consistent numerical models.

What happens after pseudo-photosphere stall-out depends on $\dot{M}(t)$. If $\dot{M}(t)$ is small then R_{ps} must also decrease and the visual L will drop, a process which will be enhanced if L_{bol} also decreases. This latter effect should be especially pronounced in novae where the peak luminosity is extremely super-Eddington. The visual pseudo-photosphere will rapidly deflate unless both $\dot{M}(t)$ and L_{bol} remain roughly constant (Kato and Iben 1992).

Visual Decline. The post-maximum evolution of classical novae is extremely complex (McLaughlin 1960; Williams 1992; Williams et al. 1994). In the UV metal line-blanketing provides an extended opacity source whose properties depend on increasingly non-equilibrium conditions in the envelope. Furthermore the shell stucture of the ejecta in some novae could lead to a limb-brightened, bullseye appearance for the UV pseudo-photosphere during the visual decline (e.g., Paresce 1994). Conditions within the ejecta are further complicated by the observed increase in ionization, a reflection of the rising color temperature of the remnant; while the visual light declines the near constant luminosity output shifts to the ultraviolet.

The spectra of novae at this phase show emission lines in the IR with P-Cygni profiles from intermediate ionization species in the optical and UV. Multiple absorption line components commonly appear at this point in the visual decline (e.g., Chochol et al. 1993). The details of multiple absorption line formation in novae is still not fully understood, but the basic model which combines features in a fast, dense stellar wind with those produced in the initial mass ejection probably describes most novae (see Kato 1994).

During this phase some novae also develop a second cool pseudo-photosphere due to the condensation of dust. The dust thermosphere is often reasonably optically thick in the UV and redistributes the bulk of the luminosity from the UV to the thermal infrared. Dust opacity also can be sufficient to obscure most of the ejecta in the optical, leading to a deep optical minimum (Gehrz 1988).

Collapse to within the Home Binary. Models and physics dictate that in the absence of a huge reservoir to support continued mass loss, a reservoir not available to novae, the pseudo-photosphere at some point must retreat to the dimensions of the binary system. This evolution in many cases is extended by mass loss driven by the common envelope phase (Livio et al. 1990) and possibly further abetted by magnetic winds (Orio et al. 1992). A personal frustration is that we have yet to identify what signatures, if any, might be associated with these mass loss phases in classical novae (see Iben and Livio 1993).

Further decreases in radius will bring the pseudo-photosphere within the white dwarf's Roche lobe. For a typical nova system with $L_{bol} \simeq 3 \times 10^4 L_\odot$

and $R_{ps} \simeq R_{RL}$, the radiation temperature will exceed 30 000 K. At this point the increasingly hot white dwarf will extensively photoionizate the ejecta. Further evolution is driven by the contraction of the remnant and very high temperatures ($T_{eff} \geq 10^5$ K are predicted and observed; e.g. Krautter and Williams 1989; Shanley et al. 1995).

Due to the rapid expansion of the ejecta and the decreasing rate of mass injection, the *mean* mass density of the ejecta must decline as the nova evolves. This trend in combination with the ultra-hot post-nova white dwarf leads to high ionization levels in nova shells. Optical lines from species such as [Ne V] or [Fe VII] are common (e.g., Barger et al. 1993), with even higher ionization state coronal lines dominating the 1-5 micron spectral region (Gehrz 1988; Greenhouse et al. 1990; Dinerstein and Benjamin 1993)

Turn Off. The amount of nuclear fuel remaining on the white dwarf after the outburst depends on the details of mass loss as well as the specifics of the pre-outburst configuration of H-rich matter on the white dwarf. The nuclear shell burning lifetimes of novae therefore are difficult to theoretically predict. Since the peak luminosity from an aging active nova will fall in the EUV or soft X-ray spectral bands, such ultra-hot novae are difficult to directly observe, and the soft X-ray radiation has been measured in only a few cases (see Orio et al. 1993; Ögelman and Orio, this conference).

The ejecta, however, give us a local probe of the radiation field within the nova system. So long as the white dwarf remains hot and luminous, the ejecta will be highly ionized. The observations of novae that are more than a decade past outburst almost invariably show low ionization levels in their nebulae. Furthermore the electron temperatures can be very low, an indication of frozen ionization occuring when the cooling time is shorter than the recombination time in a nebula. Classical novae last only a few years as luminous, hot objects (Shanley et al. 1995; Gonzalez-Riestra, this meeting).

3. Breaking Spherical Symmetry

The simple picture presented above can be criticized on numerous grounds. Two major shortcomings lie in the extreme simplification of the local physics, for example, the assumptions that strict equilibrium prevails throughout the ejecta, and spherical symmetry applies with a high degree homogeneity. Several talks in this meeting have addressed the former shortcomings, so here I examine the observational situation regarding structures in nova ejecta.

That nova ejecta are non-spherical is well known. Triple or quadruple peaked emission lines are typical and have been qualitatively modeled in terms of a ring and dual cones within the nova ejecta (e.g., Hutchings 1972).

Comparisons of line profiles for different transitions furthermore often show that conditions vary between these major components of the ejecta (see Gallagher and Anderson 1976; Chochol et al. 1993; Shore et al 1993). It is questionable to assume uniform properties for parameters such as densities or expansion velocities for nova shells, even on large scales.

Observations of nebulae surrounding old novae also demonstrate that many novae are a bit our of round. A famous example is the DQ Her nebula which has an ellipsoidal shape (Williams et al. 1978). With higher angular resolution optical/UV imaging from the *Hubble Space Telescope* and radio arrays becoming available, the evidence for large scale departures in symmetry now extends to the early phases of nearer novae. Optical interferometers offer the promise of even higher resolution imaging, as already demonstrated for nova V1974 Cyg (e.g., Quirrenbach et al. 1993).

At Wisconsin we have been applying spectropolarimetric techniques to determine the stuctures of circumstellar shells in a ground-based effort led by K. Nordsieck and in space with the ASTRO/WUPPE experiment for which A. Code is the principal investigator. A symmetrical source will have zero net polarization, while asymmetries produce intrinsic polarization. However a determination of the intrinsic polarization requires that the interstellar polarization *vector* is properly subtracted, which can now be done from long wavelength baseline, high precision spectropolarimetry.

Optical spectropolarimetry of V1974 Cyg was obtained with our 0.9-m telescope at the Pine Bluff Observatory in Wisconsin. The observations were carried out by a team consisting mainly of undergraduate students, and supported two senior thesis research projects. The results reported by Bjorkman et al. (1994) show that the initial mass ejection in V1974 Cyg was asymmetric and that the degree of asymmetry decreased and then increased in an approximately orthogonal plane during the photoionization of the nebula. Again we see a bi-symmetric is present in the ejecta.

These results agree well with radio maps obtained with MERLIN and the VLA which were reviewed in this meeting. Spectropolarimetry, however, does not depend on angular resolution and thus gives us a window into the shapes of novae in early phases of an outburst as well as for distant novae.

In summary, nova ejecta often are intrinsically aspherical with two major symmetry planes. We need to understand the origins of these planes (in the binary as often supposed?) as well as their physical implications. For example, one long standing issue is the degree to which nova speed class depends on our viewing angle, while another is whether the two shell components routinely have different densities and other physical characteristics.

4. Small Scale Clumping

Multiple absorption line components in the P-Cygni lines of novae are a standard characteristic of declining novae. This aspect of novae is now so well known that it is rarely recorded, but this is partially compensated for by a huge older literature in which such such data were featured (Payne-Gaposchkin 1957). The general trend is for the number of absorption components to increase and for their mean radial velocity blueshift to rise during the early the decline. Just before these absorptions fade in the transition spectral phase, they can become very sharp ($\delta v < 100$ km s^{-1}) despite radial velocity blueshifts often exceeding 1000 km s^{-1}. The existence of these features is immediate evidence for inhomogeneities, but what kind? One possibility is that we are seeing discreet features–clumps, filaments, or shells–within the ejecta. Another is that these features represent constant flow velocity zones in the ejecta.

One hint is provided by the time evolution of the absorption line systems. The trend to higher radial velocities suggests either that we are seeing more further into novae where wind velocities are higher due to the deeper gravitational potential, or that momentum is being transferred to ejecta over time. I favor the former model and have argued that it is also consistent with velocity oscillations which anti-correlate with optical brightness found in declining novae such as GK Per (Payne-Gaposchkin 1957). However this begs the question of the physical source for these features, and here the long term continuity of absorption line component radial velocities suggests that they are formed in matter clumps.

Discreet, narrow spectral features are also prominent in nova emission lines at all wavelengths. In HR Del we showed that some of these features matched the velocities of absorption lines–evidently the ejecta contained distinct clumps which retained their identites over long times (Gallagher and Anderson 1976). The HST images of V1974 Cyg also demonstrate that clumps are present, as in the marvelous recent images taken with the Faint Object Camera (Paresce et al. 1995).

Conclusions: Within the two major groups of ejecta, clumps exist. What is less clear is what fraction of the ejected mass is clumped, how and when condensations form within the ejecta, and the range in parameters for the condensations. However, the presence of small scale structure containing only a modest mass fraction of the ejecta can have a substantial influence on the emergent spectrum.

5. Clumping and the Nebular Spectrum

That novae have a wide range of densities is also immediately clear from their spectra, especially once the ejecta are substantially ionized by the hot post-outburst white dwarf (e.g., Barger et al. 1993). In general a range of ionization stages will be present even in a uniform density photoionized nebula, but this range will be limited by basic physical requirements. For example, a homogeneous matter-bounded nebula will contain only a trace of neutral H and because of the ionization connection between O and H via charge exchange, it will also contain very little neutral O. The only way to obtain substantial amounts of [OI] emission from such a nebula is to include density inhomogeneities which are sufficiently large as to prevent complete ionization. The [OI] will then mark the partially ionized boundaries within the inhomogeneities. The presence of pronounced [OI] emission in nova nebulae is a signature of high density clumps in the ejecta (Williams 1992).

We can also see the effects of density variations in other ways. For example in a high density nova shell the [OIII] 4363 line can become strong relative to [OIII] 4959+5007 because of collisional deexcitation of the upper state $^1D^2$ state of the latter transition (see Osterbrock 1989). Another effect of high density is to increase the H-Balmer emission line intensity decrement, as was clearly seen in our data for V1974 Cyg (Barger et al. 1993). Yet when a single density model was derived to fit these lines in V1974 Cyg, it over-predicted collisional effects in the H-recombination spectrum. This steepened the Balmer decrement so that only the invocation of non-physical interstellar "bluing" could have saved the model. Instead John Mathis resolved this difficulty by including two components with different densities such that the bulk of the H-Balmer emission comes from the less dense part of the model.

The spectra of novae in the nebular stage require clumps to produce observed emission line intensity ratios and give us the beautiful, castellated emission line profiles.

Another feature of nova nebulae is their high mean electron densities; during the early nebular stage average values of n_e routinely exceed 10^6 cm^{-3}. These densities substantially modify the nebular spectrum because collisional deexcitation is important for most of the prominent optical nebular lines . The critical densities given by Osterbrock (1989) for collisional deexcitation of the $^1D^2$ level of [OIII] is 7×10^5, for [NeIII] 8×10^6 , and for [NeV] 2×10^7. In a dense nebula the optical forbidden lines of Ne can appear stronger than those of O simply because they suffer less collisional deexcitation. Ultraviolet emission lines are mostly permitted or intercombination transitions that suffer less from these problems but can instead be subject to optical depth effects. Clumping may further enhance the impact of high densities on the emergent spectrum.

Of course modern numerical models for the spectra from ionized nova shells take these points into account, but even so they remain an area of concern. For example, the output of the models will depend on the way that mass is distributed in density as well as in radius from the central photoionization source. There are ways out of these difficulties, but they are not trivial to apply. One key approach is to fit novae over a range of times and then require that the abundances not change to constrain the models. However, the time evolution of n_e is not easily determined since the clump densities will not follow a simple scaling law. Another approach is to identify specific components of the nebula from their radial velocity or spatial position and analyze the hopefully simpler clump rather than the entire ensemble of inhomogeneities througout the ejecta.

The simplest solution of waiting until the nova ejecta becomes a "real" nebula in the low density limit is probably not available, since by the time this happens in most novae the nebula might no longer be in strict photoionization equilibrium due to the turn-off of the central source. In this case we would trade one difficulty for another (see Williams et al. 1978; Ferland et al. 1984).

6. Summary

The rich spectroscopic variety of post-outburst spectra of classical novae reflect the complex conditions which exist within nova systems. Fortunately these structural details merely obscure rather than completely hide many of the basic features of novae, such as the post-outburst constant luminosity phase. However, the specifics become more important as we seek to make measurements of key ejecta properties, such as mass and composition, and we should attempt to further quantify the magnitude of such effects on our derivations.

The bi-symmetric nature of the nova ejecta is an old empirical observation but still an interesting one. Does this structure tell us how the outburst interacts with the binary, or are the symmetry planes induced by some other factor? How do bi-symmetric novae change their properties with viewing angle?

Several lines of evidence demonstrate that matter clumps on small scales within novae. This introduces a large range in density, and we should beware of the impact of such effects on the interpretation of nebular spectra of novae. With new observational techniques classical novae in outburst will continue to illuminate an important evolutionary process and the fundamental physics of energetic mass outflows.

I thank the organizers for the opportunity to attend this workshop, which has been excellent in every regard. I would also like to express my appreciation to Karen Bjorkman, John Mathis, Ken Nordsieck, Hakki Ögelman, Marina Orio, and Mary Jane Taylor, for acting as the nova crew in Madison; Icko Iben and Jim Truran for attempts to transfer theoretical ideas, and my Wisconsin nova students, Amy Barger and Lea Shanley, for their many contributions. This research was partially supported through *Hubble Space Telescope* General Observer research program GO-5467.01-93A.

Barger, A. J., Gallagher, J. S., Bjorkman, K. S., Johansen, K. A., and Nordsieck, K. H. 1993, ApJ, 419, L85

Bjorkman, K. S., Johansen, K. A., Nordsieck, K. H., Gallagher, J. S., and Barger, A. J. 1994, ApJ, 425, 247

Chochol, D., Hric, L., Urban, Z., Kromžik, R., Grygar, J., and Papoušek, J. 1993, A&A, 277, 103

Dinerstein, D. L. and Benjamin, R. A. 1993, Rev. Mex. Ast. Ap., 27, 33

Ferland, G. J., Williams, R. E., Lambert, D. L., Shields, G. A., Slovak, M., Gondhalekar, P. M., and Truran, J. W. 1984, ApJ, 281, 194

Gallagher, J. S. and Anderson, C. M. 1976, ApJ, 203, 625

Gallagher, J. S. and Starrfield, S. G. 1978, Ann. Rev. Astron. Ap., 16, 171

Gehrz, R. D. 1988, Ann. Rev. Astron. Ap., 26, 377

Greenhouse, M. A., Grasdalen, G. L., Woodward, C. E., Benson, J., Gehrz, R. D., Rosenthal, E., and Skrutskie, M. F., 1990, ApJ, 352, 307

Greiner, J., Hasinger, G., and Kahabka, P. 1991, A&A, 246, L17

Hauschildt, P. H. , Wehrse, R., Starrfield, S., and Shaviv, G. 1992, ApJ, 393, 307

Hauschildt, P. H., Starrfield, S., Austin, S., Wagner, M., Shore, S. N., and Sonneborn, G. 1994, ApJ, 422, 831

Hutchings, J. B. 1972, MNRAS, 158, 177

Iben, I., Jr. and Tutukov, A. V. 1992, ApJ, 389, 369

Iben, I., Jr. and Tutukov, A. V. 1992, ApJ, 389, 369

Iben, I., Jr. and Livio, M. 1993, PASP, 105, 1373

Kato, M. 1994, A&A, 281, L49

Kato, M. and Iben, I., Jr. 1992, ApJ, 394, L47

Krautter, J. and Williams, R. E. 1989, ApJ, 341, 968

Livio, M., Shankar, A., Burkert, A., and Truran, J. W. 1990, ApJ, 356, 250

McLaughlin, D. B. 1960, in "Stellar Atmospheres", ed. J. L. Greenstein, p585

Mould, J. R., Cohen, J., Graham, J. R., Hamilton, D., Matthews, K., Picard, A., Reid, N., Schmidt, M., Soifer, T., Wislon, C., Rich, M. R., and Gunn, J. E. 1990, ApJ, 353, L35

Orio, M., Trussoni, E., and Ögelman, H. 1992, A&A, 257, 548

Orio, M. and Ögelman, H. 1993, A&A, 273, L56

Orio, M., Ögelman, H., Pietsch, W., Bianchini, A., Della Valle, M., Krautter, J., and Starrfield, S. G. 1993, Ann. Israel Phys. Soc. 10

Osterbrock, D. E. 1989, "Astrophysics of Gaseous Nebulae and Active Galactic Nuclei" University Science Books, Mill Valley

Paresce, F. 1994, A&A, 282, L13

Paresce, F., Livio, M., Hack, W., and Korista, K. 1995, A&A, in press

Saizer, P., Starrfield, S. G., Ferland, G. J., Wagner, R. M., Truran, J. W., Kenyon, S. J., Sparks, W. M., and Williams, R. E. 1991, ApJ, 367, 310

Payne-Gaposchkin, C., 1957, "The Galactic Novae" North Holland, Amsterdam

Prialnik, D. and Kovetz, A. 1992, ApJ, 385, 665

Quirrenbach, A., Elias, N. M., Mozurkewich, D., Armstrong, J. T., Buscher, D. F., and Hummel, C. A. 1993, AJ, 106, 1118

Shanley, L., Ögelman, H., Gallagher, J. S., Orio, M., and Krautter, J. 1995, ApJ, 438, L95

Shore, S. N., Sonneborn, G., Starrfield, S. G., Gonzales-Riestra, R., and Ake, T. B. 1992, AJ, 106, 2408

Starrfield, S. G. and Sparks, W. M. 1987, Ap&SS, 131, 379

Tomany, A. B. and Shafter, A. W. 1992, ApJS, 81, 683

Tomany, A. B. and Shafter, A. W. 1993, ApJ, 411, 640

Williams, R. E. 1992, AJ, 104, 725

Williams, R. E., Woolf, N. J., Hege, E. K., Moore, R. L., and Kopiva, D. A. 1978, ApJ, 224, 171

Williams, R. E., Phillips, M. M., and Hamuy, M. 1994, ApJS, 90, 297

DUST EVOLUTION IN NOVA AQUILAE 1993 = V1419 AQL

Boris F. Yudin
Sternberg Astronomical Institute, University of Moscow, Universitetsky prospect 13, 117234 Moscow, Russia

Ulisse Munari
Osservatorio Astronomico di Padova, Sede di Asiago, I-36012 Asiago (VI), Italy

1 Introduction

Nova Aquilae 1993 was discovered on UT May 14.6, 1993 by Yamamoto (1993) at a photographic magnitude ∼8. Inspection of U.K.Schmidt plates and Palomar Sky Survey prints led McNaught (1993) to conclude that the nova precursor was fainter than mag 22 in quiescence up to 5 years before the outburst.

The photometric evolution of N Aql 1993 during the first 220 days of the outburst in the UBV-JHKLM bands has been presented by Munari *et al.* (1994, hereafter MYKSSL). The nova belongs to the *Fe II* group, suffers from an $E_{B-V} = 0.55(\pm 0.15)$ reddening, it is at a distance $5.5(\pm 1.0)$ kpc and the speed class is *Fast* (t_3=32). N Aql 1993 entered the *transition phase* 35 days after maximum (reached on JD 2449130). At that time dust started to form in the ejecta. The dust caused an extinction of $\triangle$V=3.0 mag in the optical and a brightening of $\triangle$L=3.0 mag in the infrared (maximum reached on JD ∼2449195). High resolution echelle spectra of the emission line profiles suggest a pronunced clumpsy nature of the ejecta and an expansion velocity of ∼1000 km sec^{-1}.

Updated V, U-V, L and K-M lightcurves of N Aql 1993 are presented in Figure 1.

2 Dust properties

The dust condensed in the ejecta reached a maximum optical thickness

$$\tau_V \approx 2.2 \tag{1}$$

and a total mass of

$$M_{\text{dust}} \approx 10^{-7} \, M_\odot \tag{2}$$

Dust condensation took just ∼20 days to complete. Aside from any possible over–abundance of heavy elements in the ejecta, this may suggest that the *sticking efficiency* of condensable elements onto dust grain nucleation sites has been very high. The dust absorption efficiency turned from neutral to selective at λ ∼0.6 μm which indicates a prevalent carbon composition for grains with a diameter of the order of 0.1 μm.

The estimated gas–to–dust ratio in the nova ejecta is

$$M_{\text{dust}}/M_{\text{gas}} \approx 0.002 \tag{3}$$

295

A Bianchini et al (eds) Cataclysmic Variables 295-298

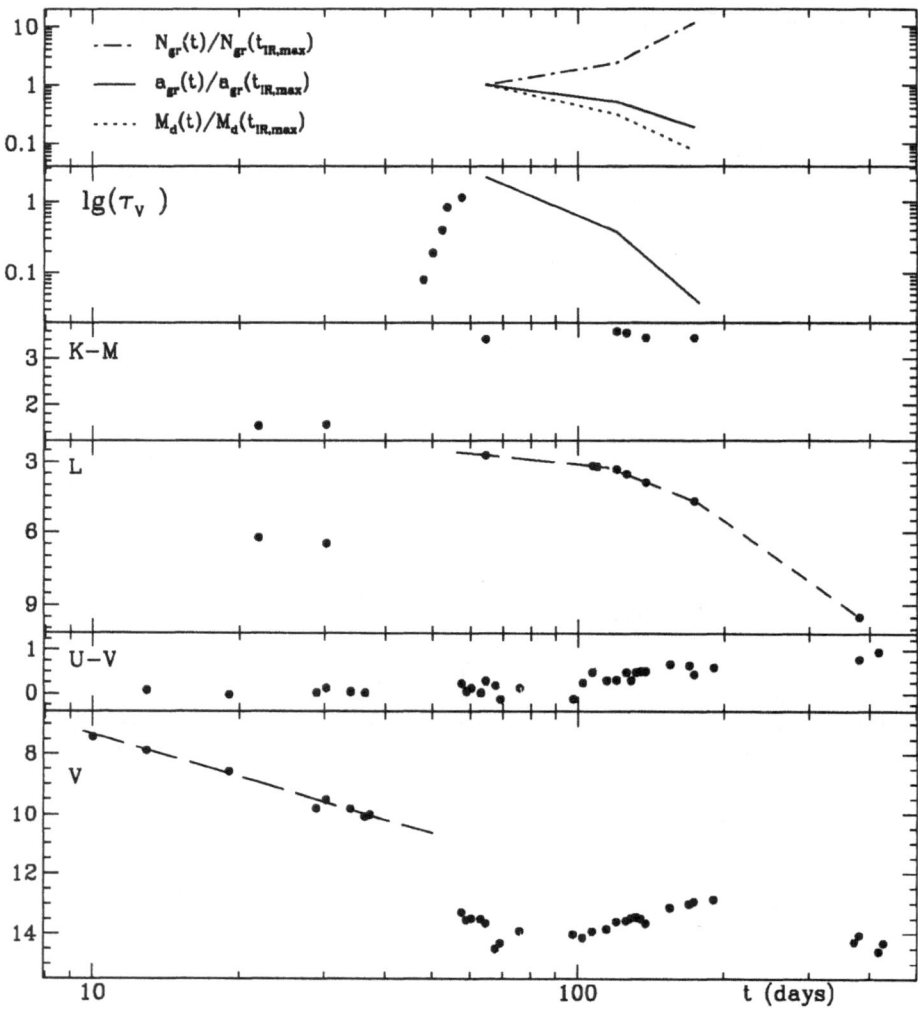

Figure 2: V, U-V, L, K-M lightcurves for Nova Aql 1993. τ = optical depth of the dust shell; a_{gr} = dust grain radius; N_{gr} = total number of dust grains; M_d = total mass of dust in the ejecta; $t_{IR,max}$ = time of maximum IR brightness.

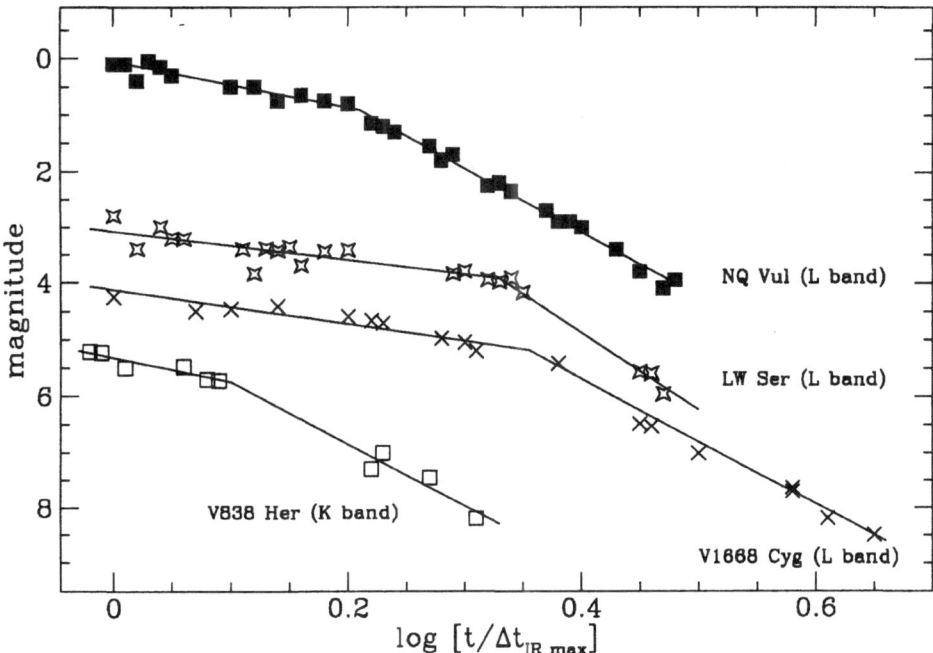

Figure 1: Light curves of NQ Vul (Nova Vul 1976), LW Ser (Nova Ser 1978), V1668 Cyg (Nova Cyg 1978) and V838 Her (Nova Her 1991). Data are from Ney and Hatfield (1978), Gehrz et al. (1980a), Gehrz et al. (1980b), and Chandrasekhar et al.(1992) respectively. Δt_{IRmax} is the time elapsed from the nova eruption to the IR maximum. Sharp changes in the IR declining rate of these dusty novae similar to that observed in Nova Aql 1993 are evident.

and the region where dust condensed covers $\leq 10\%$ of the radial extension of the ejecta. Therefore, no significant depletion of heavy elements from the gas phase is expected in a vast fraction of the ejecta themselves.

During the growth in dimension of the dust grains, the latter are expected to pass through sizes that should produce an observable reddening of the optical colors of the nova. In the case of Nova Aql 1993, this reddening phase may have lasted no longer than ~ 1 week, causing a $\Delta E_{U-\lambda} \leq 0.6$ mag . Unfortunately, our multi-band photometric coverage (see Figure 1) has a gap (due to bad weather) right at this critical phase (~ 45 after visual maximum) and the predicted effect cannot be checked.

From MYKSSL data it may be inferred that Nova Aql 1993 experienced a phase of constant bolometric luminosity (well above the Eddington limit)

$$L_{bol} \approx 2 \times 10^5 \ L_\odot \tag{4}$$

which lasted the first ~ 20 days after optical maximum, with progressive re-distribution of the energy output toward shorter wavelengths. Afterwards, the luminosity started to decline, reaching a new constant value at the time of IR maximum

$$L_{bol} \approx 2 \times 10^4 \; L_\odot \qquad (5)$$

which persisted for ~ 100 days. The mean rate of decline between the two luminosity regimes

$$L \propto t^{-2.8} \qquad (6)$$

has been faster than the decrease of the volume emission measure of the expanding gas envelope

$$n_e^2 \propto t^{-2} \qquad (7)$$

This may be the cause leading to the formation of regions where conditions were favourable for dust to condense. During the time covered by our observations in Figure 1 the decline of the dust shell luminosity was not accompanied by noticeable changes in its color temperature, as it may be inferred by the K-M color index.

When the optical depth of the dust envelope thinned to $\tau_V \sim 0.6$ (after $\sim$110 days from optical maximum), the hard radiation from the central star started to become effective in destroying the dust grains, and the rate of brightness decline in the IR consequently increased from

$$L_{IR} \propto t^{-0.9} \qquad (8)$$

to

$$L_{IR} \propto t^{-3.4} \qquad (9)$$

This abrupt change is a common feature of dusty novae (see Figure 2) and it usually appears at time

$$1.3 \leq \frac{t}{\Delta t_{IRmax}} \leq 2.3 \qquad (10)$$

where Δt_{IRmax} is the time delay between nova eruption and the maximum in the IR.

During the grain destruction phase the mass of the dust shell and the dust grain radius were decreasing at rates

$$M_{dust} \propto t^{-1.9}, \quad a(t) \propto t^{-1.1} \quad \text{(for } 1.0 < \frac{t}{\Delta t_{IRmax}} < 2.0) \qquad (11)$$

$$M_{dust} \propto t^{-3.9}, \quad a(t) \propto t^{-2.7} \quad \text{(for } 2.0 < \frac{t}{\Delta t_{IRmax}} < 3.0) \qquad (12)$$

References

Chandrasekhar T., Ashok N.M., Ragland S. 1992, MNRAS 255, 412
Gehrz R.D., Grasdalen G.L., Hackwell J.A., Ney E.P. 1980a, ApJ 237, 855
Gehrz R.D., Hackwell J.A., Grasdalen G.L., Ney E.P., Neugebaumer G.,
    Sellgren K. 1980b, ApJ. 239, 570
Mcnaught R.H. 1993 IAU Circ. 5798
Munari U., Yudin B.F., Kolotilov E.A., Shenavrin V.I., Sostero G.,
    Lepardo A. 1994, A&A 284, L9
Ney E.P., Hatfield B.F. 1978, ApJL 219, L111
Yamamoto M. 1993, IAU Circ. 5791

Results of an extended spectroscopic survey of suspected CVs

Tomaž Zwitter

Universitá di Padova, Dip. di Astronomia, Padova, Italia
University of Ljubljana, Dept. of Physics, Ljubljana, Slovenia
e-mail: tomaz.zwitter@uni-lj.si

This is a status report on some results of an extended spectroscopic survey of suspected cataclysmic variables (CVs), predominantly those without published spectra. The survey is a common effort: U. Munari and myself observe with the 1.8-m telescope of the Asiago Observatory and the ESO 1.5 m, A. Bragaglia has her hands on the 1.5-m instrument in Loiano, and Herman Mikuž from Slovenia contributes simultaneous CCD photometry when needed.

We observe with two different spectral resolutions. Observations at low resolution, so far constituting of spectra of 68 systems, are aimed predominantly at spectroscopic confirmation of faint objects without published spectroscopy. This work is important. In a recent catalogue Downes and Shara (1993, DS93) list 751 CVs and from that for 359 objects they found no published spectrum in the literature. Duerbeck (1987) published an extended catalogue for the galactic novae. Spectroscopy seems really crucial considering that 74 objects formerly considered CVs were rejected their CV status on the basis of a new spectroscopic information.

Moreover, low resolution observations are usefull to provide a clue at the type of a CV one observes, as well as to search for a signature of companion both in the red part of continuum as well as in e.g. the Ca II IR triplet. Finally, the gathered information can be used for statistical analysis of properties of continuum and lines based on a large homogeneous sample.

A. Bianchini et al. (eds.), Cataclysmic Variables, 299-301.
© 1995 Kluwer Academic Publishers.

Apart from low resolution work we observe at high resolution using Echelle spectrographs at Asiago and Loiano. So far we observed more than 20 bright CVs obtaining a small number of spectra for each system. These observations are not meant to study each system in great detail. Instead, we want to answer questions like what are typical line profiles in CVs and, in particular, what is the fraction of double peaked ones. In AGNs, which are supposed to contain an accretion disk, the double peaked BLR emission lines which can be succesfully fit with an accretion disk were observed in only 8 systems out of more than a 100. In CVs we are much better off with such an evidence for accretion disks. However we need a large homogeneous sample before we can address the question properly.

Apart from this catalogue of high resolution spectra we perform a detailed study of a few systems. So far our target was V795 Her, and we obtained also a number of spectra of SY Cnc.

In low resolution observations we finnished so far the analysis of 56 objects which lack published spectroscopy (Zwitter & Munari 1994, 1995). In 35 cases we were able to confirm their CV status, other 14 objects show continua without emission lines, and we were able to reject CV membership of 7 objects. The confirmation rate is particularly low for candidates from the Palomar–Green survey (Green et al. 1986). We have so far observed 10 objects listed as CVs in the PG catalogue and on this base included among the entries of DS93. Only one of these PG stars has shown a CV spectrum, three were clear mis–classifications (a starburst galaxy, a planetary nebula and a WD), and five have simply hot continua without emission lines ($eg.$ like those of OB subdwarfs).

Some miss-classifications are quite dramatic. PG 1136+581 turned out to be actually a spatially resolved starburst galaxy ($\equiv$Mkn 1450) receeding with a velocity of 940 km/s (Munari et al. 1995).

So, the important result is that up to 30% of objects listed in a recent catalogue as CVs may be missclassified. This is meant as a word of caution to people performing statistical analysis of CVs using such catalogues.

We postpone a throughout statistical analysis of the properties of lines and continuum till the spectroscopic survey in both hemispheres is completed. Below, however, we mention some preliminary results based on programme stars from Paper I together with bright CV templates we observed. In all cases, to our best knowledge, the results are based on objects that were observed in quiescence and outside of any possible eclipses.

Both the flux ratio of the 5876 to the 6678 He I line and the Hα to Hβ ratio require very high densities, so that standard nebular conditions do not apply.

Given the properties of Balmer, He I lines and continuum one can try to find various correlations. The first is a trivial one, namely a standard color-color diagram obtained from our spectroscopic observations. All confirmed CVs cluster around the main sequence traced by colors of a few standards we observed. Unconfirmed CVs are more scattered however mostly still consistent with a main sequence position.

There seems to be no correlation of Hα to Hβ ratio vs. color. The plot of the equivalent width of Hα vs. color is more informative. The color is estimated as the ratio of continuum fluxes at the positions of the Hα and Hβ lines. The points are quite scattered around, however a few CV classes seem to occupy a special place in this diagram. Stars classified as U Gem (without a subtype) seem to be distinctly red and with a strong Hα. SU UMa-s, on the other hand are blue and have extremely strong Hα. Also the DQ Her stars occupy a special region of very blue colors and moderately strong Hα-s.

References

[1] Downes R.A. & Shara M.M. 1993, PASP 105, 127, DS93

[2] Duerbeck H.W. 1987, *A Reference Catalogue and Atlas of Galactic Novae*, Reidel, D87

[3] Green R.F., Schmidt M. & Liebert J. 1986, ApJS 61, 305

[4] Munari U., Zwitter T., Mikuž H. 1995, A&A 296, 310

[5] Zwitter T., Munari U. 1994, A&AS 107, 503, Paper I

[6] Zwitter T., Munari U. 1995, A&AS, in print, Paper II

THE MG II LINE OF NOVA PUPPIS 1991 ON DAY 150

I. PACHOULAKIS AND P. SAIZAR
University of Pennsylvania, Dpt. of Physics & Astronomy
209 S. 33rd Street, Philadelphia, PA 19104, U.S.A.

Abstract. While line strengths are integrated measures of line formation over geometry and kinematics, high resolution line profiles encode the density and ionization structure of a shell in their shapes. The extent at which fitting of line profiles can be used in conjunction with photoionization models to disclose the spatial structure of a shell is the subject of an ongoing investigation.

To capitalize on the information content of *IUE* images related to line shapes the *SEI* method was used to model the Mg II (2800Å) resonance line doublet of Nova Puppis 1991 on day 150 after outburst. The *SEI* code (Lamers *et al.*, *Ap.J.*, **314**, 726) models winds from hot stars and PN^e, so it had to be ramified to apply to the case of a shell. Only the best fit for *IUE* high dispersion image LWP23209 is summarized below:

- The Mg II shell expands at 1,200 $km \cdot s^{-1}$ and is turbulent at a level of 30% of its coasting speed, or 360 $km \cdot s^{-1}$.
- The density profile of the Mg II shell is assumed to be gaussian. The ratio of shell radius to thickness is 40. A step-like density function did not yield significantly different results.
- The total radial integrated optical depths of the blue and the red components of the doublet are 0.8 and 1.1 respectively (the line is optically thin at 150 days).
- The line-formation mechanism is understood to be collisional excitation and radiative de-excitation. With a temperature of $T_e = 10^4$ K and $N_e = 2 \times 10^8 cm^{-3}$ we obtain $C_{12} = 8.64$ s^{-1}. A_{21} for the blue and red components are 2.61×10^8 cm^{-3}, and 2.59×10^8 s^{-1} respectively. An isothermal shell is assumed.

The implied column density of Mg II is $N = 2.7 \times 10^{14}$ cm^{-2}. Assuming constant expansion rate since outburst we obtain a shell radius of 1.5×10^{15} cm. The implied total mass of Mg II in the shell is $1.64 \times 10^{-10} M_\odot$.

303

A. Bianchini et al. (eds.), Cataclysmic Variables, 303.
© *1995 Kluwer Academic Publishers.*

5. ACCRETION INDUCED PHENOMENA

THE DISK INSTABILITY MODEL OF SU UMA STARS: CONNECTION BETWEEN THE SU UMA STARS AND THE WZ SGE STARS

Y. OSAKI

Department of Astronomy, School of Science,
University of Tokyo, Bunkyo-ku, Tokyo, 113, Japan

Abstract. Based on the disk instability model (or the thermal-tidal instability model proposed by the present author [Osaki 1989]), we demonstrate that the activity sequence among SU UMa-type dwarf novae (ranging from frequently outbursting SU UMa stars to very rarely outbursting WZ Sge stars) is understood as variation in input parameters, in particular as variation in mass transfer rate decreasing with decreasing activity. However, the very low viscosity in quiescence besides the low mass transfer rate is required for the extreme case of WZ Sge stars with extremely long outburst intervals.

1. Introduction

SU UMa stars are one sub-class of dwarf novae, whose characteristics are: (1) they exhibit two clearly distinguished types of outbursts, a short outburst called "normal outburst" which lasts for a few days and a long outburst called "superoutburst" which lasts typically for two weeks or so, (2) their orbital periods are very short, in almost all cases, below the "CV period gap", (3) they exhibit the periodic photometric light variation called "superhumps" during and only during superoutbursts. Even within the SU UMa-type dwarf novae, there exist wide differences in outburst activity: ranging from very active systems such as SU UMa and VW Hyi showing outbursts every 20 to 30 days to extremely inactive systems such as WZ Sge showing only rare large-amplitude outbursts with recurrence time of some 30 years.

307

A. Bianchini et al. (eds.), Cataclysmic Variables, 307-313.

In oder to explain superoutburst and superhump phenomenon of SU UMa stars, the present author (Osaki 1989a,b) has proposed the so-called "thermal-tidal instability model" within the general frame-work of the disk instability model of dwarf novae, in which the interplay of the two intrinsic instabilities (the thermal and the tidal instabilities) within the accretion disk plays an essential role. The main purpose of this paper is to demonstrate that this disk instability model is able to explain rich varieties in outburst activity among SU UMa-type dwarf novae.

2. Thermal-tidal instability model of SU UMa stars

Here we briefly explain the thermal-tidal instability model of SU UMa stars (Osaki 1989a,b), upon which all later discussions about the problem of various activity classes in SU UMa stars will be based. It is now well known that there are two intrinsic instabilities within accretion disks of cataclysmic variable stars (abbreviated as CVs). One of the instabilities is the thermal instability (or the ordinary disk instability) due to the partial ionization of hydrogen and this instability is now thought to be responsible for the outburst of U Gem-type ordinary dwarf novae (see, e.g.. a recent review by Cannizzo 1993).

Another is the tidal instability (or the tidally driven eccentric instability), which was first discovered by Whitehurst (1988) and was further confirmed by Hirose and Osaki (1990) and Lubow (1991a,b). When an accretion disk becomes large enough for its outer edge to reach the so-called 3:1 resonance radius (or the critical radius for the tidal instability) in the close binary system, it becomes unstable by the tidal effects of the secondary star and it is transformed to an eccentric disk whose apsidal line slowly revolves in the inertial frame of reference (a precessing eccentric disk). The "superhump" phenomenon is now believed to be explained by the periodic tidal stressing of the eccentric disk and the superhump period is explained by the synodic period of the precessing eccentric disk and the orbiting secondary star. The tidal instability occurs only in CVs with a small mass ratio where the mass ratio is defined by the ratio of masses of the secondary star to the primary star; $q = M_2/M_1$. Since the mass of the secondary star is closely related to the orbital periods in CVs, this condition requires approximately CVs below the "period gap" in order to have the tidal instability and therefore the superhumps.

The typical SU UMa stars exhibit the very characteristic pattern of outburst light curves ("super-cycle") in which several short "normal" outbursts occur between two "superoutbursts", which occur more regularly than normal outbursts. This supercycle phenomenon is explained in the thermal-tidal instability model in the following way. In this model, the mass transfer rate from the secondary star is assumed to be constant all the time and all of outburst behavior is thought to be produced by intrinsic instabilities within the accretion disk (i. e., the disk instability model). In this model, both the normal outburst and the superoutburst are produced by the thermal instability of accretion disk but the difference between the two exists in that the superoutburst is an outburst which is accompanied with the tidal instability while the normal outburst is one without the tidal instability. This difference is, in turn, produced by difference in the disk radius when an outburst occurs. That is, in the early phase of supercycle, the disk is compact and the disk's outer edge does not reach the 3:1 critical radius, although the disk expands every time when an outburst occurs. As time goes on, the disk radius gradually increases and finally it reaches the critical radius when the final normal outburst occurs. The tidal instability then takes place and it produces the superoutburst and superhumps. The long duration of the superoutburst is explained in this model by the greatly enhanced tidal torques when an eccentric disk is formed. After the end of the superoutburst, the disk becomes compact because the angular momentum of the disk is greatly removed during the superoutburst and its new supercycle begins once more.

Based on this model, we have successfully simulated light curves of SU UMa stars (see, Figures 3 and 4 of Osaki 1989a; Figures 1 and 2 of Ichikawa, Hirose, and Osaki 1993). The most important in this model is the disk radius variation during the supercycle and thus we always show it together with the light curve.

3. SU UMa/WZ Sge connection

Vogt (1993) has discussed a wide variety of SU UMa stars in their activity, ranging from very frequently erupting SU UMa stars (his group A) to the infrequent WZ Sge stars (his group C) of large amplitude, and he has suggested that the same physical process governs the activity of all of these stars as the continuous sequence of various observational parameters exist in these stars. In the Vogt (1993) activity sequence from the most active group A to the least active group C (WZ Sge stars), the frequency of the superoutburst and the number of short eruptions between two consequent superoutbursts decrease with this sequence (see his table and Table 1 of Osaki 1994a).

The present author (Osaki 1994a) has tried to explain the Vogt activity sequence as the CV evolutionary sequence below the period gap in which the binary evolution is driven by the gravitational-wave radiation and in which orbital period decreases and the mass transfer rate decreases with this sequence. In order to explain the Vogt activity sequence, it has turned out

(Osaki 1994a,b) that the most important parameters are the mass transfer rate. It has been shown that the differences between the most active group A with a typical superoutburst recurrence of 130-200 days and the intermediate group B with that of 300-500 days are explained by the difference in mass transfer rate in which group B stars have lower mass transfer rate by a factor of about two than that of group A. It has been shown there (see, figure 5 of Osaki 1994a) that a reduction of mass transfer rate reduces not only the frequency of superoutbursts but also the number of short outbursts between two superoutbursts.

However, it has been found (Smak 1993, Osaki 1994a,b) that in order to explain extremely long recurrence time of WZ Sge stars (group C) it is not enough to reduce the mass transfer rate but it is also required to reduce the viscosity parameter α_{cold} in quiescence. If we adopt a typical viscosity parameter α_{cold} but we simply reduce the mass transfer rate, the recurrence time of normal outburst becomes independent of the mass transfer rate (Ichikawa and Osaki 1994) and it is simply given by the viscous time scale of the cold disk, which is of the order of hundreds days but not of 30 years. A similar conclusion has recently been reached by Mennickent (1994 a poster paper in these proceedings) and Howell, Szkody, and Cannizzo (1994).

I have simulated the outburst light curve and the disk radius variation of WZ Sge star itself (1994b) based on the simplifying model used by the present author (Osaki 1989a), and Figure 1 illustrates the results of calculation. The most important characteristics of the outburst light curve shown in Figure 1 is a long duration of outburst (i.e., of about 30 days) which is produced by the viscous depletion of large amount of mass accumulated during the long quiescence. It must be noted that the mass transfer rate and the viscosity parameters which I used to simulate WZ Sge were much smaller than those of the typical SU UMa stars. The variation in the disk radius is as follows (not shown here because of shortage of space). Starting from the disk radius in quiescence of $R_d = 0.23A$ (which corresponds to the circularization radius or the Lubow-Shu radius), the disk expands rapidly to $R_d = 0.451A$ once the disk makes an upward thermal transition. The disk radius at this stage is still below the critical radius for the 3:1 resonance ($R_{d,crit} = 0.46A$) but the disk radius continues to increase and it reaches the critical radius soon. Then the tidal instability sets in and the disk becomes eccentric. It then produces greatly enhanced tidal torques and it keeps the disk radius within the critical radius and keeps the disk in hot viscous state for a long time. We also note that the outburst occurs every 30 years with these parameters.

In the Garching conference (Osaki 1994a), I overemphasized the connection of the Vogt activity sequence to the CV evolutionary sequence, in particular the connection between the activity group and the orbital peri-

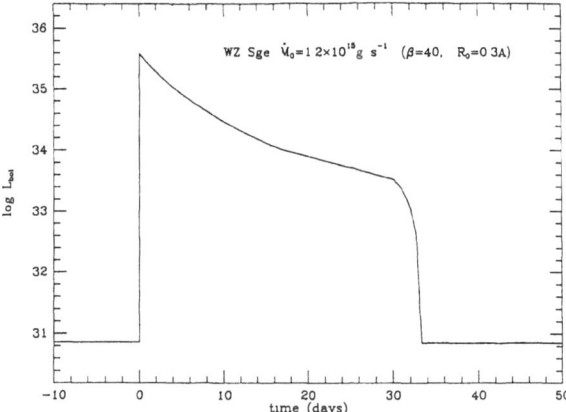

Figure 1. Model light curve of WZ Sge in which the bolometric luminosity is only due to the accretion disk. The model parameters used are: $M_1 = 1M_\odot$, the binary separation, $A = 4.46 \times 10^{10}$cm, mass transfer rate, $\dot{M} = 1.2 \times 10^{15}\mathrm{gs}^{-1}$, the viscosity parameter in hot state, $\alpha_{\mathrm{hot}} = 0.3$, and a model parameter $\beta = 40$ (which roughly corresponds to the viscosity parameter in cold state, $\alpha_{\mathrm{cold}} \simeq 0.001$)and another model parameter $R_{d,0} = 0.3A$, the disk radius at the end of the superoutburst (which represents the strength of the tidal torques in the eccentric disk).

ods. In fact, the mass transfer rate may not necessarily uniquely be given for given binary parameters as expected in the gravitational-wave radiation theory but it may likely vary secularly such as in the hibernation scenario of nova/dwarf nova alternation (see, Livio 1992). In fact, Patterson and his group (see, e.g., Skillman and Patterson 1993) have shown that there are some nova-like stars exhibiting "permanent superhumps" (hereafter called "permanent superhumpers") and this suggests that these stars should have much higher mass transfer rate than those expected from the CV evolution theory. Furthermore, Kato and Kunjaya (1994) have recently discovered that an ultraviolet excess object PG 0943+521 could be an unusual SU UMa-type dwarf nova with superoutbursts occurring every 43 days and lasting 20 days and normal outbursts occurring every four days because this star exhibits periodic photometric humps with period of 0.065 days during the long-lasting (20 days) bright phase which these authors interpret "superhumps". (It is noted here that the same star of PG 0943+521 was interpreted in this conference by Honeycutt [1994] as a nova-like star showing secular variation in mass transfer rate with a recurrence period of 43 days.) If Kato and Kunjaya's (1994) interpretation is correct, it means that the object PG 0943+521 could be an SU UMa star with a fairly high mass transfer rate, putting this star near the border-line case between nova-like stars and dwarf novae, i.e., corresponding to a "Z Cam" counterpart of cataclysmic variable stars below the period gap.

4. "Permanent Superhumpers" and a star PG 0943+521

In this section, we examine whether or not permanent superhumpers and the star PG 0943+521 can be accommodated within our thermal-tidal instability model. It will be evident that "permanent superhumpers" can be easily understood as binary systems having mass transfer rate high enough to avoid the thermal instability but suffering the tidal instability with the binary mass ratio M_2/M_1 being small enough.

If we accept Kato and Kunjaya's (1994) interpretation for the star PG 0943 +521, then this star will be the most frequently outbursting SU UMa star so far known. From the discussion presented above, such a system will most likely be explained as a system with mass transfer rate higher than those of the Vogt (1993) most active group A in our thermal-tidal instability model. In Osaki's (1989a) simplifying model, we have successfully simulated an active SU UMa star, VW Hyi, with mass transfer rate of $\dot{M} = 10^{16} \mathrm{gs}^{-1}$ and a model parameter $\beta = 0.8$, and $R_0 = 0.3A$. We have simulated the star PG 0943+521 by adopting the mass transfer rate higher by a factor 3.5 than that of VW Hyi with model parameters, $\dot{M} = 3.5 \times 10^{16} \mathrm{gs}^{-1}$ and the binary separation $A = 4.7 \times 10^{10} \mathrm{cm}$, $\beta = 0.6$, and $R_0 = 0.35A$. Our simulations have yielded the superoutburst recurrence time of 45 days, the duration of a superoutburst of 20 days, the normal outburst recurrence time of 5 days, in a good agreement with Kato and Kunjaya's (1994) observations. The long duty cycle of the superoutburst in our simulations is due to high mass transfer rate from the secondary in comparison with mass accretion rate onto the white dwarf during the superoutburst.

5. Conclusion

(1) The Vogt activity sequence of SU UMa stars is understood in the the thermal-tidal instability model as a continuous sequence of varying model parameters, in which the most important parameter is the mass transfer rate from the secondary stars.

(2) All peculiarities of WZ Sge stars, in particular of WZ Sge itself, can be understood as an extreme case of very low viscosity in quiescence and of rather low mass transfer rate.

(3) The thermal-tidal instability model can explain rich varieties of outburst behaviors of CVs below the period gap (including "permanent superhumpers" and the star PG 0943+521). Combined with the ordinary disk instability model for CVs above the period gap (i.e., the thermal limit cycle model without the tidal instability), the disk instability model can explain almost all of varieties of outburst behaviors in non-magnetic CVs.

References

Cannizzo, J. K., 1993 in *Accretion Disks in Compact Stellar Systems* ed. J. Craig Wheeler (World Scientific Publishing, Singapore), p. 6.

Hirose, M. and Osaki, Y. 1990 , *Publ. Astron. Soc. Japan*, **42**, 135.

Honeycutt, R. K. 1994, in these proceedings.

Howell, S. B., Szkody, P., and Cannizzo, J. K., 1994, *Astrophys. J.*, in press.

Ichikawa, S., Hirose, M., and Osaki, Y. 1993, *Publ. Astron. Soc. Japan* **45**, 243.

Ichikawa, S. and Osaki, Y. 1994, in *Theory of Accretion Disks-2,* eds.,W. Duschl, et al. (Kluwer Academic Publishers, Dordrecht), p. 169.

Kato, T. and Kunjaya, C. 1994, submitted to *Publ. Astron. Soc. Japan*

Livio M. 1992, in *Vina del Mar Workshop on Cataclysmic Variable Stars,* ed. N. Vogt, ASP Conference Series, **29**, p. 269.

Lubow, S. H. 1991a, *Astrophys. J.*, **381**, 259.

Lubow, S. H. 1991b, *Astrophys. J.*, **381**, 268.

Mennickent, R. 1994, in these proceedings.

Osaki, Y. 1989a, *Publ. Astron. Soc. Japan.*, **41**, 1005.

Osaki, Y. 1989b, in *Theory of Accretion Disks,* eds.,F. Meyer et al. (Kluwer Academic Publishers, Dordrecht), p. 183.

Osaki, Y. 1994a, in *Theory of Accretion Disks-2,* eds.,W. Duschl, et al. (Kluwer Academic Publishers, Dordrecht), p. 93.

Osaki, Y. 1994b, submitted to *Publ. Astron. Soc. Japan*.

Skillman D. R. and Patterson, J. 1993, *Astrophys. J.*, **417**, 298.

Smak, J. 1993, *Acta Astronomica*, **43**, 101.

Vogt, N. 1993, in *2nd Technion-Haifa Conference on Cataclysmic Variables and Related Physics* eds. O. Regev and G. Shaviv (The Israel Physical Society, Jerusalem, Israel), p. 63.

Whitehurst, R. 1988, *Monthly Notices Roy. Astron. Soc.*, **232**, 35.

IRRADIATION-DRIVEN MASS TRANSFER IN LOW MASS CLOSE BINARIES

KINWAH WU

Research Centre for Theoretical Astrophysics, School of Physics, The University of Sydney, NSW 2006, Australia

DAYAL T. WICKRAMASINGHE

ANU Astrophysical Theory Centre, School of Mathematical Science, Australian National University, ACT 0200, Australia

AND

BRIAN WARNER

Department of Astronomy, University of Cape Town, Rondebosch 7700, South Africa

Abstract. We present a formulation for irradiation-driven mass transfer in cataclysmic variables, which allows for feed-back due to radiation from an accretion-heated white dwarf. The resulting model predicts the likely mass transfer rates for CVs at different orbital periods.

1. Introduction

Mass transfer in low mass close binaries is generally believed to be driven by the orbital evolution of the binaries. However, many cataclysmic variables (CVs), a class of low mass close binaries consisting of a white dwarf and a less massive Roche-lobe filling secondary, are found to have much higher mass transfer rates than predicted by the orbital evolutionary models. Also, at the orbital periods $P_o \approx 3 - 4$ h, the spread in their absolute magnitude is very large, with $\Delta M_v \gtrsim 6$ (Warner 1987), implying that the mass transfer rates of the brightest and the faintest systems may differ by more than two orders of magnitude. Since these observations are difficult to be explained by the conventional magnetic braking (Verbunt & Zwaan

315

A Bianchini et al (eds), Cataclysmic Variables, 315-320

1981) and the gravitational radiation (Faulkner 1971) models, additional mechanisms must be invoked.

Studies (e.g. Osaki 1985, Hameury et al. 1986, Kovetz et al. 1988, Sarna 1990, Harpaz & Rappaport 1991) have shown that if the secondary in a binary is heated, mass transfer can be enhanced. In CVs, the strong UV and optical radiation emitted from the white dwarf, as a result of the accretion process, can deposit energy efficiently onto the secondary's atmosphere and heat up the star. The secondary therefore expands and further fills its Roche-lobe, causing more mass to be transferred and more radiation to be emitted. On the other hand, when mass is transferred from a low mass star to a higher mass star, the orbital separation expands. As a result, the secondary tends to detach from its Roche lobe during the mass transfer process, preventing a run-away in mass transfer. Mass transfer with irradiation heating is a feed-back process. Its stability depends on the balance of the response of the secondary to heating and the rate of radiation production due to the accretion onto the white dwarf.

In the previous studies, the heating of the secondary and the mass transfer rate were assumed to be independent. As feed-back effects are not taken into account, these studies are not self-consistent. Here we present a self-consistent formulation, in which feed-back effects due to a hot white dwarf are considered. This formulation is applied to estimate the mass transfer rates of CVs at different orbital periods.

2. Formulation

For CVs with a white dwarf of mass M_1 and a secondary of mass M_2, the mass transfer rate is $\dot{M}_2 = Q\rho c_s$, where ρ and c_s are the density and sound speed of the gas at the inner Lagrangian (L_1) point respectively, and Q the cross section of the gas stream, given by $Q = 2\pi a^3 c_s^2 / KG(M_1 + M_2)$. Here, G is the gravitational constant, a the orbital separation and K a constant depending on the mass ratio of the two stars. The density is given by $\rho = \rho_* \exp[-(\Delta r/H)^2]$, where ρ_* is the density at the isothermal atmosphere of the secondary, Δr the distance between the isothermal atmosphere of the secondary and the L_1 point, and $H = P_o c_s^2 / 2\pi\sqrt{A + 0.5}$ with $A \approx 8$. It follows that

$$\frac{\dot{M}_2}{\dot{M}_{2o}} = \left(\frac{T_2}{T_{2o}}\right)^{3/2} \exp\left[-\lambda^2\left(\left(\frac{T_{2o}\Delta r}{T_2 \Delta r_o}\right)^2 - 1\right)\right], \qquad (1)$$

where T_2 is the temperature of the secondary's atmosphere and $\lambda = \Delta r_o/H_o$. In equation (1) and hereafter, the subscript o denotes the equilibrium states (the states at which the time-derivatives of the quantities which describe the system are zero), while the quantities for the perturbed states are without a subscript. T_2/T_{2o} is given by

$$\frac{T_2}{T_{2o}} = \left[\frac{1 + (\gamma T_1^4 \dot{M}_2/T_{1o}^4 \dot{M}_{2o})}{1 + (\gamma T_1^4/T_{1o}^4)}\right]^{1/4}, \tag{2}$$

where $\gamma = \mu G M_1 \dot{M}_{2o} r_2^2 / 4 r_1 L_2 a^2$, μ the efficiency of converting the accretion energy to radiation, r_1 the white dwarf radius, T_1 the white dwarf temperature, r_2 the secondary's radius, and L_2 the secondary's intrinsic luminosity.

The rate of change in Δr as a result of mass transfer is given by

$$\frac{d}{dt}\left(\frac{\Delta r}{\Delta r_o}\right) = \left\{\left(\frac{r_{ho}}{\Delta r_o}\right)(\zeta_h - \zeta_s)\frac{\dot{M}_{2o}}{M_2}\left(\frac{\dot{M}_2}{\dot{M}_{2o}} - 1\right) + \left(1 - \frac{\Delta r}{\Delta r_o}\right)\left(\frac{r_{ho}}{\Delta r_o}\right)\right.$$

$$\left.\times\left[(\zeta_h - \zeta_s)\frac{\dot{M}_{2o}}{M_2} + \left(\frac{\partial}{\partial t}\ln r_{ho}\right)_{\dot{M}_2} - \left(\frac{\partial}{\partial t}\ln r_{2o}\right)_{\dot{M}_2}\right]\right\}, \tag{3}$$

where r_h is the radius of the Roche-lobe. $\zeta_h = (\partial \ln r_h/\partial \ln M_2)_J$ and $\zeta_s = (\partial \ln r_2/\partial \ln M_2)_S$ are the adiabatic indices of the Roche-lobe and the secondary star under irradiation heating. To the first order of $\Delta r/r_2$, the value of ζ_h and ζ_s can be taken to have the same values as in the case with no irradiation heating. The characteristic time scale for equation (3) is $\tau = [(\zeta_h - \zeta_s)(r_{ho}/\Delta r_o)(\dot{M}_{2o}/M_2)]^{-1}$. (For the derivation of equation (3), see Wu, Wickramasinghe & Warner 1994a [hereafter W^3].)

To determine T_1, we assume that: (i) the white dwarf is cooled by the emission of black body radiation, (ii) the white dwarf is heated by the gravitational energy deposited by the accretion matter, and (iii) the heating is saturated at the Eddington rate. Then, we have

$$\frac{d}{dt}\left(\frac{T_1}{T_{1o}}\right) = \frac{2}{3}\left(\frac{m_p}{\Delta M_1 k T_{1o}}\right)\left\{\frac{\mu G M_1 \dot{M}_{2o}}{r_1}\left[1 - \exp\left(-\frac{\dot{M}_2}{\dot{M}_E}\right)\right]\right.$$

$$\left.\times\left[1 - \exp\left(-\frac{\dot{M}_{2o}}{\dot{M}_E}\right)\right]^{-1} - 4\pi\, r_1^2 \sigma T_{1o}^4\left(\frac{T_1}{T_{1o}}\right)^4\right\} \tag{4}$$

where k is the Boltzmann constant, σ the Stefan-Boltzmann constant, m_p the proton mass, ΔM_1 the mass of the heated white dwarf envelope, and $\dot{M}_E$ the Eddington accretion rate.

3. Discussion and Conclusions

In our formulation, there are four variables describing the mass transfer process, and they are: the normalised degree of Roche-lobe filling $\Delta r/\Delta r_o$ $(= x)$, mass transfer rate $\dot{M}_2/\dot{M}_{2o}$ $(= y)$, the normalised temperature of the secondary's atmosphere T_2/T_{2o} (= z) and the normalised white dwarf temperature T_1/T_{1o} (= θ). Correspondingly, equations (1), (2) (3) and (4) are the four equations that govern the dynamics of the process. z can be eliminated by combining equations (1) and (2), so that the essential variables are x, y and θ.

At long orbital periods, say $P_o \approx 8$ h, the secondaries of the CVs are not very sensitive to irradiation-heating. Analysis shows that mass transfer for CVs at these orbital periods is stable. As the CVs evolve, their orbital period becomes shorter and their secondaries become less massive and cooler. Cooler stars are more sensitive to heating, so that the stability of mass transfer change when the CVs evolve. For a CV with a 1.0 $M_\odot$ white dwarf, the transition from stable to unstable mass transfer occurs at $P_o \approx 3$ - 4 h.

In Figure 1, we show the $\log(\dot{M_2}/\dot{M_{2o}})$ [= $\log y$] against $(\Delta r/\Delta r_o)$ [= x] diagram for CVs (with a 1.0 $M_\odot$ white dwarf) at $P_o = 8$ h. (The curves in the digram correspond to different T_1/T_{1o} (= θ), 0.0, 0.75, 1.0 and 1.25.) At this orbital period, the system is in a stable state. After a perturbation the system will return to its equilibrium state, represented by the point $(x, y, \theta) = (1, 1, 1)$. As a result, the mass transfer rate will not deviate much from its equilibrium value, $\dot{M_{2o}}$, and it can be determined by the orbital evolutionary models. At $P_o = 3$ h the $\log y - x$ relation is very different (Figure 2). For $\theta \gtrsim 1.25$, the $\log y - x$ relation is no longer a single-valued function of x, and the equilibrium is unstable. When the system is perturbed from equilibrium, the mass transfer rate will oscillate between a high $\dot{M_2}$ branch and a low $\dot{M_2}$ branch. The mass transfer rate on the high $\dot{M_2}$ branch can be $\sim 10 - 100$ times greater than the equilibrium value, while the mass transfer rate on the low $\dot{M_2}$ branch can be more than one to three orders of magnitude less. The typical time scale on each of these branches is $\tau \sim 10^4 - 10^6$ yr. As a system is not expected to be in the region between the high and low $\dot{M_2}$ branches for a significant time, there is a range of mass transfer rates at which very few CVs are expected be found. At the orbital periods where the transition between the stable and the unstable states occurs, the system has a wide range of $\dot{M_2}$.

In Table 1, the absolute magnitude M_V of CVs predicted by our formulation at different P_o are shown. For $P_o \leq 2.5$ h we assume that the orbital evolution is driven by gravitational radiation only and $\dot{M_{2o}}$ is given by such; for $P_o \geq 2.5$ h, by magnetic braking. We follow Warner (1987) to convert $\dot{M_2}$ to M_V. For comparison, we also show in the Table M_V as a function of P_o for the case without irradiation. The spread in the absolute magnitude (ΔM_v) predicted by our formulation is about 1.4 at $P_o = 8$ h, and its value increases when P_o decreases. At $P_o = 4$ h, $\Delta M_V = 6.7$, and at $P_o = 3.5$ h, $\Delta M_V = 8.2$. For shorter periods ($\lesssim 3$ h), there is a forbidden

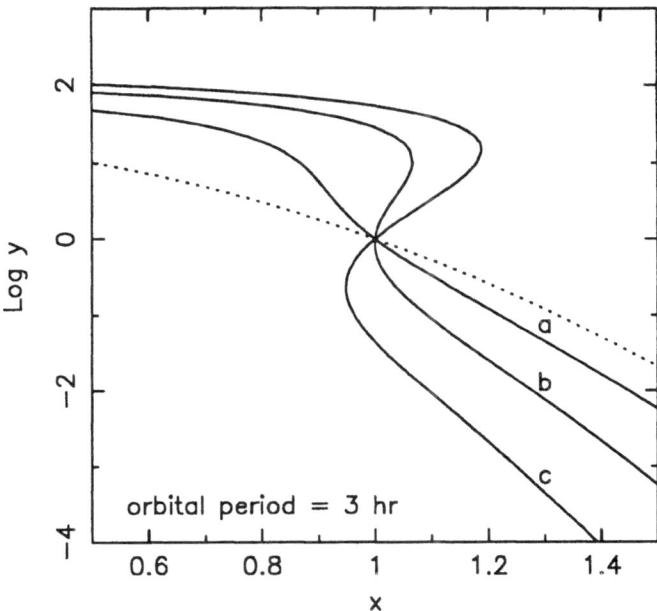

Figure 1.   The logarithm of the normalised mass transfer rate, $\log y$ $(= \log[\dot{M}_2/\dot{M}_{2o}])$, is plotted against the normalised degree of filling the Roche-lobe, x $(\Delta r/\Delta r_o)$, for a CV with $P_o = 8$ h. The dotted line corresponds to a zero white dwarf temperature, i.e. $T_1/T_{1o} = 0$. The solid lines (a), (b) and (c) correspond to $T_1/T_{1o} = 0.75$, 1.0 and 1.25 respectively.

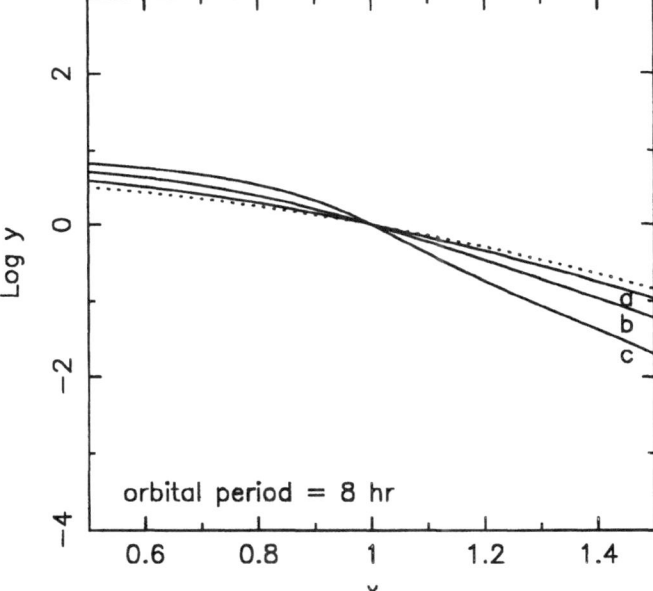

Figure 2.   The logarithm of the normalised mass transfer rate, $\log y$, is plotted against the normalised degree of filling the Roche-lobe, x, for a CV with $P_o = 3$ h.

TABLE 1. The absolute magnitude (M_V) of CVs at different orbital periods (P_o). For the cases with *, the orbital evoultion is driven by gravitational radiation; otherwise, by magnetic braking. The last column (M_V[W87]) corresponds to M_V predicted by the magnetic braking model (adopted from Warner 1987).

P_o (h)	allowed M_V	allowed M_V	forbidden M_V	M_V[W87]
1.0*	4.6 - 6.6	≥ 9.1	6.6 - 9.1	
1.5*	3.4 - 5.4	≥ 8.1	5.4 - 8.1	
2.0*	2.4 - 4.6	≥ 7.1	4.6 - 7.1	
2.5*	2.1 - 4.4	≥ 6.1	4.4 - 6.1	
2.5	3.4 - 5.3	8.4 - 17.2	5.3 - 8.4	
3.0	3.3 - 5.1	7.7 - 13.9	5.1 - 7.7	7.3
3.5	3.2 - 11.4			6.8
4.0	3.1 - 9.8			6.3
5.0	3.0 - 7.8			5.3
6.0	2.8 - 6.4			4.4
7.0	3.0 - 5.1			3.6
8.0	2.7 - 4.1			2.9

show that mass transfer is stable for systems with long orbital periods and becomes unstable when the systems evolve to very short orbital periods. The transition rough occurs at orbital periods of 3 - 4 h. When mass transfer is stable, the spread of mass transfer rate at a given orbital period is small, while at the periods when the transition between stable and unstable mass transfer occurs, the spread is the largest. When mass transfer is unstable, the mass transfer rate oscillates between a high rate branch and a low rate branch, in a time scale of $10^4 - 10^6$ yr.

References

Faulkner, J. (1971), *APJ*, **170**, L99.
Hameury, J. M., King, A. R. and Lasota, J. P. (1986), *A&A*, **162**, 71.
Harpaz, A. and Rappaport, S. (1991), *APJ*, **383**, 739.
Kovetz, A., Prialnik, D. and Shara, M. M. (1988), *APJ*, **325**, 828.
Osaki, Y. (1985), *A&A*, **144**, 369.
Sarna, M. J. (1990), *A&A*, *239*, 163.
Verbunt, F. and Zwaan, C. (1981), *A&A*, **100**, L7.
Warner, B. (1987), *MNRAS*, **227**, 23.
Wu, K., Wickramasinghe, D. T. and Warner, B. (1994a), *MNRAS*, submitted.
Wu, K., Wickramasinghe, D. T. and Warner, B. (1994b), *APJ*, submitted.

THE AM CVN STARS

D. O'DONOGHUE
Dept. of Astronomy
University of Cape Town
Rondebosch 7700
S. Africa

1. Introduction

The AM CVn stars are a class of hydrogen-deficient blue stars which are believed to be closely related to the hydrogen-rich cataclysmic variables. Yet the basic properties of these stars are poorly understood. This paper will review these properties, sketch the best model to explain them, and point the way to future work to improve our understanding.

2. Basic Properties of the AM CVn Stars

At the time of writing (late 1994), there are 6 known AM CVn stars. They are listed in Table 1 along with their most important parameters and references to recent publications.

The most important properties of the stars in the table are:

(1) All the stars are extremely hydrogen deficient. Neutral helium lines dominate the optical spectra. The optical colours are very blue.

(2) Three of the stars, the 'large amplitude variables', undergo erratic luminosity changes on a time scale of days to weeks with an amplitude of ~ 4 mag. Little change in colour is seen. The other stars remain constant in light.

(3) With the exception of GP Com, all the stars have one (or two) optical photometric period(s) between 500 and 2000 sec. The peak-to-peak amplitude of this variation varies from 1-10 %. Other periods, not simply related to those in the 500–2000 s range, are also seen in some systems (e.g. the 175-s period in V803 Cen).

(4) The pulse shapes of the photometric periods are non-sinusoidal. The stars which undergo large luminosity changes have especially complex

321

A. Bianchini et al. (eds.), Cataclysmic Variables, 321-324.
© *1995 Kluwer Academic Publishers.*

TABLE 1. The AM CVn Stars

Name (& Aliases)	P_{phot} & harmonics	Other Periods (s)	B mag. range	Refs
AM CVn (HZ29)	– , **525,** 350	1011, 289 801, 26	13.9	1,2
EC15330–1403 *New discovery*	**1119**, 560		13.6	10
CR Boo (PG1346+082)	**1490**, 745, 495, 372	1470	13.6–17.2	3,4
V803 Cen (AE-1)	**1611**, 805, 537, 402, 322, 268	175	13.4–17.4	5,6
CP Eri	**1724**, 862, 574, 431, 345		16.5–19.7	7
GP Com (G61-29)		**2790 (P_{orb})**	~15.8	8,9

References: (1) Patterson *et al.* (1993); (2) Provencal *et al.* (1994a); (3) Wood *et al.* (1987); (4) Provencal *et al.* (1994b); (5) O'Donoghue & Kilkenny (1989); (6) O'Donoghue *et al.* (1990); (7) Abbott *et al.* (1992); (8) Nather *et al.* (1981); (9) Marsh *et al.* (1991); (10) O'Donoghue *et al.* (1994).

pulse shapes. These complex pulse shapes appear as harmonics in the power spectrum (at least 7 harmonics have been seen in V803 Cen).

(5) All stars except GP Com show HeI absorption line spectra. In the large amplitude variables, this kind of spectrum appears at maximum light. When they are at their faintest, weak HeI emission lines are seen.

(6) GP Com is the 'odd man out'. It shows strong broad HeI emission lines in its optical spectrum and obvious flickering in its light curve. Its chemical composition has been shown to be N-rich (Marsh, Horne & Rosen, 1991).

(7) No eclipses are seen. Only GP Com shows radial velocity variations with short period. However the HeI absorption line profiles are generally asymmetric, and the asymmetry in AM CVn is now known to vary period-ically with P~ 13.4 hr.

3. The Model

The only model which fits all the facts is one in which a very low mass ($\sim$ 0.02 $M_\odot$) white dwarf fills its Roche lobe and transfers helium rich material to another white dwarf of normal mass via an accretion disc (Faulkner, Flannery & Warner 1972). If this model is correct, the orbital periods of the systems would be $\sim$500–2000 sec and the mass ratios extreme ($\sim$ 0.05).

The large amplitude luminosity variations, without accompanying increase in reddening, require a model involving an interacting binary. Additional support comes from the obvious flickering in GP Com. Flickering has been claimed in the other stars but the evidence is not clearcut.

Within the context of the interacting binary model, the peculiar chemical composition can then be explained by a He-rich secondary star. The fact that none of the systems show substantial radial velocity variations requires the secondary to be of very low mass. A low mass white dwarf satisfies these requirements.

The interacting binary model implies an accretion disk around the mass accreting white dwarf. The chemical composition of this accretion disk will be determined by the secondary star and therefore He rich. These features of the model account for the observed spectra of the AM CVn stars: absorption lines when the systems are bright, changing to emission lines when the systems are faint. The same kind of behaviour is seen in the hydrogen rich cataclysmic variables.

As mentioned above, the luminosity changes in the AM CVn stars require an interacting binary model. The specific mechanism for the luminosity changes then involves a change in the accretion rate on to the mass accreting white dwarf, similar to the disk instability model for dwarf novae. Recently, Warner (these Proceedings) has advanced a model to explain the luminosity changes as an instability in the mass transfer rate from the secondary, arising from irradiation by the hot mass accreting white dwarf and inner disk. In his picture, GP Com would correspond to an SU UMa dwarf nova at minimum, AM CVn and EC15330-1403 would be novalikes ('stuck in outburst') and the large amplitude variables would be VY Scl stars.

4. Unsolved Problems

Although the model outlined in the last section successfully accounts for the overall properties of the AM CVn stars, there are many details which remain to be explained.

Perhaps the most urgent problem to be solved is the mechanism that produces the 500–2000 s photometric oscillations. These oscillations are remarkable for having extremely complex pulse shapes. Some progress has been made recently: in AM CVn itself and CR Boo, the fundamental frequency of oscillation has been shown to comprise at least 2 closely-spaced components (Patterson *et al.* 1992, Provencal *et al.* 1994a,b). The period of the variability of the asymmetry in the absorption lines of AM CVn (Patterson *et al.* 1993) is the beat period between these two photometric components. Provencal *et al.* (1994b) have also shown that one of the photometric components in CR Boo is stable and maintains phase (at least over a two week period). These clues, along with the extreme mass ratio implied by the basic model, hint that one of the oscillations is a tidal phenomenon, similar to the basic model for superhumps in SU UMa-type cataclysmic variables (Hirose & Osaki 1993). The other oscillation would then be the orbital period, and arise from viewing the system at different aspects as it rotates at the binary period. An important direction for future work is to search for a 2-component structure in the other AM CVn systems, especially V803 Cen.

Another feature of the class that requires explanation is the difference between GP Com and the other members. GP Com has much stronger emission lines than the large amplitude variables at minimum. Flickering is easily seen in GP Com and has yet to be convincingly demonstrated in the other stars; yet in the hydrogen-rich CVs, flickering is seen in almost all systems, irrespective of sub-class. It is puzzling why it is not as obvious in most of the AM CVn stars.

Although it is difficult to conceive of any other model, there is no *direct* evidence that any of the stars other than GP Com are even binaries! The number of stars now known is sufficient that an eclipsing system should soon be discovered. This would provide a reassuring confirmation that the basic model is correct.

References

Abbott T.M.C., Robinson E.L., Hill G.J., Haswell C.A., 1992, *Astrophys. J.*, **399**, 680
Faulkner J., Flannery B., Warner B., 1972, *Astrophys. J.*, **175**, L79
Hirose M., Osaki Y., *Publ. Astr. Soc. Japan*, **45**, 1993.
Marsh T.R., Horne K., Rosen S., 1991, *Astrophys. J.*, **366**, 535
Nather R.E., Robinson E.L., Stover R.J., 1981, *Astrophys. J.*, **244**, 269
O'Donoghue D., Kilkenny D., 1989, *Mon. Not. R. astr. Soc.*, **236**, 319
O'Donoghue D., *et al.*, 1990, *Mon. Not. R. astr. Soc.*, **245**, 140
O'Donoghue D., *et al.*, 1994, *Mon. Not. R. astr. Soc.*, in press.
Patterson J., Sterner E., Halpern J.P., Raymond J.C., 1992, *Astrophys. J.*, **384**, 234
Patterson J., Halpern J., Shambrook A., 1993, *Astrophys. J.*, **419**, 803
Provencal J.L., *et al.*, 1994a, *Astrophys. J.*, submitted
Provencal J.L., *et al.*, 1994b, *Astrophys. J.*, submitted
Wood M.A., *et al.*, 1987, *Astrophys. J.*, **313**, 757

Irradiation-driven Mass Transfer and the Evolution of AM CVn Stars

B. Warner

Department of Astronomy, University of Cape Town, Rondebosch 7700, South Africa

Abstract. The AM CVn stars are He-transferring double-degenerate binaries. Viewed as ultra-short period systems on the increasing period branch of orbital evolution it is shown that they are analogues of the H-rich CVs, with irradiation-driven mass transfer providing an explanation of the stable high $\dot{M}$ discs and VY Scl unstable behaviour.

The late evolution of a small fraction of binaries is predicted to be a low mass He degenerate star transferring mass to the higher mass white dwarf with binary period $\sim 10^3$ s (Webbink 1979; Tutukov & Fedorova 1989; Savonije, de Kool & van den Heuvel 1986). Very low mass ratios $q = M_2/M_1$ are expected for these systems, which should lead to a precessing accretion disc as seen in SU UMa stars in superoutburst (Hirose & Osaki 1990, 1993) and permanently in the high $\dot{M}$ nova-like PG0917+342 (Skillman & Patterson 1993).

The AM CVn stars have observed properties that appear to fulfil these expectations (Warner 1994). Their observed photometric periods are best understood as superhump periods; recent spectroscopic evidence indicates that in AM CVn itself there is a 13.4h period in the shape of the line profiles, which is the beat between a superhump period of 1051s and a deduced orbital period of 1028.8s (Patterson, Halpern & Shambrook 1993). CR Boo has a coherent optical period of 1471.35s and a variable period of $\sim$1493s (Provencal et al. 1991) which are also probably orbital and superhump periods respectively. In general, therefore, we may suspect that the orbital periods of AM CVn stars are a few percent shorter than the observed principal optical photometric periods. Ordering the known AM CVn stars according to photometric (and therefore, by hypothesis, orbital) period shows a correlation with long-term optical behaviour (Table 1). The two shortest period

TABLE I

The AM CVn Stars

Star	m_v	Optical Period (s)
AM CVn	14.1–14.2	1050
EC 15330-1403	13.6	1119
CR Boo	13.0–18.0	1493
V803 Cen	13.2–17.4	1611
CP Eri	16.5–19.7	1724
GP Com	15.7–16.0	2970

325

A. Bianchini et al. (eds.), Cataclysmic Variables, 325-326.
© 1995 Kluwer Academic Publishers.

systems have stable accretion discs, analogous to the nova-like H-rich CVs above
the orbital period gap. At longer periods there are 3 systems with unstable discs,
analogous to the VY Scl behaviour seen in the 3-4h period range of H-rich CVs.
At the longest period, GP Com spectroscopically has the characteristics of a very
low $\dot{M}$ system, analogous to extreme SU UMa stars like WZ Sge.

Combining the theoretical mass transfer rate for a binary driven by gravitational radiation (Nelson, Rappaport & Joss 1986) with the evolutionary model of
Savonije, de Kool and van den Heuvel (1986) gives

$$\dot{M} \simeq 7 \times 10^{-12} P_{orb}^{-5.2}(h) \ M_{\odot}y^{-1}$$

showing that high $\dot{M}$ (and therefore stable) discs would be expected for $P_{orb} \sim$
10^3s, but low $\dot{M}$ discs with dwarf nova eruptions would be expected for $P_{orb} \geq$
2×10^3s.

However, it has been shown (Wu, Wickramasinghe & Warner 1994) that irradi-
ation of the secondary by the hot primary can sustain high $\dot{M}$ for $3 \leq P_{orb}(h) \leq 6$
in H-rich systems, and that the VY Scl star behaviour observed in the range
$3 \leq P_{orb}(h) \leq 4$ is a result of instability of this process. In the AM CVn stars the
same mechanism should occur, but in reversed direction of orbital evolution. At
low P_{orb} both GR-driving and irradiation lead to high $\dot{M}$ stable discs. By estimating the irradiative flux experienced in the VY Scl stars it can easily be shown that
the same fluxes should occur in $1300 \leq P_{orb}(s) \leq 1500$ AM CVn stars (Warner
1994). This is in reasonable accord with the observed range of unstable discs in
the AM CVn stars (Table 1). In this evolutionary scheme GP Com should be an
SU UMa star with superoutburst intervals probably of decades.

References

Hirose, M. & Osaki, Y., 1990. *Pub. Astron.Soc. Japan*, **42**, 135.
Hirose, M. & Osaki, Y ., 1993. *Pub. Astron. Soc. Japan*, **45**, 1993.
Nelson, L.A., Rappaport, S. & Joss, P.C., 1986. *Astrophys. J.*, **304**, 231.
Patterson, J., Halpern, J. & Shambrook, A., 1993. *Astrophys. J.*, **419**, 83.
Provencal, J.L. *et al.*, 1991. in *White Dwarfs*, eds. G. Vauclair & E. Sion, Kluwer, Dordrecht, p.
    449.
Savonije, G.J., de Kool, M. & van den Heuvel, E.P.J., 1986. *Astr. Astrophys.*, **155**, 51.
Skillman, D.R. & Patterson, J. 1993. *Astrophys. J.*, **417**, 298.
Tutukov, A.V. & Fedorova, A.V., 1989. *Sov. Astr.*, **33**, 606.
Warner, B., 1994. *Astr. Astrophys.* in press.
Webbink, R.F., 1979. *Int. Astr. Un. Colloq.* No. 53, p. 426.
Wu, K., Wickramasinghe, D.T. & Warner, B. 1994. *Mon. Not. R. astr. Soc.* in press.

EXTREME CIRCULAR POLARIZATION DISCOVERED
IN THE MAGNETIC CV RE 1307+535

V. PIIROLA
Tuorla Observatory of the University of Turku
FIN-21500 Piikkiö, Finland

P.J. HAKALA, O. VILHU AND D.C. HANNIKAINEN
Observatory and Astrophysics Laboratory
FIN-00014 University of Helsinki, Finland

AND

J.P. OSBORNE
Dept. of Physics and Astronomy, University of Leicester
Leicester LE1 7RH, UK

Abstract. We report the discovery of very strong variable (+50 to -20 %) circular polarization in the ROSAT EUV source RE 1307+535 in the red (600-710 nm) spectral region. The amplitude and the peak value of the broad-band circular polarization are the largest observed so far in any astrophysical object. The high degree of polarization requires that the observed red band flux of RE 1307+535 is dominated by cyclotron emission from a rather homogeneous region with little spread in the field strength and direction. Relative contributions from unpolarized (thermal) radiation sources are unusually small. We have modelled the observed circular polarization variation over the 79.7 min spin period of the white dwarf with the inclination of the spin axis $i\sim75°$, and two emission regions near the opposite magnetic poles, displaced 30-50° from the rotational poles.

1. Introduction

RE 1307+535 was detected in the ROSAT Wide Field Camera EUV survey, and subsequent optical spectroscopic and photometric observations showed that the object is most probably an AM Her system (Osborne *et al.* 1994).

327

A. Bianchini et al. (eds.), Cataclysmic Variables. 327-333.

The orbital period of 79.69 min was deduced, which is the shortest found for an AM Her system. Estimates from infrared photometry give a lower limit for the distance $d > 710$ pc and indicate that RE 1307+535 is the first AM Her system located in the galactic halo, i.e., above the galactic plane by the measured distance of $z > 630$ pc.

2. Observations

We have carried out circular polarimetry of RE 1307+535 on the nights 1994 Feb 4/5 and 5/6 at the Nordic 2.54 m telescope (NOT) on La Palma with an imaging polarimeter consisting of a superachromatic $\lambda/4$ retarder and a plane paraller calcite plate in front of the Tektronix 512x512 CCD detector. Fig. 1 gives the results from the R band (600-710 nm) measurements, folded over the 79.69 min orbital period which equals the spin period of the synchronously rotating white dwarf.

3. Results and Discussion

The circular polarization reaches about +50 %, (Fig. 1a) which makes RE 1307+535 the most strongly circularly polarized astrophysical object so far detected. Other AM Her systems show a maximum polarization which is < 40 %. Also the 70 % amplitude of the variations for RE 1307+535 (+50 to -20 %) is almost two times larger than observed in any other system. There is a reversal of the sign of the circular polarization from the positive (right handed) to negative (left handed) values in the phase interval 0.3-0.6. This indicates that cyclotron emission takes place at two opposite magnetic poles.

The R band light curve obtained during the polarization measurements is shown in Fig. 1b. The minimum brightness coincides with the maximum negative circular polarization, and the object is brightest during the phase interval of the positive circular polarization. Accordingly, the positive c.p. emission region is mostly responsible for the brightness variations. The system was in the high accretion state during our measurements in 1994 Feb 4-6 (magnitude range $R = 17.^m7$ - $18.^m6$). In the episodes of low mass transfer the brightness may fall to $R = 19.5\text{-}20.5^m$.

There is a broad dip in the bright part of the light curve at the circular polarization maximum (phase 0.0). We attribute this to a cyclotron beaming effect when the positive circular polarization region is closest to face-on. The emission intensity is largest into directions perpendicular to the field lines, and decreases to zero along the field lines. Hence, the observed brightness is reduced when we look most directly at the cyclotron region.

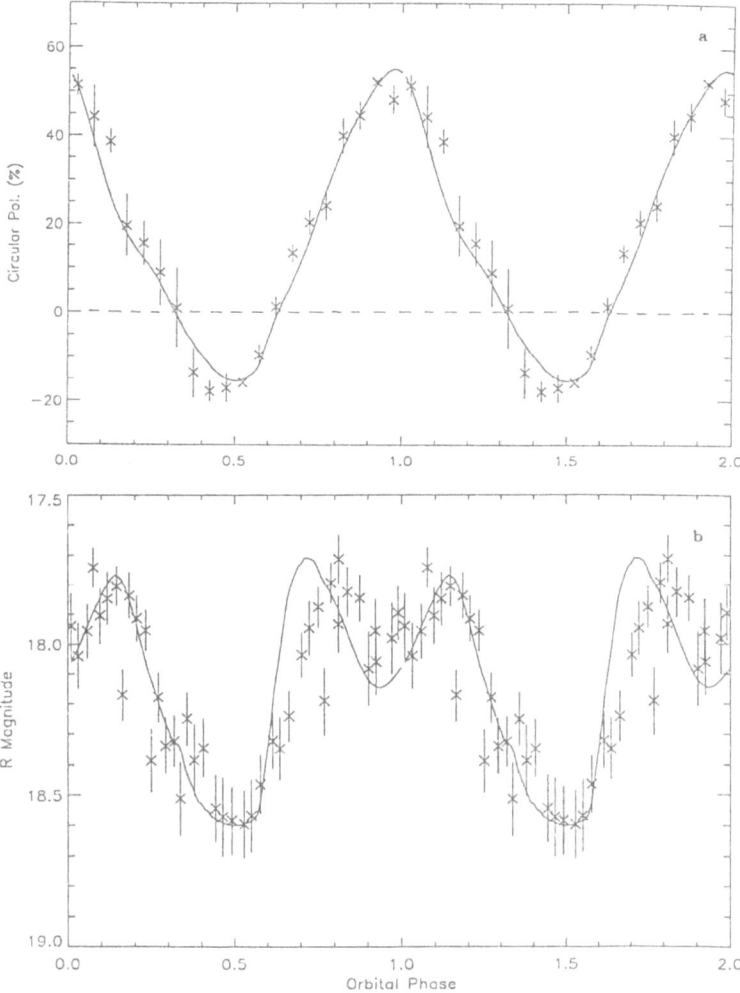

Figure 1. (a) Circular polarimetry of RE1307+535 in the red (600-710 nm) spectral region (R band), obtained at the Nordic 2.54 m Optical Telescope on La Palma on two nights, 1994 Feb 4/5 and 5/6. Each dot gives an individual measurement of 6 min duration, and the observations have been folded over the 79.69 min spin period of the magnetic white dwarf. The peak value of ∼50 % makes RE1307+535 the most strongly circularly polarized astrophysical object so far detected. (b) Modulation of the total intensity of RE1307+535 in the R band over the 79.69 min period. Note the broad dip in the light curve at the peak of the circular polarization. We attribute this to cyclotron beaming effects, which arise from the decrease of the emission intensity in viewing directions along the magnetic field lines. Continuous lines give computed curves from the model of Figs. 2-3.

Earlier geometric model considerations (Osborne *et al.* 1994) for RE 1307 +535 have been based only on light curve data in the optical and EUV, and assigned the deeper intensity minimum (phase 0.5) to cyclotron beaming effects arising when the accretion region is closest to face-on, and the weaker minimum (phase 0.0) to possible occultation of the emission region by the white dwarf body. However, such a geometry has difficulties in explaining our circular polarization curve, which shows the reversal of sign near the deeper intensity minimum, consistent with having the emission regions near the limb, one disappearing behind the white dwarf and the other coming into our view. Our model also explains the maximum in the EUV light curve as being due to the more active (positive circular polarization pole), and not to the weaker (negative circular polarization pole), as required by the earlier model.

Since there is no strong beaming effect in the X-ray intensity and the soft X-ray emission is optically thick, the soft X-ray brightness should be at the maximum when the emission region is pointing closest to us. We have compared our results with the ROSAT WFC sky survey EUV light curve (Osborne *et al.* 1994) and found that the S1 filter (90-206 eV) light curve reaches maximum near the phase 0.0, in agreement with our interpretation which gives the active accretion region closest to face-on at this phase (0.0).

There is no dip in the circular polarization curve near phase 0.0. Therefore, the contributions from the unpolarized background sources (underlying stellar photosphere and thermal emission from the circumstellar matter) must be negligible, as they have no diluting effect in the observed circular polarization. The cyclotron emission regions dominate, and totally outshine the light from other parts of the binary system. Outside the broad dip of the bright phase light curve, the shapes of the polarization and light curves suggest that geometry (visibility of the emission regions) dominates over physics (cyclotron beaming effects) there.

In order to further constrain the system geometry, we have computed simulated polarization and light curves (Fig. 2) for a model containing two extended cyclotron emission regions on the surface of the spinning white dwarf, seen at the inclination i (Fig. 3). Each strip of the regions was divided into 20 points for which all four of the Stokes parameters (Q, U, V, and I) were calculated (for details, see Piirola *et al.* 1990, 1993, 1994) taking the view angle dependences of the degree of polarization and the total flux from existing physical cyclotron models (Wickramasinghe and Meggitt 1985). We have simplified the calculations here by assuming radial field lines, which is nearly the case if the emission region is not far from the magnetic pole. For distances $< 25°$, the field lines of a centered dipole deviate $< 13°$ from the white dwarf normal.

A relatively good description of the data (continuous line in Fig. 1)

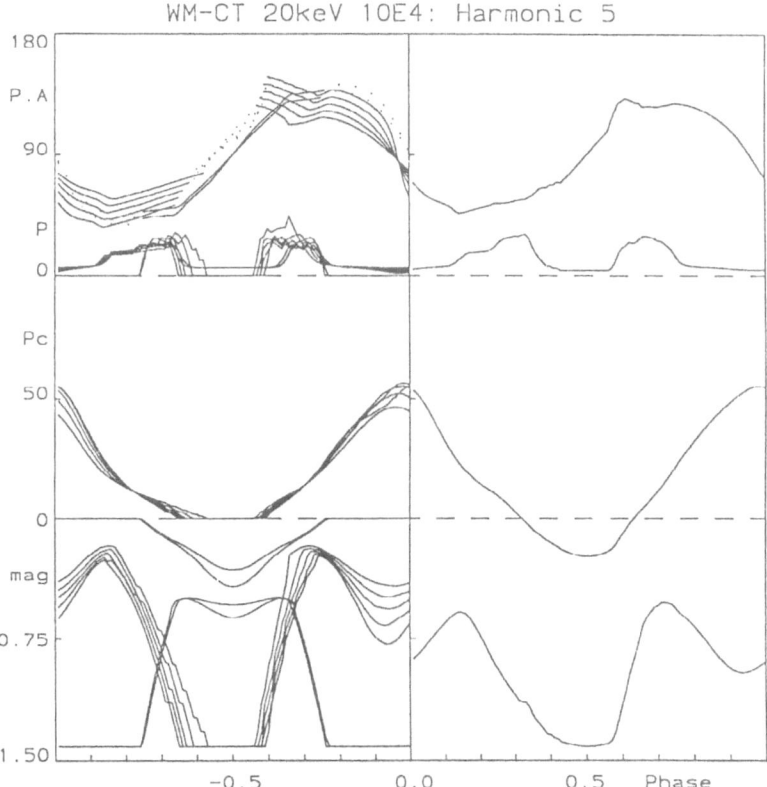

WM-CT 20keV 10E4: Harmonic 5

Figure 2.   polarization and intensity curves from a model with two extended, almost diametrically opposite, cyclotron emission regions at colatitudes β_1=30-45° and β_2=130-135°. The respective extensions in longitude are 65° and 35°. The inclination of the white dwarf spin axis is $i = 75°$. Curves from top to bottom give the position angle and the degree of linear polarization, the degree of circular polarization and the total intensity (in magnitude scale). The slightly displaced curves in the left panels correspond to individual emission strips separated by 5° in latitude (five strips for the positive and two strips for the negative circular polarization region). The right hand panels give the combined curves from the two extended regions. The kT=20 keV Λ=10⁴ cyclotron model was applied. With the magnetic field strength $B = 3\ 10^7$ Gauss the R band effective wavelength corresponds to cyclotron harmonic $n = \omega/\omega_c \sim 5$, where ω_c is the fundamental electron cyclotron frequency.

can be obtained with a simulated model with $i = 75°$ and two extended emission regions at the colatitudes (angular distance from the rotation pole pointing closest to us) $\beta_1 = 30\text{-}50°$ (positive circular pol. region) and $\beta_2 = 130\text{-}135°$ (negative c.p. region), and extended in longitude by 65° and 35°,

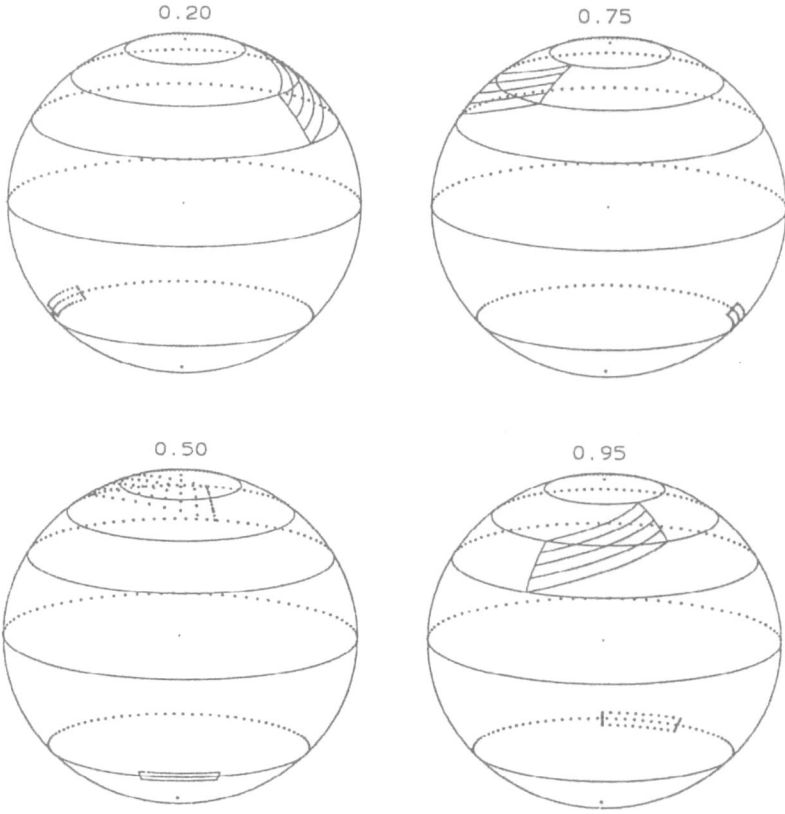

Figure 3. Illustration of the model used to compute the polarization and light curves in Figs. 1-2. The emission strips of the upper (positive circular polarization) region have a specific orientation to account for the slight asymmetry of the light curve. Values of orbital phase are given.

respectively. We have adopted for these calculations the field strength B = 30 MG (Osborne *et al.* 1994), which gives cyclotron harmonic number $n \sim 5$ for our red spectral passband (600-710 nm). We also show in Fig. 2 the linear polarization and position angle curves given by the model. Broad linear polarization pulses are predicted at the phases 0.3 and 0.7, where the magnetic field lines are nearly perpendicular to the line of sight.

The detection of such a highly variable circular polarization in RE1307 +535 indicates that the fractional contribution of the cyclotron emission to the total optical emission is significantly higher in RE 1307+535 than in any other AM Her system. It is reasonable to think that in RE 1307+535 the stream emission is significantly lower than in other AM Her stars. This

could be due to two factors: according to the estimates by Osborne *et al.*
(1994) the white dwarf field strength is relatively high (30-40 MG) and the
binary is also compact because of the 79.7 min orbital period. These two
factors could result in an unusual accretion stream geometry determined
mostly by the magnetic field. In this case the magnetic pressure would
dominate over the stream ram pressure (Hameury *et al.* 1986; Liebert and
Stockman 1985) all the way from the white dwarf to the L1 point. If we
denote the radius from the white dwarf at which the accretion stream ram
pressure equals the magnetic pressure by R_M ($\sim$2.0 10^{10} cm), and assume
a typical total mass of 0.7 $M_\odot$ for the binary, we obtain 2.4 10^{10} cm for the
distance from the L1 point to the white dwarf. This is quite comparable to
R_M, which means that in the case of RE 1307+535 the stream is possibly
not heated up due to turbulence or twisting of field lines at R_M, but instead
keeps cool and follows "nicely" the magnetic field lines. The resulting low
stream emission would explain the high polarimetric variability observed.

References

Hakala P.J., Watson M.G., Vilhu O., Hassall B.J.M., Kellett B.J., Mason K.O., Piirola
    V. (1993) *Monthly Notices Roy. Astron. Soc.* **263**, 61-68.
Hameury J.-M., King A.R., Lasota J.-P. (1986) *Monthly Notices Roy. Astron. Soc.* **218**,
    695-710.
Liebert J. and Stockmann H.S. (1985) *in Cataclysmic Variables and Low Mass X-Ray
    Binaries*, eds. D.Q. Lamb and J. Patterson, (D. Reidel Publishing Company), 151-
    177.
Osborne, J.P., Beardmore, A.P., Wheatley, P.J., Hakala, P., Watson, M.G.,Hassall,
    B.J.M., King, A.R. (1994) *Monthly Notices Roy. Astron. Soc.* in print.
Piirola, V., Coyne, G.V., Reiz, A. (1990) *Astron. Astrophys.* **235**, 245-254.
Piirola, V., Coyne, G.V., Takalo, L., Larsson, S., Vilhu, O. (1994) *Astron. Astrophys.*
    283, 163-174.
Piirola, V., Hakala, P., Coyne, G.V. (1993) *Astrophys. J.* **410**, L107-110.
Wickramasinghe, D.T., Meggitt, S.M.A. (1985) *Monthly Notices Roy. Astron. Soc.* **214**,
    605-618.

ON THE EXISTENCE OF DWARF NOVAE
WITH TREMENDOUS OUTBURST AMPLITUDES
OR TOADS IN SPACE

STEVE B. HOWELL
Planetary Science Institute 620 N. 6th Ave.
Tucson, AZ 85705 USA

AND

PAULA SZKODY
Department of Astronomy University of Washington
Seattle, WA 98195 USA

1. Introduction

During their study of high galactic latitude cataclysmic variables (CVs), Howell and Szkody (1990) discovered a group of dwarf novae which showed very large outbursts. These Tremendous Outburst Amplitude Dwarf novae (TOADs) are now known to exist at all galactic latitudes and are not just associated with high latitude, possibly metal poor CVs. The initial study concentrated on systems with minimum visual magnitudes between 16 and 22 and galactic latitudes of ≥ 45 degrees from the galactic plane. The initial assumption that these criteria would thus represent halo objects, was based on the then assumed typical absolute magnitudes of 7.5 for dwarf novae (DN) and 4.5 for novae (N) and novalikes (NL). We now realize that these values are NOT indicative of all DN, N, and NL, in particular short period systems (Sproats, Howell, and Mason 1995). Since their initial study, Howell, Szkody, and collaborators have also studied a number of low latitude systems (≤ 25 degrees from the galactic plane), as well as additional systems from the literature (see Howell et al 1995a). ¿From these developments, we now know that 1) intrinsically faint CVs seem to be abundant and at all galactic latitudes but are poorly studied and understood and 2) few of the currently well known DN and NL are actually indicitive of the majority of CVs which exist, and few are at large distances ie., halo objects.

A. Bianchini et al. (eds.), Cataclysmic Variables, 335-338.

2. The TOADs

There are three criteria a DN must possess to be a TOAD:

A) The outburst amplitude from normal minimum to maximum must be at least 6 mags. TOAD outbursts tend to be infrequent and are almost all superoutbursts.

B) TOADs all show orbital periods which are below the period gap. AR And and WW Cet are possible exceptions to this rule, but these systems need further study to determine their real quiescent magnitudes. WW Cet for example, seems to show high and low states and these may account for its listed apparent large outburst amplitude.

C) The mass transfer rate and viscosity for the TOADs appear to be unique. Mass transfer rates of $\sim 10^{15}$ g/s and viscosities of $\alpha_{cold} \sim 0.003$ are needed to explain their infrequent and tremendous outburst behavior. J. Cannizzo has developed models for the TOADs which fit the observations well (see Howell et al 1995a and Cannizzo in this volume).

Table 1 presents a list of all currently known TOADs. As one can see the listed systems are generally among the lesser known CVs and further detailed study is needed. Figure 1 shows outburst amplitudes for the TOADs with known orbital periods. HL and LL stand for high galactic latitude and low galactic latitude respectively and one can see there appears to be no correlation as to a systems position within the galactic disk. There may be a weak correlation showing larger amplitudes for shorter periods, but there are not nearly enough systems known yet to provide statistical proof of this. Howell et al (1995a) provide a detailed account of the TOADs.

Observations of TOADs in the optical and UV at outburst have so far shown that the TOADs look similar to typical DN at maximum light (Howell et al 1995b and refs therein). No differences are generally seen which provide us with clues which may account for their large outburst amplitude or why these systems have low mass transfer rates and viscosities.

The time between outbursts for the TOADs is of critical interest in order to further our understanding of these systems, predict outbursts, and produce reliable models. Currently, TOAD outbursts range from months (eg. BC UMa, AY Lyr) to decades (AL Com, WZ Sge). Some TOADs even show what appear to be regular DN type outbursts from time to time. Recent work by Howell et al (1995b) has led to the identification of four possible types of superoutburst for SW UMa and some other TOADs.

Observations at minimum, while difficult due to the faintness of the systems, are badly needed in order to provide us with values for system parameters such as masses and estimates of mass transfer rates. These systems provide a valuable laboratory for study of the evolution of short period DN. There is likely to be many more undiscovered TOADs out there,

Table 1
The Known TOADs

Star	Orb Per (min)	Outburst Amp (mags)
WX Cet	81*	8.5
WZ Sge	81.6	7.5
SW UMa	81.8	7.3
LL And	82.2*	6.0
HV Vir	83.5	7.6
AL Com	84?	9.0
VY Aqr	91	8.5
BC UMa	91.2	7.4
UV Per	93.3*	6.5
GO Com	94.8	6.9
BZ UMa	97.8	7.3
RZ Leo	102	7.7
KV And	104*	7.9
AY Lyr	105.7*	6.1
TT Boo	108*	6.9
KK Tel	112	6.2
EF Peg	123*	7.9
TV Crv	132*	7.0
AR And?	236	6.6
WW Cet?	253.2	6.4
GW Lib	?	9.5
UZ Boo	?	9.0
V592 Her	?	>7.2
AO Oct	?	7.5
V551 Sgr	?	6.5
FQ Sco	?	6.5
AK Cnc	?	6.2

* superhump period

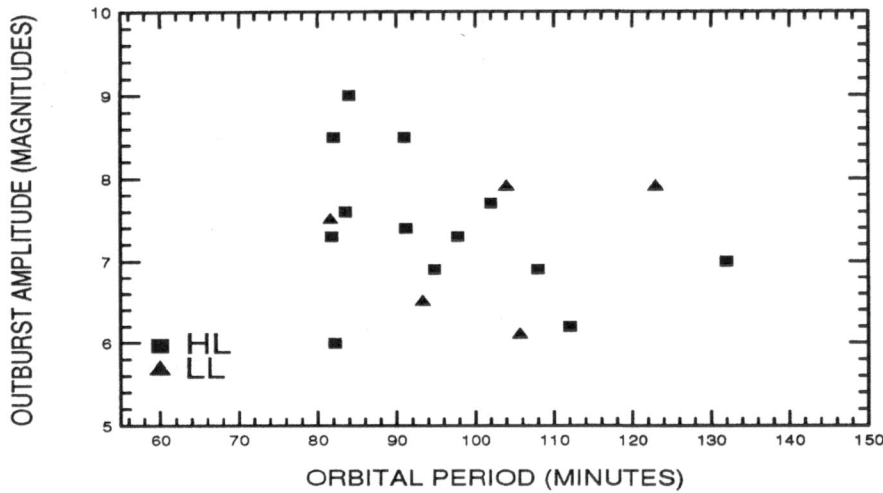

Figure 1.   Outburst amplitudes for the TOADs with known orbital periods.

and with the advent of a number of large area surveys, CCDs being available to amateurs, and the realization of their existence, no doubt many more will be found. This work was partially supported by NSF grants AST-9217971 to SBH and AST-9217911 to PS.

References

Howell S. B., and Szkody, P. 1990 ApJ 356, 623.
Howell, S. B., Szkody, P., and Cannizzo, J. K. 1995a, ApJ, 20 Jan issue.
Howell et al. 1995b, AJ, submitted.
Sproats, L., Howell, S. B., and Mason, K. O., 1995 MNRAS submitted.

ARE EXTREME DWARF NOVAE AMPLITUDES DUE TO LOW ACCRETION RATES?

PAULA SZKODY
University of Washington, Seattle, WA 98195

AND

STEVE B. HOWELL
Planetary Science Institute, Tucson, AZ 85705

Abstract. As a test of whether the extreme amplitudes observed in some dwarf novae are due to very low accretion rates associated with ultrashort orbital periods, we have estimated the accretion rates from the disk magnitudes and from ROSAT X-ray luminosities. From 8 systems with distances determined from spectral or IR measurements, four have faint disks and four have bright disks. From the ROSAT results on 7 systems, two have low accretion rate while five are normal to high. Thus, the extreme amplitudes do not appear to be solely due to low accretion rates. However, large variations in quiescent magnitudes for 3 systems may signal lower accretion values are sporadically present.

1. Introduction

Since some dwarf novae (TOADs) show outburst amplitudes comparable to that of recurrent novae (see Howell and Szkody, this volume), it is natural to ask what causes this extreme variability as compared to the normal eruption amplitudes typical of the class. Cannizzo (this volume) can obtain large, infrequent outbursts with disk instability models if he uses low mass transfer rates (near $10^{-10} M_\odot$ /yr) and very low viscosity (near $\alpha = 0.003$). At these low transfer and viscosity levels, the transport of material through the disk and subsequent accretion onto the white dwarf should be very low. This can be tested in two different ways, as half the accretion luminosity of dwarf novae is expected to be radiated by the disk and the

339

A Bianchini et al (eds) Cataclysmic Variables 339–342

other half at the boundary layer interface between the white dwarf and the disk (Lynden-Bell and Pringle 1974). If the accretion through the disk is low, the systems should show large (faint) absolute magnitudes. If the boundary layer accretion is low, the X-ray luminosity should be low. We can test these ideas by comparing the M_v and ROSAT results for TOADS with other normal dwarf novae.

2. Absolute Magnitudes

The determination of the absolute magnitudes for TOADS is dependent on obtaining distances, which in turn, is dependent on observation of the secondary star in these systems. This is best done via spectroscopy in the near-IR, but for the faintest systems, only IR photometry is available. There are only 8 systems with available distances and thus, M_v values (Table 1).

TABLE 1. Distances and M_v of TOADS

Object	d(pc)	M_v	Reference
AK Cnc	210-240	10.4-11.9	Szkody and Howell 1992
WW Cet	130	10.1	Young and Schneider 1981
AL Com	200	13.5	SHM 1994
GO Com	950	10.1	SHM 1994
RZ Leo	200	9.2	SHM 1994
WZ Sge	90	9.5	1926 parallax
BC UMa	130-400	11-13.5	Mukai et al. 1990
BZ UMa	110	12.6	Ringwald 1993

From a study of distances of normal dwarf novae, Warner (1987) gives a relation between M_v and orbital period. For systems below the gap, the range of normal M_v is between 8.4 and 9.2. From Table 1, it is apparent that 4 systems (AK Cnc, AL Com, BC UMa and BZ UMa) are more than a magnitude fainter than 10.2 (inclination effects can cause a disk to appear up to a magnitude fainter than the 9.2). Thus, half of the available data on TOADS indicate faint disks and low accretion.

3. X-ray Fluxes

ROSAT observations were obtained for 4 TOADS between June 1993-April 1994. In addition, there are 3 other measurements available in the literature. The ROSAT PSPC is sensitive to X-rays between 0.1-2.4 keV. The count rates as well as the accretion rates calculated from the Patterson and

Raymond (1985) relation between X-ray luminosity and accretion (for those systems where the spectrum and column density could be determined) are listed in Table 2.

TABLE 2. TOADS with ROSAT Observations

Object	i(deg)	ROSAT (c/s)	Reference	$\dot{M} \times 10^{-11} M_\odot$ /yr)
VY Aqr	25-50	0.12±0.02	Orio et al 1992	
TT Boo		0.008±0.001	Szkody et al 1995	0.014
WW Cet	54	0.74±0.03	Vrtilek et al 1994	4.7
WZ Sge	76	0.32±0.03	Orio et al 1992	
SW UMa	45	0.267±0.009	Szkody et al 1995	0.5
BZ UMa	<75	0.70±0.02	Szkody et al 1995	1.6
HV Vir		0.006±0.002	Szkody et al 1995	0.014

From the ROSAT count rates, known optical magnitudes and accretion rates, it appears that only 2 systems (TT Boo and HV Vir) have extremely low accretion compared to normal dwarf novae (Patterson 1984 shows a mean value of -10.5 for the log $\dot{M}$ of short period disk systems.

4. Conclusions

While half of the absolute magnitudes of the disks in TOADS which have known distances appear faint (implying very low accretion), only 2 of the 7 with known X-ray fluxes appear to have low accretion. Thus, there does not appear to be an absolute correlation between the low viscosity expected from the models and the observed properties at quiescence. However, at least 3 of these systems (WW Cet, BZ UMa and RU LMi) have sometimes been observed to be 1.5-2 magnitudes fainter than their normal quiescent values. Thus, it is possible that sporadic or interoutburst behavior may be affecting these results.

This research was partially supported by NASA grants NAGW 3158 and NAG5-1927 to PS.

References

Lynden-Bell, D. and Pringle, J. E. 1974, *MNRAS*, **168**, 603.
Mukai, K. et al. 1990, *MNRAS*, **245**, 385.
Orio, M. et al. 1992, in *Recent Results in X-Ray and EUV Astronomy*, Proc. Cospar Symp.
Patterson, J. 1984, *Astrophys. J. Suppl.*, **54**, 443.

Patterson, J. and Raymond, J. 1985, *Astrophys. J.*, **292**, 535.

Ringwald, F. 1993, *Ph.D. thesis*.

Sproats, L., Howell, S. B. and Mason, K. O. 1995, *MNRAS*, submitted.

Szkody, P. and Howell, S. B. 1992, *Astrophys. J. Suppl.*, **78**, 537.

Szkody, P. et al. 1995, in prep.

Vrtilek, S. D., Silber, A., Raymond, J. C. and Patterson, J. 1994, *Astrophys. J.*, **425**, 787.

Warner, B. 1987, *MNRAS*, **227**, 23.

Young, P. J. and Schneider, D. 1981, *Astrophys. J*, **247**, 960.

FOUR TOPICS IN ACCRETION DISK THEORY

With Application to Dwarf Novae and X-Ray Novae

JOHN K. CANNIZZO

NASA/GSFC, Laboratory for High Energy Astrophysics
Code 666, Greenbelt, MD 20771; cannizzo@stars.gsfc.nasa.gov

Abstract.

In this review I discuss four recent projects involving an application of the limit cycle theory for dwarf nova outbursts. These studies cover a range in type of object and outburst behavior. The common theme in these works is the use of observations to place constraints on the physics of accretion disks.

1. Introduction

I will touch briefly on four topics: 1) the sequencing of long and short outbursts in dwarf novae (DNe) (Cannizzo 1993), 2) the exponential decay of the outbursts in dwarf novae and X-ray novae (XNe) (Cannizzo 1994), 3) the tremendous outburst amplitude systems which have been discussed at this meeting by Steve Howell and Paula Szkody (Howell, Szkody, and Cannizzo 1994), and 4) the black hole X-ray binaries (Cannizzo, Chen, and Livio, in preparation).

2. The Sequencing of Long and Short Outbursts

In this section I discuss the long term behavior of SS Cygni which has been inferred from the AAVSO data, some of which Janet Mattei showed during her talk. The unusually complete record of SS Cyg makes possible a detailed analysis of its behavior.

SS Cyg shows a bimodal distribution of outburst durations, with the maxima in the frequency distributions occurring at about 7 and 14 days. We can find stretches of outbursts in SS Cyg where long and short outbursts simply alternate (LSLSLS...), where two short outbursts lie between two consecutive long outbursts (LSSLSSLSS...), or where three short outbursts

343

A Bianchini et al (eds) Cataclysmic Variables 343-348

lie between two consecutive long outbursts (LSSSLSSSLSSS...). Together, these three types of sequences account for about 80 percent of the observed behavior. Recent work using the accretion disk limit cycle model for dwarf nova outbursts has shown that this type of sequencing, in which n short outbursts lie sandwiched between two consecutive long outbursts, and $n \sim 1 - 3$, is fundamental to the theory (Cannizzo 1993). For outbursts which end up being short, the disk mass present at burst onset is small, and for long outbursts the stored disk mass is greater – as is the fraction of the disk which is accreted.

It is interesting to note that, if I take my favored model for SS Cyg, divide the outer disk radius by two, and divide the mass transfer rate by ten (so as to get a SU UMa system), I get sequences in which about five to ten short outbursts occur between two successive long outbursts. I believe that the limit cycle mechanism – acting alone – possesses enough complexity to account for the short and superoutbursts in the SU UMa stars, and that the thermal-tidal instablity (Osaki 1989) is not necessary (see also van Paradijs 1983).

By changing one of the control parameters of the theory in the time dependent model, we produce changes in n as well as changes in the recurrence time and quiescent disk flux. By performing very long runs with my disk instability code in which I slowly vary each parameter separately, I am able to study the correlations between different properties associated with the outbursts, and to see which correlated changes agree with those observed.

TABLE 1. Observations

	variable$_1$ $-$ variable$_2$	correlation
1.	$< t_c > - < N(L)/N(S) >$	direct
2.	$< t_q > - < N(L)/N(S) >$	direct
3.	$< t_b(L) > - < N(L)/N(S) >$	inverse
4.	$< t_b(S) > - < N(L)/N(S) >$	none
5.	$< t_c > - < f_v(\text{quiescent}) >$	inverse

Table 1 presents a summary of the observed correlations in SS Cygni. These correlations are between fluctuations seen in 1000 day moving averages of the recurrence time t_c, the quiescent interval between two outbursts t_q, the duration of long outbursts $t_b(L)$, the duration of short outbursts $t_b(S)$, the quiescent visual flux $f_v(q)$, and the ratio of number of long to number of short outbursts $N(L)/N(S)$. The most important correlations to note here are 1) a *direct* correlation between fluctuations in $< t_c >$

and $< N(L)/N(S) >$, and 2) an *inverse* correlation between $< t_c >$ and $< f_v(q) >$.

We identify five of the most obvious parameters of the theory for testing: the inner disk radius r_{inner}, the outer disk radius r_{outer}, the secondary mass transfer rate $\dot{M}_T$, the viscosity parameter α in the low state α_{cold}, and the viscosity parameter α in the high state α_{hot}. The first two questions we ask are 1) "How large would the variations in each of these parameters have to be to account for the amplitude of the fluctuations seen in the long term light curve of SS Cyg?" and 2) "On what time scale(s) would these fluctuations have to be occurring?". By numerical experimentation we find that, if r_{inner} or r_{outer} fluctuations are the ones that are producing the changes in sequencing, then these must be varying by about a factor of two. If one of the parameters $\dot{M}_T$, α_{cold}, or α_{hot} were responsible, the variations would have to be by about a factor of three. Second, from the observed variations in SS Cyg, the changes must be occurring semi-regularly on a $5 - 10$ yr time scale.

TABLE 2. Theory

	r_{inner}	r_{outer}	$\dot{M}_T$	α_{cold}	α_{hot}
1.	yes	no	no	yes	yes
2.	yes	no	no	yes	yes
3.	no	yes	no	no	yes
4.	no	no	yes	yes	no
5.	yes	no	no	yes	yes

We now discuss the correlations produced by varying each parameter separately. These are given in Table 2. A "yes" in the table indicates that the sense of the correlation is the same as that in Table 1, and a "no" indicates that it is not. The first thing we notice is that the variation of no one parameter in the theory can account for all five of the correlations listed in Table 1. There are technical reasons discussed by Cannizzo (1993), however, which may indicate that correlations listed in rows 3 and 4 of the tables may be less important than the others. That is, a "no" in columns 3 or 4 of Table 2 would not necessarily be fatal to having that parameter supplying the requisite fluctuations. Therefore r_{inner}, α_{cold}, and α_{hot} may all be viable candidates. Large variations in the mass transfer rate $\dot{M}_T$ are definitely excluded, yet it is this parameter which has been most often invoked in the recent past to account for long term variations (e.g. Hempelmann and Kurths 1990; Cannizzo and Mattei 1992). Of the three parameters r_{inner}, α_{cold}, and α_{hot}, one tends to view the viscosity parameters as fundamental physical quantities. It is not clear what mechanism could produce such large variations in α on such long time scales, especially when all the time scales intrinsic to the disk are so much shorter. We also note that, since the variations in outburst behavior produced by varying α_{cold} and α_{hot} separately are opposite, varying them together would produce only small changes (Cannizzo 1993). Finally, r_{inner} is also not without its problems. It seems unlikely that some mechanism could be operating which would give coherent, factor of ~ 2 changes in r_{inner} on a $\sim 5 - 10$ yr time scale. The Keplerian time at r_{inner} is about 10 seconds.

A strong possibility is that there is some other parameter, external to the basic limit cycle theory, which can vary so as to produce correlated changes that are in the same sense as those of r_{inner}, α_{cold}, or α_{hot}. There is a very interesting poster at this meeting by Emmi Meyer-Hofmeister, Nikolaus Vogt, and Friedrich Meyer which looks at the effect of a varying magnetic field from the secondary star which threads the outer accretion disk. A stronger field tends to push matter in the accretion disk to smaller radii, and therefore mimics a larger α_{cold}. The correlated changes in outburst properties induced by varying $\mathbf{B}_{outer}$ should therefore be in accord with the observed ones. The long time scales of $5 - 10$ yrs may be expected from magnetic cycles in the secondary star (Warner 1988, Bianchini 1990).

3. The Exponential Decay of Outbursts in DNe and XNe

For some time it has been known that the DNe outbursts decay exponentially in the visual. Bailey (1975) noted a correlation between rate of decay and orbital period. Smak (1984) showed the Bailey relation to be a natural consequence of the different size disks in different orbital period systems, and used the normalization of the Bailey relation to infer $\alpha_{hot} \sim 0.2$. Mineshige, Yamasaki, and Ishizaka (1993 = MYI) discuss the implications of the exponential decay in the X-ray nova outbursts and argue that mass and angular momentum must be extracted from the outer disk to get the observed exponential form. Moreover, the time scale for the reduction of the surface density Σ in the outer disk $-(\partial \ln \Sigma / \partial t)^{-1}$ must be constant.

Cannizzo (1994) studies the decay of the outburst in the V band for models in which $\alpha \sim r^{\epsilon}$. For runs in which ϵ is small, α decreases at small radii and the disk time scales decrease as the cooling front moves inward, leading to a faster-than-exponential decay of the optical light. For runs in which ϵ is large, the opposite is true. There is a moderate increase in disk time scales as the cooling front propagates to smaller radii, giving rise to a slower-than-exponential decay form. The decays are most nearly exponential in that $(dm_v/dt)^{-1}$ stays roughly constant for ϵ between about 0.3 and 0.4.

4. Tremendous Outburst Amplitude DNe

Why does a system like WZ Sge have a recurrence time for outbursts of more than 30 yrs? Smak (1993) and Osaki (1994) have already shown that α_{cold} must be very small in order to produce this behavior. The modeling that I have been doing with Steve Howell and Paula Szkody to attempt to provide a theoretical understanding of these systems confirms the work of Smak and Osaki. I can bracket the approximate range of observed behavior by taking $\alpha_{\text{hot}} \sim 0.2$ and $\alpha_{\text{cold}} \sim 0.001$, assuming $\dot{M}_T \sim 3 \times 10^{-11} M_\odot \text{ yr}^{-1}$. Because the ratio $\alpha_{\text{hot}}/\alpha_{\text{cold}}$ is so large, a significant fraction of the disk is accreted and only long outbursts (i.e., superoutbursts) are produced. There is a simple inverse relation between t_c and $\dot{M}_T$, as noted by Osaki (1994). If we try to get WZ Sge-like behavior by keeping $\alpha_{\text{cold}} \sim 0.02 - 0.04$ (as in "normal" DNe) and decreasing $\dot{M}_T$, we tend to get sequencing in which short outbursts lie between two consecutive superoutbursts, in contrast to the observations.

We can only speculate as to why α_{cold} should be so small for these systems. As noted by Mario Livio in his talk, in the currently favored Balbus-Hawley mechanism, the level of viscous dissipation is set by the non-linear outcome of the shearing of a weak magnetic field, and is presumably set by a balance between the generation of the poloidal field and the buoyant loss of magnetic flux to a corona overlying the disk. Since the mechanism works by amplifying an arbitrarily weak seed field, it would seem improbable that one could make the field strength in the mass stream being carried over from the secondary star so small that it would not be amplifiable. One possibility might be, however, that the disk could become thinner than the wavelength of unstable modes. This could occur if the disk were to cool to a very low temperature.

5. The Black Hole X-Ray Binaries

These systems consist of interacting binaries in which the accreting star is thought to be a $\sim 10 M_\odot$ black hole (see Mineshige and Kusunose 1993 for a recent review). The numerical treatment of the XNe accretion disks is technically more difficult than for the DNe because, for the former, $r_{\text{outer}}/r_{\text{inner}} \sim 10^4$, whereas for the latter $r_{\text{outer}}/r_{\text{inner}} \sim 10^2$. The larger dynamic range in disk radii and the accompanying disk time scales necessitates the use of many more grid points in the one-dimensional time dependent models to ensure numerical fidelity to the physics.

The XNe for which outbursts which have been well covered usually show a fast rise and slower exponential decay — $t_{\text{rise}} \sim 10$ days and $t_{\text{e-fold/decay}} \sim 30 - 60$ days (MYI). These times are much longer than for the DNe. The difference in primary mass M_1 between DNe and XNe may be able to explain, for example, the decay rate in optical flux (Cannizzo 1994). Despite the fact that strong irradiation would seem to be prevalent during the outburst, the observed exponential decay argues against irradiation preventing cooling, and also against a mass transfer event as the cause of the outburst (Lyubarskii and Shakura 1987).

6. Acknowledgement

This research was supported by the National Academy of Sciences through a National Research Council associateship at Goddard Space Flight Center.

References

Bailey, J. (1975) The Rate of Decline from Dwarf Nova Outbursts, *Journal of the British Astronomical Association*, Vol. no. **86**, pp. 30-32.

Bianchini, A. (1990) Solar-Type Cycles in Close Binary Systems, *Astronomical Journal*, Vol. no. **99**, pp. 1941-1952.

Cannizzo, J. K. (1993) The Accretion Disk Limit Cycle Model: Toward an Understanding of the Long-Term Behavior of SS Cygni, *Astrophysical Journal*, Vol. no. **419**, pp. 318-336.

Cannizzo, J. K. (1994) On the Decay of Outbursts in Dwarf Novae and X-Ray Novae, *Astrophysical Journal*, Vol. no. **435**, pp. 389-397.

Cannizzo, J. K. and Mattei, J. A. (1992) On the Long-Term Behavior of SS Cygni, *Astrophysical Journal*, Vol. no. **401**, pp. 642-653.

Hempelmann, A. and Kurths, J. (1990) Dynamics of the Outburst Series of SS Cygni, *Astronomy and Astrophysics*, Vol. no. **232**, pp. 356-366.

Howell, S. B., Szkody, P. and Cannizzo, J. K. (1995) Tremendous Outburst Amplitude Dwarf Novae, *Astrophysical Journal*, Vol. no. **439**, pp. 337-345.

Lyubarskii, Yu. É. and Shakura, N. I. (1987) Nonlinear Self-Similar Problems of Nonstationary Disk Accretion, *Soviet Astronomy Letters*, Vol. no. **13**, pp. 386-390.

Mineshige, S. and Kusunose, M. (1993) Black Hole Accretion Disk Instabilities, in *Accretion Disks in Compact Stellar Systems*, ed. J. C. Wheeler, World Scientific Publishing, Singapore, pp. 470-521.

Mineshige, S., Yamasaki, T. and Ishizaka, C. (1993) On the Exponential X-Ray Decay in X-Ray Novae, *Publications of the Astronomical Society of Japan*, Vol. no. **45**, pp. 707-713.

Osaki, Y. (1989) A Model for the Superoutburst Phenomenon of SU Ursae Majoris Stars, *Publications of the Astronomical Society of Japan*, Vol. no. **41**, pp. 1005-1033.

Osaki, Y. (1994) Disk Instability Model for SU UMa Stars: SU UMa/WZ Sge Connection, in *Theory of Accretion Disks - 2*, ed. W. J. Duschl, J. Frank, F. Meyer, E. Meyer-Hofmeister, and W. M. Tscharnuter, Kluwer Academic Publishers, Dordrecht, pp. 93-108.

Smak, J. (1984) Accretion in Cataclysmic Binaries. IV. Accretion Disks in Dwarf Novae, *Acta Astronomica*, Vol. no. **34**, pp. 161-189.

Smak, J. (1993) WZ Sge as a Dwarf Nova, *Acta Astronomica*, Vol. no. **43**, pp. 101-119.

van Paradijs, J. (1983) Superoutbursts: A General Phenomenon in Dwarf Novae, *Astronomy and Astrophysics*, Vol. no. **125**, pp. L16-L18.

Warner, B. (1988) Quasiperiodicity in Cataclysmic Variable Stars Caused by Solar-Type Magnetic Cycles, *Nature*, Vol. no. **336**, pp. 129-143.

ANGULAR MOMENTUM TRANSPORT IN ACCRETION DISKS

MARIO LIVIO
Space Telescope Science Institute 3700 San Martin Drive Baltimore, MD 21218, USA

Abstract.
Potential angular momentum transport mechanisms in accretion disks are reviewed. It is argued that the most promising mechanism is a dynamo-driven viscosity which involves the Balbus-Hawley instability as an essential ingredient.

1. INTRODUCTION

Accretion disks have been observed in a variety of astrophysical objects, ranging from young stellar objects (YSOs), to cataclysmic variables (CVs) and x-ray binaries (XRBs). Their existence is also normally postulated in most models of active galactic nuclei (AGN). Nevertheless, the nature of the angular momentum transport mechanism(s) in these disks remains elusive, in spite of extensive research. Already in the original work of Shakura & Sunyaev (1973), it has been realized that molecular viscosity is totally unimportant. For example, the Reynolds number in a typical CV disk is of order

$$\Re \sim 10^{14} \left(\frac{M_{WD}}{M_\odot}\right)^{1/2} \left(\frac{R_D}{10^{10} \text{ cm}}\right)^{1/2} \left(\frac{T}{10^4 \text{ K}}\right)^{-5/2} \left(\frac{n}{10^{15} \text{ cm}^{-3}}\right), \quad (1)$$

where M_{WD} is the white dwarf mass, R_D is the disk radius, T is the temperature and n is the number density. It was this situation which led Shakura & Sunyaev to suggest their dimensional-analysis prescription, which involves the familiar "α" viscosity parameter. The argument is based on the assumption that the origin of the viscosity is either turbulence or magnetic stresses. In the former case, the kinematic viscosity is dimensionally given by $\nu_t \sim V_t \ell_t$ (where V_t and ℓ_t are typical turbulent velocities and

349

A Bianchini et al (eds), Cataclysmic Variables, 349-358
© *1995 Kluwer Academic Publishers*

cell sizes, respectively). Since eddy sizes have to be smaller than the disk semi-thickness, H, and the turbulence is assumed to be subsonic, this can be written as $\nu_t \sim \alpha C_s H$, where C_s is the speed of sound and α is a dimensionless parameter ($\alpha \lesssim 1$). In the magnetic case, the shear stress $B_R B_\phi / 4\pi$ is assumed to be smaller than the total pressure (otherwise the field can escape in the form of magnetic bubbles). Thus, in both cases, the viscous force per unit area, $t_{\phi r}$, can be expressed as $t_{\phi r} \sim \alpha P$ (where P is the total pressure). Unfortunately, while the Shakura-Sunyaev prescription allowed many aspects of accretion disk theory to advance, until recently, no simple hydrodynamic or MHD instability that could be responsible for the turbulence or dynamo action has been identified.

In the present work, I review some of the suggested angular momentum transport mechanisms and their properties, and attempt to assess their viability. Some aspects of this problem have been recently reviewed by Hawley, Gammie & Balbus (1994) and by Vishniac & Diamond (1994).

2. NON-VISCOUS ANGULAR MOMENTUM TRANSPORT AND REMOVAL MECHANISMS

Since for a long time, attempts to identify a source for viscosity in disks have not been successful, several mechanisms of removal and transport of angular momentum which do not involve viscosity have been suggested. Among these, the most promising ones (in principle at least) have been: hydromagnetic winds and spiral shocks. I shall now review briefly these mechanisms.

2.1. HYDROMAGNETIC WINDS

Magnetically driven winds from accretion disks have been suggested as models for the collimated outflows that are observed in young stellar objects (YSOs) and in a variety of active galactic nuclei (AGN) (*e.g.*, Blandford 1993; Königl & Ruden 1993; Pringle 1993). Some authors suggested that such winds can also provide the mechanism of removal of angular momentum from the disk, which in turn, drives the accretion (*e.g.*, Blandford & Payne 1982; Pudritz & Norman 1986; Königl 1989; Lovelace, Romanova & Contopoulos 1993). The basic idea is quite simple. The disk is assumed to be threaded by reasonably ordered open field lines, gas elements are forced to corotate with the field lines up to the local Alfven radius $R_A(R)$ and are centrifugally accelerated. If we equate the rate of angular momentum removal by the wind to the rate required for the accretion to take place, we obtain

$$\frac{\dot{M}_d}{4\pi R^2} \sim \dot{m}_W [(R_A/R)^2 - 1], \tag{2}$$

where $\dot{M}_d$ is the mass flux through the disk and $\dot{m}_W$ is the rate of wind mass loss per unit area. From eq. (2) it is clear that for $R_A/R \sim 10$ it is sufficient to lose $\sim 1\%$ of the accreted mass in the wind, to drive the accretion flow. Wind mass loss rates of this order have actually been observed in CVs (*e.g.*, Drew 1990, 1995; Knigge, Woods & Drew 1994). There exists, however, both observational and theoretical evidence which suggests that magnetically driven winds may not be *the dominant* angular momentum removal mechanism. In CVs in particular, disk radii are observed (Smak 1984) to be a few times larger than the circularization radius (corresponding to the specific angular momentum of the material leaving the inner Lagrange point L_1). This suggests disk broadening by viscosity, which is not expected if angular momentum would have been efficiently removed by a hydromagnetic wind. Furthermore, the observed behaviour of the disk radius during dwarf nova eruptions (a rapid increase, followed by a slower decay, *e.g.*, Smak 1984; ODonoghue 1986) follows exactly the expectations from a viscous transport of angular momentum (*e.g.*, Livio & Verbunt 1988; Ichikawa, Hirose & Osaki 1993). On the theoretical side, Lubow, Papaloizou & Pringle (1994a,b) studied the stability of a configuration in which accretion is driven by the magnetic torques provided by the wind (which in turn is centrifugally driven). They found three equilibrium solutions. Two of these solutions were found to be stable, but unacceptable as models for the physics of accretion disks (one of them corresponds to no accretion and no outflow, and the other to a situation in which essentially all the mass is lost to the wind). The third solution, which represents the desired situation (wind removal of angular momentum with accretion; the wind mass loss rate being a small fraction of the accretion rate), was found to be *unstable*.

In a somewhat related work, Stone & Norman (1994) calculated the time evolution of an adiabatic, axisymmetric disk, which was initially threaded by a vertical magnetic field. The calculation was performed under the assumption of ideal MHD. They found that even in the case of an initially Keplerian disk, the disk *collapses on a dynamical (orbital) timescale*, both for a strong $(v_{Alfven}^2/v_{orbital}^2 \sim 0.1)$ and for a weak $(v_{Alfven}^2/v_{orbital}^2 \sim 0.001)$ magnetic field. In the case of the strong field the collapse is caused by strong magnetic braking, while in the case of the weak field it is caused by internal torques which develop as a result of the Balbus-Hawley instability (discussed in §3.3 below).

Thus, while more work on this problem is certainly needed, hydromagnetic winds do not appear at the moment to be capable of providing the *main* angular momentum removal mechanism.

2.2. COHERENT SPIRAL SHOCK WAVES

The idea behind this mechanism is based on the fact that disturbances at the outer disk edge (*e.g.*, tidal) are wound-up by differential rotation as they propagate inward, into a trailing spiral (*e.g.*, Spruit 1987; Dgani, Livio & Regev 1994). The angular momentum associated with the wave is given by (Goldreich & Tremaine 1978; Spruit 1989)

$$j \simeq -\frac{1}{2} \int \frac{\rho V_w^2}{\Omega} dV, \tag{3}$$

where V_w is the velocity amplitude of the wave, Ω is the angular velocity and the integration is carried out over the volume of the wave packet. The potential for transport of angular momentum in this mechanism is provided by the fact that two-dimensional analytical models indicated that the wave increases inward, steepening into a shock. This can lead to dissipation and to the transfer of negative angular momentum to the disk (eq. (3)). Preliminary two-dimensional numerical calculations of hot disks (Mach number $\sim$10) showed that a coherent spiral pattern is indeed obtained (*e.g.*, Rozyczka & Spruit 1989; 1993) and that angular momentum transport results, corresponding to an effective viscosity parameter of $\alpha \sim$ 0.01–0.1 (Matsuda *et al.* 1989). However, a more detailed examination of some of the aspects of this problem revealed that coherent spiral shocks may not be able to provide the required universal transport mechanism. In particular, the following problems have been identified:

-   Calculations which involved cooler disks (Mach number $\sim$ 50) showed that the spiral shocks did not reach the center (Savonije, Papaloizou & Lin 1994), thus limiting the ability of the waves to transport angular momentum. A similar problem has been identified with wave propagation in disks in general (losses due to deflection or concentration to the mid-plane of the disk; *e.g.*, Lubow & Pringle 1993).
-   It is not clear whether a coherent spiral structure is indeed obtained in three dimensions (Sawada & Matsuda 1992).
-   While the hot disk spiral-shocks calculations predict an effective (tidally induced) viscosity parameter which is proportional to the mass ratio of the binary components, $\alpha_{eff} \sim q = M_2/M_1$, the mean recurrence times of dwarf nova eruptions indicate an α_{eff} which is independent of q (Livio & Spruit 1991).
-   While it is not clear to what extent the effects of the spiral shocks are directly observable, it is worth noting that spiral patterns have not been observed in Doppler tomography studies (see also Bunk, Livio & Verbunt 1990).

As a consequence of all of the above problems it does not appear (at least to this author) that coherent spiral shocks can provide *a universal* mechanism for angular momentum transport. The potential applicability of this mechanism to particular classes of objects should be further explored.

3. ANGULAR MOMENTUM TRANSPORT BY VISCOSITY

As explained in the introduction, turbulence and magnetic stresses have been identified early on as potential sources of viscosity. What has proven, however, much more difficult than originally expected, was the identification of an actual instability which could lead to turbulence or to dynamo field generation. In the following, I review what I consider as the most promising (in principle) such instabilities.

3.1. CONVECTION

It has been known for a long time that accretion disks can have a vertical structure which is convectively unstable (*e.g.*, Livio & Shaviv 1977). One could therefore, in principle, imagine a cycle in which accretion generates a luminosity which is emitted as a vertical flux. This vertical flux drives convection which generates turbulence, which, in turn, produces the viscosity which is necessary to drive the accretion process (*e.g.*, Kley, Papaloizou & Lin 1993; Duschl 1991). It should be noted right away, that even if such a cycle were to operate, it would probably not have been applicable to all the systems which contain accretion disks (for example, it is more likely for disks around YSOs than in AGN). In addition, however, this mechanism encounters the following serious problems:

- heating from above, by the central object, may suppress convection
- $\dot{M} = 0$ is always a valid solution of the above cycle
- a linear treatment of the nonaxisymmetric convective instability showed that shear causes an *inward* transport of angular momentum (Ryu & Goodman 1992), opposite to what is required to drive accretion.

Consequently, while additional work on nonlinear transport is certainly required (a first attempt in this direction is provided by the work of Lin, Papaloizou & Kley 1993), convection does not appear as a promising mechanism for angular momentum transport in general.

3.2. THE PAPALOIZOU-PRINGLE INSTABILITY

In a remarkable series of works, Papaloizou & Pringle (1984, 1985, 1987) discovered that thick disk models are dynamically unstable to global, non-axisymmetric modes. In particular, it was found that slender annuli and tori which are characterized by a power-law angular velocity profile $\Omega(R) = \Omega_o(R/R_o)^{-q}$, are violently unstable for $q > \sqrt{3}$ (see *e.g.*, review by Narayan 1991 and references therein). However, it was found that in two-dimensional flows, the principal branch is stabilized by accretion (Blaes 1987), due to the loss of reflection at the inner disk edge (Gat & Livio 1992). The situation is less clear in three dimensions, although some indications for a potential stabilization by accretion do exist (Hawley 1991).

In spite of an enormous amount of work on this topic (e.g., Goldreich, Goodman & Narayan 1986; Glatzel 1990; Zurek & Benz 1986; Kojima 1989), the ultimate fate of three-dimensional tori (of various angular velocity distributions) *with accretion* and the potential role of the instability in driving angular momentum transport are still unclear. The interest in this instability has subsided due to the discovery of a more promising one, to be discussed in the next subsection.

3.3. LOCAL INSTABILITIES AND MAGNETIC DYNAMOS

In a recent important series of works, Balbus & Hawley (1991, 1992) and Hawley & Balbus (1991, 1992) re-examined the linear stability of an accretion disk that is threaded by a weak magnetic field and showed that such configurations are unstable against axisymmetric perturbations. While the instability itself has been discovered originally in the context of an incompressible Couette flow that is threaded by a vertical magnetic field (Velikhov 1959; Chandrasekhar 1961), its enormous potential significance for angular momentum transport in disks has only been realized recently by Balbus & Hawley.

The conditions for the instability to occur can be obtained from a highly simplified linear analysis. Imagine a vertical field line (z direction in cylindrical coordinates) that threads the disk and examine the forces that act on a fluid element attached to this field line. If the element is displaced radially outwards by an amount ζ, then, since the element is revolving with the angular velocity of its original location, it will experience an excess centrifugal force. At the same time, however, due to the bending of the magnetic field line, there will be a magnetic tension force. The net force will thus be

$$F = -\zeta k_z^2 V_A^2 - \zeta \frac{d\Omega^2}{d\ln R}, \qquad (4)$$

where k_z is the z component of the wave number vector and V_A is the Alfven speed. From expression (4) we immediately see that (for stability the net force needs to be restoring):

(a) for $\frac{d\Omega}{dR} < 0$ (as in Keplerian disks), some wavelengths are always unstable

(b) at a given wavelength, *the instability is suppressed if the magnetic field is strong enough*. The condition for instability is:

$$(\vec{k} \cdot \vec{V}_A)/\Omega \leq \sqrt{3}. \qquad (5)$$

Hawley & Balbus and collaborators embarked on a series of numerical simulations aimed at exploring in a methodical way the consequences of the instability. The following steps should be mentioned:

- First, Hawley & Balbus examined a *two dimensional flux tube* in the z direction, in an axisymmetric, uniform, Keplerian flow. They found that a radial and azimuthal field was generated, with a growth rate $\sim 0.74\Omega$ and a maximum wavelength of $\lambda_{max} \sim 2\pi V_{A,z}/\Omega$ (reconnections were found to be the amplitude limiting factor; Hawley & Balbus 1991).

- Secondly, they followed the time evolution of a *two-dimensional shearing-box* (the equivalent of a local piece of a Keplerian disk) with no vertical gravity (and hence, no buoyancy; Hawley & Balbus 1992). They explored three initial configurations: (i) a uniform B_z, (ii) a uniform B_x, (iii) a radially varying (sinusoidally) B_z.

- In case (i), they obtained a large scale radial streaming motion in effectively two channels, one near the midplane (carrying excess angular momentum outwards) and one near the disk surface (with sub-Keplerian fluid, moving in). This streaming motion was accompanied by strong magnetic fields (see also Goodman & Xu 1993). In case (ii), they obtained a structure with a small vertical scale, but a large radial scale. The mode saturated at rather modest amplitudes (with the size of the computational domain limiting the growth). In the case of a sinusoidally varying field (with a zero net field on the grid), a complex turbulent motion developed, accompanied by angular momentum transport and large field amplification.

- In the third set of simulations, Hawley, Gammie & Balbus (1995), followed the evolution of a *three dimensional shearing box* (with height $\sim C_s/\Omega$, where C_s is the sound speed) but still no vertical gravity. The results showed the following phenomena: (a) all the instabilities that were found in 2D were also present in 3D. (b) The "channel solutions" found in 2D were shown to be unstable to non-axisymmetric perturbations, consequently the coherent flow broke into *a self-sustaining turbulent state*. (c) The fluctuation amplitudes of $\delta\vec{B}$, $\delta\vec{v}$ were found to be much larger than the mean value, and in general, the flow was characterized by a *non-isotropic MHD turbulence* ($\delta B_\phi > \delta B_r > \delta B_z$). (d) Most of the power and angular momentum flux was found to be concentrated in the lowest wave numbers, namely, most of the magnetic energy was in large scale $\vec{B}$. (e) The outward flux of angular momentum was dominated by the magnetic stress $B_x B_y/4\pi$. (f) The effective viscosity parameter that was obtained was of order $\alpha_{eff} \sim 0.1$.

- A preliminary attempt to follow the evolution of *a stratified 3D box* (Stone, Hawley & Gammie 1995) resulted in MHD turbulence, in which bouyancy of strong field regions has been observed. A large fraction of the magnetic energy was found to concentrate in a magnetized corona. This preliminary calculation showed some evidence that the field strength saturates at the same strength, irrespective of the initial value (which could be indicative of dynamo action).

I should also mention that attempts have been made to follow the development of the Balbus-Hawley instability in the presence of ambipolar diffusion (which could, in principle, suppress the instability; MacLow *et al.* 1994; see, however, Regös 1995), with the inclusion of resistivity (which may introduce an instability threshold; Matsumoto & Tajima 1994) and the possible global effects of a relatively strong azimuthal field have been investigated (may require a global analysis; Gammie & Balbus 1994; Blaes & Balbus 1994; Curry, Pudritz & Sutherland 1994).

Finally, it has been attempted to tie the local results that were obtained to the global disk behaviour. Tout & Pringle (1992, 1994) suggested that ultimately, the dynamo cycle in disks may operate through the following physical processes: (i) the Balbus-Hawley instability converts an initially vertical field B_z into an horizontal one (B_r, B_ϕ). (ii) The shear in the disk generates an azimuthal component B_ϕ from the radial one, B_r. (iii) The Parker instability (buoyancy) converts B_ϕ into B_z, thus completing the dynamo cycle (see also Vishniac & Diamond 1992, 1994 for a discussion of dynamos). Curry, Pudritz & Sutherland (1994) investigated the *global* stability of a differentially rotating fluid shell, threaded by a vertical field, to linear axisymmetric perturbations. They generally found growth rates that are comparable to the local ones (see also Kumar, Coleman & Kley 1994; Gammie & Balbus 1994). They also showed that the most unstable mode is localized near the inner radius.

4. TENTATIVE CONCLUSIONS

On the basis of the work presented here one can draw the following preliminary (and clearly somewhat biased) conclusions:

- A mechanism which can generate MHD turbulence, angular momentum transport, and possibly dynamo action has been identified in the Velikhov-Chandrasekhar-Balbus-Hawley instability.
- The properties of the non-linear development of the instability in a full-scale, stratified disk, will require much future work, but all the calculations performed so far produced extremely promising results.
- The effects of additional physical processes (*e.g.*, ambipolar diffusion) will require further clarification. A recent attempt to study partially ionized disks indicates that as long as the ions-neutrals collision frequency is larger than the epicyclic frequency, the dynamo can operate (Regös 1995).

– Many questions related to other mechanisms remain open.

Among these are: (i) the stability of hydromagnetic wind solutions, (ii) the development of the Papaloizou-Pringle instability in 3D in the presence of accretion, (iii) the role of spiral shocks in 3D, realistic disks.

To end on a personal note, I feel enormously excited by the possibility that for the first time in more than 20 years, the answer to the problem of angular momentum transport in disks appears reachable. Furthermore, an examination of the potentially observable effects of anisotropic turbulence on the velocity profiles and eclipse behaviour of emission lines from disks (Horne 1995; Balbus, Gammie & Hawley 1994) shows that the shapes of the double-peaked lines of Ca, Mg, and Fe may actually serve as direct tests for the presence of anisotropic turbulence.

Acknowledgments

This work has been supported in part by NASA Grant NAGW-2678. I would like to thank Jim Pringle, Steve Balbus, John Hawley, Jim Stone, Chris Tout, Enikö Regös and Steve Lubow for extremely useful discussions.

References

Balbus, S. A., Gammie, C. F. & Hawley, J. F., 1994, MNRAS, 271, 197
Balbus, S. A. & Hawley, J. F. 1991, ApJ, 376, 214
Balbus, S. A. & Hawley, J. F. 1992, ApJ, 400, 610
Blaes, O. M. 1987, MNRAS, 227, 975
Blaes, O. M. & Balbus, S. A. 1994, ApJ, 421, 163
Blandford, R. D. & Payne, D. G. 1982, MNRAS, 199, 883
Blandford, R. D. 1993, in Astrophysical Jets, eds. D. Burgarella, M. Livio & C. O'Dea (Cambridge: Cambridge University Press), p. 15
Bunk, W., Livio, M. & Verbunt, F. 1990, A&A, 232, 371
Chandrasekhar, S. 1961, Hydrodynamic and Hydromagnetic Stability, (London: Oxford University Press).
Curry, C., Pudritz, R. E. & Sutherland, P. G. 1994, ApJ, 434, 206
Dgani, R., Livio, M. & Regev, O. 1994, ApJ, 436, 270
Drew, J. E. 1990, ApJ, 357, 573
Drew, J. E. 1995, in the Analysis of Emission Lines, eds. R. E. Williams & M. Livio (Cambridge: CUP), in press
Duschl, W. J. 1991, A&A, 247, 393
Gammie, C. F. & Balbus, S. A. 1994, MNRAS, 270, 138
Gat, O. & Livio, M. 1992, ApJ, 396, 542
Glatzel, W. 1990, MNRAS, 242, 338
Goldreich, P., Goodman, J. & Narayan, R. 1986, MNRAS, 221, 339
Goldreich, P., & Tremaine, S. 1978, Icarus, 34, 240
Goodman, J. & Xu, G. 1994, ApJ, 432, 213
Hawley, J. F. 1991, ApJ, 381, 496
Hawley, J. F., & Balbus, S. A. 1991, ApJ, 376, 223
Hawley, J. F., & Balbus, S. A. 1992, ApJ, 400, 595

Hawley, J. F., Gammie, C. F. & Balbus, S. A. 1994, in The Physics of Active Galaxies, Mt. Stromlo Observatory, in press
Hawley, J. F., Gammie, C. F. & Balbus, S. A. 1995, ApJ, 440, 742
Horne, K. 1995, A&A, in press
Ichikawa, S., Hirose, M. & Osaki, Y., 1993, PASJ, 45, 243
Kley, W., Papaloizou, J. C. B. & Lin, D. N. C. 1993, ApJ, 416, 679
Knigge, C., Woods, J. A. & Drew, J. E. 1995, MNRAS, in press
Kojima, Y. 1989, MNRAS, 236, 589
Königl, A. 1989, ApJ, 342, 208
Königl, A. & Ruden, S. P. 1993, in Protostars and Planets III, eds. E. Levy & J. Lunine, University of Arizona Press, in press
Kumar, S., Coleman, C. S. & Kley, W. 1994, MNRAS, 266, 379
Lin, D. N. C., Papaloizou, J. C. B. & Kley, W. 1993, ApJ, 416, 689
Livio, M. & Shaviv, G. 1977, A&A, 55, 95
Livio, M. & Spruit, H. C., 1991, A&A, 252, 189
Livio, M. & Verbunt, F. 1988, MNRAS, 232, 1p
Lovelace, R. V. E., Romanova, M. M. & Contopoulos, J. 1993, ApJ, 403, 158
Lubow, S. H., Papaloizou, J. C. B. & Pringle, J. E. 1994a, MNRAS, 267, 235
Lubow, S. H., Papaloizou, J. C. B. & Pringle, J. E. 1994b, MNRAS, 268, 1010
Lubow, S. H. & Pringle, J. E. 1993, MNRAS, 263, 701
MacLow, M. M., Norman, M. L., Königle, A. & Wardle, M. 1994, preprint
Matsuda, T., Sekino, N., Shima, E., Sawada, K. & Spruit, H. C. 1989, in Theory of Accretion Disks, eds. F. Meyer et al. (Dordrecht: Kluwer), p. 355
Matsumoto, R. & Tajima, T. 1994, preprint
Narayan, R. 1991, in Structure and Emission Properties of Accretion Disks, eds. C. Bertout et al. (Gif sur Yvette: Editions Frontieres), p. 153
O'Donoghue, D. 1986, MNRAS, 220, 23p
Papaloizou, J. C. B. & Pringle, J. E. 1984, MNRAS, 208, 721
Papaloizou, J. C. B. & Pringle, J. E. 1985, MNRAS, 213, 799
Papaloizou, J. C. B. & Pringle, J. E. 1987, MNRAS, 225, 267
Pringle, J. E. 1993, in Astrophysical Jets, eds. D. Burgarella, M. Livio & C. O'Dea (Cambridge: CUP), p. 1
Pudritz, R. E. & Norman, C. A. 1986, ApJ, 301, 571
Regös, E. 1995, preprint
Rozyczka, M. & Spruit, H. C. 1989, in Theory of Accretion Disks, eds. F. Meyer et al. (Dordrecht: Kluwer), p. 341
Rozyczka, M. & Spruit, H. C. 1993, ApJ, 417, 677
Ryu, D. & Goodman, J. 1992, ApJ, 388, 438
Savonije, G. J., Papaloizou, J. C. B. & Lin, D. N. C. 1994, MNRAS, 268, 13
Sawada, K. & Matsuda, T. 1992, MNRAS, 255, 17p
Shakura, N. I. & Sunyaev, R. A. 1973, A&A, 24, 337
Smak, J. 1984, Acta Astron., 34, 93
Spruit, H. C. 1987, A&A, 184, 173
Spruit, H. C. 1989, in Theory of Accretion Disks, eds. F. Meyer et al. (Dordrecht: Kluwer Academic Publishers), p. 325
Stone, J. M., Hawley, J. F. & Gammie, C. F. 1995, private communication
Stone, J. M. & Norman, M. L. 1994, ApJ, 433, 746
Tout, C. A. & Pringle, J. E. 1992, MNRAS, 259, 604
Tout, C. A. & Pringle, J. E. 1995, MNRAS, submitted
Velikhov, E. P. 1959, Soviet JETP, 35, 1398
Vishniac, E. & Diamond, P. 1994, in Accretion Disks in Compact Stellar Systems, ed. J. C. Wheeler (Singapore: World Scientific), p. 41
Vishniac, E. T. & Diamond, P. H. 1992, ApJ, 398, 561
Zurek, W. H. & Benz, W. 1986, ApJ, 308, 123

ON THE OUTBURST BEHAVIOUR OF DWARF NOVAE
ABOVE THE PERIOD GAP

NIKOLAUS VOGT

Pontificia Universidad Catolica de Chile,
Astrophysics Group, Santiago, Chile, and
Max-Planck-Institut fuer Astrophysik, Garching, Germany

Abstract.

A brief description of general tendencies in the outburst behaviour of 27 well-observed dwarf novae is given. The division between wide and narrow outbursts is very common, about 75% of all dwarf novae in the sample display such a bimodal width distribution. The total outburst width increases, the width ratio (of wide to narrow outbursts) decreases, with orbital period. Anomalous outbursts with abnormally slow rise gradients are also very common, 70% of the sample dwarf novae show them, at least occasionally. There is an exclusive distinction between normal and anomalous outbursts, as demonstrated in the case SS Cyg. Some other peculiarities, such as small peaks and micro-variability during Z Cam standstills, are also discussed.

1. Introduction

Extensive research activities have recently been dedicated to cataclysmic variables with special emphasis to the so-called "ultra-short period domain", below the striking gap between 2 and 3 hours. Dwarf novae with orbital periods $P < 2^h$ are usually SU UMa stars with spectacular super-outbursts, superhumps and associated phenomena (Vogt, 1980; Molnar and Kobulnicky, 1992). The challenge to interpret these phenomena could have been one of the reasons to neglect, to a certain extent, the study of other unexplained features in the behaviour of dwarf novae with longer periods. The present paper intends to briefly describe their most prominent outburst parameters and tendencies.

A. Bianchini et al. (eds.), Cataclysmic Variables, 359-365.
© 1995 Kluwer Academic Publishers.

Studies which compare, systematically, different dwarf novae in their long-term outburst behaviour are scarce. Szkody and Mattei (1984) analysed a total of 21 dwarf novae during a 3-years interval 1974 - 1977, based on AAVSO data. Richter and Braeuer (1989) discussed the relationships between outburst amplitude, cycle length and orbital period. More recently, the complete light curve of SS Cyg covering nearly one century (1896 - 1992) was analysed and interpreted with thorough statistical methods (Cannizzo and Mattei, 1992; Cannizzo, 1993).

2. Dwarf Nova Outburst Data

This study is based on outburst data published by the American Association of Variable Star Observers (AAVSO) and by the Variable Star Sections of the British Astronomical Association (BAA) and of the Royal Astronomical Society of New Zealand (RASNZ). Table 1 lists all dwarf novae for which relevant data could be found. They are arranged according to orbital period, taken from the latest edition of Ritter's catalogue (Ritter and Kolb, 1994). Only in four cases with undetermined orbital period (AT Ara, AH Eri, TU Ind and V 442 Cen) a rough estimate according to UBV colours in quiescent state (Vogt, 1982) is given. C is the general mean repetition cycle of outbursts, W the mean total width of outbursts, i.e., the time interval between departure from and return to quiescent state. Many dwarf novae present a bimodal distribution in the width of their outbursts (c.f. Sect. 3): W_w/W_n is the ratio of the peak values to the wide and narrow outbursts. "No" refers to cases in which no bimodal width distribution is observed. The remaining three columns refer to peculiarities such as the occurrence of anomalous outbursts (c.f. Sect. 4), small peaks and standstills (c.f. Sect. 5). The last column gives the sources used in this study (associations as mentioned above). Only in two cases (EM Cyg and AH Her) the data were derived from Beyer (1967), independently from the variable star observer associations. A detailed presentation of their numerous references is not possible here for lack of space; it will be given elsewhere.

3. Width Distribution

According to the data in Table 1 the division between wide and narrow outbursts is present in about 75% of all dwarf novae in the period range $3h < P < 10h$. This result disagrees with that of Szkody and Mattei (1984) who found evidence for a bimodal width distribution in only 2 of 16 cases with $P > 3h$. The relatively short time interval covered by them (3 years) apparently is not sufficient for determining the width distribution in a statistically significant manner.

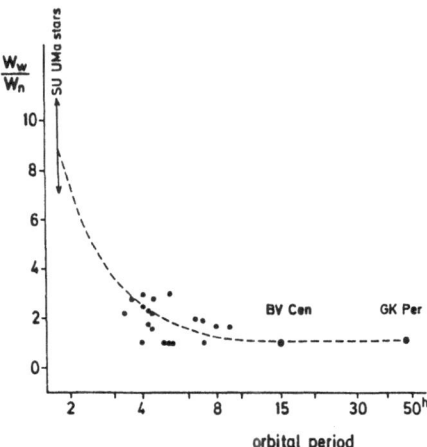

Fig. 1: Ratio of total mean widths of wide to narrow outbursts vs. orbital period.
$W_w/W_n = 1$ refers to cases without significant separation between wide and narrow
outbursts. The approximate location of SU UMa stars (superoutbursts/normal outbursts)
is shown for comparison.

The width ratio W_w/W_n is of the order 2 - 3, with an increasing tendency towards shorter orbital periods (Fig. 1). All dwarf novae with P
$> 10^h$, as well as 25% of the shorter period stars (including some Z Cam
stars) do not show a bimodal distribution of the width.

The total width of outbursts is closely correlated to the orbital period
(Fig. 2) with an approximate linear relation

$$W(days) = -18.4 + 40.8 \times logP(h)$$

valid in the range $3^h <P < 50^h$ (correlation coefficient R $= 0.9$).

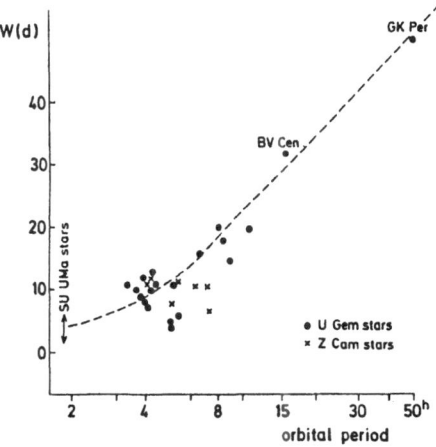

Fig. 2: The mean total width of dwarf nova outbursts vs. orbital period. The approximate location of normal outbursts of SU UMa stars is shown for comparison.

Name	Po (h)	C (d)	W (d)	Ww/Wn	anom. outb.	small peaks	Z Cam type?	Sources
GK Per	47.93	500	50	no	all			AAVSO
BV Cen	14.64	150	32	no	nearly all			RASNZ
DX And	10.59	400	20:	?	+			AAVSO
RU Peg	8.99	73	15	1.7	+			BAA
CH UMa	8.23	200	18	?	+			BAA
AT Ara	8:	65	20	1.7	+			RASNZ
EM Cyg	7.14	27	7	no	+		+	Beyer
Z Cam	6.96	23	11	1.9	+		+	BAA
SS Cyg	6.60	50	16	2.0	+	+		AAVSO,BA.
AH Her	6.19	20	11	?	+?		+?	Beyer
BV Pup	5.40	18	6	no				RASNZ
HL CMa	5.15	17	11	no?	+		+?	AAVSO
CZ Ori	5.14	31	11	3.0	+			RASNZ
RX And	5.04	13	8	no			+	BAA
AH Eri	5:	24	5	?				RASNZ
TU Ind	5:	22	4	no				RASNZ
SS Aur	4.39	56	11	2.8	+	+		AAVSO,BA.
U Gem	4.25	100	13	1.6	no			BAA
WW Cet	4.24	31	11	2.2	+	+	+	RASNZ
CW Mon	4.22	160	10	2.4	+(1)			RASNZ
UZ Ser	4.15	21	12	1.8	+	+	+	RASNZ
V 442 Cen	4:	28	8	3.0	+			RASNZ
X Leo	3.95	17	8	2.6	no			BAA
CN Ori	3.92	18	12	no	+(most)	+		RASNZ
IP Peg	3.80	65	9:	?	?			AAVSO
AB Dra	3.65	10.5	10	2.8			+?	BAA
UU Aql	3.37	55	11	2.2				AAVSO

4. Anomalous Outbursts

For several dwarf novae the occasional occurrence of "anomalous outbursts", those with an abnormally slowly rising branch in their light curve, has been reported in the literature, but no systematic study was hithero devoted to this interesting phenomenon. It turns out that anomalous outbursts are common for dwarf novae of all orbital periods $P > 3^h$. At least 19 stars in the sample (70%) display them. Only 2 of 27 cases do definitely not show anomalous outbursts (U Gem and X Leo). For dwarf novae with $P > 10^h$ anomalous outbursts seem to be the usual, if not exclusive, type of eruptions. In most of the shorter period dwarf novae anomalous outbursts happen occasionally between normal ones, often in groups at epochs of semi-chaotic behaviour.

The definition of anomalous outbursts implies that a clearly bimodal distribution in the gradients of the rising branch of the light curves is to be expected. This is, in fact, the case in SS Cyg for which the rise and decline gradients were determined and analysed for a 21 years interval (Fig. 3). The bimodal gradient distribution is only present during the rise, about 20% of all outbursts show the slow rise characteristic for anomalous outbursts. Apparently, passing maximum light, any outburst loses its memory about what happened during rise, and anomalous outbursts do not differ any more from normal ones at later stages. In particular, the distribution of decline rates is identical, and all outbursts (including anomalous ones) are either "wide" or "narrow".

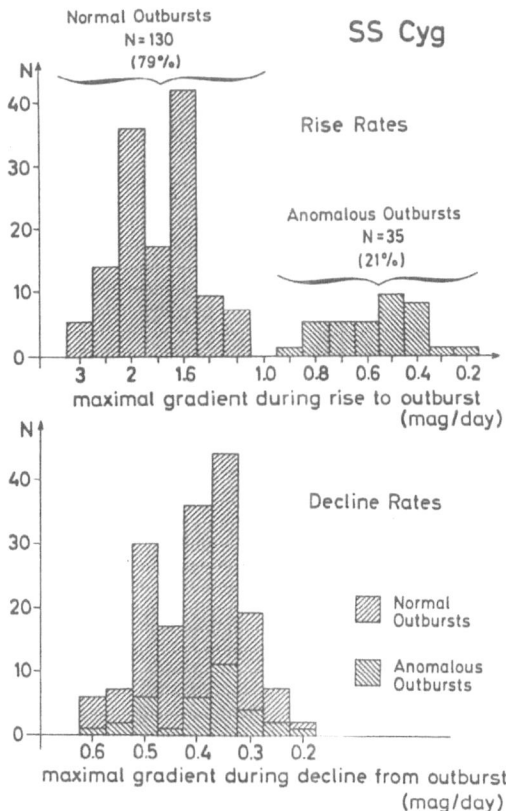

Fig. 3: Distribution of the maximal gradients during the rise branches (top) and the decline branches (bottom) of 165 outbursts of SS Cyg between 1964 and 1985, as determined from Mattei et al's (1985) light curves. Rise gradients reveal a clear separation between normal outbursts (rapid rise, 1.2 - 3 mag/d) and anomalous ones (slow rise, 0.2 - 0.9 mag/d). This difference is no longer present when considering the decline gradients.

5. Z Cam Standstills and other Peculiarities

There are 5 certain and 3 probable Z-Cam stars in the sample (30%), in the period range $3^h.6 \lesssim P \lesssim 7^h.2$. The frequency and typical duration of standstills is strongly variable from star to star, ranging from occasional short events of a few days to stages lasting for several years. During standstill the brightness is not at all constant. Irregular fluctuations, as well as minor outburst-like activities are often superimposed. There is a tendency towards "dumped oscillations" (decreasing amplitude, increasing frequency) during the last outbursts before entering a standstill (for examples see Szkody and Mattei, 1984).

As "faint peaks" we designate abnormally small outbursts with amplitudes of 1mag and a total duration of $\lesssim 1^d$. They are definitely present in 5 dwarf novae (20% of the sample), being probably a quite common phenomenon which has largely escaped detection.

Some dwarf novae with short cycle lenght show a permanent outburst activity with a nearly sinusoidal light curve and without significant quiescent state (examples: CN Ori, CY Lyr).

In general, there is no apparent correlation between the mean cycle lenght C and the orbital period. Richter and Braeuer (1989) derived such a correlation in which, however, the outburst amplitude enters as a third parameter ("Kukarkin-Parenago relationship"). Unfortunately, observed outburst amplitudes suffer many uncertainties, systematic errors and selection effects, specially when based on rather inhomogeneous data sets as the (mostly) visual magnitude estimates of dwarf novae in outburst. They are not considered here, and a direct comparison with Richter and Braeuer's (1989) results is not possible.

6. Conclusions

Many well-established details in the observed outburst behaviour of dwarf novae are hitherto unexplained, in particular the division into wide and narrow outbursts, the occurrence of anomalous outbursts, small peaks and the micro-variability during standstill. It seems rather certain that those features contain valuable information on crucial parameters, as mass transfer rate and its variability, outburst energies, mass of gas accumulated in the outer disk in quiescence, disk surface densities Sigma(r,t), alpha parameter alpha(r,t) and transition front velocities, among others.

Acknowledgments

I would like to thank Dr. H.-J. Braeuer for valuable comments on this manuscript. Part of this research was supported by the Chilean "Comision

Nacional de Investigacion Cientifica y Tecnologica" (grant FONDECYT 1940708).

References

Beyer, M.: 1967, Astron. Nachr. 290, 29
Cannizzo, J.K.: 1993, ApJ 419, 318
Cannizzo, J.K., Mattei, J.A.: 1992, ApJ 401, 642
Mattei, J.A., Salasyga, M., Waagen, R.O., Jones, C.M.: 1985, AAVSO Monograph 1
Molnar, L.A., Kobulnicky, H.A.: 1992, ApJ 392, 678
Richter, G.A., Braeuer, H.-J.: 1989, Astron. Nachr. 309, 413
Ritter, H., Kolb, U.: 1994, in "X-ray Binaries" edts. W.H.G. Lewin, J. van Paradijs and
     E.P.J. van den Heuvel, Cambridge University Press, in press
Szkody, P., Mattei, J.A.: 1984, PASP 96, 988
Vogt, N.: 1980, AA 88,66
Vogt, N.: 1982, Mitteilungen der Astron. Gesellsch. 57, 79

SPECTRALLY-RESOLVED ACCRETION DISK MAPS OF IP PEGASI DURING OUTBURST

R. BAPTISTA
Space Telescope Science Institute
3700 San Martin Drive, Baltimore, MD 21218

AND

C. HASWELL AND G. THOMAS
Columbia Astrophysics Laboratory
538 West 120th St., New York, NY 10027

Spectrally-resolved eclipse maps of the accretion disk in the dwarf nova IP Peg are presented and discussed. The analysis is based on time-resolved optical spectroscopy covering one eclipse during the 1993 May outburst.

The trailed spectrogram was divided into passbands ($\sim$ 50 Å in the continuum and $\sim$ 600 Km/s in the emission lines) and light curves were extracted for each one. Velocity-resolved light curves in Hα show minima progressively displaced towards later phases as one moves from blue to red, signaling a rotational disturbance in this emission line. Eclipse mapping techniques were used to solve for a map of the disk brightness distribution and for the flux of an additional uneclipsed component in each band. In contrast to other dwarf novae in outburst, the radial temperature profiles derived from these maps are in clear disagreement with the radial dependence predicted by the steady state disk model ($T \propto R^{-3/4}$). The flat profiles obtained are quite similar to those observed in quiescent dwarf novae, with temperatures ranging from 4,000 K to 6,500 K.

Spatially-resolved disk spectra show that the Hα and the He I λ6678 lines are in emission at all disk radii. This is in marked contrast with the results obtained for the nova-like variable UX UMa, where the lines progressively transition from absorption in the inner disk to emission in the outer disk (Rutten et al. 1994, A&A 283, 441; Baptista 1995, this volume). This result suggests that the currently accepted view that nova-like systems are similar to dwarf novae in a permanent outburst state has to be revised.

367

A Bianchini et al (eds) Cataclysmic Variables 367
© *1995 Kluwer Academic Publishers*

CESSATION OF THE OPTICAL PULSATIONS IN AE AQR

NINA BESKROVNAYA AND NAZAR IKHSANOV
Pulkovo Observatory, 196140 St. Petersburg, Russia

AND

ALBERT BRUCH
*Astronomisches Institut, Wilhelm–Klemm–Straße 10
D–48149 Münster, F.R.G.*

The $33^s.08$ oscillation and its second harmonic in the optical light of AE Aqr, detected by Patterson (1979; ApJ 234, 978) and being due to hot magnetic pole caps rotating in and out of sight, are considered a persistent characteristic of the system. Photometry performed over a time base of four weeks in July–August 1993 shows, however, that the oscillations have vanished below a detectable limit (see also Bruch et al. 1994; IBVS 3996). Power spectra of the light curves show neither a coherent signal at $33^s.08$ or $15^s.54$ nor QPO around these periods. Tests with tracer signals allowed us to put a conservative upper limit of $0^m.005$ to the amplitude of any coherent signal in the data. We investigated archive light curves of AE Aqr of 45 nights distributed (unevenly) over the years 1978 – 1992 and found the oscillations to be absent only in one the these nights. A comparable case may be V533 Her where Robinson and Nather (1983; ApJ 373, 255) reported the disappearance of the 63^s periodicity.

Eracleous et al. (1994; ApJ 433, 313) explained temporal fluctuations in the 33^s oscillation amplitude as fluctuations in the accretion rate leading to changes in the shock height above the central body. In an extreme case (very low shock height over both poles) the oscillations may cease to be detectable. We note, however, that the brightness and the colours of the system and its flaring behaviour remained normal during our observations. Hence, the mass accretion rate should not have changed significantly and the disappearance of the 33^s periodicity cannot be explained by a cessation of accretion. Other explanations might invoke geometrical factors (precession of the rotational axis of the primary), screening of the polar caps by the surrounding accretion flow, or changes of physical conditions in the caps.

A. Bianchini et al. (eds.), Cataclysmic Variables, 368.
© *1995 Kluwer Academic Publishers.*

STELLAR WEATHER PATTERNS IN IRRADIATED CV SECONDARY STARS

STEPHEN C. DAVEY[1] AND TIM J. MARTIN[2]

[1] *Astronomy Centre, University of Sussex, Brighton, UK*
[2] *British Antarctic Survey, Madingley Road, Cambridge, UK*

The effects of irradiation of the secondary by the white dwarf and disc have been seen in the light curves and radial velocity curves of many CVs. Global atmospheric motions due to the combined effects of heating and Coriolis forces seen on Earth and on Jupiter for example are surely likely on the surface of a star that is rapidly rotating and strongly heated.

Motions in a thin layer just below the photosphere extending around the equator of the secondary are considered, i.e. 2–D motions in the orbital plane. The fluid motions were computed using a finite difference scheme and Smoothed Particle Hydrodynamics. Flow vectors from the SPH model are shown in the figure below. They predict flow velocities $\sim 1.0\,\mathrm{km\,s^{-1}}$ for a temperature increase of $1000\,\mathrm{K}$ symmetrical about the L_1 point. Including the Coriolis force produces a flow pattern that is asymmetrical about the L_1 point and hot material is preferentially carried towards phase 0.75.

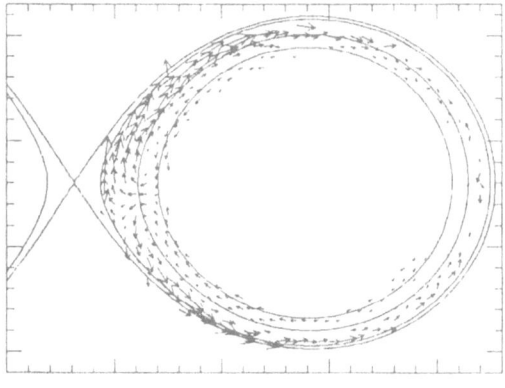

A. Bianchini et al. (eds.), Cataclysmic Variables, 369.

DETECTING CORRELATIONS IN THE LIGHT CURVES OF DWARF NOVAE

M.A. HENDRY, M.S. CATALAN, M.D. STILL AND R.C. SMITH
Astronomy Centre, University of Sussex, Brighton, BN1 9QH, UK

The linear correlation, ρ, between outburst duration (t_b), outburst amplitude (a_b), and the duration of the preceding and following quiescent interval (t_q) in dwarf nova light curves has been proposed by a number of authors as a measure of the relative importance of the mass transfer and disk instability models in describing the outburst mechanism. We have addressed the question of whether the detection / non-detection of light curve correlations may be a statistical artefact, due to poor quality data. Using a simple statistical model to generate realistic artificial light curves we have investigated the impact of sampling period and observational selection on the measured correlations.

Our results are largely insensitive to the adopted magnitude threshhold – indicating that a change of threshhold (e.g. a change of observer) will not significantly affect the measured correlation strength. Increasing the mean interval between observations, however, **does** introduce a strong negative bias in the measured correlation between outburst and quiescent timescales. For data with a mean interval longer than $\sim$ 2 days, the likelihood of detecting a significant $(\rho > 0.5)$ correlation – even when the true value is very high $(\rho = 0.9)$ – is typically less than 50%.

The mean interval in the SS Cyg database studied in Canizzo and Mattei (ApJ 401, 642, 1992) is 1.18 days. It seems likely, therefore, that their failure to detect a systematic difference between $\rho(t_b \rightarrow t_q)$ and $\rho(t_q \rightarrow t_b)$ is a statistically robust result. The SS Cyg database clearly provides the 'benchmark' against which other dwarf nova light curves must be judged.

370

A Bianchini et al (eds), Cataclysmic Variables, 370
© *1995 Kluwer Academic Publishers*

PROBING FUNDAMENTAL ACCRETION PHYSICS WITH A UNIQUE, FULLY IMPLICIT, 2D HYDRODYNAMIC EVOLUTIONARY CODE

M. HUANG AND E.M.SION
Villanova University, Villanove, PA19085, USA

W.M. SPARKS
LANL, Los Alamos, NM 87545, USA

AND

S.G. STARRFIELD
Arizona University, Tempe, AZ 85287, USA

The central accreting object in compact binaries and other accreting systems is usually hidden from direct observation, enshrouded by optically thick accretion disks or accretion columns, which are opaque to the radiation emitted by the central object as it accretes matter. However HST, EUVE, re-analyzed IUE spectra and ROSAT X-ray data have revealed an amazingly rich multi-wavelength spectroscopic database with which to study the response of the underlying degenerate stars, exposed during their quiescence or low state accretion phase, to the deposition of the mass, energy and angular momentum during their outburst or high state accretion phase. Theoretical analysis of the exposed white dwarfs with hydrodynamic accretion simulations of outburst and quiescence can deliver fundamental insight about the physics of magnetic or non-magnetic accretion onto the central object. Our 2-D, fully implicit hydrodynamic code is the first such implicit code to be applied in accretion physics. Comparing with the currently available 1-D, implicit (Sparks *et al.* 1993; Sion 1993, Proc. 2nd Technion Haifa Conf., TH2 hereafter) and 2-D, explicit codes (Kley 1993, TH2), this code simulates the heating and cooling processes on the accreting white dwarf with a more realistic geometry and *much longer time steps*. The code has been tested with the case of free fall of a pressureless gas. As the first application of this code, we will study the boundary layer in a CV system.

References

Kley, S. 1993, Proc. 2nd Technion Haifa Conf., p.156
Sparks *et al.* 1993, Proc. 2nd Technion Haifa Conf., p.96
Sion, E. 1993, Proc. 2nd Technion Haifa Conf., p.86

371

MONTE CARLO SYNTHESIS OF SPECTRAL LINE PROFILES FORMED IN CATACLYSMIC VARIABLE WINDS

CHRISTIAN KNIGGE, JOHN WOODS AND JANET DREW
Department of Astrophysics
University of Oxford, OX1 3RH, UK

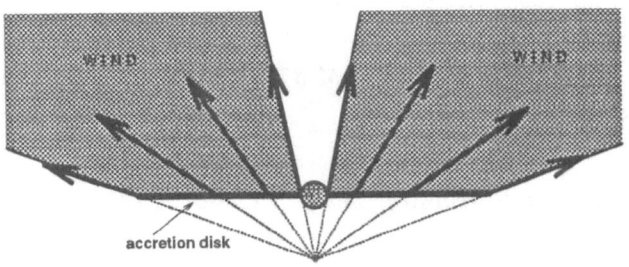

Figure 1. The geometry of our disk wind model.

A Monte Carlo line profile synthesis method is presented which allows the radiative transfer of resonantly-scattered spectral lines through a stellar wind to be solved exactly. The resulting code is designed to be used primarily in the context of non-spherical wind models, where its fully 3-dimensional nature and elimination of assumptions made by codes based on approximate analytical solutions are of particular importance.

Applying our method to the UV resonance lines formed in cataclysmic variable winds, we have constructed a new kinematic disk wind model, in which the outflow emanates from the surface of the rotating accretion disk and has a biconical geometry (Figure 1). Rotation in the wind is treated by assuming specific angular momentum to be conserved along streamlines.

All aspects of the model have been kept as simple as possible, and the outflow geometry, in particular, is specified by only two free parameters. The model is nevertheless flexible and allows a wide range of wind geometries to be explored. Such an exploration is begun here in considering the extent to which disk wind models can better explain phenomena that are difficult to accommodate within simpler, more approximate central wind models. Our results indicate that collimated accretion disk winds may indeed answer some of the questions raised by previous modelling efforts.

A Bianchini et al (eds), Cataclysmic Variables, 372
© *1995 Kluwer Academic Publishers*

ON THE OPTICAL PULSATIONS IN DQ HER STARS

PHILLIP J. MARTELL
Space Telescope Science Institute
3700 San Martin Drive, Baltimore, MD 21218 USA

AND

WILLIAM F. WELSH
Keele University, Department of Physics
Keele, Staffordshire, ST5 5BG UK

DQ Her stars are cataclysmic binary systems in which matter from a late-type near-main sequence star is accreted onto an asynchronously spinning magnetized white dwarf. X-rays generated at shocks at the magnetic poles of the white dwarf sweep the system with a "lighthouse" beam, producing optical signals modulated at the spin period of the white dwarf and/or at one or more spin–orbit sidebands. The shape of the *pulse amplitude* spectrum yields information about the emission mechanism of the pulsed component of the light. [See de Martino *et al.* in these proceedings.]

We have fitted pulse amplitude spectra derived from published data for AE Aqr, FO Aqr, BG CMi, DQ Her, AO Psc, and V1223 Sgr with power law and blackbody models. In all cases, a power law yields an excellent fit. AO Psc (805 s period) and BG CMi (913 s) each require an infinite blackbody temperature. All other systems yield blackbody temperatures ranging from $\sim$10,000 K (DQ Her) to $\sim$50,000 K (AE Aqr).

Assuming the pulsations are due to a single blackbody modulated in area, our blackbody fits yield a projected radiating area, given a distance. The results for AE Aqr are consistent with pulsed radiation originating in a small area on or near the white dwarf's surface. For FO Aqr (spin), DQ Her (spin), AO Psc (sideband), and V1223 Sgr (sideband), the projected radiating areas suggest reprocessing of X-rays in large structures. Independent evidence implies the DQ Her pulses originate over much of the disk, and in general, sidebands probably originate in X-rays reprocessed on the secondary star.

A. Bianchini et al. (eds.), Cataclysmic Variables, 373

SUPERHUMP AND SIMULATIONS OF ECCENTRIC DISKS

JAMES MURRAY
Mathematics Department
Monash University
Clayton VIC 3168 Australia [†]

Accretion disks in close binaries with mass ratios $q \lessdot 0.25$ are subject to a tidal instability. In two dimensional particle simulations (Whitehurst, Hirose and Osaki, Lubow, Murray), an eccentric disk develops. The disk is not stationary in the frame of the binary but precesses with a period a few percent longer than the orbital period of the binary itself. It has been proposed (Whitehurst) that the corresponding cyclic component to the rate of energy dissipation in these simulations is closely related to the superhump phenomenon observed in SU UMa stars in superoutburst.

For a given mass ratio, the various simulations give significantly different values for the disk precession period. We suggest the period depends upon the distribution of matter in the disk which is turn determined by the nature of the shear viscosity.

For one particular simulation (Murray) we investigated the spatial distribution of energy dissipation in the disk. The cyclic component originated in a cool extended region of the outer disk, in general agreement with observations. However the precise location and geometry of the source did not match eclipse maps of Z Cha in superoutburst (O'Donoghue). We note that the last comparison was between but a single simulation and a small set of observations. The reason for the difference is not yet clear.

References

Hirose, M. and Osaki, Y. (1990), *Publ. Astr. Soc. Japan* **42** 135
Lubow, S.H. (1991), *Ap. J.* **381** 268
Murray, J.R. (1994), *PhD Thesis*
O'Donoghue, D. (1990), *Mon. Not. R. Astr. Soc.* **246** 29
Whitehurst, R. (1988), *Mon. Not. R. Astr. Soc.* **232** 35

[†]now at CITA, University of Toronto, Toronto, Ontario M5S 1A1, Canada
email: jmurray@cita.utoronto.ca

A Bianchini et al (eds) Cataclysmic Variables 374
© *1995 Kluwer Academic Publishers*

I. ECLIPSE MAPPING IN EMISSION LINES OF HT CAS

SONJA VRIELMANN
Universitäts-Sternwarte Göttingen, Germany

KEITH HORNE
University of St. Andrews, Scotland

AND

RAYMUNDO BAPTISTA
STScI, Baltimore, USA

With the "Eclipse-Mapping" method (Horne 1985, Baptista & Steiner 1993) it is possible to reconstruct maximum-entropy continuum images of accretion disks in dwarf novae. We have calculated images in the emission lines Hβ, Hγ and Hδ. Theoretical investigations by Horne (1994) predict a symmetric X-shaped azimuthal flux distribution due to shear effects and – possibly – different asymmetric components due to anisotropic turbulence.

We used spectra of the SU UMa type dwarf nova HT Cas in quiescence first published by Young, Schneider and Shectman (1981) with a high time resolution and covering three eclipses. Since the system brightness is quite low (16^m), the spectra have a very low S/N ratio. The flux of the system appears variable, but is probably due to slit losses. Normalization of the light curves was complicated by incomplete eclipse coverage and leads to additional uncertaincies. The calculated accretion disks come out to be radially symmetric and show no hot spot for all three emission lines Hβ, Hγ and Hδ. The corresponding light curve fits have a high χ^2 of 2.5 to 3. The most prominent feature is a sharp peak in the center originating in something rather like the boundary layer than on the white dwarf.

References

Baptista, R., Steiner, J.E. (1991) *A&A*, **249**, 284
Horne, K. (1985), *MNRAS*, **213**. 129
Horne, K. (1994), *A&A*, in press
Young, Schneider, Shectman: 1981, *ApJ*, **245**, 1035

A Bianchini et al (eds), Cataclysmic Variables, 375

II. ECLIPSE MAPPING IN VELOCITY BINS OF EMISSION LINES

SONJA VRIELMANN
Universitäts-Sternwarte Göttingen, Germany

KEITH HORNE
University of St. Andrews, Scotland

AND

RAYMUNDO BAPTISTA
STScI, Baltimore, USA

The classical "Eclipse-Mapping" method (Horne 1985, Baptista & Steiner 1993) calculates maximum-entropy reconstructions of eclipsed light sources such as accretion disks in eclipsing dwarf novae in the frame of the rotating binary. We changed the method to apply it to light curves of velocity bins of emission lines originating in the accretion disk of the dwarf nova HT Cas.

A keplerian velocity field in the accretion disk leads to a dipole-like radial velocity pattern, where intermediate velocity bins correspond to crescent shaped regions in the observers frame (Horne & Marsh 1986). The additional transformation from the binary to the observers frame leads to some additional smearing (in addition to that from the maximum entropy method) for non radial symmetric structures like the hot spot. Furthermore in the default map we introduced weights depending on the "keplerian velocity distance" of the concerned pixel pair in the image.

In first order the resulting maps correspond to the expected flux distribution of a keplerian accretion disk. Only minor deviations appeared with could not yet be related with real deviation from a keplerian field.

References

Baptista, R., Steiner, J.E. (1991) *A&A*, **249**, 284
Horne, K. (1985), *MNRAS*, **213**, 129
Horne, K., Marsh, T. R. (1986), *MNRAS*, **218**, 761

A. Bianchini et al. (eds.), Cataclysmic Variables, 376.
© 1995 Kluwer Academic Publishers.

Two Modes of Gas Streams in Binaries

Zdislav Šíma, Astronomical Institute of AV ČR , Boční II 1401,
141 31 Praha 4, Czech Republic; E-mail: sima@ig.cas.cz

1 Introduction

Two types or "modes" of gas rings around a primary star were described by Smak (1985), that is a hot case – geometrically thick, and a cold case – geometrically thin. Šíma and Hadrava (1987) suggested that there are four possible ways to compute gas streams flowing from L_1 towards the ring. Only two of the models are physically meaningful: the adiabatically expanding stream (cool case, geometrically thin, higher density) and the stream in a state of radiative equilibrium (hot case, geometrically large, lower density).

2 Scenario

If there are two modes of gaseous streams there must be also a limiting case between them. The physical background of this is as follows. When there is only a small mass exchange $\dot{M}$, the transferring gas can be heated up by the hotter primary towards which it flows. This means that the stream will be hot and consequently it will be geometrically large, with small density. When $\dot{M}$ increases, the radiative field in the binary (mainly from the hotter primary star) is no longer able to heat up the streaming gas which will adiabatically expand along the trajectory of the stream. The temperature in the stream will fall dramatically, and so the stream will be cold and dense. As a consequence, it will be of small geometrical dimensions (about 30 % of the preceeding case). There are no other possibilities between these two cases. A small mass transfer $\dot{M}$ creates a large stream while a bigger $\dot{M}$ results in a geometrically small stream.

Critical mass transfer $\dot{M}_s$ between these two modes of gas streaming is a function of the radiation field in the binary, i.e. 1) of the stars' temperature; and 2) of the geometry of the system.

The heat equilibrium equation for the stream is as follows:

$$\sigma \dot{M}_s \Delta T = \dot{Q}_1 + \dot{Q}_2 - \dot{Q}_e \qquad (1)$$

where σ is the specific heat of the gas, $\dot{Q}_1$ is heat energy from the primary star absorbed by the stream, $\dot{Q}_2$ is the same from the secondary, $\dot{Q}_e$ is heat energy emitted by the stream and

$$\Delta T = T_r - T_0 \qquad (2)$$

where T_r is the lowest temperature necessary for the stream to be in the "radiative equilibrium" state and T_0 is the original temperature of the stream when close to L_1.

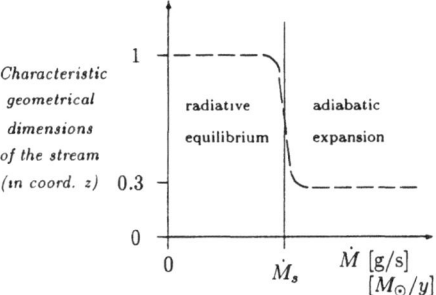

Fig. 1. $\dot{M}_s$ – critical mass transfer of the stream

These two modes of gas streams are combined with Smak's two modes of gas rings making together several possible means to create a hot spot from which two are of the greatest importance: the case of a "central punch" and that of a "pincers". From here several possibilities for forming the S-wave follow, and these may be spectroscopically distinguished one from the other.

References

[1] Smak J.: 1985, Acta Astron. **35**, 351.

[2] Šíma Z., Hadrava P.: 1987, Astroph. & Space Sci. **130**, 151.

377

A. Bianchini et al (eds), Cataclysmic Variables, 377.
© *1995 Kluwer Academic Publishers.*

6. THE ROLE OF MAGNETIC FIELDS

POLARS IN THE ROSAT ERA I: OVERVIEW

K. BEUERMANN, V. BURWITZ, K. REINSCH
Universitäts-Sternwarte Göttingen,
Geismarlandstr. 11, 37083 Göttingen, Germany

A.D. SCHWOPE
Astrophysikalisches Institut Potsdam, Germany

AND

H.-C. THOMAS
MPI für Astrophysik, Garching, Germany

1. Introduction

Polars (or AM Herculis binaries) and intermediate polars (or DQ Herculis binaries) are subgroups of Cataclysmic Variables (CVs) which contain a mass-losing late-type main-sequence star and an accreting magnetic white dwarf. Orbital periods range from 78 min to 8 hours or 48 hours for the two subclasses, respectively. The defining property is the degree of synchronism between white-dwarf rotation and orbital motion: the white dwarf rotates synchronously in polars and freely in intermediate polars (IPs). Synchronization occurs when the magnetic torque between primary and secondary overcomes the accretion torque (e.g. Campbell 1985). We include systems which are only slightly out of synchronism into the subgroup of polars.

The magnetic field of the white dwarf determines the accretion flow and thereby the observational appearence of these systems including the physical properties of the accretion plasma and the radiative transfer of the released energy.

2. ROSAT's Impact on the Field

Ritter's (1990) pre-ROSAT catalogue lists 18 polars (categories AM + AM?) and 13 IPs (categories IP + DQ + DQ?). The principal impact of the ROSAT mission on the study of magnetic CVs was to substantially increase the number of known systems: 27 new soft X-ray detected po-

A Bianchini et al (eds), Cataclysmic Variables, 381-388

lars (and polar candidates) were so far identified among the RASS sources. Two further new polars are from pointed ROSAT observations: RX J0524 = "Paloma" (P = 143 min, Hasinger, private communication) and RX J1940 near NGC6814 (Madejski et al. 1993, Staubert et al. 1994). Together with some additional optically discovered systems, the number of known polars has surpassed 50. The significantly improved statistics now allows to proceed from the study of individuals to the study of the properties of the class. Table 1 lists the newly discovered soft X-ray selected polars and polar candidates ordered according to orbital period. Period searches using X-ray data from the ROSAT All-Sky-Survey (RASS) suffer from severe aliasing and, hence, periods determined from RASS data only are not yet definitive. They are indicated by a :: sign in Table 1. Time-resolved optical or X-ray follow-up observations have, in several cases, resolved these uncertainties.

Not surprisingly, the distribution of the soft X-ray selected systems over the sky is far from uniform. The galactic plane is largely devoid of the less luminous short-period systems. Also at high galactic latitudes, regions characterized by nearby atomic hydrogen as, e.g. the North-Polar Spur, contain few polars. In the south galactic polar cap, several polars were found with PSPC countrates $> 0.5\,\mathrm{cts\ s^{-1}}$, but complete identification of the very soft RASS sources down to 0.2 cts s^{-1} did not add a single new polar to our list. This result suggests that ROSAT sees practically all polars at high galactic latitudes provided they are in their 'ON' state. With appropriate corrections for detection losses, we estimate the local space density of short-period polars to be $n_o \simeq 5\,10^{-7}\mathrm{pc^{-3}}$ and the standard deviation of a Gaussian density distribution perpendicular to the galactic plane to be $\sigma_g \simeq 240\,\mathrm{pc}$. The space density of long-period polars is lower by a factor of $5 - 10$, as expected from the shorter evolutionary time scale.

Since IPs dominate among the known longer-period magnetic CVs and polars among the short-period ones, King et al. (1985) suggested that the former evolve into the latter as they become synchronized. This appears to be a viable hypothesis, save for the fact that the few IPs with known magnetic field seem to have lower polar field strengths than typical polars. The discovery of IPs with X-ray spectra as soft as those of polars re-opens this discussion (Mason et al. 1992, Haberl et al. 1994, Haberl and Motch 1995). It is also noteworthy that the space density of IPs is comparable to that of long-period polars ($n \sim 7\,10^{-8}\,\mathrm{pc^{-3}}$) which suggests an evolutionary connection.

Among the long-period systems ($P_{\mathrm{orb}} > 3\,\mathrm{h}$), synchronism need not be strictly enforced. V1500 Cyg ($P_{\mathrm{orb}} = 201\,\mathrm{min}$) lost synchronism in the common-envelope phase of the 1975 nova eruption and is currently re-establishing it on a time scale of 170 yrs (Schmidt et al. 1994). BY Cam (Silber et al. 1992, Piirola et al. 1994) and RX J1940 (Staubert et al. 1994)

TABLE 1. New polars (AM Herculis binaries) and polar candidates from the ROSAT All Sky Survey. Given are the system designation, the mean PSPC countrate from the RASS, the hardness ratio HR1, an indication for a detection by the WFC on board of ROSAT (y = yes), a typical visual magnitude, the type, and the orbital period in min.

System	PSPC cts/s	HR1	WFC	Mag	Type	Period min	Ref.
RX J0132	0.55	−0.90		19	AM:	78:	1,2
RX J1307	2.50	−0.85	y	17	AM	80	3
RX J1015	1.32	−0.74		17	AM	80	1,2
RX J0153	0.44	−0.92		17	AM:	89	1
RX J1149	2.15	−0.95	y	17	AM	90	4
RX J1844	1.39	−0.57	y	16	AM	90	5
RX J1957	1.12	−0.89		17	AM:	99	1,6
RX J0453	1.70	−0.89	y	19	AM	102:	7
RX J0953	0.25	−0.53		19	AM	102	1
RX J1002	0.63	−0.84		17	AM:	107	1,2
RX J1802	2.4				AM	113	8
RX J1724	0.33	−0.97			AM?	120::	1,8
RX J2107	1.29	−0.91	y	15	AM	125 ecl.	9
RX J1846	0.21	−0.93			AM?	129::	1,8
RX J0531	1.20	−0.70	y	16	AM	133	10
RX J1938	8.91	−0.94	y	14	AM	140	6,11,12
RX J0501	0.29	−0.75		15	AM:	171 ecl.	1
RX J0929	0.4	−0.83		17	AM	203 ecl.	13
RX J1007	2.11	−0.91		18	AM:	208:	1,6
RX J2316	1.47	−0.74		18	AM:	209:	1,6
RX J1313	2.09	−0.79		16	AM	255	1,6
RX J0203	0.55	−0.67		17	AM:	275	1,2
RX J0515	0.50	−0.82		15	AM	480 ecl.	6,14,15,16
RX J0512	0.24	−0.79		17	CV		1
RX J0600	0.49	−0.84		19	CV		1
RX J0859	0.34	−0.69		18	CV		1
RX J2022	0.48	−0.83		19	CV		1

(1) This work, (2) Beuermann and Schwope 1994, (3) Osborne et al. 1994, (4) Mittaz et al. 1992, (5) Bailey et al. 1995, (6) Beuermann and Thomas 1993, (7) Burwitz et al. 1995, (8) Greiner, private communication, (9) Schwope et al. 1993a, (10) Reinsch et al. 1994, (11) Buckley et al. 1993, (12) Schwope et al. 1995b, (13) Buckley et al. 1994, (14) Garnavich et al. 1995, (15) Shafter et al. 1995, (16) Walter et al. 1995.

also appear to be slightly asynchronous and other long-period systems may be so, too. The process of synchronization is far from being understood and a systematic study of the degree of synchronism among the long-period systems would clearly be useful. Presently, the systems with the longest periods

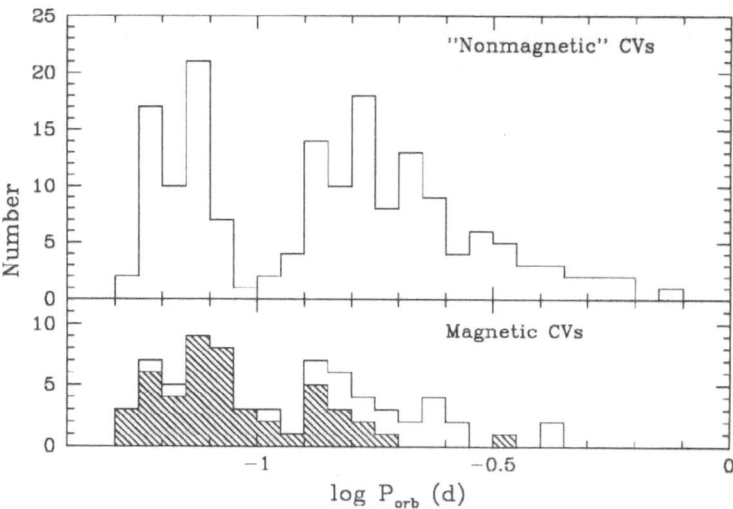

Figure 1. Orbital-period distributions of non-magnetic and magnetic CVs. In case of the magnetic systems, the shaded histogram refers to polars (AM Her stars) and the open section to intermediate polars (DQ Her stars). Data are from Ritter and Kolb (1993) from Haberl and Motch 1995 and from Table 1. A paucity of systems with periods between 2 and 3 hours (period gap) is indicated for the non-magnetic and for the magnetic systems.

are RX J1313-32 (P_{orb} = 255 min), RX J0203+29 (P_{orb} = 275 min) and RX J0515+01 (P_{orb} = 480 min) (Beuermann and Schwope 1994, Garnavich et al. 1995, Shafter et al. 1995, Walter et al. 1995).

The secular evolution of non-magnetic CVs is thought to be governed by two different angular-momentum loss processes, magnetic braking (Verbunt and Zwaan 1981) and gravitational radiation, being dominant above and below $P_{orb} \simeq 3$ h, respectively. Thermal evolution causes the secondary to loose contact with its Roche surface between $P_{orb} \simeq 3$ h and 2 h which produces the well-known period gap (Fig. 1). The evolution of polars may differ from that of non-magnetic CVs because angular-momentum losses by a stellar wind may depend strongly on the strength and geometry of the magnetic field. Wickramasinghe and Wu (1994) suggested that the magnetic configuration of polars prevents the wind loss, causing them to evolve merely by gravitational radiation, in which case there would be no gap. The observational situation seems to refute that suggestion, although the number of systems in the gap is still too small to arrive at a final conclusion. If we define the gap as extending from $\log P_{orb}(d) = -0.90$ to -1.05 (2.1 to 3.0 h) and define a bin immediately above the gap from $\log P_{orb} = -0.80$ to -0.90 (3.0 to 3.8 h) then the numbers below the gap, in the gap, in the bin immediately above the gap, and at $P_{orb} > 3.8$ h are 57, 7, 26, 76 and 32, 7, 13, 19 for the non-magnetic and the magnetic systems, respectively.

While the relative number in the gap is higher for the magnetic systems by a factor of ~2, the difference is not statistically significant. At $P_{orb} > 3.8$ h, however, the relative number of magnetic systems clearly drops, an effect which is even more pronounced for the polars. The low number of polars is a consequence of the fact that a typical magnetic CV synchronizes only near $P_{orb} \sim 4$ hours. The relatively low number of IPs can be a selection effect or a genuine result caused by different evolutionary histories of magnetic and non-magnetic systems. We conclude that the presence of a period gap for magnetic systems indicates that the process which produces the gap in non-magnetic CVs is acting also in magnetic ones. Hence, magnetic and non-magnetic CVs seem to evolve under the same angular-momentum loss processes but differences in the relative efficiencies of these processes for the two subclasses can not presently be excluded.

3. Magnetic Field of the White Dwarf

The strength and structure of the white-dwarf magnetic field determine not only the evolution of a system but also the accretion geometry and the radiative properties. Part of the luminosity is released as cyclotron radiation which displays the higher harmonics of the electron gyrofrequency as severely broadened emission lines from the $\sim 10^8$ K post-shock plasma. After minor special-relativistic corrections, the separation of these lines directly measures the field strength B_{cyc} in the accretion spot. Cool matter surrounding the accretion spot may lead to Zeeman lines in the cyclotron quasi-continuum. This effect measures the field strength B_h in a gaseous halo around the accretion spot and possibly at some height above the photosphere. Finally, photospheric Zeeman lines may be detected if accretion ceases or the spot is hidden behind the white dwarf. In this case, a flux-weighted mean field $\bar{B}_{phot}$ over the visible hemisphere is measured.

In lucky circumstances, a combination of such measurements provides insight into the field structure of the white dwarf. We may expect $B_h \simeq B_{cyc} \simeq B_{pole}$ because emission and absorption occur near one of the poles and nearby in space; furthermore, in a dipole geometry B_{pole} would exceed $\bar{B}_{phot}$ by a factor of up to 2. Surprisingly, however, in some systems $\bar{B}_{phot}$ is similar to B_{cyc} (or B_h) or even significantly exceeds it, a result which indicates substantial deviations of the surface field from a dipole geometry.

Table 2 provides an overview of currently available field measurements of the white dwarfs in 22 polars. Field strengths for the main accretion region cluster around 30 MG with a standard deviation of 15 MG. There are no field strengths below ~7 MG which probably indicates that even short-period systems do not synchronize below this value. There are also no field strengths in the main accretion spot in excess of 61 MG. This is

TABLE 2. Field strengths of magnetic white dwarfs in polars (AM Herculis binaries). $B_{cyc,1}$ and $B_{cyc,2}$ refer to field strengths obtained from the spacing of cyclotron lines, B_h to the Zeeman effect in the cool halo of the accretion spot, and $\bar{B}_{phot}$ to the flux-weighted mean photospheric field from Zeeman lines.

System	P_{orb} (min)	$B_{cyc,1}$ (MG)	$B_{cyc,2}$ (MG)	B_h (MG)	$\bar{B}_{phot}$ (MG)	Ref.
EF Eri	81			13		1
DP Leo	90	31	59			2,3,4
RXJ1149+28	90	43				5
RXJ1957-57	99			14		6
VV Pup	100	31	54			3,7
V834 Cen	101	23		23	22	8,9
RXJ0453-42	102:	36				10
V2301 Oph=H1752	113			7		11
MR Ser	113	24		24	27	12
BL Hyi	114			12	22	13
ST LMi	114	12			19:	14,15
EK UMa	114	35:/47:				16
AN UMa	115	29				3,17
HU Aqr=RXJ2107	125	35				18
EU Cnc	125	42:				19,20
UZ For	126	53	75(113)			21
RXJ0531-46	133	19				22
QS Tel=RXJ1938	140	47	70(80)			23
AM Her	186	14			13	24,25
BY Cam	202	28				3,16
QQ Vul	222	36:				3
RXJ0515+01	480	61:				26

(1) Östreicher et al. 1990, (2) Cropper et al. 1990a, (3) Schwope 1991, (4) Cropper and Wickramasinghe 1993, (5) Schwope (in prep.), (6) Thomas et al. (in prep.), (7) Wickramasinghe et al. 1989, (8) Schwope and Beuermann 1990, (9) Ferrario et al. 1992, (10) Burwitz et al. 1995, (11) Ferrario et al. 1995, (12) Schwope et al. 1993b, (13) Schwope et al. 1995b, (14) Schmidt et al. 1983, (15) Ferrario et al. 1993, (16) Cropper et al. 1990b, (17) Cropper et al. 1988, (18) Schwope et al. 1993a, (19) Pasquini et al. 1994, (20) This work, (21) Schwope et al. 1990, (22) Reinsch et al. 1994, (23) Schwope et al. 1995a, (24) Young et al. 1981, (25) Bailey et al. 1991, (26) Shafter et al. 1995.

surprising because single white dwarfs show a much broader distribution which extends to beyond 500 MG (Chanmugam 1992). A viable hypothesis for the absence of high-field magnetic CVs involves a strong wind which couples onto the white-dwarf field lines and causes rapid evolution of such systems with a correspondingly small space density and chance of discovery (Hameury et al. 1988).

For some systems, two values are given for B_{cyc}. In these cases, two systems of cyclotron lines were observed which refer to two accretion regions. If these are located essentially opposite to each other, field strengths different by a factor of ~2 are incompatible with a centered dipole. In addition, there is evidence for non-dipolar fields in other systems: e.g., BL Hyi displays pronounced Zeeman absorption lines in the cyclotron continuum which indicate a field strength of 12 MG in the accretion region. In low states, however, shallow Zeeman absorption lines appear which indicate a mean photospheric field of 22 MG with a substantial spread in field strength (Schwope et al. 1995b). This is impossible in a dipole geometry. A somewhat less extreme example is MR Ser with a field of 24 MG in the main accretion spot and a mean photospheric field of 27 MG (Schwope et al. 1993b). Accepting direct and indirect evidence, there are 9 systems with significantly different field strengths at two 'poles'. In all cases, the main accretion spot is located at the low-field 'pole'. For neutron stars, Romani (1993) suggested that accretion may cause the horizontal component of the field to be submerged and carried away from the pole. Unfortunately, no quantitative treatment exists and the applicability to white dwarfs remains open because the total accreted mass is uncertain. Also, nova explosions may counteract submersion. Alternatively, the non-dipolar field structure could be a property of the internal field-generating process in the progenitor star and the magnetic orientation of the white dwarf a consequence of the mechanism of synchronisation (Wu and Wickramasinghe 1993).

Field strengths in many IPs are uncertain because they are lower than those of polars and not easily measured. Several new ROSAT-discovered IPs with X-ray spectra as soft as those of polars, however, are of interest in this respect because they represent a new subclass as possible progenitors of polars (Haberl and Motch 1995). At least the field strength of one of these, PQ Gem ($=$RX J0751+14$=$RE0751+14), is comparable to that of low-field polars: $B \simeq 8 - 18\,\text{MG}$ (Piirola et al. 1993).

4. Conclusions

The ROSAT mission has led to the discovery of many new polars and of an important subclass of intermediate polars with X-ray spectra as soft as those of polars. These findings allow to proceed from the study of individuals to the study of the properties of the class, start a new attempt to understand the evolution of these systems, and to investigate the magnetic-field structure of white dwarfs and, maybe, more generally of compact stars.

Acknowledgements

This work was supported in part by the DARA under project number 50OR9210 and 50OR9403

References

Bailey, J., Ferrario, L. and Wickramasinghe, D.T. (1991) *MNRAS*, **251**, 37P
Bailey, J., et al. (1995) *MNRAS*, **272**, 579
Beuermann, K. and Thomas, H.-C. (1993), *Adv. Space Res.*, **13**, (12)115
Beuermann, K. and Schwope. A.D. (1994), *ASP Conf. Ser.*, **56**, 119
Buckley, D. et al. (1993), *MNRAS*, **262**, 93
Sekiguchi, K., Nakada, Y., and Basset, B. (1994), *MNRAS*, **266**, L51
Burwitz, V. et al. (1995), *A&A*, in press
Campbell, C.G. (1985), *MNRAS*, **215**, 509
Chanmugam, G. (1992), *Ann. Rev. Astron. Astrophys.*, **30**, 143
Cropper, M. et al. (1988), *MNRAS*, **236**, 29P
Cropper, M., Mason, K.O. and Mukai, K. (1990a), *MNRAS*,**243**, 565
Cropper, M. et al. (1990b), *MNRAS*, **245**, 760
Cropper, M. and Wickramasinghe, D.T. (1993), *MNRAS*,**260**, 696
Ferrario, L. et al. (1992), *MNRAS*, **256**, 252
Ferrario, L., Bailey, J. and Wickramasinghe, D.T. (1993), *MNRAS*, **262**, 285
Ferrario, L. et al. (1995), *MNRAS*, in press
Garnavich, P.M. et al. (1995), *ApJ Letters*, in press
Haberl, F. et al. (1994), *A&A*, **291**, 171
Haberl, F. and Motch, Ch. (1995), *A&A*, in press
Hameury, J.M., King, A.R. and Lasota, J.P. (1988), *MNRAS*, **237**,848
King, A.R., Frank, J. and Ritter, H. (1985), *MNRAS*, **213**, 181
Madejski,G.M., et al. (1993), *Nature*, **365**, 626
Mason, K.O. et al. (1992), *MNRAS*, **258**, 749
Mittaz, J.P.D., Rosen, S.R., Mason, K.O., and Howell, S.B. (1992), *MNRAS*, **258**, 277
Östreicher, R., Seifert, W., Wunner, G. and Ruder, H. (1990), *ApJ*, **350**, 324
Pasquini, L., Belloni, T. and Abbott, T.M.C. (1994), *A&A*, **290**, L17
Piirola, V., Hakala, P., and Coyne, G.V., S.J. (1993), *ApJ*, **410**, L107
Piirola, V., et al. (1994), *A&A*, **283**, 163
Reinsch, K., Burwitz, V., Beuermann, K., Schwope, A.D. and Thomas, H.-C. (1994), *A&A*, **291**, L27
Ritter, H. (1990), *A&A Suppl.*, **85**, 1179
Ritter, H., and Kolb, U. (1993), MPA preprint 757
Romani, R. (1993), *Nature*, **347**, 741
Schmidt, G.D., Stockman, H.S. and Grandi, S.A. (1983), *ApJ*, **271**, 735
Schmidt, G.D., Liebert, J., and Stockman, H.S. (1994), *ApJ*, in press
Schwope, A.D. and Beuermann, K. (1990), *A&A*, **238**, 173
Schwope, A.D., Beuermann, K. and Thomas, H.-C. (1990), *A&A*, **230**, 120
Schwope, A.D. (1991) PhD thesis, Technical University of Berlin
Schwope, A.D., Thomas, H.-C. and Beuermann, K. (1993a), *A&A*, **271**, L25
Schwope, A.D., Beuermann, K., Jordan, S. and Thomas, H.-C. (1993b), *A&A*, **278**, 487
Schwope, A.D., Thomas, H.-C., Beuermann, K., Burwitz, V., Jordan, S. and Haefner, R. (1995a), *A&A*, **293**, 764
Schwope, A.D., Beuermann, K., and Jordan, S. (1995b), *A&A*, in press
Shafter, A.W., Reinsch, K., Beuermann, K., Misselt, K.A., Buckley, D.A.H., Burwitz, V. and Schwope A.D. (1995), *ApJ*, in press
Silber, A. et al. (1992), *ApJ*, **389**, 704
Staubert, R., et al. (1994), *A&A*, **288**, 513
Verbunt, F. and Zwaan, C. (1981), *A&A*, **100**, L7
Walter, F.M., Wolk, S.J., and Adams, N.R. (1995), *ApJ*, in press
Wickramasinghe, D.T., Ferrario, L. and Bailey, J. (1989), *ApJ*, **342**, L35
Wickramasinghe, D.T., and Wu, K. (1994), *MNRAS*, **266**, L1
Wu, K., and Wickramasinghe, D.T. (1993), *MNRAS*, **260**, 141
Young, P., Schneider, D.P. and Shectman, S.A. (1981) *ApJ*, **245**, 1043

POLARS IN THE ROSAT ERA II: THE INDIVIDUALS

A.D. SCHWOPE

[1]*Astrophysical Institute Potsdam, An der Sternwarte 16
D-14482 Potsdam (ASchwope@aip.de)*

K. BEUERMANN[2,3], V. BURWITZ[2]

[2]*Universitäts-Sternwarte Göttingen,* [3]*MPE Garching*

AND

K.-H. MANTEL[4], R. SCHWARZ[1]
Universitäts-Sternwarte München

1. Introduction

Polars (or AM Herculis stars) are magnetic cataclysmic binaries with synchronously rotating white dwarfs. The accretion energy is released in small regions on the white dwarf surface as optical cyclotron radiation, hard X-ray bremsstrahlung and soft X-ray radiation from the heated photosphere.

The impact of ROSAT on the field is (at least) twofold: (1) a large number of new systems were detected and (2) detailed studies of individual objects (new and already known) became possible. Here we highlight both aspects by presenting the results of pointed ROSAT observations with the PSPC of well-established AM Her stars (BL Hyi, VV Pup, EK UMa) as well as follow-up observations of one of the most exciting new discoveries, HU Aqr (=RX J2107.9-0518). More general aspects of the AM-Her sky as seen with ROSAT are described by Beuermann & Schwope (1994) and Beuermann et al. (this volume).

The systems presented here have in common that their main emission regions are self-occulted by the white dwarf for extended phase intervals. This advantageous situation allows to derive constraints on the geometry of the hot spots, to search for emison from a secondary region or from an active chromosphere of the secondary stars. The full content of the data will be presented elsewhere in the near future (Schwope et al., in preparation).

389

A. Bianchini et al. (eds.), Cataclysmic Variables, 389-396.

2. EK UMa

Only 2 observations of this polar were reported before ROSAT (Morris et al. 1987, Cropper et al. 1990). Clayton & Osborne (1994) performed a first soft X-ray study with the PSPC in May 1992. Optical and X-ray data displayed an eclipse-like event in the bright phase interpreted as absorption in the transient accretion flow streaming outside the orbital plane.

We performed two further pointed observations of EK UMa in Nov. 1992 (4.2 ksec) and May 1993 (12.0 ksec) and, in addition, extracted the photons originating from this system also from the ROSAT All Sky Survey (Beuermann & Thomas 1993). Using all available data we constructed a linear ephemeris for the center of the absorption dip

$$\text{HJD} = 244\,8756.5865(11) + E \times 0.079544032(16)$$

which is the first long-term ephemeris for this system.

The phase-folded soft X-ray light curves of EK UMa and details of the absorption dip are shown in Fig. 1. The main results of the ROSAT observations can be summarized as follows. The light curve has a roughly symmetric shape, a bright phase lasting from $\phi \simeq 0.62$ to 1.21, and a faint phase with no detection in X-rays. The bright phase displays large amplitude intensity variations due to flaring and/or absorption. The average bright-phase intensity varyied by about a factor of 2 when comparing the different ROSAT observations. The eclipse is variable in shape, length and absolute phase. It always displays a well-defined ingress followed by a phase of totality. The second part of the dip, $\phi \simeq 0.00 - 0.05$, is affected by a variable amount of absorption. At these later phases we are looking through matter out of the plane further upstream. Hence, the different observations show that matter is stripped off from an extended range of the horizontally flowing stream (in the orbital plane). The horizontal stream is stopped at some location in the magnetosphere and the vertical stream then becomes opaque. The width of this part of the stream is variable by a factor of ~2. The temperature of the soft, bona-fide blackbody, component was 57 eV in 1992 but only about 25 eV in 1993. No other AM Herculis system has shown such huge temperature changes.

3. VV Pup

VV Pup, the 'soft X-ray machine' (Patterson et al. 1984) was observed during the reduced pointing phase of ROSAT in October 1991 for a total of 17.3 ksec. It was in a high accretion state reaching a peak countrate of about ~80 cts/s, thus becoming the second brightest polar. The phase-folded light curve is shown in Fig. 2 (top and bottom panel with different scaling at the ordinate), the hardness ratio $\text{HR1} = \frac{H-S}{H+S}$ with H and S

being the counts above and below 0.5 keV, respectively, is shown in the middle panel of Fig. 2.

The light curve of the mainly accreting pole is asymmetric in soft and hard X-rays with a slow increase and a fast decrease. The bright phase lasts from $\phi = 0.72$ to 1.16 with pre- and postcursors at low light level of length $\sim$0.06 phase units. Hence, the total length of the bright phase is in the range 0.44 – 0.56. Strong flaring is observed throughout the whole bright phase, but definite conclusions about the real time signature are difficult to be reached because the object was located in the center field of the PSPC where the entrance window support structure may introduce spurious signals. At one occasion we tentatively identified in a 500 sec interval QPOs with $f \sim 0.011\,$Hz. At 10 sec time resolution the X-ray flux during the bright phase is always positive. In particular, we could not detect the pronounced absorption dip at phase $\sim$0.95 due to a transient accretion stream reported by Patterson et al. (1994).

VV Pup is known to be a twopole accretor from optical cyclotron spectroscopy and polarimetry (e.g. Liebert & Stockman 1979, Schwope 1990). Here we report the clear identification of the secondary, fainter accretion region in X-rays. It is detected during the faint phase at an average countrate of 0.26 PSPC counts s^{-1} ($\phi = 0.3 - 0.6$). There is a clear trend in the X-ray brightness during the faint phase becoming monotonically fainter. Again, at 100 sec time resolution VV Pup is never completely off.

The bright pole in VV Pup is extremely soft with HR1($\phi = 0.75 - 1.15$, binsize 100 sec)$= -0.964 \pm 0.008$. The flux in the soft bona-fide blackbody ($kT = 32$ eV) and the hard bremsstrahlung component ($kT = 20$ keV) is $\geq 10^{-10}$ and $\sim 2.2 \times 10^{-12}$ erg cm^{-2} s^{-1}, respectively. The corresponding values for the faint phase are $F_{bb} \simeq 10^{-12}$ and $F_{br} \simeq 2 \times 10^{-13}$ erg cm^{-2} s^{-1}, respectively. Hence, the soft X-ray excess is more pronounced in the low-field accretion region which gains the higher mass accretion rate.

Polarimetry has shown that the main accretion region is an extended structure (e.g. Piirola et al. 1990). This becomes evident also from the asymmetric shape of the X-ray lightcurve. Our data show in addition that there is structure along the accretion ribbon. At start of the bright phase HR1 becomes significantly larger indicating the presence of a hard knot at the leading edge of the accretion arc.

4. BL Hyi

BL Hydri was observed with the PSPC in late spring 1991 for a total of 18.2 ksec (on-axis) and in October 1993 for 3.8 ksec (off-axis). Phase-resolved optical photometry was obtained occasionally in October 1992 and September 1993. A brief report on the 1991-observations has been given in Schwope & Beuermann (1993). During that observation BL Hyi was encountered in a state previously not seen in an AM Her binary. Episodes with vanishing accretion (zero detection at X-rays) were disrupted by strong accretion flares reaching 75 s^{-1} peak countrate. This mode of completely irregular accretion alternated with a term of regular accretion at the former (during the EXOSAT era) main accretion pole at a low level. The two most prominent examples of flares are shown in Fig. 3 (top left), and the light curve in the mode of regular accretion in the bottom panel.

Soft and hard X-ray light curves of the 1993 PSPC observation (incomplete phase coverage) are shown together with an optical lightcurve taken 24 days apart in the right of Fig. 3. These data suggest that BL Hyi was in a stable high state with more or less regular accretion at that time. X-rays were detected troughout the faint phase suggesting either two-pole accretion or accretion along a more extended region than at lower accretion states (e.g. Beuermann & Schwope 1989) or a latitudinal migration of the accretion spot. The X-ray spectrum became softer in the faint phase whereas in e.g. VV Pup the reversed behaviour was observed. In both systems the faint pole has the higher fieldstrength. These findings are relevant to the recent discussion of the relation between soft X-ray excess and fieldstrength (Beuermann & Schwope 1994, Ramsay et al. 1994) and show that more than one parameter, the fieldstrength, determines the size of the soft X-ray excess.

Onset of the bright phase occurs earlier in the 1993 high state than predicted by our ephemeris (Schwope & Beuermann 1989), which indicates a migration of the accretion spot at least in longitude. The bright phase starts earlier in hard X-rays than in soft X-rays which, as in VV Pup, gives clues to a temperature structure along the accretion region.

5. HU Aqr = RX J2107.9-0518

HU Aqr was discovered by us as bright eclipsing polar in the course of an optical identification programme of soft ROSAT survey sources (Schwope et al. 1993). Intensive optical follow-up observations took place in fall 1992 and summer 1993 using several telescopes at Calar Alto and La Silla. Here we describe briefly the results of high-speed (1min), high-resolution (1Å FWHM) spectroscopy and of high-speed (0.5 sec) photometry obtained in the 1993 high state and compare them with our intermediate state measurements in 1992.

Our updated eclipse ephemeris is
$$\mathrm{HJD} = 244\,8896.543678(7) + E \times 0.086820450(3)$$
The period derived by Hakala et al. (1993) is incorrect.

The slit-loss corrected, continuum-subtracted and phase-folded spectrograms of HeII 4686 are shown in Fig. 4. The profile consists of a narrow line (NEL, FWHM $\sim$2.2 Å), which is most intensive around phase 0.5, and a broad underlying structure with higher radial velocity amplitude. The NEL has zero radial velocity around conjunction of the secondary star. Phasing, width and lightcurve of this component are compatible with an origin on the X/UV illuminated hemisphere of the companion star. This component, often discussed but mostly barely resolved, is seen here in unprecedented detail. Its lightcurve is somewhat skewed and its velocity is nonzero at conjunction which is indicative of asymmetric illumination of the secondary star (shadowing by the stream). The broad underlying component has a phase-dependent asymmetric profile. It reaches maximum velocity amplitude of $\geq$ 1000 km/s and has a width corresponding to $\simeq$ 700 km/sec. Most of the flux in this component is emitted in the accretion stream which has left the orbital plane.

The shape of the optical lightcurve has changed remarkably between 1992 (intermediate accretion state) and 1993 (high state, Fig. 5). In the high state the system has a double-humped bright-phase lightcurve which is naturally explained by cyclotron beaming. Two eclipse-like features can be recognized inbetween. The first, we call it dip, at $\phi = 0.88$ is due to the transit of the out-of-plane accretion stream and the second is the genuine eclipse by the secondary star. During the dip only the cyclotron spot is occulted, the remaining light originates in the luminous accretion stream. The shape of the dip is well described with a Gaussian profile (FWHM $\cong$ 12°). The eclipse is two-stage due to the subsequently occulted accretion spot and the accretion stream. The eclipse of the white dwarf wasn't clearly resolved due to its faintness. The ingress and egress times of the spot measured in the B-filter are 2.9 ± 0.2 sec, equivalent to $\sim 10^8$ cm on the white dwarf. The ingress time of the stream is $\sim$2 min, egress takes somewhat longer because the stream is seen then better in its whole elongation from the side.

In the 1992 intermediate state neither the double cyclotron emission hump nor the pronounced absorption dip is recognizable and ingress of the accretion stream takes much longer, $\sim$5 min. Instead of one well-shaped macro-dip we observed an extended phase interval with partly reduced intensity just before eclipse which we assign to the random occurence of micro-dips, $\phi = 0.82-0.96$. This implies that the accretion stream was much more fragmented and that the volume in the magnetosphere which could be occupied by accreting matter was larger in the intermediate brightness state (longer ingress and egress times). This probably causes an emission region on the white dwarf with higher internal structure as well as the dominance of the signatures of instationary accretion over cyclotron beaming in the shape of the light curve.

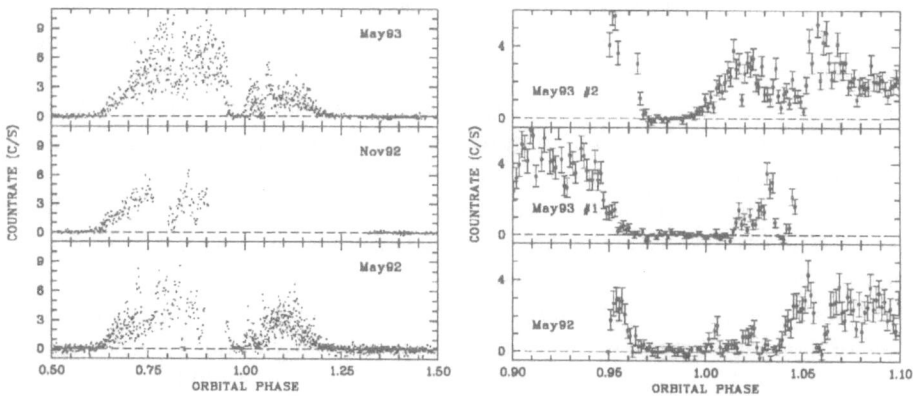

Figure 1. EK UMa: *(left)* Phase-folded PSPC data obtained at 3 occasions in 1992 and 1993. *(right)* Uninterrupted trains of photons completely covering the absorption dip.

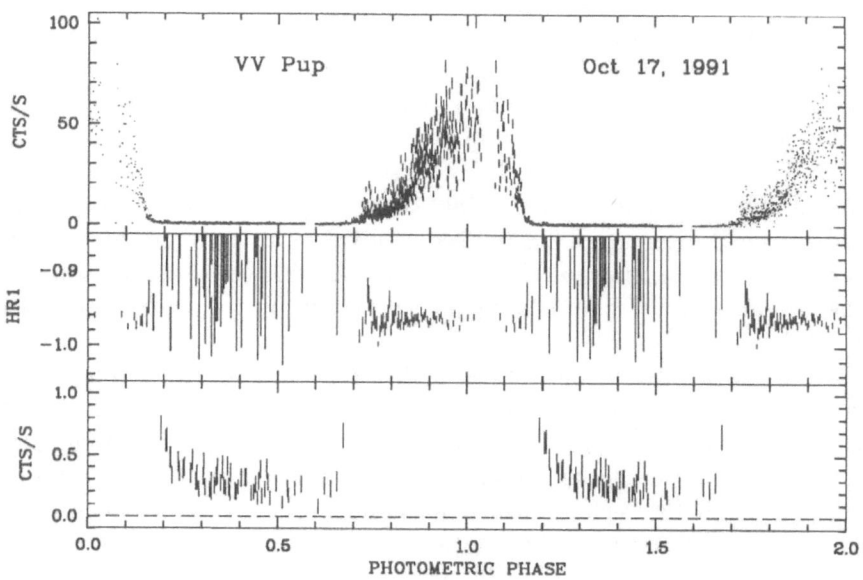

Figure 2. Phase-folded PSPC data of VV Pup.

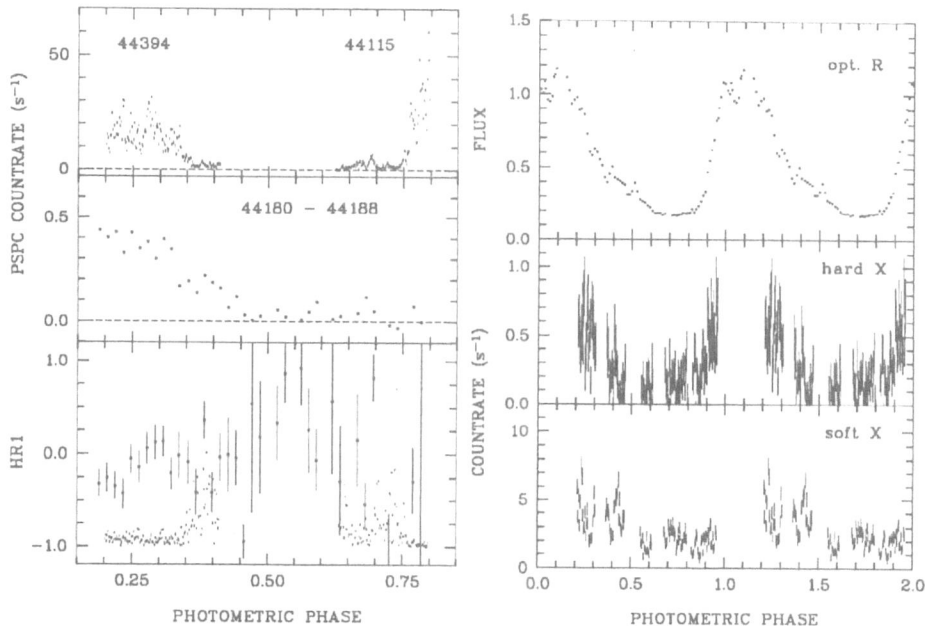

Figure 3 BL Hyı *(left)* Phase folded PSPC data obtained in 1991 (top) Soft flares observed at given cycles, binsize 10 sec, (middle) superposition of counts (binsize 100 sec) obtained in 5 individual observational slots in the given interval of cycles adding up to a light curve in the regular mode of accretion, (bottom) hardness ratio for the light curves in the upper panels *(right)* Phase folded PSPC data and optical lightcurve obtained in 1993 (binsize 30 sec)

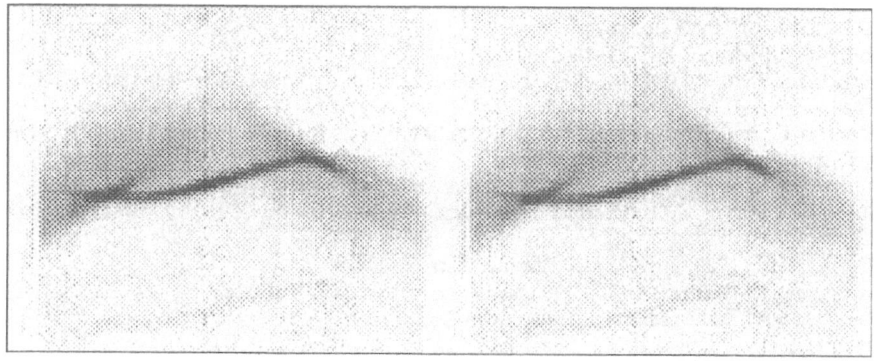

Figure 4 HU Aqr Phase-folded continuum-subtracted spectrograms of HeII 4686 Time (phase) moves along the y-axis and all data are shown twice for clarity The wavelength range displayed is 4651 – 4722 Å

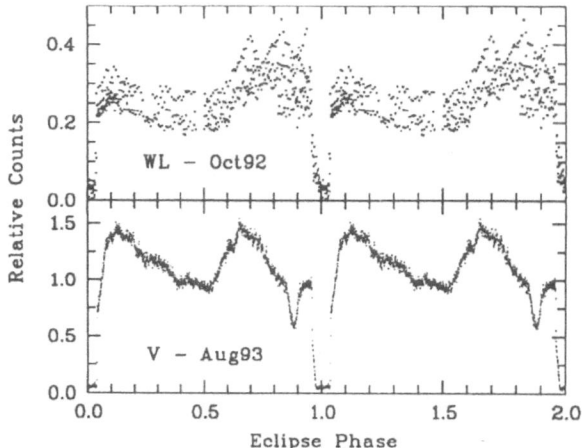

Figure 5. HU Aqr: Phase-folded continuum lightcurves in different accretion states. The data in the upper panel were not averaged because of an overall decline in brightness by ~1.5mag over the period of observation (Oct. 1 – 4, 1992). The light curve obtained in the high state (Aug. 16 –18, 1993, lower panel) contains phase-averaged brightness values of data in equivalent to ~8 orbital cycles.

References

Beuermann K., Schwope A.D., 1989, A&A 223, 179

Beuermann K., Schwope A.D., 1994, in *Interacting Binary Stars*, ed. A.W. Shafter, ASP Conf. Ser. 56, 119

Beuermann K., Thomas H.-C., 1993, Adv. Space Res. 13(12), 115

Clayton K.L., Osborne J.P., 1994, MNRAS 268, 229

Cropper M., Mason K.O., Mukai K., 1990, MNRAS 243, 565

Hakala P.J., Watson M.G., Vilhu O., Hassall B.J.M., Kellett B.J., Mason K.O., Piirola V., 1993, MNRAS 263, 61

Liebert J., Stockman H.S., 1979, ApJ 229, 652

Morris S.L., Schmidt G.D., Liebert J., Stocke J., Gioia I., Maccacaro T., 1987, ApJ 314, 641

Patterson J., Beuermann K., Lamb D.Q., Fabbiano G., Raymond J.C., Swank J., White N.E.: 1984, ApJ 279, 785

Piirola V., Coyne G.V., Reiz A., 1990, A&A 235, 245

Ramsay G., Mason K.O., Cropper M., Watson M.G., Clayton K.L., 1994, MNRAS 270, 692

Schwope A.D., 1990, Reviews in Modern Astronomy 3, 44

Schwope A.D., Beuermann K., 1993, Ann. Isr. Phys. Soc. 10, 234

Schwope A.D., Beuermann K., 1989, A&A 222, 132

Schwope A.D., Thomas H.-C., Beuermann K., 1993, A&A 271, L25

ADS, KHM: Visiting astronomers, German-Spanish Astronomical Center, Calar Alto, operated by the Max-Planck-Institut für Astronomie, Heidelberg, jointly with the Spanish National Comission for Astronomy

This work has been supported by the BMFT under grant 50 OR 9403 5.

MAGNETIC CVS OBSERVED WITH IUE

D. DE MARTINO
IUE Observatory VILSPA-ESA, Madrid, Spain
On leave from Osservatorio di Capodimonte, Naples, Italy

1. Introduction

Magnetic CVs contain a magnetized white dwarf which accretes material from a late type companion via Roche Lobe overflow. These comprise of the Polar systems, or "AM Her" stars, and the Intermediate Polars (IPs), or "DQ Her stars". In Polars, the magnetic field is strong enough ($\sim 10 - 60MG$) to synchronize the white dwarf rotation (spin) with the orbital period and to channell accretion towards its polar regions (Cropper 1990). IPs are asynchronous systems, with $P_{spin} \ll P_{orb}$. Their magnetic fields are believed to be lower than those of Polars , which can allow the formation of an accretion disc (Berriman 1988).

The accretion geometry strongly influences the emission properties at all wavelengths and its variability. The knowledge of the behaviour in all energy domains can allow one to locate the different accreting regions. The UV domain represents the link between the high energy X-ray and EUV emissions, which map regions close to the white dwarf surface, and the optical/IR regimes mapping regions up to the orbital distance. However the UV flux is not single component but has different contributions. In Polars, sources of emission can be the accretion stream and column towards and above the white dwarf, the heated polar cap(s) and the intrinsic emission of the white dwarf itself. In IPs, possible sources of UV flux can be the accretion curtain above the white dwarf polar caps, the impact region of the accretion stream with the disc (hot-spot), the inner disc regions and X-ray reprocessed radiation at the impact site (bulge). In the latter systems, a hot corona above the accretion disc can also be an additional source of the UV emission lines.

In sect. 2 the observed characteristics in both UV continua and emission lines are described, while in sect. 3 the main results of UV variability studies which allowed the identification of different emitting regions are summarized.

A Bianchini et al (eds) Cataclysmic Variables 397-403

2. UV properties of Magnetic CVs

Up to now and thanks to the long life of the IUE satellite about one third of magnetic CVs has been observed and the UV picture is currently improving with observations of newly discovered systems from the recent ROSAT Sky Survey. Both Polars and IPs are relatively strong UV emitters ($L_{UV} \sim$ 2 10^{30} $-$ 3 10^{32} erg s^{-1}). Their UV spectra are characterized by a blue continuum and strong emissions of resonance lines of NV (1240 Å), CIV (1550 Å), SiIV (1397 Å), HeII (1640 Å and 2733 Å) and MgII (2800 Å). The continuum luminosity is $\sim$ 2 $-$ 10% the line luminosity (Bonnet-Bidaud & Mouchet 1988). UV emission line and continuum measures presented in this section are based on orbital averaged spectra or on spectra whose length is comparable with the orbital period, which have been taken from the IUE data bank.

Polars and IPs have similar ranges of UV line luminosities ($\sim$ 2 10^{29} $-$ 5 10^{30} $ergs^{-1}$). In Fig. 1 line ratios, L_{NV}/L_{SiIV} and L_{NV}/L_{CIV} are shown for both classes. For Polars (Fig. 1a) these ratios have been derived using high accretion state spectra, whereas for IPs quiescent spectra have been used (see sect. 3). The luminosity ratios are compatible with collisional excitation with ion abundances from photoionization of black-body continua at 15-45 eV (Kallman 1983). Exceptions have been found in the Polar BY Cam (Bonnet-Bidaud & Mouchet 1987) and the IP TX Col (Mouchet et al. 1991), which display anomalous line ratios suggesting nonsolar abundances. Though UV spectra of the two classes might appear similar there are differences between them. Polars display a wider range of ionization states: NV (1250Å), NIV (1718Å) and NIII (1748Å); CIV (1550Å), CIII (1175Å) and CII (1335Å); SiIV (1397Å) and SiIII (1300Å) and AlIII(1859Å). IPs instead show only marginally emission lines of lower ionization states. Indication of a higher ionization efficiency in IPs also comes from emission line luminosities, being on average $L_{SiIV} < L_{NV} < L_{CIV}$ whilst in Polars $L_{SiIV} > L_{NV} < L_{CIV}$ (Fig 1). The effect of increasing ionization efficiency tends to suppress SiIV and increase NV and CIV. The NV and SiIV ratio is very sensitive to the ionization parameter, $\propto L/Nh^2$, where L is the luminosity of the photoionizing continuum, N is the densitiy in cm^{-3} and h the scale height of the ionized material (Kallman 1983). A higher ionization in IPs could be due to a strong absorption of the soft X-ray/EUV component which in these systems is marginally detected.

Differences between the two classes are also found in their UV continuum emission. IPs are systematically found at higher continuum luminosities (by a factor $\sim$ 10) with respect to Polars (Bonnet-Bidaud & Mouchet 1988), indicating that IPs experience a higher mass accretion rate, $\sim 10^{-16} - 10^{-17}$ g s^{-1}. The UV continuum slopes in both systems

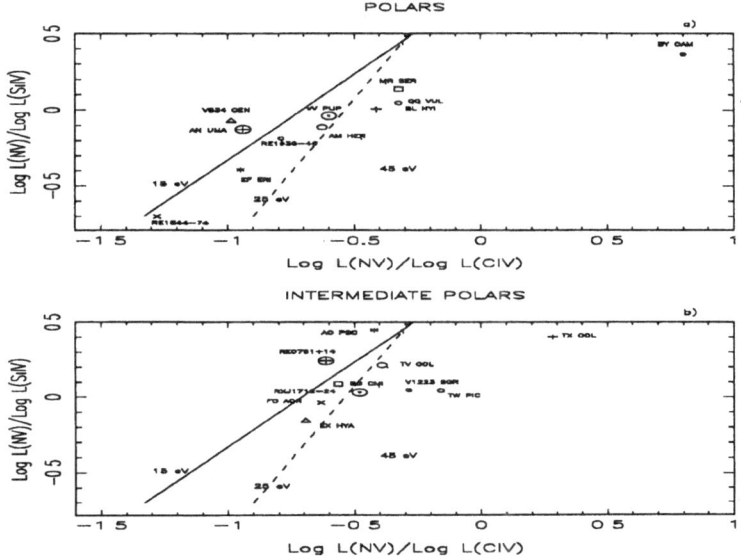

Figure 1. NV/SiIV and NV/CIV luminosity ratios compared with predictions from photoionization models of Kallman (1983) for polars a) and for IPs b).

are generally described by power laws ($\propto \lambda^{-0\ 0-3\ 0}$), while black-bodies do not efficiently represent the observed spectral shapes in the whole range covered by IUE ($\lambda\lambda 1000 - 3400$) (Fig. 2). Continuum fluxes in Polars (Fig. 2a) have been measured on high state spectra. For two systems, AM Her and V834 Cen, low state spectra in the IUE archive are of good quality to permit adequate continuum measures (Fig.2a). For IPs quiescent (cfr. sect. 3) UV spectra have been used. The tendency for Polars to show a steeper energy distribution in the far-UV (SWP range, $1000 - 1990\text{Å}$) with respect to the near-UV (LWP range, $1700 - 3400\text{Å}$) is seen in Fig. 2a but no λ^{-4} tail expected from a hot soft-X-ray/ EUV black-body continuum is observed. The different behaviour in the two spectral ranges indicates that the UV continuum emission is multicomponent (cfr.sect.3).

3. Variability

3.1. LOW AND HIGH STATES

One of the major characteristics is a strong variability with high and low states observed in both classes. IPs generally are found in quiescence at a relatively stable luminosity level. For these systems, few examples are

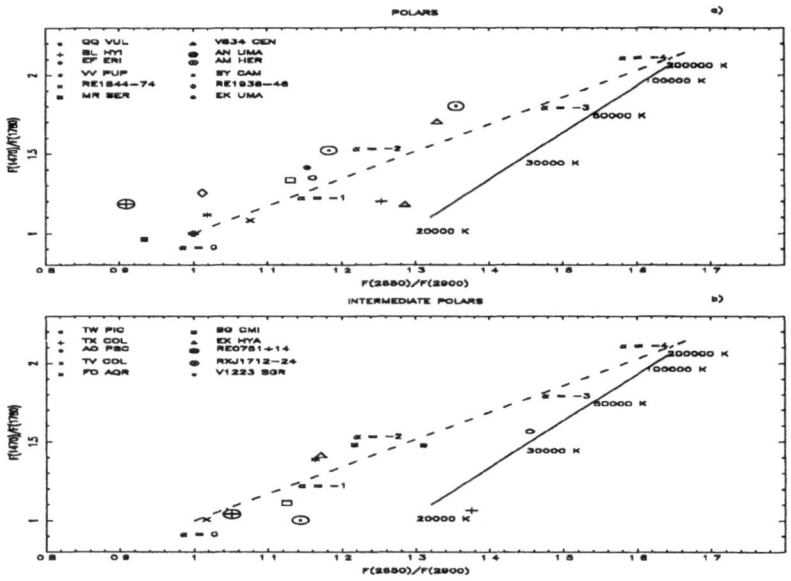

Figure 2. Far-UV versus near-UV continuum flux ratios for Polars **a)** and for IPs **b)**

known during activity phases. Except for the old nova GK Per ($P_{orb} = 2^d$) and the peculiar system AE Aqr ($P_{orb} = 9.9^{hrs}$), outbursts fall in the dwarf novae case but their phenomenology, mainly observed in the optical, is different from case to case (Hellier 1993). Magnitude variations are $\sim 2-4mag$ and time scales of occurrence are different from system to system as well as the duration of the outbursts. TV Col (Szkody & Mateo 1994), EX Hya (La Dous 1990) and the IP candidate H0857-24 (Alvarez et al. 1994) were observed with IUE during activity. The main UV spectral characteristics are changes in spectral slope and the appeerence of P-Cygni profiles in the resonance UV lines indicating events of mass loss. The wind profiles disappear when re-entering in quiescence.

Worthnoticing is the case of the peculiar IP TX Col (Mouchet et al. 1991), showing a long term UV variability in both continuum and emission line intensity by a factor of 3. This variations have been interpreted in terms of changes in the mass accretion rate.

Polars, instead, do not show abrupt variations, but mostly a continuum sequence of luminosity states in a range of 1-3mag on time scales of months. UV observations of AM Her (Heise & Verbunt 1988; Gansicke et al. 1994; Silber et al. 1994; La Dous 1995), V834 Cen (Szkody et al. 1982, Maraschi et al. 1984), MR Ser and AN UMa (Szkody et al.1988) have shown that during low states the white dwarf dominates the UV continuum emission. Interesting cases are those of V834 Cen and AM Her for which several UV observations during low state are available in the IUE archive. For one of the faint states of V834 Cen, a black-body temperature of 26500 K was derived (Maraschi et al. 1984) while on other occasions UV continuum spectra were compatible with a black-body at 17000 K (La Dous 1995), thus revealing a cooling of the white dwarf. On the other hand, from L_α absorption and far-UV continuum, a temperature of ~ 20000 was derived for AM Her during its different low state levels (Heise & Verbunt 1988; Gansicke et al. 1994; Silber et al. 1994; La Dous 1995), suggesting that the white dwarf does not cool down when accreting at a low rate. Szkody et al. (1988) pointed out that the white dwarf in Polars remains hot at low accretion rates but the issue appears to be more complex.

3.2. PERIODIC VARIATIONS

The contribution to the UV emssion during high states comes from different regions whose indentification can be made by studying periodic variations. The orbital and white dwarf clocks can allow one to identify fixed regions in the binary and white dwarf frames. Polars are variable at the orbital period (hours) while IPs vary on a wide range of frequencies and mainly at the orbital, spin and sideband periods (tens of mins). Except for EX Hya ($P_{spin} = 67^{mins}$) IUE phase resolved studies in IPs have been performed at the orbital period.

As for the continuum in Polars, early UV observations of AM Her (Raymond et al. 1979) and V834 Cen (Maraschi et al. 1984) were described with a two component distribution: a the steep λ^{-4} responsible for the far-UV modulation and a flatter and constant λ^{-1} component. The steep contribution was ascribed to the black-body tail of the soft X-ray component. Nevertheless this interpretation found strong difficulties when the so-called "X-ray reverse mode" in Polars was osberved (Heise & Verbunt 1988; Osborne et al. 1987). The far-UV modulation in AM Her during the X-ray "reverse mode" (Heise & Verbunt 1988) was not in phase with the soft X-rays indicating that previous interpretations that the UV flux is the tail of the soft X-ray component was not correct. The UV emission was compatible with a large area of a heated white dwarf at $\sim 30000K$. In QQ Vul, the orbital variable UV continuum component is in excess of the extrapolated soft X-ray emission (Mukai et al. 1986). Also in the bright EUV Polar RE1938-46, the far-UV emission is variable with the orbital period but does not show any λ^{-4} distribution (de Martino et al. 1994). This suggests that the white dwarf heated region is large and possess a temperature distribution with the cooler outer regions mainly contributing in the UV.

UV line emission studies have been carried out for few Polars. In V834 Cen (Takalo & Nousek 1988), QQ Vul (Mukai et al. 1986) and RE1938-46 (de Martino et al. 1994), the prominent emission lines (NV, CIV, SiIV and HeII) show orbital variations in phase with the broad component seen in the optical lines suggesting a common origin. Flux variations (a factor of 3-4), velocity amplitudes ($\sim 400 km s^{-1}$), and FWZI ($\pm 2600 km s^{-1}$) are consistent with the magnetically confined accretion stream towards the white dwarf. Also, in V834 Cen (Takalo & Nousek 1988), RE1938-46 (de Martino et al. 1994) and in high resolution spectra of AM Her (Raymond et al. 1979), a blue shifted broad emission component in CIV line was observed indicating a multicomponent structure. HST high temporal and spectral resolution spectra are needed to resolve these components.

In IPs, the nature of the UV emission is more complex since they are characterized by periodic variations at different frequencies. The identification of the continuum and emission line regions can be done studying the modulations at the main periods P_{orb}, P_{spin} and P_{beat} (de Martino 1993). While only for EX Hya (Krautter & Buchholz 1990) a rotational modulation study could be performed, IPs have been observed with IUE along their orbital period. In that system, the UV continuum pulsation is quite strong ($\sim 30\%$ in the SW range) and slightly red ($\sim 38\%$ in the LW range). From optical studies (de Martino et al. 1994; 1995; Welsh et al. 1994) it also appears that the spectrum of the spin pulsation is not very hot (≤ 20000 K), indicating that the responsible regions of the accretion curtain are not confined close to the white dwarf surface.

UV continuum orbital modulations in FO Aqr and BG CMi are strong, $\sim 30\%$ and $\sim 70\%$ respectively, with a remarkable similar orbital phasing and with temperatures $\geq 20000 K$ and $\sim 18000 K$ respectively which have been interpreted in terms of X-ray reprocessing at the bulge (de Martino et al 1994; 1995). Also, in FO Aqr the UV/optical continuum unmodulated component is compatible with an accretion disc which contributes to $\sim 8\%$ of the total accretion luminosity (de Martino et al. 1994).

UV emission lines in IPs have also different origins but studies are still scarce in this respect. EX Hya is again the unique IUE example for which the main emissions (NV, CIV and SiIV) are found to be strongly variable ($\sim 50\%$) with the spin period. Trailed spectra during eclipse phases have been used to infer an emission component arising from the accretion stream (Krautter & Buchholz 1990) thus indicating a multicomponent structure. FO Aqr is also an interesting example. NV and CIV line fluxes are stable along the orbital period, while HeII (1640Å) flux is in phase with the UV continuum modulation (Buckley et al. 1994). This behaviour suggests that the orbital modulated flux in HeII and UV continuum originate in the same region, whilst the resonance lines of NV and CIV are formed in a large volume region, possibly a hot corona above the disc.

References

Alvarez, R. et al. (1994), this conference.

Berriman, G. (1988) in *Polarized radiation of circumstellar origin*, Vatican Press, p281.

Bonnet-Bidaud, J.M. & Mouchet, M. (1987) *A&A*, **188**, 89.

Bonnet-Bidaud, J.M. & Mouchet, M. (1988) *ESA SP-281*, p271.

Buckley, D. et al. (1994), in preparation.

Cropper, M. (1990) *Space Science Reviews*, **54**, 195.

de Martino, D. (1993) in *Cataclysmic Variables and Related Physics*, eds. O.Regev & G.Shaviv, p201.

de Martino, D. et al. (1994), *A&A*, **284**, 125.

de Martino, D. et al. (1994), in preparation.

de Martino, D. et al. (1995), *A&A*, in press.

Gansicke, B.T. et al. (1994), in 9th *European Workshop on White Dwarfs*, in press.

Heise, J. & Verbunt, F. (1988), *A&A*, **189**, 112.

Hellier, C. (1993) in *Cataclysmic Variables and Related Physics*, eds. O.Regev & G.Shaviv, p205.

Krautter, J. & Buchholz (1990) 11th *American Workshop on CVs and LMXBs*, p229.

La Dous, C. (1990), *Space Science Rev.*, **52**, 203.

La Dous, C. (1995), *A&A*, in press.

Mouchet, M. et al. (1991) *A&A*, *A&A*, **250**, 99.

Maraschi, L. (1984), *ApJ*, **285**, 214.

Mukai, K. et al. (1986), *MNRAS*, **221**, 839.

Osborne, J.P. et al. (1987), *ApJ Let.*, **315**, L123

Raymond, J.C. et al. (1979), *ApJ Let.*, **230**, L95

Silber, A. et al. (1994), this conference.

Szkody, P. & Mateo, M. (1984), *ApJ*, **280**, 729.

Szkody, P. et al. (1982), *ApJ*, **257**, 686

Szkody, P. et al. (1988), *PASP*, **100**, 362.

Welsh, W. et al. (1994), this conference.

A BOUNDARY LAYER DYNAMO MODEL FOR THE SECONDARY COMPONENTS OF CATACLYSMIC VARIABLES

ANTONIO BIANCHINI AND LUCA ZANGRILLI
Dipartimento di Astronomia, Università di Padova
vicolo Osservatorio 5, 35122, Padova, Italy.

Abstract.

A simplified boundary layer $\alpha - \omega$ dynamo model operating at the base of the convective envelope of the secondaries of Cataclysmic Variables can produce the magnetic field intensities required for the magnetic braking mechanism. We show that the dynamo model can explain the observed orbital period distribution of CVs better than previous braking laws. This model spontaneously produces the period gap between 2 and 3 hr. In particular, it might account for the observed lack of dwarf nova systems at orbital periods between 3 and 3.5 hr. We show that if the α mechanism is free to operate in a larger convective zone than ω, and non linear effects are introduced, we obtain more realistic intensities of the magnetic field and periods of the magnetic cycles similar to those observed in CVs.

1. Introduction

It is now largely recognized that mass transfer in Cataclysmic Variables (CVs) is mainly maintained by continuous loss of orbital angular momentum by the binary system. In most cases the main mechanism has been recognised to be the magnetic braking of the tidally coupled secondary by its own magnetic stellar wind (Verbunt & Zwaan 1981; Rappaport, Verbunt & Joss 1983). An approach for modelling the magnetic braking mechanism in CVs has been developed by Hameury *et al.* (1988). This model is based on the theory of magnetic stellar wind developed by Mestel & Spruit (1987) and requires the knowledge of the intensity of the dipolar magnetic field, which is usually derived by the simple relation $B = 650/P_{orb}(hr)G$, where $P_{orb}(hr)$ is the orbital period in units of an hour. The validity of this equa-

A Bianchini et al (eds), Cataclysmic Variables, 405-414

tion is based on the observations of single G and K dwarf stars in galactic clusters, which show an approximate linear correlation between their $B_{surface}$ and Ω, with Ω the angular velocity (see Vilhu 1984). This choice is however questionable for at least two reasons. First, the chromspheric activity, due to the eruption of toroidal magnetic flux tubes, could not be linearly correlated with the dipolar magnetic field, which is responsible for the magnetic braking. Secondly, it could be difficult to extrapolate such a correlation to the extreme rotational regimes of the secondaries of CVs. Saar (1991) and Vilhu (1984) found that the chromospheric activity itself tends to saturate in rapidly rotating stars. The saturation is reached for the Sun at $\Omega > 7\Omega_\odot$.

In spite of the many stimulating results achieved by the different magnetic braking models, Shafter (1992) has demonstrated that they are all unable to properly explain the observed distribution of the orbital periods of CVs and, in particular, that of dwarf nova systems (DN) alone. He noticed that theoretical $\dot{M}(P)$ tracks are unacceptably steep while decreasing towards shorter periods, and offer no explanation for the apparent lack of DN between 3 and 3.5 hr, where almost only classical novae (CN) and nova like systems (NL) are seen. Another observation concerns the usual explanation of the period gap. In fact, the turn off of the mass transfer is usually obtained by artificially decreasing the efficiency of the braking law as soon as the star becomes fully convective.

In this paper we derive the intensity of the magnetic field of the secondary using a simplified boundary layer $\alpha - \omega$ dynamo model operating at the base of the convective region. To test how this may change the evolution of CVs we have computed evolutionary tracks using a bipolytrope model of the secondary. This choice offers a great computational speed while testing different braking laws with a large number of varying input parameters.

In section 2 we briefly describe the input parameters for standard bipolytrope model of the secondary. The dynamo model is presented in section 3. In section 4 the wind model is rapidly described. The results are discussed in section 5.

2. The bipolytrope model for the secondary

Numerical evolutionary computations have been performed using for the secondaries of CVs a double polytrope code as described by Rappaport et al. (1983) with the improved interpolated formulae given by Hameury et al. (1988). The chemical composition has been assumed to be $X_c = 0.4$ and $Z_c = 0.02$ for the radiative nucleus, and $X_e = 0.7$ and $Z_e = 0.02$ for the envelope. The assumed mean molecular weight in the photosphere is $\mu_{ph} = 1.25$. Photospheric opacities have been computed by interpolating with Bspline functions the tables of Alexander *et al.* (1983). For those at the core boundary we have used the interpolated formulae given by Stellingwerf (1975) modified as prescribed by Hameury *et al.* (1988), adopting values of the corrective parameter δ, (the α parameter in Hameury *et al.*), between 1.5 and 4.0.

3. The dynamo model for the secondaries of CVs

3.1. THE STANDARD BOUNDARY LAYER DYNAMO MODEL

The $\alpha - \omega$ dynamo equations can be expressed in the form:

$$\frac{\partial A}{\partial t} = \alpha B_\phi + \eta \nabla^2 A \tag{1}$$

$$\frac{\partial B_\phi}{\partial t} = [\nabla \Omega \times \nabla (Ar \sin \theta)]_\phi + \eta \nabla^2 B_\phi \tag{2}$$

$$\vec{B}_p \equiv \nabla \times (0, 0, A) \tag{3}$$

where A is the vector potential, B_ϕ is the toroidal component of the magnetic field, B_p is the poloidal field, Ω is the angular velocity and θ is the polar angle. α is the poloidal component regenerating term, while that of the toroidal field is the differential rotation (the ω part of the dynamo). η is the turbolent diffusivity and operates as a linear dissipative term. For a derivation and a complete discussion of dynamo equations, see, for example, Yoshimura (1975a,b), (1978), and Parker (1979).

It is currently assumed that the dynamo operates mainly in a thin layer (about a scale heigh pressure, or even smaller) located at the base of the convective zone or even in the overshoot region. This would explain the latitudinal migration of solar spots towards the equator (see C.A.Morrow *et al.* 1988), as well as helioseismological results, which suggest that the angular velocity gradient is confined in a thin shell near the overshoot region (Libbrecht & Morrow 1991). This confinement is however restricted to the ω part of the dynamo, which is responsible for the toroidal magnetic field amplification. Instead, there seems to be no *a priori* reason to confine the α mechanism inside a thin boundary layer. Actually, recent results suggest that it may operate in the whole convection zone (Vainshtein *et al.* 1993). In the following we will explore both these possibilities.

Robinson & Durney (1982) (hereafter RD) obtained boundary layer dynamo equations considering only local latitude dependent solutions. To take into account non linear effects these authors assumed Yoshimura's prescription (Yoshimura 1978), multiplying α and $\nabla \Omega$ by the factor

$$nlin = \exp\left(-\gamma \frac{B_\phi^2}{4\pi\rho V_c^2}\right)$$

where the constant γ might assume different values for the two dynamo terms.

3.2. THE SIMPLIFIED BOUNDARY LAYER DYNAMO

A more schematic form of dynamo equations for the boundary layer (bl) model, in which all the essential features of dynamo action are represented, can be given as:

$$\frac{dA}{dt} = \alpha B_\phi - \frac{A}{\tau_p} \tag{4}$$

$$\frac{dB_\phi}{dt} = \frac{\Delta_L\Omega}{L}A - \frac{B_\phi}{\tau_\phi} - \frac{B_\phi}{\tau_{Par}} - \frac{u}{L}B_\phi \tag{5}$$

where $\Delta_L\Omega$ is a typical value of the radial shear in the angular velocity in a pressure hight scale. The last two terms of equation (5) represents non linear dissipative mechanism, which do not appear in the original $\alpha - \omega$ equations. The first is due to the magnetic buoyancy, u being the magnetic flux tubes rise velocity; the second corresponds to Parker's instabilty, with $1/\tau_{Par} = 0.3V_{A,\phi}/L/$ and $V_{A,\phi}$ the Alfvèn speed (Parker 1979).

Expressions of the regenerative terms α and $\Delta_L\Omega$ are given by Durney & Robinson (1982) (hereafter DR) as

$$\alpha = c_1\frac{L^2\Omega}{R_{core}} \quad and \quad \Delta_L\Omega = c_2\left(\frac{L}{R_{core}}\right)^2\Omega$$

where c_1 and c_2 are constants properly chosen in order to reproduce the $1G$ surface polidal field and the $11yr$ periodicity in the solar model.

We note that in equations (3) of RD the time scale for turbolent dissipation of both the magnetic field and the vector potential is equal to $1/(\eta K^2)$, where $\eta = 1/3V_cL$, and $K = 2./R_{core}$, L being the local pressure hight scale, V_c the mixing length convective velocity, and R_{core} the radius of the radiative nucleus. So, we may write:

$$\tau \equiv \tau_\phi = \tau_p = \frac{3}{4}\frac{R_{core}^2}{V_cL} \tag{6}$$

If we define

$$t_A \equiv \left(\frac{1}{A}\frac{dA}{dt}\right)^{-1} \quad and \quad t_{B_\phi} \equiv \left(\frac{1}{B_\phi}\frac{dB_\phi}{dt}\right)^{-1}$$

as the amplification time scale for the potential vector, A, and the toroidal field, B_ϕ, respectively, from (4) we obtain $A = \alpha t_A \tau B_\phi / (t_A + \tau)$. Since $B_p = \nabla \times A \simeq A / R_{core}$, we finally derive the simple expression for the poloidal magnetic field

$$B_p = \frac{1}{R_{core}} \frac{\alpha t_A \tau}{t_A + \tau} B_\phi \qquad (7)$$

In the *bl* model, we shall assume that the dynamo operates in a one pressure scale height layer at the base of the convective zone. According to the mixing lenght theory, the dimensions of the convective eddies will also be taken equal to the local pressure scale height, with the prescription that they can never become larger than the half width of the convective envelope.

A way to solve the simplified dynamo equations is to assume that the amplification time scales of the magnetic field and of the vector potential are equal to the time rise of the flux tubes, $t_{rise} = L/u$ (Parker 1979). This assumption requires that in eq. (5) we drop out the dissipative term due to the buoyancy, since we have now changed from an Eulerian to a Lagrangian view of the dynamo action. Thus, defining $t \equiv t_A = t_{B_\phi} = t_{rise}$, and $c_3 \equiv \alpha t_A \tau / [R_{core}(t_A + \tau)]$, after simple calculations, dynamo equations yeld

$$\left(\frac{1}{t}\right)^2 = \frac{dA}{dt} \frac{1}{A} \frac{1}{B_\phi} \frac{dB_\phi}{dt} = \left(\frac{\Delta_L \Omega}{L} - \frac{1 + \tau/\tau_{Par}}{\tau R_{conv} c_3}\right) \left(\alpha - \frac{L c_3}{\tau}\right)$$

and, finally

$$\left[\left(\frac{\Delta_L \Omega}{L} - \frac{1 + \tau/\tau_{Par}}{\tau R_{core} c_3}\right) \left(\alpha - \frac{L c_3}{\tau}\right)\right]^{0.5} \frac{L}{u} = 1 \qquad (8)$$

Numerical solution of this equation gives the toroidal magnetic field intensity. Clearly, when the argument of the square root becomes negative, dissipative terms overcome the regenerative ones, and field amplification ceases.

It must be noted that in Parker's formulation the rise velocity of the flux tubes requires the knowledge of their radius, R_{tube}. Assuming a filling factor of about 0.1 for B_ϕ, we obtain radii of about $0.05L$, which might be consistent with the observed dimensions of sunspots.

A mechanism which plays against buoyancy is provided by the tensions along the flux tubes. According to Weiss (1964), we may write the ratio of the inward to the buoyancy force as:

$$\frac{F_{sink}}{F_{buoyant}} = \left(1 - c_3^2\right) \frac{R_{Ryd} T}{g R_{curv}}$$

where R_{Ryd} is the Rydberg constant, g the gravity, T the local temperature, and R_{curv} the curvature radius of flux tubes. The buoyant velocity is then reduced by the factor $[1 - R_{Ryd}T/(gR_{curv})]^{0.5}$. For strong magnetic filds the inward force may be larger than the buoyancy, thus turning off the dynamo process.

3.3. THE BOUNDARY LAYER PLUS ENVELOPE DYNAMO

If α is assumed to operate in the whole convective envelope, the amplification time scales of the toroidal magnetic field and of the potential vector will be different. In the boundary layer plus envelope (ble) model we treat the ω mechanism exactly as in the previous pure bl model. To evaluate the α term, we tentatively adopt values of thermodynamical parameters calculated in the middle of the shell where α is efficient. Following Canuto & Mazzitelli (1991), we assume that typical dimensions of the convective eddies in the envelope are equal to half the width of this active region. The evolution of a magnetic flux tube in the remaining envelope is obtained by integrating the equation of motion and applying the flux and mass conservation laws inside the tube (for detailed calculations see Zangrilli & Bianchini, work in progress). The buoyancy time in this region, t_{env}, results

$$t_{env} = \frac{1}{C}\left[\frac{1}{R_{b.l.}} - \frac{1}{R_{star}}\right] \tag{9}$$

with constant C given by

$$C = B^2_{\phi,b.l.}\frac{3}{32\pi}\frac{QR^2_{tube}}{DV_cL^2\rho_{b.l.}R^2_{b.l.}} \tag{10}$$

where all quantities with index $b.l.$ are evaluated at the upper edge of the boundary layer, $R_{b.l.}$; R_{tube} is the radius of a flux tube, R_{star} is the stellar radius, and D is a constant of the order of unity. The constant Q is defined in Parker (1979).

The mean value of the toroidal magnetic field produced by the boundary layer is then obtained by numerically solving the equation

$$\left[\left(\frac{\Delta_L\Omega}{L} - \frac{1 + \tau/\tau_{Par}}{\tau R_{core}c_3}\right)\left(\alpha - \frac{Lc_3}{\tau}\right)\right]^{0.5}[t_{rise}(t_{rise} + t_{env})]^{0.5} = 1 \tag{11}$$

3.4. MAGNETIC ACTIVITY CYCLES

DR, neglecting dissipative effects, obtained that the amplification time scale for the magnetic field coincides within a factor 2π with the period of the dynamo cycle. For this reason, we may assume that even in presence of dissipative effects the periods are $t_{cycle} = 2\pi t$. In both bl and ble models the cycles are determined by the rise time of the flux tubes through the boundary layer alone.

4. The magnetic wind model

The model we adopt here is that of Mestel & Spruit (1987), also discussed by Hameury *et al.* (1988). The only difference is that the magnetic field is now explicitly derived from our dynamo models. In the ble model the calculated angular momentum losses have been multiplied by a factor $J_{fac} = 0.6$. One might speculate that a factor lower than unity could account for the presence in the binary system of the white dwarf, which is suspected to reduce the efficiency of the magnetic braking.

5. Results and conclusions

Our evolutionary sequences start with a $1M_\odot$ secondary star and a $1M_\odot$ white dwarf which is assumed to periodically loose all the accreted material through nova outbursts. In Table 1 we give some relevant input parameters for three selected models. Models BL1 and BL2 represent pure bl dynamos; model BLE represents a ble dynamo. The coefficients γ_1 and γ_2 determine the efficiency of non linear effects on α and $\Delta_L\Omega$, respectively. The last column specifies when shear saturation at $\Omega > 7\Omega_\odot$ is assumed. The evolutionary tracks for mass transfer rate and poloidal magnetic field are shown in Fig. 1 and 2, respectively.

Table 1

model	δ	γ_1	γ_2	c_1	c_2	J_{fac}	R_{tube}/L	shear saturated
BL1	1.5	0.03	0.03	5.5×10^{-4}	7.1×10^3	1.0	0.05	yes
BL2	1.5	1.00	4.00	9.0×10^{-4}	5.5×10^4	1.0	0.05	no
BLE	4.0	5.00	5.00	1.6×10^{-3}	9.0×10^4	0.6	0.05	no

As we can see, similar behaviours can be obtained using different sets of parameters and/or models. If we take very small γ values, as, for example, in model BL1, the dynamo shows a general tendency to produce a peak in the

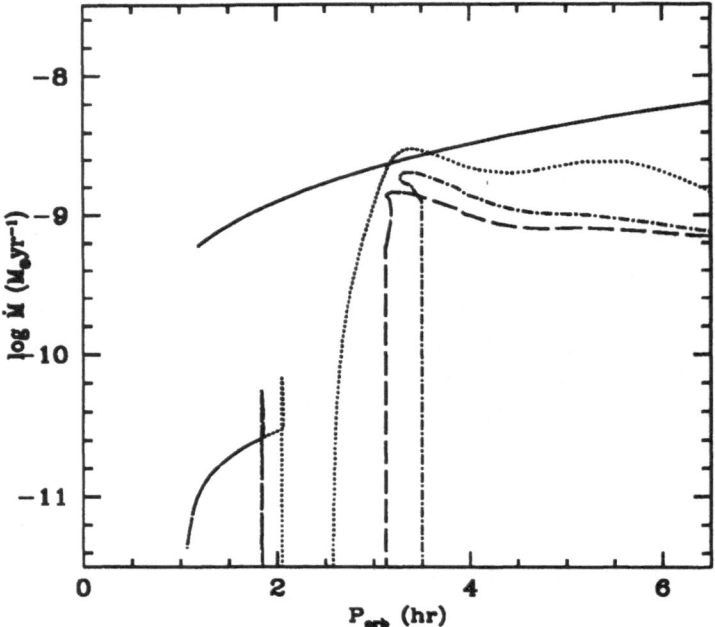

Figure 1.   Mass transfer rate *vs.* orbital period for models BL1 (dashed), BL2
(dashed-dotted) and BLE (dotted). The full line represents the critical mass transfer
rate for accretion disc stability.

mass transfer rate, when the binary system approaches the $3hr$ upper limit
of the period gap. This appears to be an encouraging feature of the dynamo
model, because it might explain the observed lack of unstable systems in
the range $3 - 3.5hr$. It must however be reminded that in this model, a
saturation in the shear has been introduced. Unfortunately, unacceptable
high toroidal fields, well above the energy equipartition limit ($B_\phi \simeq 16 \times B_{eq}$), and too short magnetic cycles (few months) are produced.

Model BL2 shows that we can obtain evolutionary tracks similar to the
previous one, using larger γ values (of the order of unity) to reproduce
shear saturation, provided that γ_1 is chosen smaller than γ_2. The periods
of magnetic cycles are now of the order of a few years, in agreement with
the observational results of Bianchini (1990) and Warner (1988). The field
intensities also satisfy the equipartition limit.

Since a larger efficiency of α with respect to $\Delta_L \Omega$ can be obtained by
letting α free to act in a larger portion of the convection zone, we have
investigated also model BLE, which, at the moment, seems to be the most
promising approach.

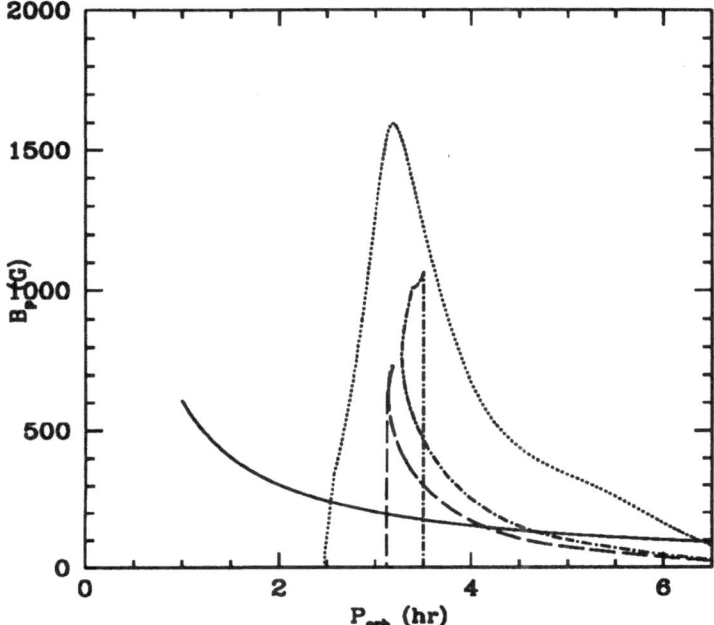

Figure 2. Poloidal magnetic field intensity *vs.* orbital period. The full line represents the empirical law for the magnetic field. The other lines are as in Figure 1.

It is worth noticing that dynamo models spontaneously produce a period gap. In models BL1 and BL2 the rapid decrease of mass transfer rate is due to both the decrease of the dynamo efficiency and to the increase of the tensions along toroidal field lines. In model BLE, the toroidal field is much weaker, and tensions are unimportant, although the poloidal field appears stronger. The gap is produced because angular momentum losses decrease so rapidly that the star looses contact with its Roche lobe.

We finally comment that more detailed calculations might demonstrate that some magnetic field can be still produced even in very low mass secondaries below the period gap, as it is required by the presence of several Polar systems.

References

Alexander, D.R., Johnson, H.R., Rypma, R.L. 1983, *Ap. J.*, **272**, 773.
Bianchini, A. 1990, *A. J.*, **99**, 1941.
Canuto, V.M., Mazzitelli, I. 1991, *Ap. J.*, **370**, 295.
Durney, B.R., and Robinson, R.D. 1982a, *Ap. J.*, **253**, 290.
Hameury, J.M., King, A.R., Lasota, J.P. and Ritter, H 1988, *M. N. R. A. S.*, **231**, 535.

Libbrecht, K.G. & Morrow, C.A. 1991, in *Solar Interior and Atmosphere*, ed. A.N. Cox, W.C. Livingston & H.M.S. Matthews (Tucson: Univ. of Arizona), 479

Mestel, L.,and Spruit, H.C. 1987, *M. N. R. A. S.*, **226**, 57.

Morrow, C.A., Gilman, P.A., DeLuca, E.E. 1988, *Seismology of the Sun & Sun-Like Stars*, ESA SP-286, 109

Parker, E.N. 1979, *Cosmical Magnetic Fields* (Oxford University Press)

Rappaport, S., Verbunt,F. and Joss, P.C. 1983, *Ap. J.*, **275**, 713.

Robinson, R.D. and Durney, B.R. 1982b, *Astron. Ap.*,**108**, 322.

Shafter, A.W. 1992, *Ap. J.*, **394**, 268.

Stellingwerf, R.F. 1975, *Ap. J.*, **195**, 441.

Stix, M. 1976, *I.A.U. Symp.***71**, 367

Vainshtein S.I., Parker E.N., and Rosner R. 1993, *Ap. J.*, **404**, 773.

Verbunt, F., Zwaan, C. 1981, *Astron. Ap.*,**100**, L7.

Vilhu, O. 1984, *Astron. Ap.*,**133**, 117.

Warner, B. 1988, *Nature*,**336** 129.

Weiss, N. O. 1964,*M. N. R. A. S.*, **128**, 225.

Yoshimura, H. 1975 a, *Ap. J.*, **201**, 740.

Yoshimura, H. 1975 b, *Ap. J. Suppl.*, **294**, 467.

Yoshimura, H. 1978, *Ap. J.*, **220**, 692.

THE EVOLUTION OF MAGNETIC CVS

G. CHANMUGAM
Department of Physics and Astronomy, Louisiana State University
Baton Rouge, LA 70803, USA

1. Introduction

About one-fourth of all cataclysmic variables (CVs) contain white dwarfs with magnetic fields that are sufficiently strong ($B \gtrsim 5 \times 10^4$ G) to disrupt the accretion disk, either partially or completely, and are known as magnetic CVs (Chanmugam 1992, Ritter & Kolb 1995). Two classes of magnetic CVs have been distinguished: the synchronous systems in which the spin period of the white dwarf P_{spin} and the orbital period P_{orb} are equal to within about 2 %, and the asynchronous systems where the spin and orbital periods are more widely different (Sparks et al 1995). The former are also known as AM Herculis binaries or polars (Cropper 1990). Their magnetic field strengths B have been determined from Zeeman and cyclotron spectroscopy, and from cyclotron emission models for the optical/IR polarization they emit (Chanmugam and Dulk 1981; Meggitt and Wickramasinghe 1982). With the recent discovery of the probably synchronized system 1H1752+08 with $B \approx 7$ MG (Ferrario et al. 1994), the range of their magnetic fields is 7 MG - 70 MG (Chanmugam 1992). In addition, most of these binaries have $P_{orb} \lesssim 4$ hours. The resulting short orbital separation and strong magnetic field, enforces the synchronous rotation of the white dwarf (e.g., Joss, Katz and Rappaport 1979; Campbell 1990; King et al. 1990; Wu & Wickramasinghe 1993). The number of AM Her binaries has more than tripled in the last few years to over 50 (see e.g., Beuermann 1995) as result of the large number of them discovered by ROSAT from their softer X-ray spectra.

The asynchronous systems are also known as DQ Her binaries or Intermediate Polars (IPs) but in an inconsistent manner. Thus, some authors refer to all of them as DQ Her binaries (e.g., Patterson 1994). Others desig-

A Bianchini et al (eds), Cataclysmic Variables, 415-421
© *1995 Kluwer Academic Publishers*

nate systems with $P_{spin}/P_{orb} \sim 0.1$ as IPs (Warner 1985), while those with $P_{spin}/P_{orb} \ll 0.1$ are called DQ Her binaries. The asynchronous rotators, in general, do not emit polarized optical radiation, so there is no direct evidence of a magnetic field.

In this paper we review first early models for the evolution of magnetic CVs; and then we critically discuss rapid and slow evolution models.

2. Early Models for the Evolution of Magnetic CVs

Cataclysmic Variables evolve by losing angular momentum due to emission of gravitational radiation, and probably magnetic braking caused by a stellar wind along the magnetic field lines from the tidally locked red dwarf star (Verbunt and Zwaan 1981; Spruit and Ritter 1983; Rappaport, Verbunt and Joss 1983). This mechanism was applied to magnetic CVs by Chanmugam and Ray (1984), who argued that as the orbital separation a and mass transfer rates decrease, the magnetospheric radius r_m of the white dwarf becomes comparable to a and the white dwarf synchronizes if, typically, $B \gtrsim 3$ MG or more correctly the magnetic moment $\mu \sim BR^3 \gtrsim 10^{33}$ G cm^3 (see also Lamb & Melia 1987). The model provided an explanation for why most of the synchronous rotators had relatively short orbital periods ($P_{orb} \lesssim 4$ hr), and why most asynchronous rotators had $P_{orb} \gtrsim 3$ hr. The model failed to explain why there is a dearth of asynchronous rotators at short orbital periods. King, Frank and Ritter (1985) proposed therefore that the two classes of magnetic CVs had similar magnetic moments. This implied that all the asynchronous rotators would become synchronized as the evolution proceeded. But these models failed to explain why none of the then known asynchronous rotators were progenitors of the AM Her binaries, as none were known to emit polarized radiation characteristic of a strong magnetic field (Lamb & Patterson 1983, Wickramasinghe, Wu & Ferrario 1991).

In these models, as in the case for the evolution of CVs, it was assumed that a 2 - 3 hr period gap arose because of the suppression of magnetic braking when the Roche lobe filling secondary becomes convective, at the mass corresponding to $P_{orb} \approx 3$ hr, and ceased to be magnetically active. Nevertheless, Chanmugam and Ray (1984) predicted that when CVs are discovered in the period gap a higher proportion of them would be AM Her binaries, because of the reduced accretion rates which would lead to increased synchronization of systems. Over 5 AM Her binaries have since then been detected to lie in the period gap, a significantly higher proportion than for the general CV population (King 1994).

3. Rapid Evolution Model for High Field CVs

A puzzling feature of magnetic CVs is that there are none with field strengths $B \gtrsim 70$ MG, whereas isolated magnetic white dwarfs have field strengths that extend up to ~ 500 MG (Schmidt and Norsworthy 1991; Chanmugam 1992). Liebert and Stockman (1985) therefore suggested that magnetic CVs with strong magnetic fields would evolve several times more quickly than those with weaker fields, because the wind from the secondary would attach onto the stronger magnetic field of the white dwarf leading to greater magnetic braking and more rapid evolution (Schmidt, Stockman and Grandi 1986). Hence, the secondary would be whittled down quickly to a small mass and become detached. The system would then not undergo mass transfer and appear as an isolated strongly magnetic ($B \gtrsim 70$ MG) white dwarf which rotates with a period characteristic of the synchronous rotators. This is consistent with the existence of several apparently isolated magnetic white dwarfs with the expected rotation periods of several hours (Schmidt and Norsworthy 1991). A serious difficulty with the model is, however, that the enhanced magnetic braking leads to a broader period gap than is observed (Hameury et al 1987).

4. Slow Evolution Model for High Field CVs

Motivated by the discovery of five AM Her binaries in the period gap, Wickramasinghe and Wu (1994, hereafter WW) proposed a model where the evolution proceeds more slowly than for non-magnetic CVs so that there is no gap. They hypothesized that magnetic braking by the wind from the red dwarf is suppressed or reduced after the strongly magnetized white dwarf gets synchronized. The first serious attempt to understand the interaction between the magnetic fields of the two stars was subsequently made by Li, Wu and Wickramasinghe (1994), who studied the simple case where the magnetic dipoles of the stars are oriented perpendicular to the orbital plane but are anti-parallel. They estimated the strength of magnetic braking, once synchronization of the white dwarf has occurred, for a 2D model in which the magnetohydrostatic equilibrium is determined in the plane containing the two dipoles. They then determined the regions where wind pressure exceeds the magnetic pressure and the equilibrium breaks down. They find that the field lines from the red dwarf can connect with those of the white dwarf, or form closed loops. Thus, two dead zones are formed near the red dwarf where magnetic braking is suppressed compared to that expected from single red dwarfs. Hence the strong field AM Hers evolve mainly by emitting gravitational radiation rather than by magnetic braking. This they claim explains the presumed low luminosities in the AM Her binaries compared to the asynchronous rotators and non-magnetic CVs.

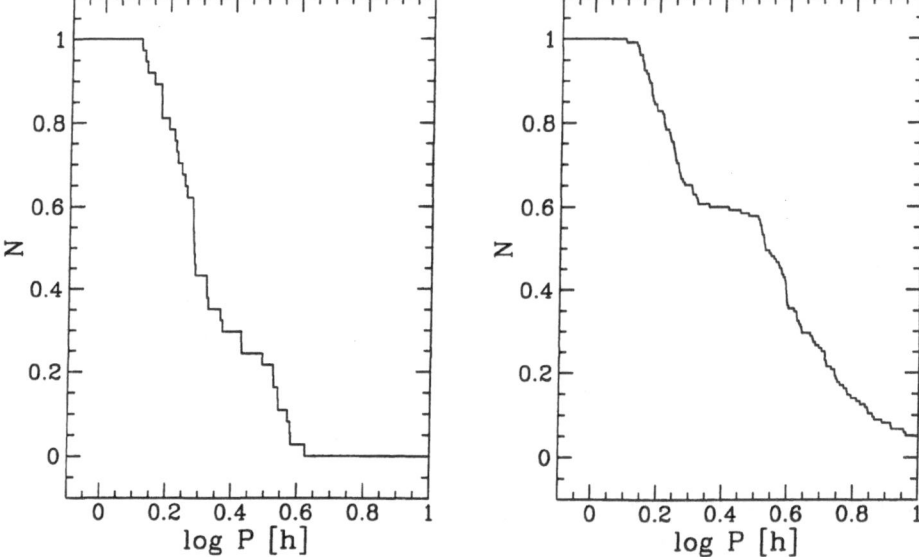

Figure 1.   Integrated period distribution of the AM Her binaries on the left and for non-magnetic CVs on the right taken from Kolb and de Kool (1993). P is the orbital period.

Furthermore the model implies that the period gap is less pronounced for the AM Hers than for CVs, and provides an explanation for their relatively high proportion in the period gap.

Kolb and de Kool (1993) have discussed the orbital period distribution of 37 AM Her binaries assuming that they evolve without magnetic braking as proposed by WW. Instead of plotting the observed period distribution $n(P)$ they consider the integrated period distribution

$$N(P) \propto \int_P^\infty n(P')dP',$$

which is independent of binning. In this case the existence of a period gap is evident for the 141 non-magnetic CVs considered whereas evidence for a period gap for the AM Her binaries is much weaker (see Fig. 1).

It was also argued by WW that magnetic braking is suppressed in synchronized detached systems, and that such systems would not be able to start mass transfer, as the time required for the red dwarf of mass M_2 to reach contact at $P_{orb} \approx 9\ M_2/M_\odot$ hr, is longer than the Hubble time. Li et al (1994), also show that magnetic braking is cut off sharply for $B \gtrsim 70$ MG for a white dwarf mass of 0.7 $M_\odot$ and $P_{orb} = 5$ hr. This explains why no high field ($B \gtrsim 70$ MG) CVs have been detected. On the other hand, King et al.

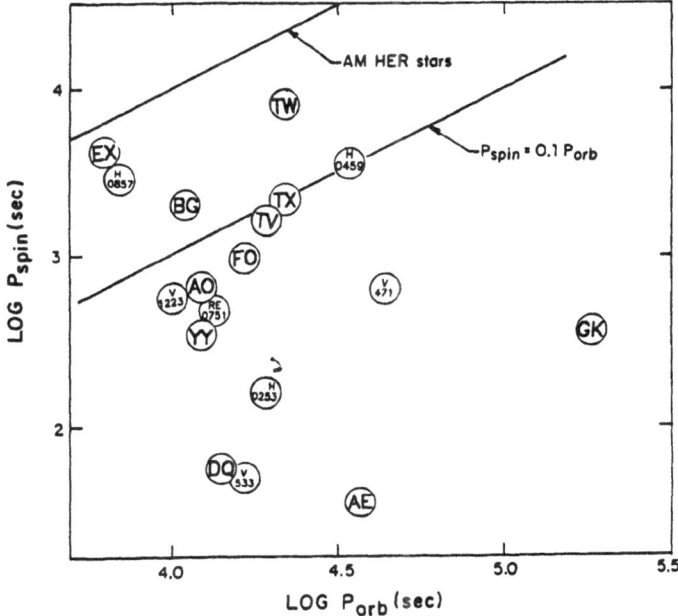

Figure 2. Plot of the spin-period vs the orbital period of magnetic CVs taken from Patterson (1994). The upper straight line corresponds to the AM Her binaries. The lower straight line corresponds to $P_{spin} = 0.1 P_{orb}$.

(1994) show that over 80 % of pre-CVs emerge from the common-envelope phase with periods $\lesssim$ 8 hr and evolve into contact within the lifetime of the Galaxy. They deduce that 80 % of AM Her binaries would be born at short enough periods for the limit given by Li et al. (1994) not to apply. If King et al. (1994) are correct then one should expect to see some CVs with $B \gtrsim 70$ MG which is not the case. On the other hand, the WW model requires the added hypothesis that AM Her binaries emerge from the common-envelope phase at long periods.

5. $P_{spin} - P_{orb}$ Relation in Asynchronous Rotators

A possible interesting correlation between the spin and orbital periods of the IPs, was first noted by Barrett, O' Donoghue and Warner (1988), who proposed that for these systems $P_{spin} \approx 0.1 P_{orb}$. Explanations for this relation were subsequently proposed by King and Lasota (1991) and Wickramasinghe, Wu and Ferrario (1991). A difficulty with proposing such a relation is that one needs to have an unbiased list of asynchronous rotators. Thus, those asynchronous rotators with $P_{spin} << P_{orb}$ were not included in the plot and only the IPs which have $P_{spin} \sim P_{orb}$ were. Recently, Patterson (1994) has presented a list of what he considers to be a carefully selected list of asynchronous rotators and then plotted all of them in a $P_{spin} - P_{orb}$ diagram. From his figure (see Fig. 2) one notes that there is no obvious correlation between P_{spin} and P_{orb} (see also Patterson and Thomas 1993), at least at present.

6. Conclusions

Magnetic CVs with $\mu \gtrsim 10^{33}$ G cm^3 are likely to synchronize at some point in their orbital evolution. Those with sufficiently high magnetic moments $\mu \gtrsim 10^{34}$ G cm^3 typically synchronize above the period gap and emit strong optical/IR polarized radiation and form the AM Her binaries. Systems with 10^{33} G cm^3 $\lesssim \mu \lesssim 10^{34}$ G cm^3 are likely to synchronize at shorter orbital periods and emit weak optical/IR polarized radiation. BG CMi was the first asynchronous rotator observed to emit weak circularly polarized radiation (Penning, Schmidt and Liebert 1986). Fits to the polarized radiation imply magnetic field $B \sim 4$ MG so that it is a possible example of such a binary (Chanmugam et al 1990). RE 0751+14 which emits stronger optical/IR linear and circular polarization is a more convincing example as it has a magnetic field $B \sim 10$ MG (Piirola, Coyne and Reiz 1993 ; Väth 1994; Väth et al. 1995). This value is larger than that of the AM Her binary 1H1752+08 with $B \approx 7$ MG (Ferrario et al 1994). While this supports the argument that the synchronous rotators were asynchronous in their past, it seems puzzling that more of them are not seen. It should be noted, however, that at longer orbital periods the evolution of such binaries is typically a factor of 10 or so faster than those at short orbital periods (Chanmugam and Ray 1984), and hence there may not be a dearth of such long period binaries considering uncertainties in observational selection effects (Patterson 1994).

Significant progress has been made by Li et al (1994) who have presented a detailed model for the resulting magnetic field configuration for the interaction of the white dwarf and red dwarf magnetic fields, in a simplified case, and which leads to reduced magnetic braking. But, more realistic field configurations for the AM Hers need to be studied in order to make the conclusions more convincing.

The absence of high field CVs with $B \gtrsim 70$ MG has not yet been satisfactorily explained by fast (WW) or slow evolution models (Liebert and Stockman 1985). We suggest instead that their absence may be either intrinsic, or because there are observational biases which prevent them from being seen.

This research was supported by NASA grant NAGW 2447. We thank U. Kolb for providing Fig. 1.

References

Barrett, P. E., O'Donoghue, D., and Warner, B. 1988, *MNRAS*, **233**, 759.

Beuermann, K. 1995, these proceedings.

Campbell, G. C. 1990, *MNRAS*, **244**, 367.

Chanmugam, G. 1992, *ARA&A*, **30**, 143.

Chanmugam, G., Dulk, G.A. 1981, *ApJ*, **244**, 569.

Chanmugam, G., Frank, J., King, A.R., & Lasota, J.-P. 1990, *ApJ*, **350**, L13.

Chanmugam, G., & Ray, A. 1984, *ApJ*, **285**, 252.

Cropper, M. 1990, *Space Sci. Rev.*, **54**, 195.

Ferrario, L., Wickramasinghe, D. T., Bailey, J., and Buckley, D. 1994, preprint.

Hameury, J. M., King, A.R., & Lasota, J.-P. 1991, *ApJ*, **316**, 275.

Joss, P. C., Katz, J. I., & Rappaport, S. A. 1979, *ApJ*, **270**, 176

King, A. R. 1994, in *Interacting Binary Stars*, ed. A. W. Shafter ASP Conference Series.

King, A. R., Frank, J., & Ritter, H. 1985,*MNRAS*, **213**, 181

King, A. R., Frank, J., & Whitehurst, R. 1990 ,*MNRAS*, **244**, 731.

King, A. R., Kolb, U., de Kool, M., & Ritter, H 1994,*MNRAS*, in press.

King, A.R., & Lasota, J.-P. 1991, *ApJ*, **378**, 674.

Kolb, U., & de Kool, M. 1993, in *Interacting Binary Stars*, ed. A. W. Shafter, A. S. P. Conf. Series, **56**, 338.

Lamb, D. Q., Melia, F. 1987. Ap. Space Sci. 131, 511.

Lamb, D. Q., Patterson, J. 1983, in *Cataclysmic Variables and Related Objects*, eds. M. Livio and G. Shaviv, Dordrecht: Reidel pp. 229.

Liebert, J., Stockman, H. S. 1985. in *Cataclysmic Variables and Low-Mass X-Ray Binaries* eds. (Dordrecht: Reidel), eds. Lamb, D. Q., and Patterson, J. pp. 151.

Li, J., Wu, K., & Wickramasinghe, D.T. 1994,*MNRAS*, **268**, 61.

Meggitt, S.M.A., & Wickramasinghe, D.T. 1982,*MNRAS*, **198**, 71.

Patterson, J. 1994, *PASP*, **106**, 209.

Patterson, J. & Thomas, G. 1993, *PASP*, **105**, 59.

Penning, W.P., Schmidt, G.D., & Liebert, J. 1986, *ApJ*, **301**, 881.

Piirola, V., Hakala, P., & Coyne, G.V. 1993, *ApJ*, **410**, L107.

Rappaport, S., Verbunt, F., Joss, P. C. 1983, *ApJ*, **275**, 713.

Ritter, H., & Kolb, U. 1995, in *X-ray Binaries*, eds. W. G. H. Lewin & van den Heuvel, Cambridge University Press, Cambridge, in press.

Sparks, W. M., Starrfield, S. G., Sion, E. M,. Shore, S. N., Chanmugam, G., and Webbink, R. F. in *Astrophysical Quantities*, eds. A. N. Cox and W. M. Sparks, AIP, New York, in press.

Schmidt, G. D., Norsworthy, J. E. 1991. *Ap. J.*, **366**, 270.

Schmidt, G. D., Stockman, H. S., and Grandi, 1986. *Ap. J.*, **300**, 804.

Spruit, H. C., Ritter, H. 1983. *A&A*,**124**,267.

Verbunt, F., Zwaan, C. 1981. *A&A*, **100**, L7

Väth, H. 1994, Ph.D. thesis, Louisiana State University

Väth, H., Chanmugam, G., & Frank, J. 1995, in preparation

Warner, B. 1985, in *Interacting Binaries*, ed P.P. Eggleton, J.E Pringle, Reidel, Dordrecht, pp 367.

Wickramasinghe, D.T.& Wu, K. 1994,*MNRAS*, **266**, L1 (WW).

Wickramasinghe, D.T., Wu, K., & Ferrario, L. 1991,*MNRAS*, **249**, 460.

Wu, K., & Wickramasinghe, D. T. 1993,*MNRAS*, **260**, 141.

Modelling MHD with Particle Methods

An Overview of SPMHD

Joe Morris

Monash University

Melbourne, Australia

email: jpm@fermat.maths.monash.edu.au

Smoothed particle magnetohydrodynamics (SPMHD) is a particle method for modelling magnetohydrodynamical fluids which has been applied to many astrophysical problems. This presentation gives a brief overview of the relevance of MHD to cataclysmic variables. The advantages of using particle methods for modelling astrophysical fluid dynamics are explored. Smoothed particle hydrodynamics (SPH) is introduced and the extensions to SPMHD are outlined. In order to demonstrate the versatility and accuracy of the method, SPMHD is then applied to problems involving wave propagation, a Rayleigh-Taylor instability and perpendicular magnetic shocks.

Magnetic fields can have a profound effect upon various processes in cataclysmic variables. Magnetic stresses could operate as the mechanism for angular momentum transport, particularly as the shear present in disks could be a natural environment for dynamo generation of magnetic fields. Hydromagnetic winds from the disk could be responsible for some degree of angular momentum transport in cataclysmic variables. Magnetic fields will funnel accreted matter onto specific areas on the white dwarf surface. Magnetic fields about a compact object may disrupt the accretion disk within some radius. The expanding envelope from a nova may be collimated and funnelled by magnetic fields.

In Smoothed particle hydrodynamics (SPH) each particle is, effectively, an interpolation point, moving with the fluid velocity. The value of a fluid property at a given point is obtained by summation interpolation over the particles. The equations of energy and motion for fluid dynamics then become ordinary differential equations which describe the evolution of the particle properties. The particles provide a Lagrangian description of the fluid, advecting contact discontinuities, preserving translational and rotational invariance, and reducing the computational diffusion of various fluid properties including momentum. It is possible to formulate the particle equations such that mass, momentum and energy are conserved exactly. It is a method which encourages insight into the physics of the fluid behavior. The method is also very robust and is easily extended to a wide range of astrophysical problems.

There are several approaches which can be taken when trying to develop a SPH-MHD method. One could develop a hybrid particle-grid method where the magnetic field is solved for on a grid (possibly using an already established and tested code) while the particles are used to evaluate the hydrodynamical forces. Another approach is to take all of the equations of MHD and reformulate SPH so as to solve them in a fashion analogous to the purely hydrodynamical case. This latter approach is used here to derive and test an implementation of smoothed particle magnetohydrodynamics (SPMHD).

A. Bianchini et al. (eds.), Cataclysmic Variables, 423.

FORMATION OF CATACLYSMIC VARIABLES BY MAGNETIC FIELDS IN COMMON-ENVELOPES

E. REGŐS AND C. A. TOUT
Institute of Astronomy
Madingley Road, Cambridge CB3 0HA, UK

A common envelope will be formed when a red-giant star in orbit with a low-mass main-sequence star fills its Roche lobe at a time when mass transfer is dynamically unstable. The degenerate core of the red giant and the relatively dense main-sequence star find themselves together inside the giant's expanded envelope. If some kind of frictional process can transfer angular momentum and energy from the orbit of these cores to the envelope the cores will spiral together and the envelope can be ejected. Depending on how much of the released orbital energy goes into driving away the envelope the result will be either a wide main-sequence plus white-dwarf binary, a pre-cataclysmic variable (that is close enough for magnetic braking to initiate mass transfer within the age of the galaxy) or a coalesced object that looks like a rapidly rotating red giant. Livio and Soker introduced a parameter α_{CE} that expresses the fraction of the released orbital energy that goes into driving off the envelope.

Differential rotation and convection, driven by both the energy loss from the orbit and nuclear burning, can drive a strong $\alpha - \Omega$ magnetic dynamo. We have found that the resulting magnetic fields by their tendency to reduce differential rotation can both transfer angular momentum from the cores to the envelope and drive stellar winds all on a timescale of about 15 years, as fast or faster than alternative dynamical friction possibilities.

We have investigated the wide range of possible initial conditions (envelope mass, initial separation, etc.) using a self consistent dynamo model and found α_{CE} to be large, about 0.7, on average. The remnant magnetic fields provide a simple explanation for the considerable range in white dwarf fields from polars to non-magnetic cataclysmic variables.

A. Bianchini et al. (eds.), Cataclysmic Variables, 424.
© *1995 Kluwer Academic Publishers.*

DISC VISCOSITY FROM A MAGNETIC DYNAMO

C. A. TOUT AND J. E. PRINGLE
Institute of Astronomy,
Madingley Road,
Cambridge
CB3 0HA
England.

We demonstrate (Tout and Pringle 1991) how differential rotation and two simple magnetic instabilities that operate in accretion discs combine to create a self-sustaining dynamo.

1) Differential rotation can easily convert radial field lines to azimuthal lines by magnetic induction.

2) Magnetic buoyancy converts field in the plane of the disc to perpendicular (vertical field).

3) The Balbus–Hawley instability occurs whenever there is a weak vertical field and $\frac{d\Omega}{dR} < 0$. A vertical field line that is perturbed azimuthally attempts to straighten and tends to slow down associated orbiting material which, consequently, falls inwards where it is stretched further and pulls back even harder. If the field is strong enough the perturbation will be stabilized but otherwise material falls in further still creating radial field lines from the vertical. It is this same process that leads to transfer of angular momentum and hence viscosity in the disc. Following these processes we find that vertical field builds up until the Balbus–Hawley instability turns off. Vertical field then decays by reconnection at the surface of the disc until it becomes unstable again. There is no stable equilibrium and indeed $B = 0$ is unstable. The α parameter in the usual viscosity prescription, $\nu = \alpha C_S^2/\Omega$, where C_S is the local sound speed and Ω the Keplerian angular velocity, is about 0.4. The time taken for fields to build up may account for colour changes in the disc of TT Ari (Tout, la Dous and Pringle 1993).

References

Tout, C. A., Pringle, J. E. and la Dous, C., 1993, *Mon. Not. R. astr. Soc.*, **265**, L5.
Tout, C. A. and Pringle, J. E., 1992, *Mon. Not. R. astr. Soc.*, **259**, 604.

A Bianchini et al (eds), Cataclysmic Variables 425

BY CAM: A MULTIPOLE MAGNETIC FIELD MODEL

PAUL A. MASON
Department of Astronomy, Case Western Reserve University,
Cleveland, OH, 44104, U.S.A.

G. CHANMUGAM
Department of Physics and Astronomy
Louisiana State University, Baton Rouge, LA, 70803, U.S.A.

I. L. ANDRONOV AND S. V. KOLESKINOV
Odessa Astronomical Observatory
Odessa University, Shevchenko Park, 270014 Odessa, Ukraine

AND

E. P. PAVLENKO AND N. M. SHAKOVSKOY
Crimean Astrophysical Observatory
P/O Nauchny, 334413, Crimea, Ukraine

We explain the complex photo-polarimetry variations seen in BY Cam in terms of a multipole magnetic field model (Wu and Wickramasinghe 1992). Based on 54 nights of photometry and 12 nights of polarimetry a new tentative rotational ephemeris is $HJD = 2446586.2559417(\pm 10) + 0.13842380E(\pm 58) + 2.89 \times 10^{-10}E(\pm 25)$. A photometric beat period of 14.11 ± 0.02 days is observed. From narrow emission-line data we derive an orbital period of $P_{orb} = 201.30 \pm 0.03$ min (c.f., Silber et al. 1992). We show that there is a satisfying consistency among the five observationally determined quantities $P_{spin} = 199.3303 \pm 0.0004$ min (JD=2446586), P_{orb}, P_{beat}, $dP_{spin}/dt = 0.132 \pm 0.04$ sec/year, and $dP_{beat}/dt = 0.0125 \pm 0.0008$ days/year. We thank L. Staufer, R. Remillard and A. Silber for new orbital period and light curve data. This research was supported by LaSpace and NASA grant NAGW 2447 to LSU.

References

Silber, A., Bradt, H. V., Ishida, M., Ohashi, T., Remillard, R. A., 1992, *ApJ*, **389**, 704.
Wu, K, Wickramasinghe, D. T., 1992, in *Vina Del Mar Workshop on Cataclysmic Variable Stars*, ed. N. Vogt, ASP conference series, **Vol 29**, p. 203.

A. Bianchini et al. (eds.), Cataclysmic Variables, 426.
© *1995 Kluwer Academic Publishers.*

7. SUPERSOFT X–RAY SOURCES AND THEIR RELATION TO CVs

THE "ZOO" OF SUPER-SOFT X-RAY SOURCES

MARINA ORIO
Dept. of Physics, University of Wisconsin,
1150 University Ave., Madison, WI 53706, USA
and Osservatorio Astronomico di Torino, I-10025
Pino Torinese (TO), Italy.

The definition of super-soft X-ray sources (SUSO's) appears with slightly different versions in the literature; it generally indicates black-body like emitters at temperatures kT<100 eV (usually 20-40 eV) and at very high luminosities, $\gtrsim 10^{36}$ ergs s^{-1} and often $\simeq 10^{38}$ ergs s^{-1} (Eddington luminosity for a $1 M_\odot$ star). These stars are therefore among the hottest in the universe. As Table 1 shows, the X-ray flux was observed to be variable in five sources, on different time scales (days to years). Therefore the sources might fall below the detection threshold of the present instruments and be hidden for part of their life. The X-ray fluxes are still not well determined because fitting a black body instead of realistic and complex atmospheric model the flux is usually overestimated.

Two were the paths leading to the discovery of these sources. The first was serendipitous discoveries in the Magellanic Clouds, were the low reddening makes the sources easier to find (e.g. Cowley et al, 1984), and the second path was the search for hot post novae in the Galaxy (e.g. Ögelman et al, 1992). Great impulse was given by the *ROSAT* PSPC, thanks to its sensitivity and moderate spectral resolution in the 0.1-0.5 KeV energy band. Classical novae emitting "super-soft" X-rays after the outburst were searched to estabilish whether the nova white dwarfs retain material after the outburst and thereby can become type Ia supernovae or undergo an accretion induced collapse. The theory predicted a "super-soft" phase lasting tens or hundreds years but up to now only three hot hydrogen burning post-novae were detected by *ROSAT* (GQ Mus, see Shanley et al. 1994, this conference), N Cyg 1992 (Krautter 1994, this conference) and the symbiotic nova RR Tel (which might be however *not* burning hydrogen, see Jordan et al, 1993). In Table 2 I include also QU Vul and PW Vul among the Galactic sources, but these were detected by *Exosat* without the neces-

429

A Bianchini et al (eds) Cataclysmic Variables 429-433

TABLE 1. Super-soft Sources in the Magellanic Clouds

Source	Optical Id.	P_{orb}	Optical Spectrum	X-ray variability	WD ?	H burning?
LMC sources						
Cal 83 (1,4,24)	V≃17.3	1.04 d.	em.lines: strong HeII (wind?), CIII CIV, OVI		likely	likely
Cal 87 (2,18,21)	V≃20.1	10.6 hrs.	em. lines: weak Balmer strong HeII	eclipse and other changes	likely	likely
RXJ0513.9-6951 (3,19,20)	B≃17 binary		em. lines: Balmer He II, OVI	outburst with F_x decreasing by factor 20	likely	likely
RXJ0527.8-6954 (6)	?			F_x in 3 y. increased by factor 3		
RXJ0550-72 (3,8)	?			variable		
RXJ0537.7-7034 (16)	?			variable timescale 1 year		
RXJ0439-68 (7)	?			constant for 14 months		
SMC sources						
N67 (25)	PN	NOT binary	PN	no	yes	yes
1E0035-72 (10,26)	B≃ 20	≥ 1.5 hrs.	em. lines: weak HeII	yes, possibly associated with P_{orb}	likely	likely
RXJ0058-71 (10)	?					
RXJ0112-72 (10)	?					
SMC 3 (10)	symbiotic		em. lines		yes	very likely

TABLE 2. Super-soft Sources in the Galaxy

Source	Optical Id.	P_{orb}	Optical Spectrum	X-ray variability	WD ?	H burning?
GQ Mus (14,15,22)	Nova	1.42 hrs.	rich nebular spectrum with strong HeII	"ON" for 9.5 yrs.	yes	yes
RR Tel (9)	Symb. Nova		em. lines		yes	?
PW Vul ? (14)	Nova	5.13 hrs.	em. lines	"ON" for few years	yes	yes
QU Vul ? (14)	Nova	2.68 hrs.	em. lines	"ON" for few years	yes	yes
N Cyg 1992 (11)	Nova	1.95 hrs.	em. lines	"ON" for 1 year	yes	yes
RX J0019+21 (8)	B = 12 blue star	15.8 hrs.?	em. lines: Balmer, HeII			
RXJ0925-48 (8)	B = 17 blue star	3.5 d.?	em. lines: H, HeII NVII CIII			
H1504+65 (13, 27)	V=16.24 PG 1159	NOT binary	absorption: OVI, CIV	no	yes	no
RXJ2117.1+3412 (12)	V≃13.2 PG 1159	NOT binary	absorption: OVI, CIV	no	yes	no

TABLE 3. Optical properties of identified sources

Source	ref	M_v	<V-R>	<B-V>	<U-B>	Δ(m)
1E0035.4-7230	17	+1.31±0.05	-0.04±0.10	-0.34±0.07	≳-0.50±0.10	≥0.2
CAL 83	4	-1.43±0.03		-0.090±0.034		≃0.2
	24	-1.84±0.1 -2.39±0.01				0.22
CAL 87	2	+0.248±0.196	+0.090±0.043	+0.075±0.043	-0.700±0.075	1.2
	18	+0.19±0.300	+0.09±0.06	+0.24±0.06		
RXJ0513.9-6951	19	-2.01		-0.16±0.04		
GQ Mus	5	+2.66		-0.3	-1.1	≃0.5

sary spectral resolution for an unambiguous interpretation (Ögelman et al 1987). The monitoring of post-novae with *ROSAT* is described in more detail by Ögelman (1994) in this conference. The number of serendipitously discovered sources is instead rapidly growing. Tables 1 and 2 list all sources defined as "super-soft" X-ray emitters by various authors and compares their basic properties. The references are indicated in the tables under the name of each source. RXJ0537.7-7034 , rejected as spurious by Hasinger, 1993, is included; new *ROSAT* data obtained by us show that its X-ray flux is variable on time scales of a year, therefore the suspected identification with a new, close-by field white dwarf seems ruled out. I would like to draw attention on 1E0035.4-7230 , a source in the SMC recently identified by us (Orio et al., 1994); it seems to bear remarkable similarities with Cal 83 and Cal 87.

Table 2 shows two unusually hot PG 1159 stars. The existence of such hot white dwarfs was not thought possible previously to their discovery. They can be better defined as "white dwarfs being made", hot degenerate cores of former AGB giants, depleted of hydrogen and in the case of H1504+65 also of helium. H1504+65 and RXJ2117+3412 were detected by *ROSAT* at effective temperatures of 170000 K and 150000 K respectively, at the lower end of the temperature distribution of super-soft sources. Also a planetary nebula core, N67, was detected as a super-soft X-ray source by *ROSAT* (most likely it is burning hydrogen in a shell). It will be very important to estabilish how many of the super-soft sources in the Magellanic Clouds belong to this class of isolated stars and how many are close binaries instead. It seems that in a variety of situations a white dwarf appears as a super-soft X-ray source: at "birth" under particular conditions, or while burning hydrogen in a shell in different types of close binary systems. Table 3 (from Orio et al. 1994), compares the optical properties of the candidates optically identified up to now. A thorough discussion can be found in the identification paper. I would like to remark that, as we see in the Table, even before detection of the orbital period the analysis of the magnitudes and colour indexes helps to distinguish between old, not recorded novae and systems with longer periods and more massive secondary stars.

References

Cowley, A.P., *et al.*, 1984, *Ap J* , **286**, 196 (1)
Cowley, A.P., *et al.*, 1990, *Ap J* , **350**, 288 (2)
Cowley, A.P., *et al.*, 1993, *Ap J* , **418**, L63 (3)
Crampton, D., *et al.*, 1987, *Ap J* , **321**, 745 (4)
Dias, M.P., Steiner, J.E., 1989, *Ap J* , **339**, L41 (5)
Greiner, J., Hasinger, G. and Kahabka, P., 1991, *A & A*, **246**, L17 (6)
Greiner, J, Hasinger, G., Thomas, H.C., 1994, *A & A*, **281**, L61 (7)

Hasinger, G., 1994, in " *Proc. of Maryland Conf., Evolution of X-Ray Binaries*", ed. S.S. Holt & C.S. Day (New York:AIP), 611 (8)

Jordan, S., Mürset, U., Werner, K., 1994, *A & A*, **283**, 475 (9)

Kahabka, P., Pietsch, W., Hasinger, G., 1993, *A & A*, **288**, 538 (10)

Krautter, J., 1994, *this conference* (11)

Motch, C., Werner, K., Pakull, M.W., 1993, *A & A*, **268**, 561 (12)

Nousek, J. A., et al., 1986, *Ap J*, **309**, 230 (13)

Ōgelman, H., Krautter, J., Beuermann, K., 1984, *A & A*, **177**, 110 (14)

Ōgelman, H., Orio, M., Krautter, J., Starrfield, S., 1993, *Nature*, **361**, 331 (15)

Orio, M., Ōgelman, H., 1993, *A & A*, **273**, L56 (16)

Orio, M., Della Valle, M., Massone, G., Ōgelman, H., 1994, *A & A*, **289**, L11 (17)

Pakull, M.W., et al., 1988, *A & A*, **203**, L27 (18)

Pakull, M.W., et al., 1993, **278**, L39 (19)

Schaeidt, S., Hasinger, G., Trümper, J., 1993, *A & A*, **270**, L9 (20)

Schmidtke, P.C., Mc Grath, T.K., Cowley, A.P., Frattare, L.M., 1993, *PASP*, **105**, 863 (21)

Shanley, L., Ōgelman H., Gallagher, J., Orio, M., 1994, *this conference* (22)

Sion, E.M., Starrfield, S., 1993, *Ap J*, **421**, 261 (23)

Smale, A.P., et al., 1988, *MNRAS*, **233**, 51 (24)

Wang, Q., 1991, *MNRAS*, **252**, 47p (25)

Wang, Q., Hamilton, T., Helfland, D.J., Wu, X., 1991, *Ap J*, **374**, 475

Wang, Q., Wu, X., 1992, *Ap J*, **78**, 391 (26)

Werner, K., 1991, *A & A*, **251**, 147 [27]

SUPERSOFT X-RAY SOURCES IN THE LMC AND SMC

P. KAHABKA

Astronomical Institute, University of Amsterdam
Kruislaan 403, NL-1098 SJ Amsterdam,
Max-Planck-Institut für extraterrestrische Physik
D-85740 Garching

1. Introduction

There are predictions from stellar evolution calculations, that binaries or components of binaries may show up peculiar during certain phases of interaction. In the case of white dwarfs (WDs) in a binary system experiencing high mass overflow rates of hydrogen rich matter of $\sim 10^{-7} M_\odot \ yr^{-1}$ (c.f. Fig.1) stationary nuclear burning was predicted to occurr on the WD surface [14], [8]. Stationary nuclear burning of helium-rich matter needs mass overflow rates of $\sim 10^{-6} M_\odot \ yr^{-1}$. These hot and luminous WDs were found to show up as bright EUV (*supersoft*) sources which may be difficult to be detected due to substantial interstellar absorption. These EUV sources have been found in the regime of ultraviolet radiation in the symbiotic systems (which count as CV like systems) and termed the *hot component* of symbiotics. Some discussion has been about their nature in terms of either nuclear burning WDs or accretion events. Observational facts were in many systems favoring the first scenario [27]. Symbiotic binaries have wide orbits in order to engulf the big giant star within its Roche lobe. It was much more difficult to detect the close binary systems due to the faintness of the optical counterpart, which is supposed to be an evolved main sequence star [44]. What tuns out in these systems to be predominant is the much brighter accretion disk. It was the unique chance of the satellite born X-ray imaging instruments (*Einstein* and *ROSAT*) to discover these EUV and soft X-ray bright sources.

435

A Bianchini et al (eds.) Cataclysmic Variables 435-442
© *1995 Kluwer Academic Publishers*

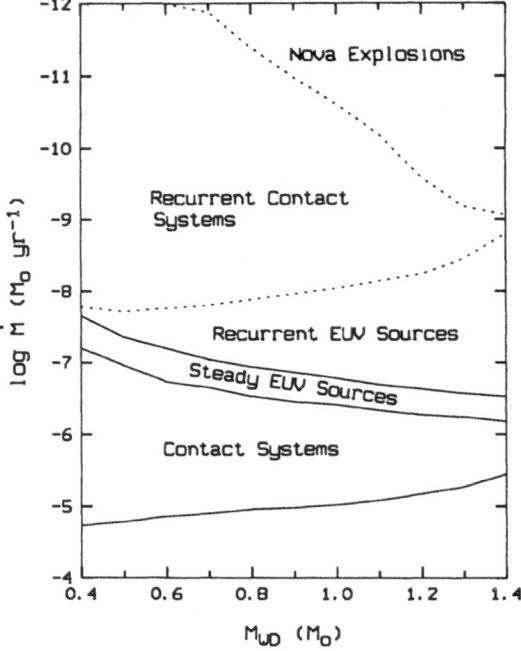

Figure 1. Characteristics of hydrogen shell burning in the $\dot{M}$-M_{WD} diagram (from [8]).

2. Observations

Starting with the *first light* observation on the LMC during the calibration phase of *ROSAT* originally dedicated to discover X-ray emission from the supernova 1987A the first EUV sources were already covered [43] including one new discovery. This led to term them *supersoft* (or *ultrasoft*). *ROSAT* performed after the calibration phase a complete all-sky Survey [47] and numerous pointed observations. This led to discover additional *supersoft* sources in the Large and Small Magellanic Cloud and to confirm the *supersofts* seen by the *Einstein* satellite. In total 11 such systems are presently known in the Clouds (c.f. Tab.1 and references given there).

One important finding was that *supersofts* are of transient or even recurrent nature (c.f. J.Greiner, these proceedings). This makes their real number at the present epoch quite uncertain.

3. The BH-NS-WD origin

A variety of explanations has been proposed in the literature starting from black holes (BHs) [41],[2] to neutron stars (NSs) [9],[23] contained in low

TABLE 1. Supersoft X-ray sources - the Magellanic Clouds sample

Name	PSPC [c/s]	kT [eV]	opt. ID	opt. magni.	Orb. Per.[d]	Remarks	Ref.[a]
LMC							
RX J0439.8-6809	1.38	18	-	B>19			1,15
RX J0513.9-6951	<0.06-2	40	HV 5682	V~17	0.43?	recurrent	2-4,15
RX J0527.8-6954	0.03-0.3	30				transient	4-8,15
Cal 83	0.98	40	+	V~17	1.04	persistent	6,9,10
Cal 87	0.12	31	+	V~19	0.44	eclipsing	9-13
RX J0550.0-7151	0.9	32	-				4,15
SMC							
1E0035.4-7230	0.38	41	+	V~20.2	0.17	variable	14,16,18
RX J0048.4-7332	0.19	20	SMC3	V~15.5		symb. nova	13,19
RX J0058.6-7146	0.025	42	-			outburst	13
1E0056.8-7154	0.33	28	N67	V~16.6		PN	14,17
RX J0122.9-7521	0.81	15	+	V~15.5			13,18

[a] Ref.:
(1) Greiner, Hasinger & Thomas 1994; (2) Schaeidt, Hasinger & Trümper 1993; (3) Pakull et al. 1993; (4) Cowley et al. 1993; (5) Trümper et al. 1991; (6) Greiner, Hasinger & Kahabka 1991; (7) Orio & Ögelman 1993; (8) Hasinger 1994; (9) Long, Helfand & Grabelski 1981; (10) Smale et al. 1988; (11) Pakull et al. 1988; (12) Schmidtke et al. 1993; (13) Kahabka, Pietsch & Hasinger 1994; (14) Wang & Wu 1992; (15) Cowley et al. 1995; (16) Orio et al. 1994; (17) Heise et al. 1994; (18) Schmidtke et al. 1994; (19) Vogel & Morgan 1994.

mass X-ray binary (LMXB) systems up to WDs [44]. The BH model was originated during the *Einstein* period and referred to Cal 83 and Cal 87 as their spectrum was claimed to be soft but not well constrained [1]. It was known from BH candidates that they show a bright and soft state [42] but the *supersofts* were later proofed to be much softer [9], [37], [19]. Cal 87 was considered to contain a BH due to mass determination arguments for the compact object [2],[38], [22]. But this can be doubted [44]. The NS or LMXB nature was claimed as the optical appearance of a *supersoft* source is quite similar to that what is known from LMXBs in terms of a bright accretion disk with emission lines due to excitation by the strong radiation field of the central compact object. The main problem was to explain the very soft characteristic of the X-ray spectrum. Either a scattering accretion disk corona [7] or a large cocoon engulfing the NS [9], [23] was invoked. But arguments were collected against the reliability of such scenarii [44], [36].

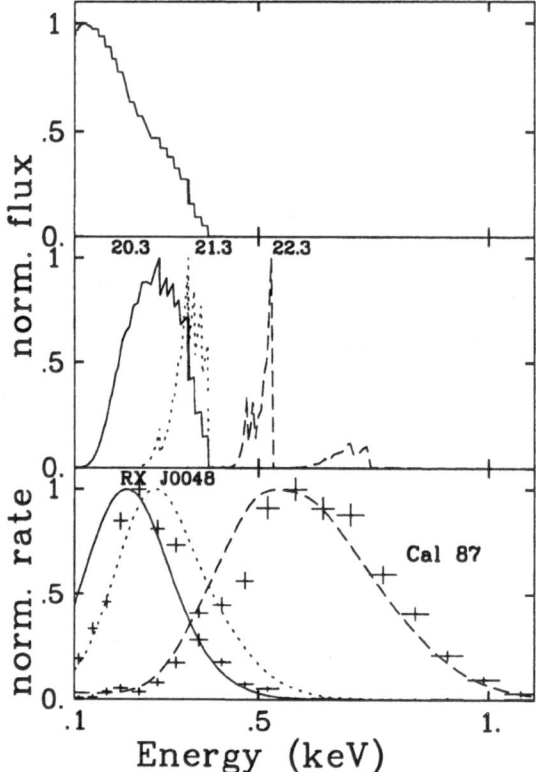

Figure 2. Formation of supersoft spectra. Upper panel: WD model atmosphere flux distribution for an effective temperature of $4 \times 10^5 K$, middle panel: same spectrum folded with three absorbing column densities ($logN_H = 20.3, 21.3, 22.3 \; atoms \; cm^{-2}$), lower panel: spectra folded with the ROSAT PSPC detector response. For comparison the observed spectra of RX J0048.4-7332 and Cal 87 are given.

The WD model got support from the observational [48], [30], [37], [10], [19] and theoretical [36], [40], [14] side and is presently the main field of study and testing [12], [45], [26]. A review of the observations is given in [11], [20], [4].

4. The Sample

6 *supersoft* sources were reported in the LMC and 5 in the SMC (c.f. Tab.1 and Fig.3). Each system is very specific. What they have in common is the low effective temperature $log(T[K]){\sim}5.4$-6.0 and the high (bolometric) luminosity $log(L[erg \; s^{-1}]){\sim}36.9$-$38$ [45], although these parameters are

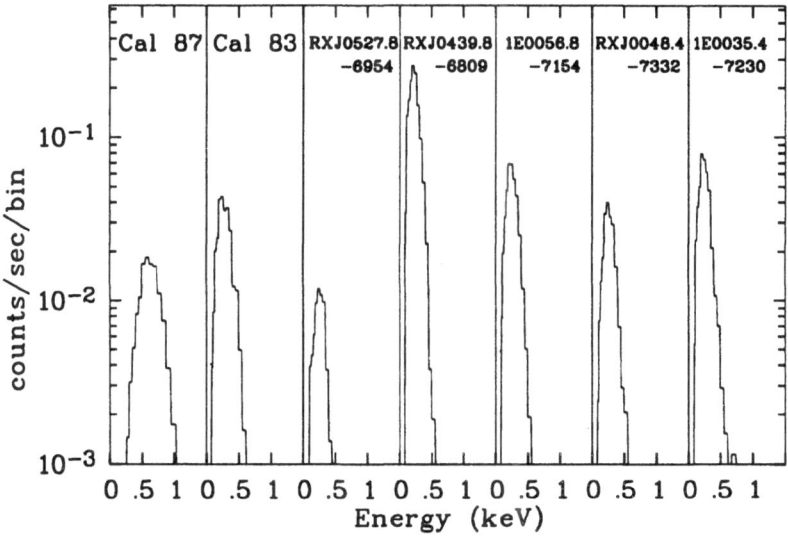

Figure 3. ROSAT PSPC spectra of the LMC & SMC supersoft sources.

dependent on the spectral model used (c.f. chapter 5). For 7 systems an optical counterpart has been found. For 4 sources binary orbital periods in the range of 0.2-1.0 days have been determined, these are besides the classical prototypes Cal 83 and Cal 87, RX J0513.9-6951 and 1E 0035.4-7230, only one system shows optical (and X-ray) eclipses, 3 are transients, 1 is a symbiotic nova and 1 coincides with the nucleus of a planetary nebula. Do we know such systems in the Galaxy ? RX J0048.4-7332 may be similar to the galactic symbiotic novae RR Tel [18] and AG Dra which have been found in X-rays as *supersoft* sources. 1E0056.8-7154 is an extremely hot nucleus of a planetary nebula [48], a similar galactic counterpart has not yet been detected. The PG1159 stars, hydrogen deficient and hot pre-WDs, may be of a similar nature. RX J2117.1+3412 is such an object and has been discovered as a *supersoft* source [28]. Cal 83, Cal 87, RX J0513.9-6951 and 1E0035.4-7230 may have galactic counterparts in RX J0925.7-4758 [29] and RX J0019+21 [35] for which the close binary nature could be established. RX J0527.8-6954 resembles a slow nova as it gradually fades out in X-rays with a timescale of a few years [11]. The *supersoft* transient RX J0058.6-7146 may be a recurrent EUV source [19] and is probably very similar to the LLGCXS 1E1339.8-2837 found in the galactic globular cluster M3 [13]. RX J0513.9-6951 was found to be a close binary system [4] and recurrent in X-rays [11]. It is much X-ray brighter than the forementioned candidates.

An interesting question is, whether there are (nearly) pure helium ac-

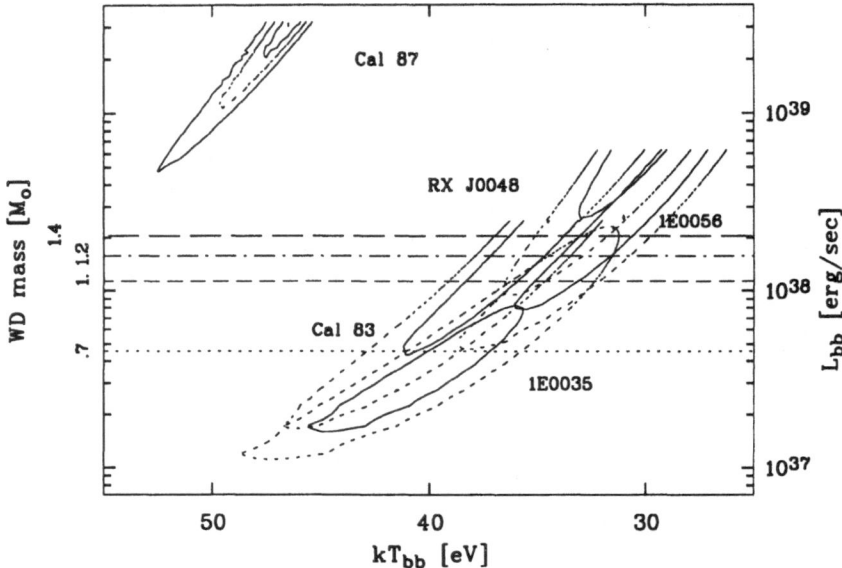

Figure 4. Hertzsprung-Russel diagram of supersoft sources. 68% and 95% parameter contours are given. The plateau luminosity [14] (maximum luminosity for which steady-state solutions exist) is given as dashed lines for different WD masses. A blackbody description of the spectral energy distribution has been used which most probably overestimates the true bolometric luminosity by a factor of ∼10 [12],[45]. Cal 87 is in the blackbody description not consistent with a nuclear burning WD. A similar diagram deduced from WD atmosphere spectral fits is presented in [45].

creting and burning systems [16], [17]. In Cal 87 [33], [2] and RX J0925.7-4758 [29] H_β is missing in the optical spectra. The recently discovered (possible) orbital period of ∼4 hours in 1E0035.4-7230 [39] would not fall into the predicted range of orbital periods for hydrogen burning binaries [36] and would favor a helium star as donor involved. *Supersoft* systems could also harbor magnetic WDs. Then *supersoft polars* (SSP) or *supersoft intermediate polars* (SSIP) could be observed. The existence of such systems may be justified by the finding, that nuclear burning may not be suppressed by magnetic fields below ∼ 10^8 G [24]. SSPs do not form accretion disks. The galactic *supersoft* source GQ Mus [30] may be a SSP as it is of the AM Her type [6].

5. X-ray spectra

The first X-ray spectra of *supersofts* were observed from Cal 83 and Cal 87 with the *Einstein* satellite. But the limited energy resolution did not allow to constrain the spectral parameters quite well [1]. The spectra also suffer from substantial absorption due to the Galaxy and the Clouds (c.f. Fig.2). A major improvement was achieved by *ROSAT* observations. [9] have applied blackbody spectral models to Cal 83 and RX J0527.8-6954 and deduced effective temperatures of 20-50 eV. The luminosities were poor constrained by the X-ray data alone but assuming an upper bound for the absorbing column about Eddington luminosities for a $1M_\odot$ compact object were deduced [9]. Such an analysis has also been applied to the other *supersofts* in the Clouds [37], [10], [19]. But there was a systematic trend that blackbody spectra gave luminosities above the Eddington value and for Cal 87 it was shown, that the luminosity was at least $3\times$Eddington for a $1M_\odot$ compact object [19] (c.f. Fig.4). This apparent problem was recognized by [12] and it was shown, that WD atmosphere model spectra give substantially lower luminosities (by a factor of$\sim$10) of $\sim 10^{37} - 10^{38}$ erg s^{-1} consistent with what is predicted in the steady nuclear burning WD model [14],[36]. Reanalysing all available spectra of *supersofts* it was shown [45], that all sources fit into the nuclear burning model. Cal 87 was shown to be consistent with a very hot nuclear burning WD [45].

6. Conclusions

Supersoft sources have been discussed as progenitors of supernovae type Ia and as systems that eventually may undergo accretion induced collapse (AIC). This topic has extensively been discussed in [25] and references therein. It is concluded, that *supersoft* systems having CO degenerates of mass 0.7-$1.2M_\odot$ will, if they grow due to accretion beyond the Chandrasekhar limit, experience carbon deflagration and henceforth a type Ia supernova event but are not a case for the AIC (initial masses in excess of $1.2M_\odot$ are required). As type Ia supernovae comprise a rather inhomogenous class [5], the progenitors may be found among the *supersofts*, recurrent novae but also among WD mergers. In late type galaxies CV like systems like the *supersofts* and recurrent novae are favoured and in early type galaxies double degenerates [5]. It is interesting to note, that *supersofts* in which the WD does not grow beyond the Chandrasekhar limit, will become double degenerates [15].

References

1. Brown T., Cordova F., Ciardullo R., et al., 1994, ApJ 422, 118
2. Cowley A.P., Schmidtke P.C., Crampton D., et al., 1990, ApJ 350, 288
3. Cowley A.P., Schmidtke P.C., Hutchings J.B., et al., 1993, ApJ 418, L63
4. Cowley A.P., Schmidtke P.C., Crampton D., et al., 1995, *Supersoft X-Ray Sources in the LMC*, in: IAU Symposium 165, Compact Stars in Binaries
5. Della Valle M., Livio M., 1994, ApJ 423, L31
6. Diaz M.P., Steiner J.E., 1989, ApJ 339, L41

7. Fabian A,C., et al., 1987, MNRAS 255, 29p
8. Fujimoto M.Y., 1982, ApJ 257, 767
9. Greiner J., Hasinger G., Kahabka P., 1991, A&A 246, L17
10. Greiner J., Hasinger G., Thomas H.-C., 1994, A&A 281, L61
11. Hasinger G., 1994, *Supersoft X-Ray Sources*, in: AIP Conference Proceedings 308,
    The Evolution of X-Ray Binaries, eds. S.S. Holt, C.S. Day, 611
12. Heise J., van Teeseling A., Kahabka P., 1994, A&A 288, L45
13. Hertz P., Grindlay J.E., Bailyn C.D., 1993, ApJ 410, L87
14. Iben I., 1982, ApJ 259, 244
15. Iben I., Tutukov A., 1984, ApJS 54, 335
16. Iben I., Tutukov A.V., 1993, ApJ 418, 343
17. Iben I. & Tutukov A.V., 1994, ApJ 431, 264
18. Jordan S., Mürset U., Werner K., 1994, A&A 283, 475
19. Kahabka P., Pietsch W., Hasinger G., 1994, A&A 288, 538
20. Kahabka P., Trümper J., 1995, *Supersoft ROSAT Sources in the Galaxies*, in: IAU
    Symposium 165, Compact Stars in Binaries
21. Long K.S., Helfand D.J., Grabelski D.A., 1981, ApJ 248, 925
22. Khruzina T.S., Cherepashchuk, 1994, Astr.Rep. 38 (3), 386
23. Kylafis N.D., Xilouris E.M., 1993, A&A 278, L43
24. Livio M., Shankar A., Truran J.W., 1988, ApJ 330, 264
25. Livio M., 1994, *On the Fate of Accreting White Dwarfs in Cataclysmic Variables
    and Related Systems*, in: Millisecond Pulsars: A Decade of Surprise
26. Meyer F., Meyer-Hofmeister E., 1994, (preprint)
27. Mikolajewska J., Kenyon S., 1992, MNRAS 256, 177
28. Motch C., Werner K., Pakull M.W., 1993, A&A 268, 561
29. Motch C., Hasinger G., Pietsch W., 1994, A&A 284, 827
30. Ögelman H., Orio M., Krautter J., 1993, Nat 361, 331
31. Orio M., Ögelman H., 1993, A&A 273, L56
32. Orio M., Della Valle M., Massone G., et al., 1994, A&A 289, L11
33. Pakull M.W., Beuermann K., van der Klis M., et al., 1988, A&A 203, L27
34. Pakull M.W., Motch C., Bianchi L., et al., 1993, A&A 278, L39
35. Reinsch K., Beuermann K., Thomas H.-C., 1993, in: Astronomische Gesellschaft,
    Abstract Series No. 9, 41
36. Rappaport S., Di Stefano R., Smith J.D., 1994, ApJ 426, 692
37. Schaeidt S., Hasinger G., Trümper J., 1993, A&A 270, L9
38. Schmidtke P.C., McGrath T.K., Cowley A.P., et al., 1993, PASP 105, 863
39. Schmidtke P.C., Cowley A.P., McGrath T.K., 1994, IAU Circ, No 6107
40. Sion E.M., Starrfield S.G., 1994, ApJ 421, 261
41. Smale A.P., Corbet R.H.D., Charles P.A., 1988, MNRAS 233, 51
42. Tanaka Y., 1994, *Black-Hole Binaries*, in: X-Ray Binaries, eds. W.H.G. Lewin, J.
    van Paradijs & E.P.J. van den Heuvel
43. Trümper J., Hasinger G., Aschenbach B., et al., 1991, Nat 349, 579
44. van den Heuvel E.P.J., Bhattacharya D., Nomoto K., et al., 1992, A&A 262,97
45. van Teeseling A., Heise J., Kahabka P., 1994, (preprint)
46. Vogel M., Morgan D.H., 1994, A&A 288, 842
47. Voges W., 1992, *The ROSAT All-Sky-Survey*, in: Proc. European ISY meeting,
    Symposium *Space Sciences with particular emphasis on High Energy Astrophysics*,
    p 223
48. Wang Q., 1991, MNRAS 252, 47
49. Wang Q., Wu X., 1992, ApJS 78, 391

VARIABILITY OF SUPERSOFT X-RAY SOURCES

J. GREINER

*Max-Planck-Institut für Extraterrestrische Physik,
85740 Garching, FRG, jcg@mpe-garching.mpg.de*

1. Introduction

Supersoft X-ray sources (SSS) are characterized by very soft X-ray radiation of high luminosity. The *ROSAT* spectra are well described by blackbody emission at a temperature of about kT≈ 25–40 eV and a luminosity close to the Eddington limit (Greiner et al. 1991, Heise et al. 1994). After the discovery observations of supersoft X-ray sources with *Einstein* , the *ROSAT* satellite has discovered more than a dozen new SSS. Most of these have been observed in nearby galaxies. These SSS are in particular: CAL 83 (Long et al. 1981, Pakull 1984, Cowley et al. 1984, Crampton et al. 1987, Smale et al. 1988, Greiner et al. 1991), CAL 87 (Long et al. 1981, Callanan et al. 1989, Cowley et al. 1990, Kahabka et al. 1994), 1E 0035.4–7230 (Jones et al. 1985, Wang & Wu 1992, Kahabka et al. 1994, Orio et al. 1994), RX J0045.4+4145 (White et al. 1994, Supper et al. 1995), RX J0048.4–7332 (Morgan 1992, Kahabka et al. 1994), RX J0058.6–7146 (Kahabka et al. 1994), 1E 0056.8–7154 (Wang & Wu 1992, Kahabka et al. 1994), RX J0122.9–7521 (Kahabka et al. 1994, Schmidtke et al. 1994) RX J0439.8–6809 (Greiner et al. 1994), RX J0513.9–6951 (Schaeidt et al. 1993), RX J0527.8–6954 (Greiner et al. 1991, Orio & Ögelmann 1993, Cowley et al. 1993), RX J0537.7–7034 (Orio & Ögelmann 1993), RX J0550.0–7151 (Cowley et al. 1993),

Galactic SSS are hard to detect due to the high interstellar absorption in the galactic plane. Besides the early discovery of the old nova GQ Muscae as SSS there are now two luminous galactic SSS known with the classical CAL 83 like properties, namely RX J0925.7–4758 (Motch et al. 1993) and RX J0019.8+2156 (Beuermann et al. 1995, Greiner & Wenzel 1995). The *ROSAT* X-ray observations and optical identifications of a few of them have established all these SSS to form a new class of accreting binary systems.

A Bianchini et al (eds) Cataclysmic Variables 443-451
© *1995 Kluwer Academic Publishers*

TABLE 1. Variable supersoft X-ray sources

Source name	Variability X-ray (X) / Optical (O)	ROSAT PSPC countrate (cts/sec)	m_B (mag)	Time scale	References
RX J0019.8+2156	O	2.0	12.6	hr—yr	1, 2, here
RX J0048.4–7332 (SMC3)	O	0.19	15.5–17	months	3, 4
RX J0058.6–7146	X	<0.001–0.7		days	3
RX J0527.8–6954	X	0.02–0.25	> 19	yr	5, here
RX J0513.9–6951 (HV 5682)	X/O?	<0.06–2.0	16.7	days	6–8, here

[1]Beuermann et al. 1995, [2]Greiner & Wenzel 1995, [3]Kahabka et al. 1994, [4]Morgan 1992, [5]Hasinger 1994, [6]Schaeidt et al. 1993, [7]Pakull et al. 1993, [8]Schaeidt et al. 1995

2. Variability among selected supersoft X-ray sources

Though SSS are established as a new class of objects the data on individual systems are still sparse, mainly because of the lack of optical identifications for nearly half of the objects. Excluding orbital variations of X-ray/optical intensity (such as eclipses in CAL 87) there are 5 systems out of the 15 SSS which show variable emission (either in X-ray and/or optical) by more than a factor 2 (see Table 1). The study of this variability is very useful because the time scales tell us about the physical processes involved.

In the following, I want to discuss three of these variable systems in more detail: 1) RX J0513.9–6951, which was discovered due to its strong X-ray variability, 2) RX J0527.8–6954, which was 10 times fainter during *Einstein* observations than during the first *ROSAT* observations, and which is declining since 1991 (a fact first noted by Orio & Ögelmann 1993), and 3) RX J0019.8+2156, a galactic SSS which shows intriguing optical variability.

2.1. X-RAY VARIABILITY OF RX J0513.9–6951

RX J0513.9–6951 was discovered during a variability study of X-ray sources near the south-ecliptic pole (Schaeidt et al. 1993). This source increased in X-ray intensity by a factor of 30 within two days. Several follow-up pointed *ROSAT* observations have been performed to study the temporal evolution of this enigmatic source. Interestingly, RX J0513.9–6951 was found to be alternately on and off on a few months timescale (Schaeidt et al. 1995). These observations suggest that the variability occurs on timescales which are short compared to the nuclear burning timescale of novae or the limit cycle length as proposed by van den Heuvel et al. (1992).

RX J0513.9–6951 was optically identified with a 16.7 mag object (HV 5682) (Pakull et al. 1993). The optical spectrum exhibits striking similarities with CAL 83. Due to photographic exposures taken within the EROS *Macho* project (Aubourg et al. 1993) there is near-simultaneous optical coverage of RX J0513.9–6951 around the times of the X-ray observations. According to these optical observations there is probably no optical brightening with the X-ray outburst unless the optical decay was less than <10 days.

2.2. X-RAY VARIABILITY OF RX J0527.8–6954

RX J0527.8–6954 was discovered in the *ROSAT* first light observations of the Large Magellanic Cloud (LMC) in June 1990, but is not yet optically identified. The X-ray spectral properties are quite similar to CAL 83 (Greiner et al. 1991). A few months later the source was scanned during the All-Sky-Survey (end of October 1990) for 21 days. Since then, several *ROSAT* pointings have been performed, partly on different targets and partly also on N132D for calibration purposes (1991–1994). Unfortunately, in many pointings RX J0527.8–6954 is affected by the support structure of the PSPC entrance window. We have corrected for this obscuration by folding the exposure map with the wobbled instrument map. However, there certainly remains a systematical error making the count rate estimate a lower limit of the actual intensity because during times of complete obscuration of the source by the support structure no photons are detected which could be corrected for. Details of this procedure and attempts for improvement will be presented elsewhere (Greiner et al. 1995).

Fig. 1 shows the lightcurve deduced from 14 *ROSAT* PSPC pointings, two *ROSAT* HRI pointings and the All-Sky-Survey between 1990 and 1994, revealing a gradual decline over the past 4 years. The total X-ray amplitude between maximum and minimum observed count rate is a factor 10. It was not detected during *Einstein* observations when it was in its field of view. Thus, RX J0527.8–6954 must have brightened up by at least a factor of 10 between *Einstein* observations and summer 1990. At the present (1994) X-ray intensity the source would have been invisible for the *Einstein* observatory.

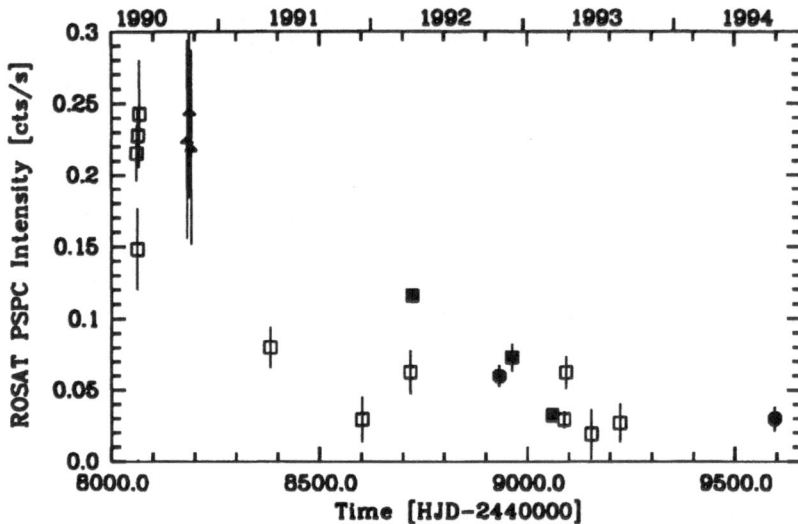

Figure 1. Steady decrease of the X-ray intensity of RX J0527.8–6954 over the last four years. The filled squares represent PSPC observations with the source at an off-axis angle less than 15′, whereas open squares represent off-axis angles between 15–40′. Due to obscuration effects in many observations by the detector entrance support structure, there is a systematical error in addition to the statistical errors plotted here. The triangles are 5 day means of the scan data during the *ROSAT* All-Sky-Survey, and the circles represent HRI pointings with the observed count rates transformed into PSPC rates (factor 7.8 was applied).

2.3. OPTICAL VARIABILITY OF RX J0019.8+2156

RX J0019.8+2156 was found during the search for galactic SSS in the *ROSAT* All-Sky-Survey data and optically identified with a V=12.2 mag object (Beuermann et al. 1995). Its optical spectrum is very similar to that of CAL 83. As the optically brightest SSS it is the best member suited for a study of the long-term optical behaviour of this class of objects.

RX J0019.8+2156 was measured on 689 Sonneberg Observatory sky patrol plates and 667 Harvard College Observatory plates (series I, Q, AC, MC, RH, DB) resulting in a temporal coverage of >100 years. In addition to the optical variability already described in Beuermann et al. (1995), namely

1. a cyclic variation of 15.8 h period with an amplitude of about 0.3 mag,
2. quasiperiodic "pulsations" of roughly 2 hours length each and a range of <0.1 mag,

two further variability components were discovered (Greiner & Wenzel 1995):

3. long-term variations, seemingly non-periodic, with timescales of up to 20 years and an amplitude of 1 mag (Fig. 2) and
4. irregular fluctuations of timescales of weeks to months and small range (Fig. 3).

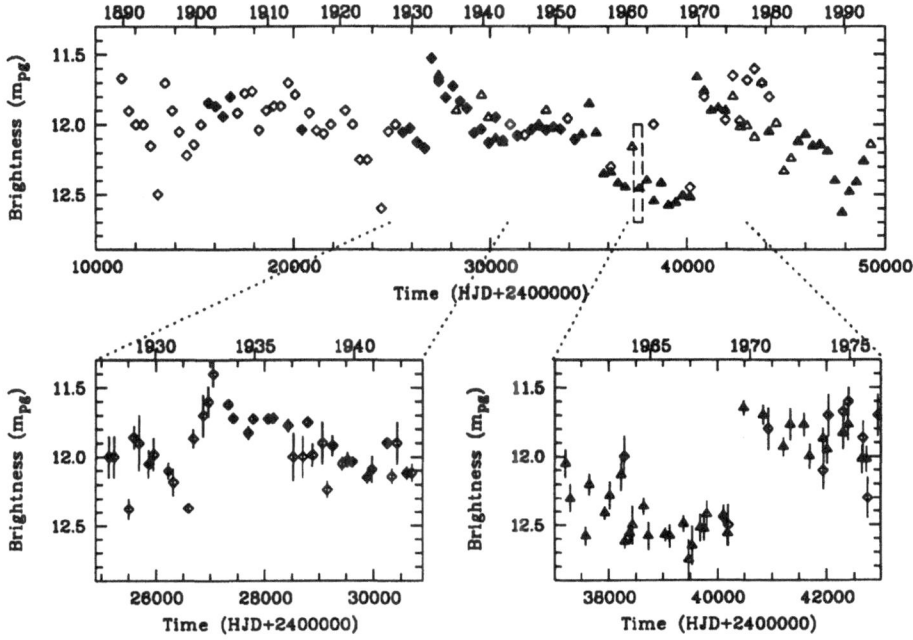

Figure 2. Optical lightcurve of RX J0019.8+2156 derived from photographic plates at Harvard and Sonneberg observatory. Each data point represents the mean of one season. Filled symbols contain more than 9 individual measurements, whereas open symbols contain 9 or less. Lozenges denote Harvard plates and triangles Sonneberg plates. The 1σ statistical error for filled symbols is less than 0.05 mag. The dashed box marks the part shown in more detail in Fig. 3. The lower panels are blow-ups of the two time intervals covering the sudden intensity jumps. Here, the data are shown in bins of three months together with 1σ error bars. (Adapted from Greiner & Wenzel 1995.)

In the best-covered sections of the Sonneberg material, component 1 can be easily traced despite of the other components. The observations present no contradictions to the period given by Beuermann et al. (1995), allow however, a substantial improvement of the lightcurve elements (ephemeris):

$$\text{Min.(hel.)} = 243\ 5799.247 + 0\overset{d}{.}6604565 \times E$$

which is valid at least for the years 1955–1993 (Greiner & Wenzel 1995).

The stability of the period over many years is most easily explained by orbital motion. The very broad minimum, and the obvious vanishing of the "pulsational" feature (above item 2) during the central parts of the minimum point towards an eclipse of a bright accretion disk. The dimension of the circular orbit (≈ 4 solar radii) fits well with the half amplitude of the observed radial velocity curve (Beuermann et al. 1995), giving ≈ 2 solar masses for the sum of the compact object and companion. This excludes any high mass component or a giant secondary star (classic symbiotic binary) and suggests the applicability of an accreting white dwarf model.

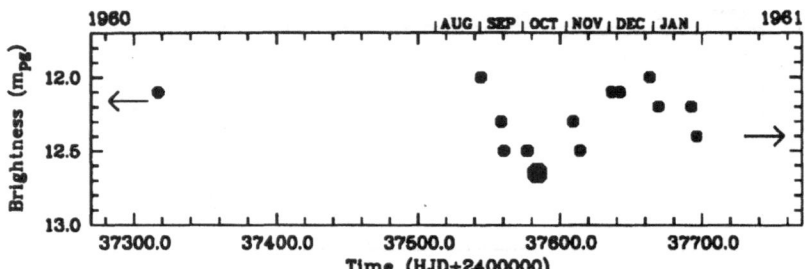

Figure 3. Typical variation on the timescale of weeks. The large symbol represents 9 magnitude determinations, the remaining are single measurements (all Sonneberg plates). Data of the phase interval 0.9–0.1 have been discarded. Arrows mark the mean of the preceding and following seasons. (Adapted from Greiner & Wenzel 1995.)

The behaviour of RX J0019.8+2156 during the last years is worth to be commented in particular: Given an X-ray intensity of of 2.1 cts/sec during the All-Sky-Survey (summer 1990), and 2.0 cts/sec in pointed *ROSAT* observations performed in 1992 and 1993 (Beuermann et al. 1995), there seemingly is no variation in X-ray luminosity over this 3 year period. Also, no temperature changes could be found. During the same time interval the mean optical intensity has increased by nearly 0.5 mag. Unfortunately, there are no exactly simultaneous optical measurements, which could provide the optical intensity during the pointed *ROSAT* observations. Given the short-term variability (timescale of weeks, see Fig. 3), no firm conclusion on the relation of X-ray and optical luminosity can be drawn so far.

3. Discussion

The most popular model involves steady nuclear burning on the surface of an accreting white dwarf (WD) (van den Heuvel et al. 1992). Hydrogen burning on the WD surface is stable for $\dot{M} \geq 10^{-7}$ $M_\odot$/yr (Paczynski & Zytkow 1978). This steady burning is processing hydrogen into helium at the rate of accretion. The burning timescale is of the order of 100–300 yrs which is considerably longer than the timescales of X-ray and optical variability observed so far in all five variable SSS.

The short-term outbursts of RX J0513.9–6951 have been ascribed to the direct response of the WD on the accretion rate (Pakull et al. 1993). Kato (1985) had shown that considerable expansion and conforming cooling of the WD surface occurs at accretion rates slightly below the Eddington rate. Thus, a slightly reduced mass transfer would rise the WD temperature into a range which is observable with *ROSAT* , and give rise to the observed rapid increase in X-ray luminosity (Pakull et al. 1993). In the context of this scenario the X-ray constant SSS would accrete at a lower rate.

TABLE 2. Timescales and luminosity amplitudes for various accretion scenarios

low accretion rate	high accretion rate	very high accretion rate	
$\dot{M} < \dot{M}_{ex}$	$\dot{M}_{ex} < \dot{M} < \dot{M}_{RG}$	$\dot{M} > \dot{M}_{RG}$	
shell flashes	stable burning	expansion of	
		envelope	WD surface
Iben 1982, Fujimoto 1982		van den Heuvel et al. 1992	Kato 1985, Pakull et al. 1993
1–100000 yrs	years	≈ 100 yr limit cycle	instantaneous
$\triangle L \approx 10^4$	$\triangle L \leq 3$	$\triangle L =$??	

The non-detection of RX J0527.8–6954 by *Einstein* and the decline to a similar intensity level during the last two years sets the variability timescale to less than ≈ 10–15 years. Due to the lack of observations inbetween the timescale could also be half or a third of that value. Extensive steady state calculations of massive white dwarfs accreting at high rates (10^{-7} $M_\odot$/yr) have shown (Iben 1982, Fujimoto 1982) that the WD masses have to be larger than 1.1 $M_\odot$ in order to achieve on/off timescales of H flashes shorter than 10 years.

If the observed 1 mag jumps in the optical intensity of RX J0019.8+2156 are interpreted in terms of shell flashes, this amplitude puts constraints on shell flash models. Typical flashes produce optical intensity amplitudes as large as 4 orders of magnitude. Such a phenomenon is certainly not occuring in SSS which show optical flux amplitudes of about a factor 2–3. Table 2 summarizes the standard picture and gives the timescales and luminosity amplitudes (last two lines) in dependence of the accretion rate.

Hydrodynamic calculations have shown (Livio et al. 1989) that the luminosity in the off state increases with increasing $\dot{M}$, thus resulting in smaller outburst amplitudes. Thus, one could imagine that (possibly varying) mass transfer near the burning rate drives the WD only into mild flashes. After having burned the rather small hydrogen layer, the system returns to instability. In order to avoid large amplitude flashes, the subsequent cooling (along the constant radius track in the HR diagram) has to be stopped by resuming accretion at a high rate. Thus, hydrogen ignition on massive WDs would not necessarily be connected with large amplitude intensity variations and the optical variability of RX J0019.8+2156 can possibly understood by temporarily reduced mass transfer which leads to long-term "oscillations" between stable and unstable H burning.

An interesting complication in the above scenario is that though no mass is ejected at accretion rates of about 10^{-7} $M_\odot$/yr the WD would expand to ≈ 80 $R_\odot$ (Livio et al. 1989) being much larger than the binary dimension. Sion & Starrfield (1994) therefore proposed to consider extremely hot but low-mass WDs (≈ 0.5–0.7 $M_\odot$) which remain at compact radii ($R_{WD} \leq 0.1$ $R_\odot$ for 10^{-8} $M_\odot$/yr). However, it remains unclear how the intensity variations and amplitudes as observed in RX J0019.8+2156 can be explained. Certainly, shell flashes in such systems are much too violent, and the deduced cycle lengths are a few thousand years (Sion & Starrfield 1994).

The above scenario has attributed the cause of the variability to changing mass transfer rates which in turn lead to variable nuclear burning rates and hence departures from a limit cycle recurrence time. Consequently, one should also ask for the possible reasons of mass transfer variations. First, if the donor star is more massive than the compact object, mass transfer is unstable on a thermal timescale. Second, if indeed a F or G type dwarf star (as suggested for CAL 83 and CAL 87) is the mass sponsoring secondary component, our hypothesis is that variable median and long-term intrinsic activity of the secondary ("solar cycle activity", depending on the size of the convective atmosphere) might give rise to a changing mass transfer, and heating of the accretion disk, and therefore to an optical variability. Spectroscopic investigation of the mass transfer and accretion rate during low and high states can clarify this point. Third, also intensity and/or spectral variations of the irradiation of the secondary can affect the mass transfer.

4. Summary

The observational properties of the above discussed three supersoft X-ray sources RX J0513.9–6951, RX J0527.8–6954 and RX J0019.8+2156 and their consequences can be summarized as follows:

RX J0513.9–6951 Recurrent X-ray outbursts, optically constant. The on/off changes occur very fast and with large amplitude, thus excluding $\dot{M}$ changes along stable burning track; no 'scaled' nova.

RX J0527.8–6954 X-ray long-term variable, optically unidentified yet. The timescale of decline is still compatible with burning luminosity changes caused by changing $\dot{M}$ (factor 10 within several years).

RX J0019.8+2156 X-ray constant over the past 3 years, optically variable. There are four different timescales of variability, and all intensity changes are faster than the burning timescale.

While the observed X-ray and optical variability are believed to be consequences of unstable mass transfer onto a WD which sporadically is burning hydrogen, some of the details of the proposed models are not in line with the observed amplitude and timescale of variability. In addition, the applicability of some of the previously applied assumptions such as accretion via a disk instead of spherical accretion, the effect of the expected strong radiation driven wind, the angular momentum loss carried by the escaping gas and tidal effects in these tight binaries might be explored more carefully.

Acknowledgements

I thank G. Hasinger and R. Schwarz for help in the *ROSAT* data analysis of RX J0527.8–6954, and I. Iben for an enlightening discussion on unstable H burning. JG is supported by the Deutsche Agentur für Raumfahrtangelegenheiten (DARA) GmbH under contract FKZ 50 OR 9201.

References

Aubourg E., Bareyre P., Brehin S., et al. 1993, The Messenger 72, 20
Beuermann K., Reinsch K., Barwig H., et al. 1995, A&A (in press)
Cowley A.P., Crampton D., Hutchings J.B., et al. 1984, ApJ 286, 196
Cowley A.P., Schmidtke P.C., Crampton D., Hutchings J.B., 1990, ApJ 350, 288
Cowley A.P., Schmidtke P.C., Hutchings J.B., et al. 1993, ApJ 418, L63
Crampton D., Cowley A.P., Hutchings J.B., et al. 1987, ApJ 321, 745
Greiner J., Hasinger G., Kahabka P., 1991, A&A, 246, L17
Greiner J., Hasinger G., Thomas H.-C., 1994, A&A 281, L61
Greiner J., Wenzel W., 1995, A&A (in press)
Greiner J. et al. 1995 (in prep.)
Heise J., van Teeseling A., Kahabka P., A&A 288, L45
Jones L.R., Pye J.P. McHardy I.M., Fairall A.P., 1985, Space Sci. Rev. 40, 693
Kahabka P., Pietsch W., Hasinger G., 1994, A&A A&A 288, 538
Kato M., 1985, PASJ 37, 19
Long K.S., Helfand D.J., Grabelsky D.A., 1981, ApJ 248, 925
Morgan D.H., 1992, MNRAS 258, 639
Motch C., Werner K., Pakull M.W., 1993, A&A 268, 561
Callanan P., Machin G., Naylor T., Charles P.A., 1989, MNRAS 241, 37p
Orio M., Ögelman H., 1993, A&A 273, L56
Orio M., Della Valle M., Massone G., Ögelman H., 1994, A&A 289, L11
Paczynski B., Zytkow A., 1978, ApJ 222, 604
Pakull M.W., 1984, IAU Symp. 108, 317
Pakull M.W., Motch C., Bianchi L., et al. 1993, A&A 278, L39
Schaeidt S., Hasinger G., Trümper J., 1993, A&A 270, L9
Schaeidt S., et al. 1995 (in prep.)
Schmidtke P.C., Cowley A.P., McGrath T.K., et al. 1994, IAU Circ. No. 6107
Smale A.P., Corbet R.H., Charles P.Q., et al. 1988, MNRAS 233, 51
Supper et al. 1995 (in prep.)
van den Heuvel E.P.J., Bhattacharya D., Nomoto K., et al. 1992, A&A 262, 97
Wang Q., Wu X., 1992, ApJS 78, 391
White N.E., Giommi P., Angelini L., Fantasia S., 1994, IAU Circ. No. 6064

ON THE NATURE OF THE SOFT X-RAY EMISSION FROM NOVA GQ MUSCAE 1983

J.W. TRURAN
Department of Astronomy and Astrophysics
University of Chicago
5640 S. Ellis Avenue
Chicago, Il 60637 USA

AND

S.A. GLASNER
Racah Institute of Physics
Hebrew University of Jerusalem
Jerusalem, Israel

Abstract. A model is proposed for the interpretation of the extended phase of soft X-ray emission observed for Nova GQ Muscae 1983. It is argued that such a phase is a forced consequence of the evolution of classical novae in outburst (MacDonald, Fujimoto, & Truran 1985). The early appearance of the soft X-ray emission (Ögelman, Krautter, & Beuermann 1987) is interpreted to be a consequence of common envelope (CE) driven mass loss, occurring over the first few hundred days of the nova outburst. The subsequent turn-off of GQ Muscae as a soft X-ray source occurred between early 1992 and late 1993. The approximately 10 year phase of soft X-ray emission thus defined is found to be consistent with the nuclear burning timescale for the exhaustion of the residual (post common envelope phase) hydrogen envelope on a 1 $M_\odot$ white dwarf, in a binary system of period 85.5 s (Diaz & Steiner 1989).

1. Introduction

Ögelman *et al.* (1993) reported the detection of Nova GQ Muscae 1983 at soft x-ray wavelengths, nine years after outburst, with the ROSAT satellite. The observed spectrum was very soft, and was determined to be consistent

453

A Bianchini et al (eds) Cataclysmic Variables, 453-461

with black body emission from a $\sim$ $M_\odot$ white dwarf, burning at a near Eddington luminosity and an effective temperature $\sim$ $3.4x10^5$ K. Such a soft X-ray signature is entirely consistent with expectations for the long term evolution of classical novae in outburst (MacDonald, Fujimoto, and Truran 1985), as it can arise as a natural consequence of the hardening of the radiation from novae, as the photospheric radius decreases during an extended phase of shell hydrogen burning at approximately constant bolometric luminosity. Nova GQ Muscae was observed to turn off as a super soft X-ray source over the next year (Krautter et al. 1995; Shanley et al. 1995). In this paper, we (1) briefly review the observational situation for Nova GQ Muscae 1983 (the prototype event of this nature?); (2) discuss factors which allow us better to understand the timescales for the onset and duration of the observed phase of soft x-ray emission, and (3) address the obvious question as to why it is that GQ Muscae is one of only a few of the recent classical novae to have been found to exhibit this behavior.

It is now generally accepted that the outbursts of classical novae are the results of thermonuclear runaways occurring on accreting white dwarfs in low mass close binary systems. The early evolution of a nova in response to runaway involves the rapid achievement of a near Eddington bolometric luminosity (a super-Eddington luminosity, for fast novae (Truran 1982)) and the expansion of the envelope to red-giant dimensions. A significant fraction of the envelope can be ejected in this early phase if, as is the case for fast novae, the expansion occurs on a dynamical timescale. Subsequently, the nova settles into a phase of shell hydrogen burning of the residual envelope matter at a (*constant bolometric*) luminosity consistent with the core mass-luminosity relation for degenerate cores (Paczynski 1971; Iben 1977). The retreat of the photospheric radius during this phase, which occurs as a direct consequence of the progressive depletion of the residual hydrogen envelope, gives rise to an increase in the effective temperature and a concomitant hardening of the radiation to UV, EUV, and ultimately soft x-ray wavelengths. The two critical issues here with regard to the interpretation of the observed behavior of GQ Muscae in outburst are those concerned with (1) the understanding of the rapid timescale on which novae appear to evolve to the onset of soft X-ray emission, and (2) the identification of the factors which define the duration of the phase of soft X-ray emission.

2. Nova GQ Muscae 1983: The Observational Situation

The variety of observations of Nova GQ Muscae in outburst provide important constraints on the characteristics of the underlying nova system and the nature of its outburst. The rate of decline t_3 $\sim$ 40 days and expansion velocity v$\sim$ 800 km s^{-1} provided by Krautter et al. (1984) suggest that Nova GQ Muscae may reasonably be classified as a moderately fast novae. The visual magnitude at maximum was compatible with shell hydrogen burning at a near Eddington luminosity on a $\sim$ 1 $M_\odot$ white dwarf, at a distance $\sim$ 4.7 kpc (Ögelman, Krautter, & Beuermann 1987). Similarly, the timescale of the early evolution of the visual light curve (for this moderately fast nova) was consistent with expectations for a nova of metal enrichment Z $\sim$ 0.40 (Hassall et al. 1990), on an $\sim$ 1 $M_\odot$ white dwarf. Spectroscopic observations of GQ Muscae in the nebular phase (Krautter & Williams 1989) revealed evidence for "photoionization from a hot radiation source with T $\sim$ 4×10^5 K whose temperature increases with time."

We are concerned specifically with the onset and duration of the period of (super) soft X-ray emission from GQ Muscae. The initial discovery of soft X-ray emission (Ögelman, Beuermann, & Krautter 1984) approximately one year into outburst was followed by EXOSAT observations over a period of $\sim$ two years, which revealed an essentially constant level of soft X-ray emission. Subsequently it was detected with ROSAT (Ögelman, Orio, Krautter, & Starrfield 1993), nine years years into the outburst, at a luminosity comparable with the Eddington luminosity for a 1 $M_\odot$ white dwarf. It turned off as a soft X-ray source between early 1992 and late 1993. The observations thus identify a phase of super soft X-ray emission that began (less than ?) approximately one year into outburst and continued for a period of eight or nine years. An interesting and perhaps distinguishing feature of this exhibited soft X-ray behavior is the fact that the period for which GQ Muscae appeared as a soft X-ray source was significantly longer than the timescale for the onset of soft X-ray emission.

3. The Hydrogen Envelope Depletion Timescale

The timescale for the outburst of a classical nova and its return to minimum is expected to be equivalent to the effective timescale for the exhaustion of the available hydrogen fuel and the associated cessation of nuclear burning. There exist several possible mechanisms of envelope mass depletion that can contribute: (1) rapid (burst) ejection accompanying the early dynamic stage of the outburst; (2) thermonuclear burning; (3) wind driven mass loss; and (4) common envelope driven mass loss. A critical challenge with respect to our understanding of nova explosions is the determination of the roles played by each of these mechanisms. As we will emphasize below, observations of the outburst of GQ Muscae 1983 may help considerably our efforts to identify the relative magnitudes of their respective contributions.

3.1. THE NUCLEAR BURNING AND WIND DEPLETION TIMESCALES

Given the condition that observed novae typically involve white dwarfs in the mass range $\sim$ 1.1-1.2 $M_\odot$, it is immediately apparent that evolution on a purely nuclear timescale (e.g. assuming no mass loss) is inconsistent with observations. The nuclear burning timescale may be written

$$\tau_{nuclear} \sim 300 \ \ \text{yrs} \ \ (\frac{M_{envl}}{10^{-4} M_\odot})$$

where we have normalized to conditions approximating those appropriate to a $M_\odot$ white dwarf. For comparison, a white dwarf of 1.25 $M_\odot$ would have, rather, $M_{envl} \sim 2 \times 10^{-5}$ $M_\odot$ and $L \sim 4.4 \times 10^4$ $L_\odot$, yielding a timescale $\tau_{nuclear} \approx 40$ yrs. We also note that burning of the entire 10^{-4} $M_\odot$ accreted envelope required to trigger thermonuclear runaway on a $M_\odot$ white dwarf would yield a total energy output $\sim 10^{48}$ erg, approximately two orders of magnitude larger than is observed for classical novae.

It thus appears that some alternative mechanism of mass loss must be effective in removing a substantial fraction of the envelope. Dynamical ejection may, for the fastest novae, be capable of effecting loss of at most perhaps fifty percent (and, more typically, less than $\sim$ ten percent) of the envelope matter. This is not sufficient, in general, to account for the shorter observed timescales for novae in outburst. Radiation pressure driven mass loss can contribute to envelope depletion on a more rapid timescale, particularly for the cases of more massive white dwarfs, for which hydrogen shell burning proceeds at higher timescales. Recent calculations of the effects of such winds by Kato & Hachisu (1989, 1994) have provided a measure of their potential contributions to nova envelope mass depletion. A crude estimate of the appropriate timescale may follow from consideration of the theory of winds from main-sequence OB stars (Abbott 1982). For matter of solar composition

$$\dot{M}_{wind} \approx 5.4 \times 10^{-7} \ (\frac{L}{10^4 \ L_\odot})$$

where Γ is the ratio of the luminosity to the Eddington luminosity

$$\Gamma = 0.26 \ (\frac{L}{10^4 \ L_\odot}) \ (\frac{M}{M_\odot})^{-1}$$

The dependence on mass is clear from these expressions. For a 1 $M_\odot$ white dwarf, $L \sim 2.9 \times 10^4 L_\odot$, $\Gamma \sim 0.75$, and we thus obtain $\dot{M}_{wind} \approx 4 \times 10^{-6} M_\odot yr^{-1}$. Alternatively, for a 1.25 $M_\odot$ white dwarf, $L \sim 4.4 \times 10^4 L_\odot$, $\Gamma \sim 0.90$, and we thus obtain $\dot{M}_{wind} \approx 2 \times 10^{-5} M_\odot yr^{-1}$. Noting that the hydrogen envelope masses for the cases of $M_\odot$ and 1.25 $M_\odot$ white dwarfs are, respectively, 1.4×10^{-4} and $2 \times 10^{-5} M_\odot$ (Truran & Livio 1986), the implied envelope wind depletion timescales are ~ 35 yrs and ~ 1 yr. It is thus clear that a relatively reliable determination of the underlying white dwarf mass is critical to the understanding of the role of winds in mass depletion.

3.2. THE COMMON ENVELOPE DEPLETION TIMESCALE

It is well known that a common envelope phase is an unavoidable stage in the development of a classical nova outburst: the photospheric radii at visual maximum ($\sim 10^{12}$ cm) are much greater than the typical orbital separations ($< R_\odot$). For such conditions, the frictional interaction between the expanded envelope on the nova white dwarf and the companion star can be expected to be important (MacDonald 1980: MacDonald, Fujimoto, & Truran 1985). Indeed, exploratory numerical studies of this common envelope phase (Livio, Shankar, Burkert, & Truran 1990; Shankar, Livio, & Truran 1991) have demonstrated that the associated interactions can have a significant impact on the evolution of a nova system in outburst. In particular, these authors have found that drag experienced in the common envelope phase can cause mass loss at high rates ($\dot{M}_{outflow} > 10^{-6} M_\odot yr^{-1}$), at velocities of order 1000-2000 km s^{-1}. We note here again that the occurrence of such common envelope driven mass loss as a major factor in the depletion of nova envelopes (and the rapid turn off-off of novae) is entirely consistent with the fact that the symbiotic novae, due to the absence of a common envelope phase, tend to return to minimum on a much longer timescale. Livio *et al.* also find that, as a consequence of the asymmetric loss of mass during the common envelope phase, a prolate morphology is predicted for nova shells. The existing observations of nova shells (e.g. for DQ Her, RR Pic, T Aur, and more recently Nova V1974 Cygni 1992) seem to be consistent with this prediction.

It would therefore appear reasonable to assume that common envelope driven mass loss is an important mechanism of envelope depletion. The magnitude of $\dot{M}_{outflow}$ is, however, somewhat uncertain. We note that over the past two years, ROSAT has recorded both the turn-on and the turn-off of soft X-ray emission from Nova V1974 Cygni 1992 (Krautter *et al.* 1995), and the turn-off of Nova GQ Muscae 1983 (the turn-on of GQ Mus was seen by EXOSAT (Ögelman, Beuermann, & Krautter 1984). If we assume that it is the common envelope mechanism that is responsible for the rapid onset of soft X-ray emission from both GQ Muscae and V1974 Cyg, then we can estimate the rates of mass loss that are necessary to explain the evolution of the light curves of these two novae in outburst. Mass estimates for the envelopes of both of these novae are in the range $\sim 10^{-4}~M_\odot$, and both evolved to temperatures consistent with soft X-ray emission (for which condition, the radii will have retreated to values comparable with the Roche dimensions and only a small fraction of the initial envelope will remain) on timescales of order one year. This implies a mass loss rate approaching $\sim 10^{-4}~M_\odot~yr^{-1}$, quite in excess of the values expected for wind mass loss alone, and supports the view that common envelope evolution must play a critical role.

4. The Evolution of Nova GQ Muscae in Outburst

In light of the considerations of possible mechanisms of envelope hydrogen depletion discussed above, we find that the observed soft x-ray emission from GQ Muscae can be understood, in a straightforward manner, in the context of the standard thermonuclear runaway model for the outbursts of classical novae. Observations to date are consistent with a white dwarf mass of order $M_\odot$. For a white dwarf of this mass, calculations of the accretion phase leading to runaway indicate that a mass $\sim 10^{-4}~M_\odot$ of accreted (envelope) matter is required to initiate runaway. For a degenerate dwarf of one $M_\odot$, the nuclear burning luminosity defining the constant bolometric luminosity phase of evolution in outburst is $\sim 3 \times 10^4~L_\odot$. The timescale for exhaustion of the (accreted) hydrogen-rich envelope matter by nuclear burning alone, as shown above, is approximately 300 years.

Such a long timescale is quite clearly inconsistent with the observations of GQ Muscae 1983, for which the visual magnitude was observed to fall as the system evolved quite rapidly to high photospheric temperatures. This rapid evolution of the nova system can most readily be attributed to a rapid depletion of the hydrogen envelope. Following our discussion of the last section, for a nova of the speed class of GQ Muscae 1983, it is unlikely that more than $\approx 50~\%$ of the envelope was ejected in the early dynamic phase. The question that remains is whether to attribute the rapid envelope depletion to winds or to common envelope effects. Expectations for wind driven mass loss accompanying burning on a $\sim M_\odot$ white dwarf also appear to fall short.

We argue that, for the case of Nova GQ Muscae 1983, the dominant mechanism of mass loss on the relevant timescale (~ 1 year) is common envelope evolution. The numerical calculations by Livio et al (1990) and Shankar et al (1991) demonstrated that such mass loss can be expected to occur on a reasonably rapid timescale. The observations indicate that this must have occurred on a timescale of order one year. Note that the Roche dimension of the GQ Muscae system, for a period of 85.5 seconds, is $\approx 0.1\ R_\odot$, for which condition the effective photospheric temperature is $\approx 200{,}000\ K$, and a significant fraction of the emission in a black body spectrum occurs at soft X-ray wavelengths. The rapid onset of the phase of soft X-ray emission is thus an expected consequence of rapid common envelope driven mass loss, which operates efficiently down to dimensions comparable to the Roche lobe.

Subsequently, the evolution of the system is expected to proceed on the timescale for the exhaustion of the hydrogen fuel by nuclear burning. This factor serves to provide perhaps the most accurate measure of the mass of the underlying white dwarf in the GQ Muscae system. Calculations by Iben (1982), for accretion rates in the range $1.5 \times 10^{-9} - 1.5 \times 10^{-8}\ M_\odot\ yr^{-1}$, indicate a timescale for the duration of the phase of soft X-ray emission of $\sim 10-15$ years, for a $M_\odot$ white dwarf, consistent both with mass estimates for white dwarf in the GQ Muscae system and with the observed timescale for soft X-ray emission from GQ Muscae. This is essentially the time required for the hydrogen burning remnant to evolve through the termination of shell hydrogen burning. During this phase of evolution at temperatures in the range $\sim 200{,}000 - 400{,}000\ K$, the contributions to envelope depletion from wind driven mass loss are less significant. The general compatibility of the nuclear burning evolution timescale calculated by Iben (1982), for the case of a $M_\odot$ white dwarf, with the observed timescale for the duration of the soft X-ray emission for GQ Muscae, would thus appear to confirm the earlier estimate of the white dwarf mass for this system.

It is interesting to consider the question as to whether a somewhat larger mass could be appropriate. We note that, for a mass of say $1.2\ M_\odot$, the rate of wind mass loss would be considerably higher, and could explain the rapidity of the evolution to the onset of the soft X-ray emission. For this mass, however, the timescale for the *duration* of the phase of soft X-ray emission - equivalently the timescale for the nuclear burning of the mass remaining within the Roche lobe - would be significantly shorter, less than of order one year. This is consistent with the observed duration of the phase of soft X-ray activity for the case of Nova V1974 Cygni 1992 (Krautter *et al.* 1995), suggesting a mass for the white dwarf in this system of order 1.2 $M_\odot$.

5. Summary and Conclusions

In the context of the thermonuclear runaway model for classical novae, and based on the considerations of the hydrogen envelope depletion timescales reviewed in this paper, the following conclusions can be drawn:

1. All classical novae may be expected to be strong EUV and soft X-ray emitters at some intermediate to late stage of their outbursts. This is a forced consequence of the fact that the evolution of the hydrogen shell burning remnants continues to temperatures exceeding $\sim 300,000 K$, for white dwarfs of masses $\sim 1 - 1.2 M_\odot$ characteristic of novae observed in outburst (Truran and Livio 1986).

2. The level of energy output can be expected to approach Eddington. This follows from the fact that the long term evolution of the hydrogen burning remnant proceeds at an approximately constant bolometric luminosity, given by the core mass-luminosity relation (Paczynski 1971; Iben 1977; Iben & Tutukov 1994), that is $\sim 75 - 90\%$ of the Eddington limit.

3. The appropriate timescales for both the *onset* and the *duration* of these late phases of EUV and soft X-ray emission are strong functions of the white dwarf mass, the accretion rate, and the effectiveness of envelope exhaustion mechanisms.

4. The observed soft X-ray emission for Nova GQ Muscae 1983 can be understood in a straightforward manner: following outburst, mass loss driven by common envelope evolution brought the system to dimensions comparable to the Roche dimensions of the system and the temperature to a value exceeding $\sim 200,000 K$ on a timescale of ~ 1 year; the subsequent evolution of this $\sim M_\odot$ white dwarf occurred on a nuclear timescale of order 10 years, consistent with the calculations by Iben (1982), during which phase Nova GQ Muscae remained a super soft X-ray source.

A significant amount of discussion of the soft X-ray emission from GQ Muscae has arisen from the fact that it would appear to be such an exception. Of the approximately 26 Galactic novae that had been observed in outburst in the 10 years prior to the ROSAT all-sky-survey (Orio *et al.* 1993), only two were found to appear as luminous super soft X-ray sources: GQ Muscae 1983 and V1974 Cygni 1992. Concern with this factor, and an awareness of the similarity between the X-ray properties of GQ Muscae and those of the supersoft LMC sources (see e.g. the discussion by van den Heuvel *et al.* 1992), led Ögelman *et al.* (1993) to consider how GQ Muscae might be understood in some other manner. While we too are concerned with this issue, we nevertheless believe that the model we have sketched for the evolution of GQ Muscae through outburst is entirely consistent with the available calculations of the hydrodynamics of classical nova explosions and the observed and inferred properties of the GQ Muscae system. We call attention to the fact that, since the mean mass of the white dwarfs for novae observed in outburst is expected to be $\sim 1.2 M_\odot$ (Truran and Livio 1986; Ritter *et al.* 1991), the period over which the evolution will proceed at temperatures in excess of $\sim 200,000 K$ and luminosities near Eddington can be less than or comparable to a year, hence many of the novae observed in the all-sky-survey might simply have been missed. We hope that future observations will be able to resolve this issue.

6. Acknowledgements

This research was supported in part by the National Science Foundation, under grant AST 93-96039 at the University of Chicago, and by the National Aeronautics and Space Administration, under grants NAG 5-2081 at the University of Chicago, and NAGW 2421 at the University of Illinois.

7. References

References

Abbot, D.C. 1982, ApJ, 259, 282
Diaz, M.P. & Steiner, J.E. 1989, ApJ, 239, L41
Diaz, M.P., & Steiner, J.E. 1990, Rev. Mex. Astron. Astrof., 21, 369
Hayes, J. 1993, Ph.D. thesis, University of Illinois
Hassall, B.J.M., et al. 1990, in *Physics of Classical Novae*, ed. A. Cassatella & R. Viotti (Berlin: Springer-Verlag), p. 202
Iben, I., Jr. 1977, ApJ, 217, 788
Iben, I., Jr. 1982, ApJ, 259, 244
Iben, I., Jr. & Tutukov, S. 1994, ApJ, 431, 264
Kato, M. & Hachisu, I. 1989, ApJ, 336, 424
Kato, M. & Hachisu, I. 1994, ApJ, 437, 802
Krautter, J., et al. 1984, A&A, 137, 307
Krautter, J., Ögelman, H., Wichmann, R., Starrfield, S., & Trümper, J. 1995, ApJ, in press
Krautter, J., & Williams, R.E. 1988, in *Coordination of Observational Projects in Astronomy*, ed. C. Jaschek and C. Sterken (Cambridge: Cambridge University Press), p. 141
Krautter, J. & Williams, R.E. 1989, ApJ, 341, 968
Livio, M., Shankar, A., Burkert, A., & Truran, J.W. 1990, ApJ, 356, 250
Livio, M. & Truran, J.W. 1994, ApJ, 425, 797
MacDonald, J., Fujimoto, M.Y., & Truran, J.W. 1985, ApJ, 294, 263
Ögelmann, H., Beuermann, K., & Krautter, J. 1984, ApJ, 287, L31
Ögelmann, H., Orio, M., Krautter, J., & Starrfield, S. 1993, *Nature*, 361, 331
Ögelmann, H., Krautter, J., & Beuermann, K. 1987, A&A, 177, 110
Orio, M., et al. 1993, Adv. Space Res., 13, 351
Paczynski, B. 1971, Acta Astr., 21, 427
Ritter, H., Politano, M.J., Livio, M., & Webbink, R.F. 1990, ApJ, 376, 177
Shankar, A., Livio, M., & Truran, J.W. 1991, ApJ, 374, 623
Shanley, L., Ögelman, H., Gallagher, J.S., Orio, M., & Krautter, J. 1995, ApJ, in press
Sparks, W.M., Starrfield, S., & Truran, J.W. 1978, ApJ, 220, 1063
Truran, J.W. 1982, in *Essays in Nuclear Astrophysics*, ed. C.A. Barnes, D.D. Clayton, & D.N. Schramm (Cambridge: Cambridge University Press), p. 467
Truran, J.W. 1990, in *Physics of Classical Novae*, ed. A. Cassatella & R. Viotti (Berlin: Springer-Verlag), p. 373
van den Heuvel, E.P.J., Bhattacharya, D., Nomoto, K., & Rappaport, S.A. 1992, A&A, 262, 97

THE VISUAL LIGHT IN SUPERSOFT X–RAY SOURCES

F. MEYER AND E. MEYER-HOFMEISTER
Max-Planck-Institut für Astrophysik,
Postfach 1523, D-85740 Garching, Germany

Abstract. We study the sources of visual light in supersoft X-ray sources. Our investigation concerns systems in which mass flows in a close binary system from a main-sequence star via an accreting disk over onto the white dwarf at a rate such that steady nuclear burning produces the supersoft X-rays (as suggested be van den Heuvel *et al.* 1992). Among the supersoft sources for which periods are known the eclipsing system CAL 87 is of special interest.

1. Introduction

Luminous X–ray sources with very soft spectra had already been found with the IPC detector of the EINSTEIN X–ray observatory. The ROSAT observations now led to the discovery of many new sources of this kind, and a separate class, the super–soft sources (SSS), has been established. The SSS are luminous X–ray objects with very soft spectra (more than 90% of the observed photons below 0.5 keV). This corresponds to a black-body temperature of $10^5 - 10^6$K. The bolometric luminosity is close to the Eddington limit for $1M_\odot$, $\approx 10^{38}$erg/s. About 25 to 30 objects have been found up to now, in our galaxy as well as in the Magellanic Clouds and the Andromeda Nebula.

Such luminosities and temperatures correspond to blackbody surfaces of typical white dwarf dimension, $r \approx 10^9$cm. The very soft spectra might originate from different kinds of objects where we might expect a hot and possibly accreting white dwarf as in novae, planetary nebulae, symbiotic binaries but possibly even from neutron stars surrounded by an appropriate cocoon of gas (for recent reviews see Hasinger (1994), Kahabka *et al.* (1994)). ¿From the different behaviour of the observed SSS (variability, on–

463

A Bianchini et al (eds) Cataclysmic Variables 463-468

and off–states) one might conclude that they do not form one unique class of objects.

Optical counterparts of several SSS were found, mainly blue objects. It was argued that the irradiated accretion disk around the compact object would be the dominant optical light source (van den Heuvel *et al.* 1992, later referred to as HBNR 1992). The aim of our investigation is to clarify how much visual light can originate from the disk and the secondary star. For three systems orbital periods are known: CAL 83, CAL 87 and RX J0019±21. The periods give constraints on the geometry of the binary system. Further important information comes from the eclipse light curve of CAL 87. Based on the model of van den Heuvel *et al.* (HBNR 1992) for steady nuclear burning on accreting white dwarfs we have a narrow range of mass accretion rates, white dwarf masses and secondary star masses.

2. Contributions to the visual light

In such a binary system we expect visual light from the secondary star and the accretion disk, both irradiated by the hot white dwarf. The latter contributes only a very small part due to the large bolometric corrections.

The contribution from the secondary star changes with the photometric phase as we successively see more of the irradiated side of the unheated backside. For the comparison with the eclipse curve of CAL 87 we determined the visual light from of a 1.5 $M_\odot$ main sequence star, which is irradiated by a central light source of 10^{38}erg/s at a distance according to the orbital period. We have computed how much light we get from such a star at each photometric phase, seen under different aspect angles. For this one computes the effective temperature of each surface element resulting from irradiation plus the star's own luminosity and determines its contribution to the visual light as seen from the observers direction. We used bolometric corrections of Allen (1973). For high inclinations we find a flat bottom when we look towards the unheated side. This is due to the fact that the point source is close to the secondary ($R_{2,R}/a = 0.44$ for the geometry of CAL 87, following from orbital period) and less than a hemisphere is irradiated. Maximum light corresponds to that photometric phase where we look towards the irradiated side.

The observed light curve of CAL 87 (Schmidtke *et al.* 1993) shows a deep narrow primary minimum when we look towards the unheated side of the secondary. Since this narrow minimum cannot be related to the change of light from the secondary it must be connected with the occultation of the disk by the secondary star and this disk needs to be bright. We show in the following how we determined the disk luminosity.

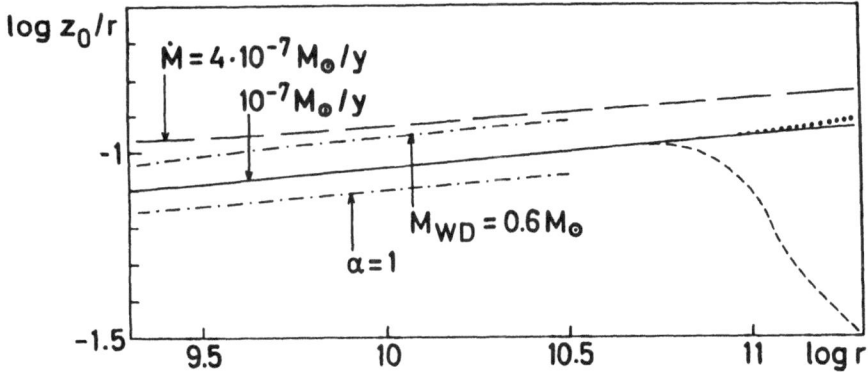

Figure 1. Shape of the irradiated disk: z_0 height above midplane, r distance from central object in cm. Parameters taken: mass of white dwarf $1 M_\odot$, luminosity 10^{38} erg/s, $\alpha = 0.3$ (except where indicated otherwise). $\dot{M}$ mass flow rate. Small dashes: without irradiation. Dots: white dwarf luminosity $2 \cdot 10^{38}$ erg/s

3. Properties of the accretion disk, irradiated by a point source (steady burning white dwarf)

3.1. SHAPE OF THE DISK

The radiation of the central source heats the surface of the accretion disk. The resulting effective temperature T_e is determined by both, frictional heating, $T_{e(0)}$, and irradiation, $T_{e(\text{irr})}$:

$$T_e^4 = T_{e(0)}^4 + T_{e(\text{irr})}^4 \quad . \tag{1}$$

The irradiation term depends on the shape of the disk (z_0 height of photosphere above midplane, r distance from central object)

$$T_{e(\text{irr})}^4 = \frac{\eta L}{4\pi r^2 \sigma} \left(\frac{dz_0}{dr} - \frac{z_0}{r} \right) = \frac{\eta L}{4\pi r^2 \sigma} \left(\frac{z_0}{r} \frac{d\ln z_0/r}{d\ln r} \right) \tag{2}$$

with L luminosity of the central object, σ Stefan–Boltzmann constant. Terms of order $\left(\frac{z_0}{r} \right)^2$, $(dz_0/dr)^2$ have been neglected compared to 1. This approximates the central source by a point source. It is uncertain which part of radiation intercepted by the disk is reprocessed to thermal radiation. We introduce an efficiency parameter η.

We show in Figure 1 how z_0/r varies with r for different parameters M (mass of white dwarf), $\dot{M}$ (mass overflow rate) and viscosity paramter α

(Shakura and Sunyaev 1973). Approximation by a power law gives

$$z_0/r = 10^{-0.95} r_{11}^{0.093} \left(\frac{M}{M_\odot} \right)^{-0.38} \dot{M}_{-7}^{0.17} \alpha_{0.3}^{-0.12} \quad , \tag{3}$$

where $r_{11} = r/10^{11}$, $\dot{M}_{-7} = \dot{M}/(10^{-7} M_\odot/y)$, $\alpha_{0.3} = \alpha/0.3$.
This leads to

$$T_{e(\mathrm{irr})} = \eta^{0.25} \, 10^{4.3} r_{11}^{-0.47} \left(\frac{M}{M_\odot} \right)^{-0.10} \alpha_{0.3}^{-0.03} \dot{M}_{-7}^{0.04} L_{38}^{0.25} \tag{4}$$

with $L_{38} = L/10^{38} \mathrm{erg/s}$. Relation (4) shows the weak dependence on α and $\dot{M}$.

3.2. EFFECTIVE TEMPERATURE

We show in Figure 2 temperatures of a disk with mass flow rate $\dot{M} = 10^{-7} M_\odot/y$ ($4 \cdot 10^{-7} M_\odot/y$) and the effect of irradiation taken as described in formulae (4). For all computations we took $M = M_\odot$ and $\alpha = 0.3$ and maximal efficiency $\eta = 1$.

3.3. LUMINOSITY AND MAGNITUDES

In Figure 3 we give the bolometric luminosity L_d integrated from an inner disk edge at $r = 10^9 \mathrm{cm}$ outward to the disk radius r_d. L_d is the luminosity of both sides of the disk together. A star with the same luminosity has absolute bolometric magnitude M_{bol} given on the right scale. When we apply bolometric corrections $BC(T_{\mathrm{eff}})$ (Allen, 1973) to the individual contributions before integrating we obtain the absolute visual magnitude M_V as shown by the lines labelled "visual". Due to the fact that the corrections are hign in the hot inner disk the visual contribution increses more steeply with r than the bolometric one.

The absolute magnitude of a disk depends on the viewing angle, and equals that of a star of the same luminosity if seen under $i = 60°$. This is easily understood by comparing the projected "stellar disk" of $\frac{1}{4}$ the stellar light with the projected halfside of an inclined disk of $\frac{1}{2} \cos i$ of the disk light.

4. Results

If we compute the theoretical light curve for CAL 87 from contributions of irradiated secondary star and irradiated accretion disk including the occultation of the disk by the star we find that we have two discrepancies: (1) the primary minimum is not deep enough (which means the unocculted

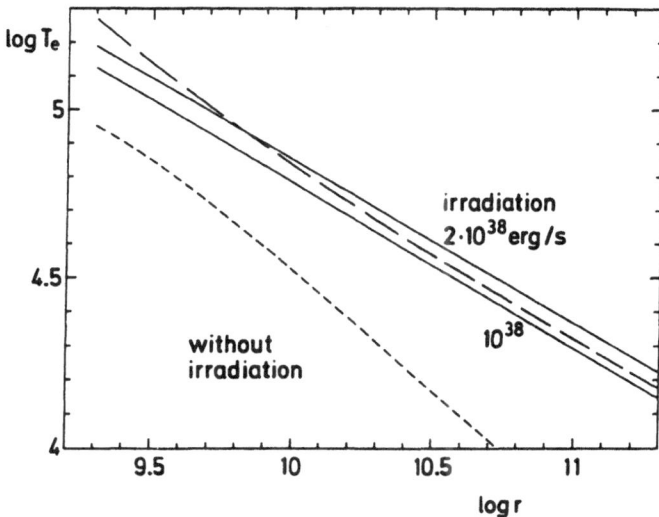

Figure 2. T_e effective temperature of the disk at distance r in cm, $\dot{M} = 10^{-7} M_\odot/\mathrm{y}$ except long dashes $\dot{M} = 4 \cdot 10^{-7} M_\odot/\mathrm{y}$. $\eta = 1$

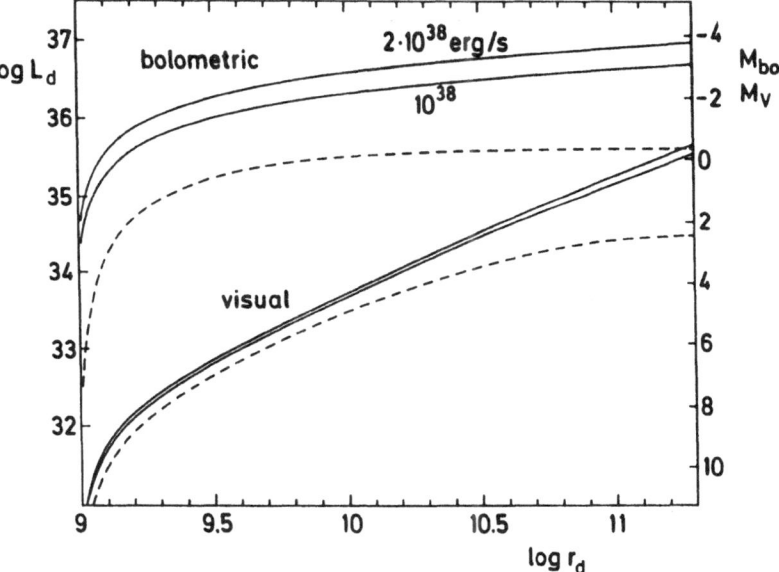

Figure 3. Bolometric luminosity L_d and absolute magnitudes M_{bol}, M_V (for $i = 60°$) for a disk extended from inner edge at 10^9cm outward to r_d. Solid lines: with irradiation ($\eta = 1$), dashed lines: without irradiation.

disk is not bright enough) and (2) the theoretical eclipse curve is symmetric around the minimum due to the azimuthally symmetric disk. Both problems might be solved by allowing for a bright disk rim connected with the hot spot and therefore not azimuthally symmetric.

5. References

Allen C.W., 1973, *Astrophysical Quantities*, 3rd edition, Univ. London, The Athlone Press, p.205

Hasinger G., 1994, Reviews in Modern Astronomy, Astron. Gesellschaft

Kahabka P., Pietsch W., Hasinger G., 1994, *A&A*, **288**, 538

Reinsch K., Beuermann K., Thomas H.C., 1993, *Astron. Ges. Abstract Series* Vol. 9, p.41

Schmidtke P.C., McGrath T.K., Cowley A.P., Frattare L.M., 1993, *PASP* **105**, 863

Shakura N.I., Sunyaev R.A., 1973, *A&A* **24**, 337

van den Heuvel E.P.J., Bhattacharya D., Nomoto K., Rappaport S.A., 1992, *A&A* **262**, 97 (HBNR 1992)

ROSAT OBSERVATION OF THE OLD CLASSICAL NOVA CP PUPPIS

SÖLEN BALMAN[1], MARINA ORIO[1,2], HAKKI ÖGELMAN[1]
1 *Dept. of Physics, University of Wisconsin,*
1150 University Ave., Madison, WI 53706, USA
2 *Osservatorio Astronomico di Torino, I-10025*
Pino Torinese (TO), Italy.

CP PUPPIS is an old classical nova, that had an outburst in 1942 as one of the fastest and brightest novae ever recorded (Δ m=17 mag, t_3 = 6.5 days ; $V_{ej} \geq$ 1000 km/sec, Payne-Gaposhkin 1957). We observed it in X-rays with the ROSAT Position Sensitive Proportional Counter in April 1993. The source was detected with a count rate of 0.067 ± 0.004 c/s. The total X-ray flux between 0.1 and 2.4 keV was $\sim 2.88 \times 10^{-12}$ ergs s^{-1} cm^{-2}. The spectrum is best fitted by a thermal bremsstrahlung emission model with temperature ≥ 1 keV. Temperatures as high as 30 keV are allowed within 2σ errors. The X-ray count rates are modulated with the $\simeq$1.5 hrs. spectroscopic and photometric orbital periods (Bianchini et al. 1985; Warner 1985; O'Donoghue et al. 1989; Diaz & Steiner 1991, White et al. 1993). The significance of the modulations at the spectroscopic period detected by White et al., (1993) (P $\sim$ 5295 s) was tested by a phase dispersion method and found to be at the 3σ confidence level. Considering the variation of X-ray flux with the orbital period and the possibility of high plasma temperatures in the emitting medium, we favor an intermediate polar interpretation for this system.

References

Bianchini, A., Friedjung, M., and Sabbadin, F., 1985, Inf. Bull. Var. Stars No. 2650
Diaz, M.P.,and Steiner, J.E., 1991, PASP, 103, 964
O'Donoghue, D., and Warner, B., Wargau, W., Grauer, A.D., 1989, MNRAS, 240, 41
Payne-Gaposchkin, C. 1957, The Galactic Novae (Amsterdam: North-Holland)
Warner, B., 1985, MNRAS, 217, 1P
White, J.C., Honeycutt, K.R., Horne, K., 1993, ApJ, 412, 278

A Bianchini et al (eds), Cataclysmic Variables, 469

VARIATIONS OF THE SOFT X-RAY SPIN AMPLITUDE IN TV COL

P. BARRETT AND E. SCHLEGEL
Universities Space Research Association &
NASA/Goddard SFC, Greenbelt, MD 20771, USA

Over 27 ks of ROSAT PSPC observations were made during two periods in late 1992 and early 1993. Contemporaneous optical observations were made using the UCT photometer and polarimeter module on the 1.9 m telescope at SAAO during late Nov. 1992. The X-ray data was divided into 4 phases for each of the three photometric periods and three energy ranges (soft: channels 20–50, medium: channels 51–101, and hard: channels 102–201). We have nearly complete phase coverage for the orbital and 5.2-hr photometric periods, but incomplete coverage for the 4-day period. The result of folding each 4-day phase period on the 1911 s spin period shows that the amplitude of the spin period varies with the phase of the three photometric periods. In addition changes in the amplitude of the spin pulse are more noticeable in the soft energy band that in the medium or hard energy bands, which is indicative of photo-electric absorption.

Hellier et al. (1993) has reanalyzed the EXOSAT data for TV Col and several other IPs. By folding the CMA (0.05–2.0 keV) and ME (1.4–4.5 and 4.5–8.5 keV) data on the orbital period, they find that there is preferentially a minimum in the light curve at phase 0.8 and that the energy dependence is characteristic of photo-electric absorption. An eclipse at this orbital phase may be due to obscuration of the white dwarf by the accretion flow or hot spot). Our ROSAT observations support their result and extend it to the other periods in the system, most importantly to the 4-day period.

Barrett et al. (1988) have shown that among several different models the most plausible is a precessing accretion disk. The period of such a disk for a typical CV is about 4 days, lending strong support that precession is the correct mechanism. The current dispute is whether the precession is due to a tilted accretion disk as proposed for Her X-1 or to an elliptical disk as proposed for those CVs showing superhumps. Further analysis is required before any definite conclusion can be made.

470

A. Bianchini et al. (eds.), Cataclysmic Variables, 470.
© *1995 Kluwer Academic Publishers.*

THE GINGA HARD X-RAY SPECTRUM OF AM HERCULIS

A.P. BEARDMORE

Dept. of Physics, The Open University, Walton Hall, Milton Keynes MK7 6AA, UK

C. DONE AND J. P. OSBORNE

X-ray Astronomy Group, Dept. of Physics & Astronomy, University of Leicester, Leicester LE1 7RH, UK

AND

M. ISHIDA

Institute of Space and Astronautical Science 3-1-1 Yoshinodai, Sagamihara, Kanagawa 229, Japan

We present an analysis of the phase-resolved 2 − 30 keV X-ray spectrum of AM Herculis obtained with the *Ginga* satellite. The bremsstrahlung flux varies by more than a factor of 8 as a function of orbital phase demonstrating that the X-ray orbital intensity variation is due to partial occultation which varies with the viewing angle. The spectrum is hardest when the source is brightest in its orbital cycle, and the phase resolved spectra are not well fit by simple models with a narrow line plus continuum. The derived high and variable bremsstrahlung temperature cannot account for the observed line emission, and the residuals to these fits indicate complex behaviour at high and low energies. The latter is shown to be consistent with a complex absorber, and either partial covering or partial ionisation both give a good description of the low energy spectrum. The residuals above 6 keV are well modelled by reflection of the incident bremsstrahlung flux from the white dwarf surface, where the amount of reflection varies with phase as predicted by the changing inclination of the white dwarf surface. The inclusion of this hard and variable spectral component gives a temperature for the post shock region of $\sim$ 13.5 keV which is constant with phase. This value is considerably lower than previous estimates, allowing the high equivalent width of the iron line to be explained as a combination of a 6.8 keV thermal blend and 6.4 keV fluorescent components. This new low bremsstrahlung temperature suggests that the hard X-ray luminosities of AM Her systems may have been over estimated, exacerbating the soft X-ray problem. The detailed modelling of the complex low energy spectrum affects the derived ionisation state of the reflector: with partial covering of cold material the reflection spectrum is significantly ionised, but with an ionised absorber the reflecting surface is cold. Our results clearly predict that both a 6.4 keV and 6.8 keV iron line blend should be observed in X-ray observations of AM Her made with instrumentation of greater energy resolution in the 2 − 10 keV range (eg ASCA and JET-X). Also, together with ROSAT data, we should be able to determine the relative importance of the cold and ionised absorption.

471

A. Bianchini et al. (eds.), Cataclysmic Variables, 471.

THE GALACTIC SUPERSOFT X-RAY SOURCE RXJ0019+21

K. REINSCH AND K. BEUERMANN
Universitäts-Sternwarte Göttingen
Geismarlandstr. 11, D-37083 Göttingen, Germany
(reinsch@usw050.dnet.gwdg.de)

We report the discovery (Beuermann *et al.* 1995) of a galactic supersoft X-ray source, RX J0019.8+2156, which we have identified with a 12.2 mag blue star. The system shows quasi-sinusoidal variations of the optical, UV, and X-ray fluxes and systematic radial velocity variations with a period of 15.85 hours which reveals the binary nature of the source. The optical/UV continuum is characteristic of a bright accretion disk and a heated secondary with strong He II and Balmer emission lines superimposed. The Balmer lines display time-variable P Cygni type profiles indicative of a fast wind. The source is slightly reddened with $E_{B-V} \simeq 0.1$, consistent with the total galactic neutral hydrogen column density at the position of RX J0019.8+2156, $N_H = 4 \times 10^{20}$ cm^{-2}. This suggests that the object is located outside the galactic hydrogen layer. Lack of any spectral signatures of the expected F-type secondary it is likely that the distance is $d > 1.5$ kpc. At an adopted distance of $d = 2$ kpc, the system has $M_V = +0.4$ and $L_X \simeq 6 \times 10^{36}$ ergs/s. The hot component probably is a white dwarf which accretes at a rate of $\sim 10^{-7} M_\odot$yr^{-1} near the limit for stable hydrogen burning. The X-ray luminosity and temperature of RX J0019.8+2156 seem to be lower compared to other supersoft sources suggesting that the system currently is in a low state between weak shell flashes. This picture is supported by the optical long-term behaviour of RX J0019.8+2156 (Greiner and Wenzel 1995).

References

Beuermann, K., Reinsch K., Barwig, H., Burwitz V., de Martino, D., Mantel, K.-H., Pakull, M.W., Robinson, E.L., Schwope A.D., Thomas H.-C., Trümper, J., van Teeseling, A., and Zhang, E. (1995) Discovery of a galactic supersoft binary X-ray source, *Astron. Astrophys.* submitted

Greiner, J. and Wenzel, W. (1995) Optical long-term variability of the supersoft X-ray source RX J0019.8+2156, *Astron. Astrophys.* in press

A. Bianchini et al. (eds.), Cataclysmic Variables, 472.

THE SOFT X-RAY TURN-OFF OF NOVA MUSCAE 1983

L. SHANLEY AND J. S. GALLAGHER
Dept. of Astronomy, Univ. of Wisconsin-Madision
475 N. Charter St., Madison, WI 53706-1582

AND

H. ÖGELMAN AND M. ORIO
Dept. of Physics, Univ. of Wisconsin-Madison
1150 University Ave., Madison, WI 53706-1582

Nova GQ Muscae 1983 was unique among recent novae in maintaining nearly constant bolometric luminosity for nearly a decade after outburst. During most of this time GQ Mus maintained a high temperature and was observed as a "super-soft" x-ray source (e.g., Ögelman et al. 1993). ROSAT observations in 1992 showed that the super-soft source was still present, and thus GQ Mus offered the opportunity to investigate a classical post-nova where the theoretically anticipated, multiple-year nuclear burning phase appears to have occurred. To this end, follow up observations were obtained. ROSAT PSPC observations taken in January and September 1993 were complimented with B band photometry taken in January 1993. By January 1993, the X-ray count rate had declined by a factor of 17, yet no appreciable decrease in the optical magnitude nor change in the amplitute of modulation was evident. By September 1993 GQ Mus was no longer detected by ROSAT, requiring a decrease in soft X-ray flux by a factor of ≥ 30. This decline may be attributed to a nuclear burning shutdown resulting from the complete consumption of the residual hydrogen rich envelope. However, observations indicate that the optical luminosity is not simply coupled to the X-ray luminosity (e.g. through reprocessing).

References

Ögelman, H., Orio, M., Krautter, J., and Starrfield, S. G. (1993) *Nature*, 361, 331.
Shanley, L., Ögelman, H., Gallagher, J. S., Orio, M, and Krautter, J. (1995) The Soft X-ray Turn-off of Nova Muscae 1983, *Astrophysical Journal Letters*, in press.

A Bianchini et al (eds) Cataclysmic Variables, 473
© *1995 Kluwer Academic Publishers*

X-RAY AND OPTICAL OBSERVATIONS OF TWO NEW
AM HERCULIS SYSTEMS DISCOVERED BY ROSAT

V. BURWITZ AND K. REINSCH
Universitäts-Sternwarte,
Geismarlandstr. 11, D-37083 Göttingen
(burwitz@usw051.dnet.gwdg.de)

Our optical identification program of bright soft high-galactic latitude sources from the ROSAT all-sky survey ($|b| > 20, > 0.2$ PSPC cts/s, HR1 < 0) led to the discovery of 23 new CV's most of these are AM Her stars. We present the results of detailed X-ray and optical follow-up studies of two ROSAT discovered polars: the V~17 mag system RX J0531.5–4624 (Reinsch *et al.* 1994) and the V~19 mag system RX J0453.4–4213 (Burwitz *et al.* 1995). Optical spectroscopy reveals that the orbital period of RXJ0453 is either 102 min or 94 min. For RXJ0531 we find an orbital period of 133.5 min placing it well within the 'period gap' of cataclysmic variables. Both objects display cyclotron emission lines from which their magnetic field strengths in the accretion region are determined as B~19 MG for RXJ0531 and B~36 MG for RXJ0453. In the case of RXJ0453 the analysis of the phase dependent movement of the cyclotron emission line maxima contrains the accretion geometry. It gives an angle ~85° for the sum of the orbital inclination i and the co-latitude of the accretion region β. The X-ray lightcurve in RXJ0531 is ~ 80% modulated and has two maxima per orbital cycle whereas in RXJ0453 it is dominated by flaring.

References

Burwitz, V., Reinsch, K., Schwope, A.D., Beuermann K., Thomas, H.-C., Greiner, J., (1995) RX J0453.4-4213: a new ROSAT discovered polar with strong cyclotron emission, *Astron. Astrophys.* submitted

Reinsch, K., Burwitz, V., Beuermann, K., Schwope, A.D., Thomas, H.-C. (1994) RXJ0531.5-4624: a new ROSAT-discovered AM Herculis binary in the 'period gap', *Astron. Astrophys.* L27-L30

A. Bianchini et al. (eds.), Cataclysmic Variables, 474.

ROSAT OBSERVATIONS OF 3 CATACLYSMIC VARIABLES

ERIC M. SCHLEGEL AND K. MUKAI

NASA-GSFC and Universities Space Research Association

AND

J. SINGH

NASA-GSFC and National Research Council

We studied the cataclysmic variables MV Lyr, KR Aur, and EP Dra using the *ROSAT* PSPC; we report on each.

EP Dra (=H1907+690) is an eclipsing polar. The PSPC exposure lasted 14.6ksec; the observed count rate was $\sim$0.02 counts s^{-1}. The spectrum is that of a classic polar. The fitted spectrum yields a temperature of $\sim$15eV for the blackbody portion; the results are insensitive to the assumed temperature of the 10-30keV bremsstrahlung component. The measured column of 1-20x10^{20} cm^{-2} is reasonable based upon an estimate of the column from the mean extinction and the distance. The 0.1-0.5 keV flux for the blackbody component is $\sim$8x10^{-14} ergs s^{-1} cm^{-2}. The light curve in the 0.1-2.0keV band shows a strong pulse cut by the eclipse. The off-phase may be relatively brief, as only 1 or 2 phase bins are consistent with zero counts. The hardness ratio (H-S/H+S, with H = energies 0.5-2.1keV, S = 0.1-0.5keV) is relatively constant at $\sim$0.2 except during the eclipse phase when it becomes $\sim$-0.2.

MV Lyr and KR Aur have insufficient phase coverage for any phased study. The spectrum of MV Lyr can be fit by a pure blackbody of temperature $\sim$0.45keV, in contrast to the *Einstein* IPC results in which a thermal bremsstrahlung spectrum of kT $\sim$5 keV provided the best fit. An explanation for the difference is not clear at present.

The spectrum of KR Aur can be equally well fit by a blackbody of temperature $\sim$0.55keV or by a power law of index 0.6-0.7. The KR Aur IPC spectrum is consistent with the PSPC results. The *EXOSAT* results are not, however: a 30keV bremsstrahlung and an 80eV blackbody were necessary for a good fit. Note, however, that the phase coverage of all three spectra of KR Aur is not identical.

A Bianchini et al (eds), Cataclysmic Variables, 475
© *1995 Kluwer Academic Publishers*

8. EVOLUTION AND EVOLUTIONARY LINKS

THE REACTION OF LOW–MASS STARS TO ANISOTROPIC IRRADIATION AND ITS IMPLICATIONS FOR THE SECULAR EVOLUTION OF CATACLYSMIC BINARIES

H. RITTER, Z. ZHANG AND U. KOLB
Max-Planck-Institut für Astrophysik
Karl-Schwarzschild-Str. 1
Postfach 15 23
85740 Garching, Germany

Abstract. A semi–analytic model for the reaction of a low–mass star to anisotropic irradiation of low incident flux is presented. By applying this model to the donor star of cataclysmic binaries (CBs) it is shown that CBs are likely to be unstable against irradiation–driven runaway mass transfer. The implications of this instability for the long–term evolution of CBs are examined. The possibility is discussed that because of this instability CBs evolve through a limit cycle in which phases of high and low mass transfer rate alternate on a time scale short compared to the evolutionary time scale.

1. Introduction

The possible importance of the reaction of low–mass stars to external irradiation for the long–term evolution of compact binaries has been realized only rather recently in the context of the evolution of low–mass X–ray binaries (LMXBs). So far, most papers have dealt with this problem in spherical symmetry and in the limit of the very high incident fluxes relevant for LMXBs (Podsiadlowski 1991; Harpaz and Rappaport 1991; Frank et al. 1992; Hameury et al. 1993). The limit of low flux, but still in spherical symmetry, has been considered by D'Antona and Ergma (1993). Anisotropic irradiation in the limit of high flux has been studied by Gontikakis and Hameury (1993), Hameury et al. (1993) and by Ritter (1994). To the best

A Bianchini et al (eds) Cataclysmic Variables 479-486

of our knowledge, anisotropic irradiation in the limit of low flux has not been considered so far. In this paper we present a first attempt to do so. For this we develop a simple semi–analytical model for the reaction of a low–mass star to anisotropic irradiation and apply this model mainly to CBs.

2. The reaction of a low–mass star to anisotropic irradiation

The following discussion applies only to low–mass stars ($M \lesssim 1 M_\odot$) on or near to the main sequence. Such stars have either a deep outer convective envelope and a radiative core ($M > M_{conv} \approx 0.35 M_\odot$) or they are fully convective ($M < M_{conv}$). We model the influence of anisotropic irradiation on the structure of such stars as follows: We assume that a fraction s of the stellar surface is exposed to an incident flux F_{irr} and that the remaining fraction $(1 - s)$ of the surface remains in the shadow. Denoting by R_s the radius of the star and by T_1 and T_2 respectively the effective temperatures on the unlit and the irradiated part, the stellar luminosity (energy lost by the star from its interior per unit time) is

$$L = 4\pi R_s^2 \{(1 - s)\sigma T_1^4 + s\sigma T_2^4 - sF_{irr}\} \quad . \tag{1}$$

In order to use the modified Stefan–Boltzmann law (1) as an outer boundary condition for stellar evolution calculations one has to know T_2. This, in turn, is determined by the details of how irradiation affects the energy loss through the irradiated surface.

Now it is important to realize that because the stars under investigation have a deep outer adiabatic convective envelope their mechanical and thermal structure remains spherically symmetric to a very high degree of accuracy out to the onset of the superadiabatic convection zone despite anisotropic irradiation. It is only the very thin superadiabatic layer (mass $\sim 10^{-10} M_\odot$), where energy is mainly transported via radiation, which is strongly affected by irradiation. So the asphericity of the star is restricted to a very thin surface layer, and it is this property which allows us to determine T_2 in (1) with a simple model. In this model we assume that convection is adiabatic out to a point where the pressure is $P = P_B$ and the temperature $T = T_B$ (see sketch in Fig. 1). The point (P_B, T_B) is the base of the superadiabatic convection zone which extends to the photosphere where the pressure is the photospheric pressure P_{ph} and the temperature the effective temperature T_{eff}. Because in this zone convection is ineffective, we assume the energy transport to be only radiative, i.e. $\nabla_{rad} = \nabla$ (where $\nabla \equiv \partial \ln T / \partial \ln P$). Because on the irradiated part of the star the effective temperature is $T_2 > T_1$, but T_B is the same for both parts, we see that irradiation, by raising the effective temperature, reduces the temperature

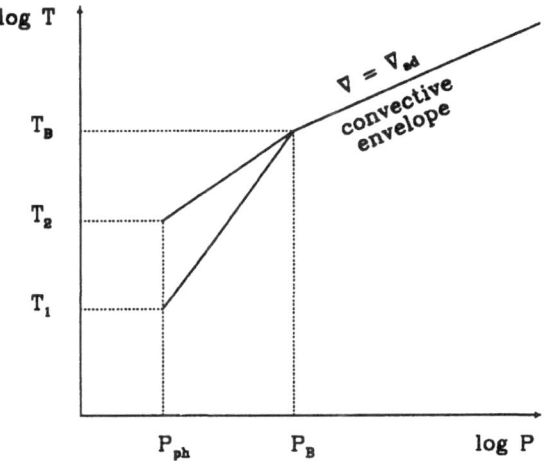

Figure 1. Sketch of the assumed run of temperature versus pressure in the outer layer of our model star.

gradient ∇ and thus the radiative energy loss through these layers. As is sketched in Fig. 1 we assume for simplicity that P_{ph} is the same on both parts of the star, though this is of course not strictly true. Using now the radiative diffusion approximation

$$F_{rad} = -\frac{ac}{3\kappa\varrho}\frac{dT^4}{dr} \quad , \tag{2}$$

where ϱ is the density, κ the opacity, and the other symbols have their usual meaning, and a power law approximation for κ

$$\kappa = const. P^a T^b \quad , \tag{3}$$

a one zone model for the superadiabatic layers yields (for details see Ritter et al. 1995)

$$\frac{F_{rad,1}}{F_{rad,2}} \equiv \frac{\sigma T_1^4}{\sigma T_2^4 - F_{irr}} = \begin{cases} \frac{T_B^n - T_1^n}{T_B^n - T_2^n} \quad , & n = 5 - b \neq 0 \\[2mm] \frac{\ln T_B - \ln T_1}{\ln T_B - \ln T_2} \quad , & n = 5 - b = 0 \end{cases} \tag{4}$$

Eq.(4) can be solved for $T_2(T_1, T_B, F_{irr})$ which together with Eq.(1) can be used as an outer boundary condition for numerical calculations.

3. Stability against irradiation–induced mass transfer

Let's now examine the situation in which a low–mass star transfers mass to a compact companion and, in turn, is irradiated by the accretion light source. For reasons which will become clear later, we restrict our discussion mainly to CBs, i.e. the compact star is a white dwarf (of mass M_{WD} and radius R_{WD}). The average flux with which the donor star is irradiated is

$$\langle F_{irr} \rangle = \frac{\eta}{8\pi} \frac{GM_{WD}(-\dot{M}_s)}{R_{WD}\, a^2} \quad , \tag{5}$$

where $(-\dot{M}_s)$ is the mass transfer rate, a the orbital separation and $\eta \lesssim 1$ a dimensionless "efficiency" factor which absorbs such factors as the albedo of the donor and takes into account that the irradiating light source does not necessarily radiate isotropically. Now, from Eqs.(1) and (4) one derives that

$$\frac{\partial L}{\partial F_{irr}} = 4\pi R_s^2 sg(T_1, T_2, T_B) < 0 \quad , \tag{6}$$

where

$$g(T_1, T_2, T_B) = \begin{cases} \dfrac{nT_1^4 T_2^{n-1}}{nT_1^4 T_2^{n-1} + 4T_2^3(T_B^n - T_1^n)} \, , & n = 5 - b \neq 0, \ T_B > T_2 \\[3mm] \dfrac{T_1^4}{T_1^4 + 4T_1^4(\ln T_B - \ln T_1)} \, , & n = 5 - b = 0, \ T_B > T_2 \ . \end{cases} \tag{7}$$

Thus the main effect of irradiating the star is to reduce its luminosity. This, in turn, means that part of the energy which the star generates in its interior is prevented from leaking through the irradiated surface and, therefore, is stored as internal and gravitational energy with the result that the star swells. In a mass–transferring binary, the swelling of the donor drives additional mass transfer and in this way generates even more irradiating flux. Therefore, we need now to examine under which conditions such a situation is stable against irradiation–induced runaway mass transfer. For this we write the temporal change of the stellar radius as

$$\frac{d\ln R_s}{dt} = \zeta_S \frac{\dot{M}_s}{M_s} + \left(\frac{\partial \ln R_s}{\partial t}\right)_{ml} + \left(\frac{\partial \ln R_s}{\partial t}\right)_{irr} \quad , \tag{8}$$

where ζ_S is the adiabatic mass radius exponent, $(\partial \ln R_s/\partial t)_{ml}$ the thermal relaxation term due to mass loss and $(\partial \ln R_s/\partial t)_{irr}$ the one due to irradiation. The condition for (dynamical) stability against mass transfer can be derived in the same way as e.g. in Ritter (1988). The result is

$$\zeta_S - \zeta_R > \zeta_{irr} \equiv -M_s \frac{\partial}{\partial M_s}\left(\frac{\partial \ln R_s}{\partial t}\right)_{irr}$$

$$= -M_s \frac{\partial}{\partial L}\left(\frac{\partial \ln R_s}{\partial t}\right)_{irr} \frac{\partial L}{\partial F_{irr}} \frac{\partial F_{irr}}{\partial M_s} \quad , \tag{9}$$

where ζ_R is the mass radius exponent of the Roche radius, and ζ_{irr} is a dimensionless number which measures how sensitively the stellar radius changes in response to irradiation. In order to evaluate $\partial^2 \ln R_s / \partial L \partial t$ we use the bipolytrope model (e.g. Kolb and Ritter 1992) and obtain

$$\frac{\partial}{\partial L} \left(\frac{\partial \ln R_s}{\partial t} \right)_{irr} =$$

$$-\frac{R_s}{GM_s^2} \left\{ \frac{\partial H}{\partial Q} \frac{1}{H_2(Q,n_1)} + \frac{1}{f} \frac{\partial f}{\partial Q} \right\} \left\{ \frac{\partial H}{\partial Q} + \frac{H}{f} \frac{\partial f}{\partial Q} \right\}^{-1} ,$$

(10)

where the quantities in the curly brackets $\{\}\{\}^{-1}$ define a dimensionless number which depends only on the relative size $Q = r_{core}/R_s$ of the radiative core and the polytropic index n_1 in it. In particular, for a single polytrope $n = 3/2$, i.e. $Q = 0$, one has $\{\}\{\}^{-1} = 7/3$. Taking now $F_{irr} = \langle F_{irr} \rangle$, Eqs.(5,6,7) yield

$$\zeta_{irr} = \frac{\eta}{2} \frac{M_{WD}}{M_s} \frac{R_s}{R_{WD}} \left(\frac{R_s}{a} \right)^2 s \{\}\{\}^{-1} g(T_1, T_2, T_B) .$$

(11)

Since in a normal CB $\zeta_S - \xi_R \approx 1$, systems in which $\zeta_{irr} \gtrsim 1$ are unstable against irradiation–induced runaway mass transfer. Under which conditions is $\zeta_{irr} > 1$? Using ZAMS models for the donor (to determine $R_s(M_s)$ and $T_B(M_s)$), Nauenberg's (1972) mass radius relation for white dwarfs and $T_2 = T_1$ (initial state is an unirradiated star) we find that $\zeta_{irr} > 1$ is reached for surprisingly small values of η (see Fig. 2), i.e. $\eta \gtrsim 0.04 \ldots 0.1$ is sufficient for instability. The conclusion to be drawn from this is that at least at large fractions of CBs is likely to be unstable against irradiation–induced runaway mass transfer.

Before we proceed to discuss the long–term evolution of CBs under this instability, we can now make a brief remark regarding the LMXBs. Replacing the white dwarf by a neutron star means that, if everything else is equal, ζ_{irr} is larger by a factor $R_{WD}/R_{NS} \gtrsim 10^3$ and that, therefore, $\zeta_{irr} \gtrsim 1$ is reached for much smaller η, i.e. $\eta \gtrsim 10^{-3}$. Thus LMXBs are extremely unstable against irradiation of the donor star.

4. Long–term evolution of CBs under the irradiation instability

Since there is no observational evidence showing that the known CBs are in a phase of runaway mass transfer and since a number of well–observed properties of CBs such as the period gap can only be explained if their evolutoin on the long–term average follows closely a standard evolution without taking into account irradiation (e.g. Kolb and Ritter 1992, Kolb 1993), we must conclude that the instability is either not relevant for CBs (because η is too small) or, if it is, that the instability must be quenched

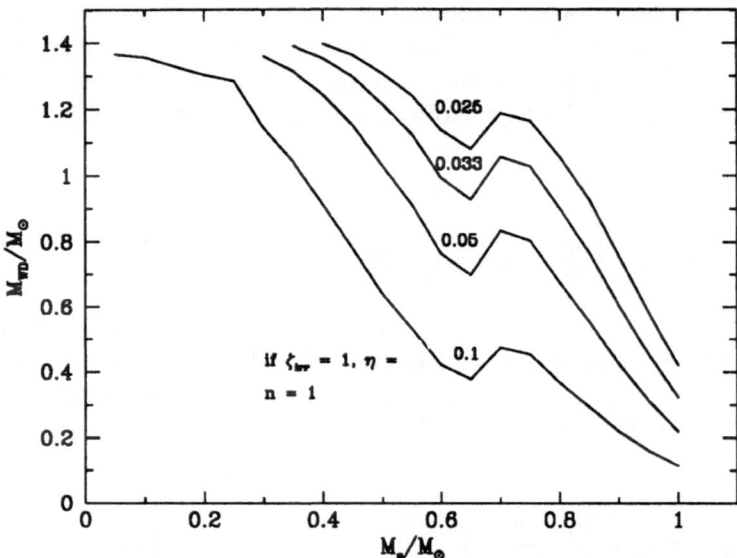

Figure 2.   Contour lines for η = const. in the $M_s - M_{WD}$ plane along which $\zeta_{irr} = 1$. The donor star is assumed to be on the main sequence with $T_2 = T_1$, $s = 0.5$ and $n = 1$ ($\hat{=} b = 4$, i.e. H^- opacity).

such as to prevent the mass transfer from running away too strongly. If the latter is the case, it is then conceivable that CBs go through a limit cycle in which phases of high and low states alternate on a time scale short compared to the evolutionary time scale. The high state would correspond to a phase during which the donor swells in response to irradiation and transfers mass at a rate above the secular mean. Mass transfer above the secular mean, however, cannot be maintained indefinitely and sooner or later the system has to go into a low state during which the mass transfer rate is below the long–term mean and irradiation is correspondingly unimportant. In this context we note that King (this volume) discusses the possibility of such limit cycles from a more general point of view. In his paper he arrives at the conclusion that they are possible only if the thermal relaxation term of the secondary's radius (cf. Eq.8) depends explicitly on the mass transfer rate and that the only conceivable way how this could be is via irradiation of the secondary. The question is now whether there are such quenching mechanisms which could give rise to limit cycles? In fact there are and in the following we are going to describe two of them.

One quenching mechanism operates only for donor stars with a mass $M_s \gtrsim 0.6 M_\odot$. In this case quenching is possible because with increasing T_2 not only the temperature gradient decreases but also the optical depth

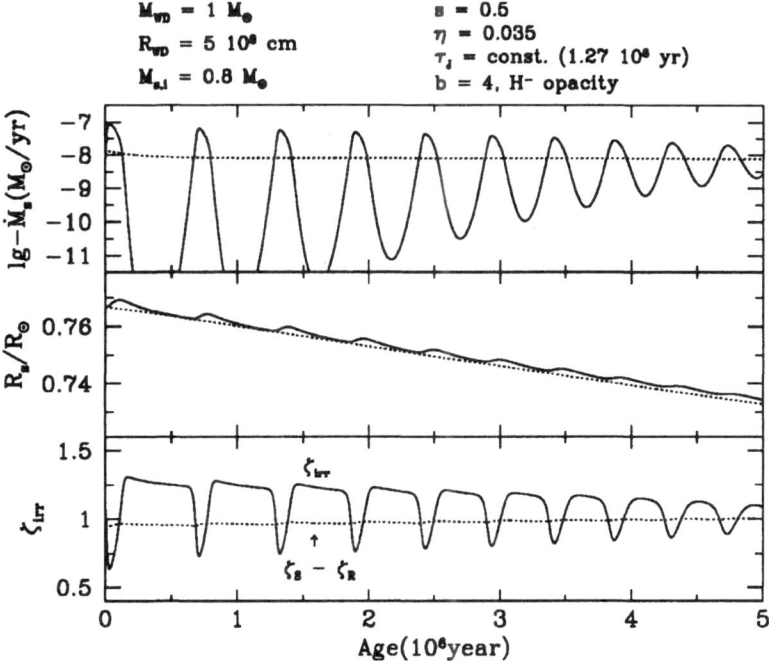

Figure 3. Temporal evolution of $-\dot{M}_s$, R_s and ζ_{irr} in a CV with parameters as listed at the top of the figure.

in the superadiabatic layer becomes larger (note that the relevant opacity source is H^- and the corresponding value of b is $b \approx 4-5$). As a consequence of this $\partial\zeta_{irr}/\partial T_2 < 0$. To illustrate this we show in Fig.3 the evolution of a system starting with an $0.8M_\odot$ ZAMS secondary and a $1M_\odot$ white dwarf. The calculations which have been performed with an appropriately modified bipolytrope programme (see Kolb and Ritter 1992) show in fact that such a system can go through a limit cycle, though in this case the oscillations are damped on a time scale of $\sim 10^7$yr. This behaviour is rather typical for systems with a secondary in the mass range $0.6M_\odot \lesssim M_s \lesssim 0.9M_\odot$, though the oscillations in $\dot{M}_s$ can also be growing, depending on the run of T_B/T_1 with M_s. More details will be presented elsewhere (Ritter et al. 1995).

The other quenching mechanism is of more fundamental nature. Quenching sets in as soon as $T_2 = T_B$. Since for stars with a mass $M \lesssim 0.6M_\odot$ already T_1 is only slightly lower than T_B, $T_2 = T_B$ is reached for rather small mass transfer rates. As soon as $T_2 \geq T_B$, the star's luminosity L does no longer depend on F_{irr}, i.e. $\partial L/\partial F_{irr} = 0$ and so $\zeta_{irr} = 0$. This is because for $T_2 \geq T_B$ an isothermal layer is formed on the irradiated side of the star which extends from the photosphere ($P = P_{ph}$) down to $P = P_B(T_2/T_B)^{1/\nabla_a}$, and through which no energy is transported, i.e. $\sigma T_2^4 = F_{\imath rr}$. The mass transfer rate for which quenching sets in, i.e. when $T_2 = T_B$ is

$$\dot{M}_q = \frac{1}{\zeta_{\imath rr}} \frac{s}{1-s} \frac{R_s L}{GM_s^2} \{\}\{\}^{-1} \left(\frac{T_B}{T_1}\right)^4 g(T_1, T_2 = T_B, T_B)$$

$$\lesssim \frac{s}{1-s} \frac{R_s L}{GM_s^2} \{\}\{\}^{-1} . \tag{12}$$

When compared with the mass transfer rates for a standard CB evolution (e.g. Kolb and Ritter 1992) we find that $\dot{M}_q$ is less than $\dot{M}_{tr}$, but by not more than a factor of two, if $M_s \lesssim 0.5 M_\odot$. This, in turn, would mean that no limit cycle is possible below $\sim 0.5 M_\odot$. Rather such systems are stable and the iradiation instability permanently quenched, i.e. $T_2 > T_B$ at all times. However, as far as stars below $0.5 M_\odot$ are concerned, our conclusions may be premature. This is because application of our model to stars with a mass $M \lesssim 0.6 M_\odot$ is problematic. The reason is that Eq.(4) was derived in the diffusion approximation which requires that the optical depth in the superadiabatic layer is sufficiently high. This, however, is not the case if T_B is only slightly above T_1 as in main sequence stars with $M \lesssim 0.6 M_\odot$.

Thus it remains to be demonstrated whether the irradiation instability described here, in combination with a suitable quenching mechanism, allows for a stable and large amplitude limit cycle which, in turn, could account for the observed large scatter of the mass transfer rates of CBs around the long-term mean.

5. References

D'Antona, F., Ergma, E. 1993, A&A 269, 219

Frank, J., King, A.R., Lasota, J.P. 1993, ApJ 385, L45

Gontikakis, C., Hameury, J.–M. 1993, A&A 271, 118

Harpaz, A., Rappaport, S. 1991, ApJ 383, 739

Hameury, J.–M., King, A.R., Lasota, J.P., Raison, F. 1993 A&A 277, 81

Kolb, U. 1993, A&A 271, 149

Kolb, U., Ritter, H. 1992, A&A 254, 213

Nauenberg, M. 1972, ApJ 175, 417

Podsiadlowski, Ph. 1991, Nature 350, 136

Ritter, H. 1988, A&A 202, 93

Ritter, H. 1994, in: Evolutionary links in the zoo of interacting binaries, F. D'Antona, V. Caloi, C. Maceroni, F. Giovanelli (eds.), Mem. Soc. Astron. It. 65, 173

Ritter, H., Zhang, Z., Kolb, U. 1995, in preparation

CATACLYSMIC BINARIES WITH HE-RICH SECONDARIES, SUPERNOVAE AND ACCRETION-INDUCED COLLAPSES

L.R. YUNGELSON AND A.V. TUTUKOV

Institute of Astronomy

48 Pyatnitskaya Str., 109017, Moscow, Russia

Abstract. We model the galactic population of ultra-short period cataclysmic binaries. Their total number is $(6 - 7) \cdot 10^7$. 'Observable' sample of systems with $V \leq 13^m$ contains 120 objects with 6 min $\lesssim P_{orb} \lesssim 1$ hr, $0.01 \lesssim M_2/M_\odot \lesssim 0.1$. The rate of accretion-induced collapses in these systems is $\sim 10^{-5}$ yr^{-1}. The rate of Supernova-scale events due to accretion of He onto white dwarfs in systems with degenerate and nondegenerate He donors may be as high as $\sim 6 \cdot 10^{-3}$ yr^{-1}.

1. Introduction

Some cataclysmic binaries (CB) have orbital periods below $P_{min} \approx 49$ min (Paczyński and Sienkiewicz 1981) for CB with hydrogen-rich secondaries: GP Com (46.5 min), CP Eri (28.7 min), V803 Cen (26.9 min), CR Boo (24.8 min), AM CVn (17.2 min). With exception of GP Com, all these systems exhibit several periodicities close to ones cited above and it is still uncertain, if they really reflect the orbital motion (Ulla 1994). The absence of hydrogen features in their spectra and the presence of helium ones strongly syggests that their secondaries are He-rich, because, e. g., for accretion disks Balmer emission lines are stronger than He I lines even when He/H is ~ 100 in number (Williams and Ferguson 1982). A group of low-mass X-ray binaries: 4U 1915-05 (50.0 min), 4U 1627-67 (41.4 min) and 4U 1820-303 (11.4 min) may be related to the ultra-short period CB.

We address the interpretation of these objects as close binaries with He-rich secondaries (Faulkner *et al.* 1972, Nather *et al.* 1981, Wood *et al.* 1987, O'Donoghue *et al.* 1987, O'Donoghue & Kilkenny 1989, Ulla 1994, Warner this volume). They deserve a special attention because accretion

A Bianchini et al (eds) Cataclysmic Variables 487-493

of pure He may result in the accretion-induced collapses (AIC) and double detonations comparable in power with Supernovae.

We use a population synthesis code which combines the statistical data on the birthrates of binaries depending on the masses of components and their separations with the data on stellar evolution. Under assumption of constant star formation rate through galactic lifetime, we generate a population of zero-age CB with degenerate He-rich secondaries. Then, we compute analytically evolution of each 'new-born' system for Hubble time, $(15 \cdot 10^9$ yr), as described, e. g., in Tutukov & Yungelson (1979).

We assume that He white dwarfs obey the mass-radius relation for zero-temperature degenerate objects (Nauenberg 1972)

$$R = 7.84 \cdot 10^8 \left[(M/M_{Ch})^{-2/3} - (M/M_{Ch})^{2/3} \right]^{1/2} \quad cm, \qquad (1)$$

with M - mass in $M_\odot$, $M_{Ch} = 1.433 M_\odot$. For additional details of computations see, e. g., Tutukov and Yungelson (1994) and references therein.

2. Results

Theoretically, three types of He-rich secondaries in semidetached systems with WD primaries are possible: (i) He degenerate dwarfs; (ii) He-burning nondegenerate stars; (iii) He-mantle stars with CO cores. While orbital periods of all these systems overlap with the shortest observed CB periods, types (ii) and (iii) may be excluded as candidate CB because of high luminosity. Their counterparts are, most probably, hot subdwarfs. However, mass exchange rates in all three types of systems allow accumulation of massive helium layer at the surface of WD.

Progenitors of systems with He dwarf secondaries in most cases have $1.5 \lesssim M_{10}/M_\odot \lesssim 6$, $0.8 \lesssim M_{20}/M_\odot \lesssim 2.5$, $30 \lesssim A_0/R_\odot \lesssim 1000$. Primaries experience type C mass exchange, while secondaries type B one. Typical mass of primary white dwarf is 0.6 $M_\odot$, while masses up to M_{Ch} are possible. Condition of stable mass exchange $\frac{d \ln R}{d \ln M_2} \leq \frac{d \ln R_{cr}}{d \ln M_2}$ limits the masses of secondaries from above by $\sim 0.3\ M_\odot$. The lower limit is the lowest possible mass of He dwarf $\sim 0.13\ M_\odot$. Time interval between formation of secondary WD and the RLOF by the latter is between 10^4 and 10^{10} yr. Initial orbital periods of He-dwarf CB are about 3 min and they evolve to longer periods (Tutukov & Yungelson 1979; Fig. 1).

The Table lists the birthrates of systems with He secondaries of all types and of H-rich CB, as given by scenario code. If the lifetime of CB of both types is limited by Hubble time, CB with He dwarf donors have to be about twice more numerous than H-rich CB, because most of the latter must be the systems which passed through P_{min}. The birthrate of systems with He dwarf donors only slightly depends on the common envelope parameter α_{ce}.

TABLE 1. Birthrates and numbers of low-mass semidetached systems and rates of related events

Type of system	Birthrate, yr^{-1}	Number	AIC, yr^{-1}	He-flashes, yr^{-1}
Hydrogen donors. $\alpha_{ce}= 1$	7.1(-3)	6.2(7)	2.6(-6)	0.0
Hydrogen donors, $\alpha_{ce}= 0.5$	1.1(-2)	7.0(7)	0.0	0.0
Subgiant donors, $\alpha_{ce}= 1$	9.8(-6)	1.6(3)	0.0	0.0
Subgiant donors, $\alpha_{ce}= 0.5$	3.0(-6)	7.7(2)	0.0	0.0
He dwarf donors, $\alpha_{ce}= 1$	1.3(-2)	1.5(8)	2.3(-5)	0.0
He dwarf donors, $\alpha_{ce}= 0.5$	1.6(-2)	1.8(8)	1.8(-6)	0.0
He dwarf donors, doubled R_{wd}	1.3(-2)	2.4(7)	2.6(-5)	1.0 (-3)
He dwarf donors, WW	1.3(-2)	1.5(8)	2.6(-5)	0.0
Nondeg. He-donors, $\alpha_{ce}= 1$	4.9(-3)	1.4(6)	1.2(-4)	1.0(-3)
Nondeg. He-donors, $\alpha_{ce}= 0.5$	3.0(-3)	1.1(6)	5.9(-6)	3.5(-3)
Nondeg. He-donors, WW	4.9(-3)	1.4(6)	1.4(-4)	5.6(-3)
He-mantle donors, $\alpha_{ce}= 1$	8.5(-4)	1.9(2)	9.9(-6)	0.0
He-mantle donors, $\alpha_{ce}= 0.5$	2.4(-4)	5.7(2)	3.0(-7)	0.0

Note: Numbers in parentheses give power of decimal exponent. In WW case the flash occurs on dwarfs of $M \geq 0.6 M_\odot$ after accretion of $0.2 M_\odot$ of He

Lower panel of Fig. 1 gives the distribution of all systems with He dwarf donors over P_{orb}. As mass loss rate decreases with time, systems accumulate at largest periods attainable in Hubble time. This distribution is independent of α_{ce}. In at least a half of all systems secondaries overfill Roche lobes in less than $\sim 10^8$ yr after formation. This time is certainly too short to attain mass-radius relation for completely degenerate cold configurations. We computed an additional model, assuming that R_{wd} is two times higher than given by Eq. (1). If evolution is supported by AML via GWR, $\dot{M} \propto R_{wd}^{-3}$. This results in lower $\dot{M}$ at the instant of RLOF. Hence, secondaries with masses higher than in 'standard' case can stably transfer matter to the companions with the rate lower than $\dot{M}_{Edd}$.

Because M_2 in He-rich CB does not decrease in Hubble time below $\sim 0.005 M_\odot$ (Fig. 2), secondaries are probably stable against tidal disruption due to the finite time of transfer of angular momentum to the disk and from the disk back to the orbit, which ensues when $M_2 \lesssim 0.002 - 0.004\ M_\odot$ (Ruderman and Shaham 1983, 1985, Hut and Paczyński 1984).

Our model gives the birthrate ~ 0.002 yr^{-1} for H-rich CB that pass through the periods between 2 hr and 80 min. If $\dot{M} \lesssim 5 \cdot 10^{-11}$ $M_\odot$ yr^{-1} and donor loses about 0.2 $M_\odot$ prior to P_{min}, there have to be $\sim 8 \cdot 10^6$ such systems in the Galaxy. Ritter and Kolb (1993) list ~ 90 systems below the

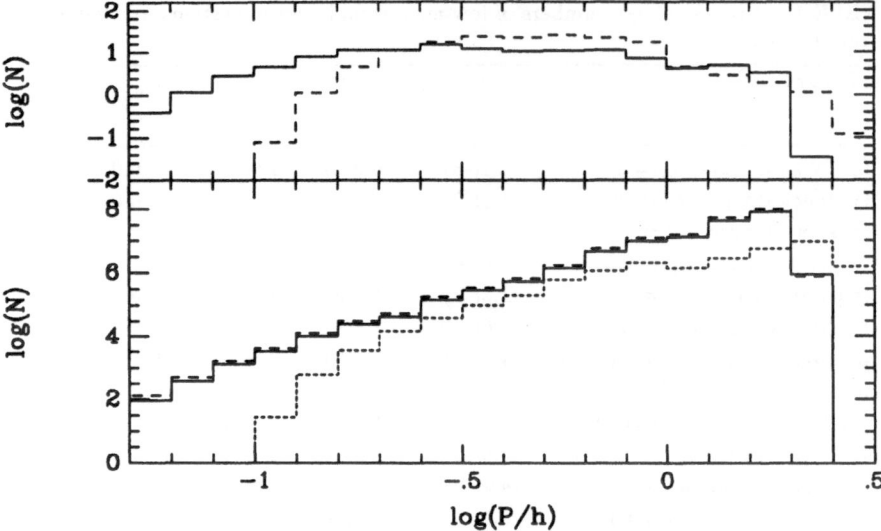

Figure 1. Distribution of systems with He dwarf donors over orbital periods for total sample (lower panel) and the sample limited by V=13^m (upper panel). Solid line corresponds to 'standard' α_{ce}=1, dotted one - to the model with doubled R_{wd}, dashed one - to the model with α_{ce}=0.5.

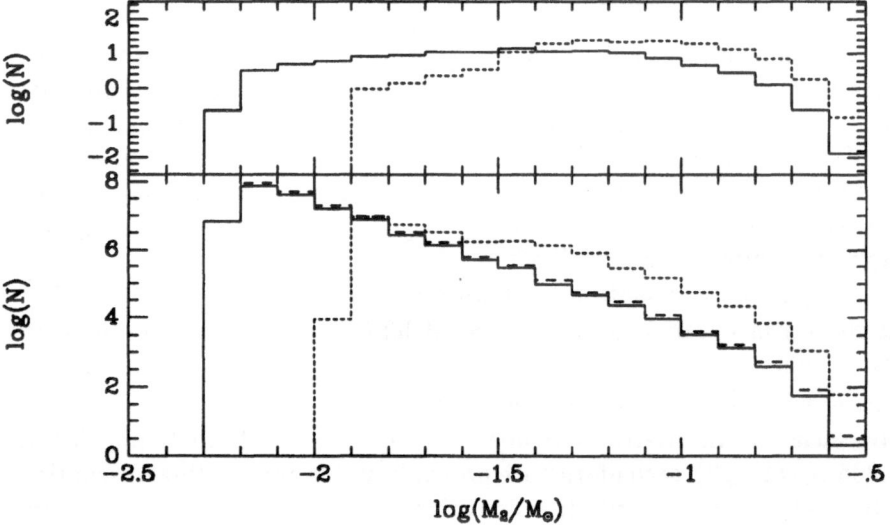

Figure 2. Distribution of systems with He dwarf donors over masses of secondaries. Panels and lines as in Fig. 1.

period gap, i. e. we observe $\sim 10^{-5}$ of all of them. In our 'population' there are $\sim 3 \cdot 10^6$ systems with He dwarf donors with $10^{-11} \leq \dot{M} \leq 10^{-10}\,\mathrm{M_\odot\,yr^{-1}}$. This gives the proportion of observed systems equal to 10^{-6}. The order of magnitude difference in proportions of observed ordinary CB and He-dwarf CB may be ascribed to still unknown properties of systems with He donors (e.g., bolometric corrections of the disks) and unknown selection effects.

Considering radiation of accretion disk as the main source of the luminosity of CB in quescence, we computed the numbers of systems and their distributions over P_{orb} and M_2 for samples limited by $V_{lim} = 13^m$ which is suggested by brightness of V803 Cen and CR Boo in maximum. If orbital inclination of the disk $i = 90°$, then $M_{Vmax} = -9.48 - \frac{5}{3} \log(M_1 \dot{M})$, where M_1 is in $M_\odot$ and $\dot{M}$ in $M_\odot$ yr^{-1} (Webbink et al. 1987). For $V_{lim} = 13^m$ the number of 'observable' stars reduces to ~ 120. For a sample with doubled R_{wd} and $V_{lim} = 13^m$ it contains ~ 1200 stars. Flat maximum of distribution of limited sample over P_{orb} between 15 min and 48 min, covers the range of periods of He-rich CB (Fig. 1).

We considered an event as AIC, if in the course of evolution M_{wd} grew up to 1.39 $M_\odot$. In fact, this growth may result in a collape or an explosion depending on $\dot{M}$, chemical composition, and M_{wd} at the onset of accretion (Nomoto & Kondo 1991, Dominguez et al. 1993). We assumed, after Iben & Tutukov (1985) and Dominguez et al. (1993), that all components of close binaries with initial masses 9 - 11.4 $M_\odot$ evolve into oxygen-neon-magnesium (ONe) WD which experience AIC. This assumption overestimates the rate of AIC by an order of magnitude, if masses of progenitors of ONe WD are confined to 10.3 - 10.6 $M_\odot$, as originally was guessed by Iben & Tutukov (1985). However, AIC are also likely for CO WD, if solidification of their interiors occurs (Nomoto & Kondo 1991).

If accreting dwarf is cold and $\dot{M} \lesssim (3-5) \cdot 10^{-8}$ $M_\odot$ yr^{-1}, at least 0.1 - 0.15 $M_\odot$ of He has to be accumulated at the surface of WD prior to He ignition (Fujimoto & Taam 1982, Limongi & Tornambé 1991, Iben et al. 1987, Tutukov & Khokhlov 1992, Woosley & Weaver 1994). If mass of the dwarf exceeds ~ 0.8 $M_\odot$ (Tutukov & Khokhlov 1992) or 0.6 $M_\odot$ (Woosley & Weaver 1994), double detonation occurs and energy release has a Supernova scale. The computed light curves resemble SN Ia in the shapes, but the objects are by a half less luminous than SN Ia (Woosley & Weaver 1994).

If $\dot{M}$ exceeds $(3-5) \cdot 10^{-8}$ $M_\odot$ yr^{-1}, the growth of He shell is limited by mass loss in thermal flashes, formation of common envelopes, and radiatively driven stellar wind. We took all these effects into account.

The Table shows the rates of AIC and double-detonation Supernovae induced by He shell flashes in systems with H and He donors. Systems with He nondegenerate donors and systems with He-mantle donors are potentially the site of most numerous AIC. For H accretors Table gives the upper limits of the rates of AIC, estimated under assumption of conversion of all accreted H into He for steady hydrogen burning WD and massive ($M \gtrsim 1.25$ $M_\odot$) white dwarfs, accreting at $\dot{M} \geq 10^{-8}$ $M_\odot$ yr^{-1}, as suggested by observations of recurrent Novae, which do not show indications of mass loss during outbursts. If AIC result in formation of neutron stars and systems with nondegenerate He donors become LMXB and continue their life until donor becomes WD, we arrive to $\sim 10^3$ bright LMXB. It is not clear what effects extinguish them from the sample of observed LMXB.

The number of LMXB from AIC in systems with He degenerate donors is $\sim 3.5 \cdot 10^6$. Their typical X-ray luminosity is $0.2 - 2$ $L_\odot$. Our model predicts existence of about 100 sources with $10^3 \lesssim L_x/L_\odot \lesssim 10^4$. Dimmer sources may contribute to the X-ray background. In bright sources most of X-rays are probably absorbed by the matter flowing from the donor and radiatively induced stellar wind. X-ray burster 4U 1820-30 with the shortest known orbital period of 665 s may be a representative of this class of objects.

Only systems with He nondegenerate donors produce double detonations with the rate that is comparable to the observational estimate of the frequency of SN Ia in the Galaxy ~ 0.003 yr^{-1}(van den Bergh and Tammann 1991). If we assume, that they occur only on the dwarfs that are more massive than 0.8 $M_\odot$ after accretion of 0.1 $M_\odot$, $\nu_{He} \approx 10^{-2}$ yr^{-1}. If after Woosley and Weaver (1994) we allow to explode 0.2 $M_\odot$ He shells on all dwarfs which are more massive than 0.6 $M_\odot$, $\nu_{He} \approx 0.006$ yr^{-1}. However, systems with nondegenerate He donors are not older than 10^9 yr, and systems with He dwarf donors are younger than $6 \cdot 10^9$ yr (Tutukov & Yungelson 1994). Thus, they can not challenge merging double degenerates as pre-SN Ia in early-type galaxies.

3. Conclusions

1. The systems with helium degenerate and nondegenerate donors cover the interval of orbital periods of the ultra-short period CB. The systems with He degenerate dwarf donors are their most probable counterparts.

2. The systems with helium nondegenerate donors may be the site of most numerous AIC. If AIC produce neutron stars, these systems have to generate a population of bright LMXB with 20 min $\lesssim P_{orb} \lesssim$ 30 min.

3. Double-detonations induced by burning of He in the shell in systems with He nondegenerate donors may provide a significant contribution to the SNIa rate from populations with the age below $\sim 10^9$ yr.

Acknowledgments
This study was partially supported by International Science Foundation grant MPT000, ESO C&EE Programme grant A-01-019, and Russian Foundation for Fundamental Research grant 93-02-2893. We acknowledge travel grants from SOC and ISF. L.Y. acknowledges financial support from MPI für Astrophysik (Garching). We thank Mrs. S. Sushko for her help with the preparation of the manuscript.

References

Dominguez, I., Tornambé, A. & Isern, J., 1993. *ApJ*, **419**, 268.

Faulkner, J., Flannery, B. P. & Warner, B., 1972. *ApJ*, **175**, L79.

Fujimoto, M.Y. & Taam, R.E., 1982. *ApJ*, **260**, 249.

Hut, P. & Paczyński , B., 1984. *ApJ*, **284**, 685.

Iben, I. Jr. & Tutukov, A. V., 1985. *ApJSS*, **58**, 661.

Iben, I. Jr., Tutukov, A. V. & Yungelson, L. R., 1994. *ApJ*, submitted.

Iben, I. Jr., Nomoto, K., Tornambé, A. & Tutukov, A. V., 1987. *ApJ*, **317**, 717.

Limongi, M. & Tornambé, A., 1991. *ApJ*, **371**, 317.

Nauenberg, M., 1972. *ApJ*, **175**, 417.

Nather, R. E., Robinson, E. L. & Stover, R. J., 1981. *ApJ*, **244**, 269.

Nomoto, K. & Kondo, Y., 1991. *ApJ*, **367**, L19.

O'Donoghue, D., Menzies, J. W. & Hill, P. W., 1987. *MNRAS*, **227**, 347.

O'Donoghue, D. & Kilkenny, D., 1989. *MNRAS*, **236**, 1989.

Paczyński, B. & Sienkiewicz, R., 1981. *ApJ.*, **248**, L27.

Ritter, H. & Kolb, U., 1993. In 'X-ray Binaries', eds. W. H. G. Lewin, J. van Paradijs, and E. P. J. van den Heuvel, (Cambridge: Cambridge Univ. Press).

Ruderman, M. & Shaham, J., 1983. *Nature*, **304**, 425

Ruderman, M. & Shaham, J., 1985. *ApJ*, **289**, 244.

Tutukov, A. V. & Khokhlov, A. M., 1992. *AZh.*, **69**, 754.

Tutukov, A. V. & Yungelson, L. R., 1979. *Acta Astron.*, **29**, 665.

Tutukov, A. V. & Yungelson, L. R., 1994. *MNRAS*, **268**, 871.

Ulla, A., 1994. *Space Sci. Rev.*, **67**, 241.

van den Bergh, S. & Tammann, G. A., 1991. *ARAA*, **29**, 363.

Webbink, R. F., Livio, M., Truran, W. & Orio, M., 1987. *ApJ*, **314**, 653.

Williams, R. E. & Ferguson D. H., 1982. *ApJ*, **257**, 672.

Wood., M. A. et al., 1987. *ApJ*, **313**, 757.

Woosley, S. E. & Weaver, T.A., 1994. *ApJ*, **423**, 371.

FORMATION RATES OF SUPERNOVAE IA AND CATACLYSMIC BINARIES AS A FUNCTION OF THE AGE OF STELLAR SYSTEM

A.V. TUTUKOV AND L.R. YUNGELSON

Institute of Astronomy of Russian Ac. of Sci.
48 Pyatnitskaya Str., 109017, Moscow, Russia

Abstract. A model of populations of merging binary degenerate dwarfs and cataclysmic binaries (CB) was constructed by means of a scenario code. Formation rates of both populations decrease with the age of progenitor stellar population. This explains the observed trends in SNe Ia and novae rates. Higher observed ratio of frequencies of novae and SNe Ia in elliptical galaxies as compared with spiral galaxies is explained by the model as a consequence of faster decrease with the age of the rate of SNe Ia relative to the formation rate of CB. The model shows, that the mean dwarf mass in zero-age CB changes with the age of the stellar population from $\sim M_\odot$ for ages below $\sim 10^8$ yr to $\sim 0.3 M_\odot$ for the oldest CB with the age about 10^{10} yr. Thus, the observed difference between 'disk' and 'bulge' novae may be a result of different average ages of their progenitors.

1. Introduction

Supernovae of type Ia (SNe Ia) show evidences of photometric and spectroscopic diversities among them (e.g., Ruiz-Lapuente and Fillipenko 1993, Canal and Ruiz-Lapuente 1993). Branch and van den Bergh (1993) also have found an evidence of a dependence of the expansion velocity of the SNe Ia envelope on the galaxy type or, probably, on the average age of stars. The velocity drops from 10000-14000 km s^{-1} for Sb-Sc-Sd-Ir galaxies to 9000 km s^{-1} for E-S0 galaxies. The reason of this variation with the simultaneous threefold decrease of the SNeIa rate per unit of luminosity (Cappellaro et al. 1993) is unclear. We propose that the decrease of

495

A. Bianchini et al. (eds.), Cataclysmic Variables, 495-501.
© *1995 Kluwer Academic Publishers.*

the total mass of merging double dwarfs, which are progenitors of SNe Ia, with their age may be responsible for the change of the expansion velocity with the age of population. An additional source of variation of SNe Ia properties with the age of stellar population may be the nonuniformity of their progenitors. Helium shell explosions leading to detonation of carbon (Iben et al. 1987; Tutukov & Khokhlov 1992; Woosley & Weaver 1994) may represent a part of SNe Ia among young stars ($t \lesssim 10^9$ yr).

Novae form a rather nonuniform family as well. They were divided into two families (Duerbeck 1990; Della Valle et al. 1992). Fast novae in the galactic disk have mean $z \simeq 130$ pc. Bulge slow novae, concentrated towards the galactic centre, have mean $z \simeq 500$ pc. Disk novae in LMC and M 33, according to Della Valle et al. (1994), have $1.2 - 1.4 M_\odot$ dwarfs, and bulge novae in M 31 and M 81 contain dwarfs with masses $0.7 - 0.9 M_\odot$. Della Valle et al. (1994) have found that the novae rate per unit of luminosity is about four times higher for 'blue' galaxies like LMC and M 33, as compared to the rate for 'red' galaxies like M 31 and M 81.

In the present study we model the age dependences of cataclysmic binaries (CB) and SNe Ia formation rates by means of a numerical scenario code for the population synthesis of binary stars. Some details of the model are given in Tutukov and Yungelson (1992, 1994), Tutukov et al. (1992).

2. A simple model of the history of star formation in a galaxy

All galaxies may be divided into three groups according to the history of star formation (SF). In elliptical galaxies most stars were formed during first 5 billions of years of their life. Irregular galaxies with masses below $\sim 10^{10} M_\odot$ had stationary SF during the Hubble time T_H, and more massive irregular and spiral galaxies show enhanced rates of SF at the begining of their life, as compared to the current rate (Gallagher et al. 1984). Since the formation rate of binaries of certain type changes with the age of stellar population, this rate appears to be related to the integral color of the galaxy, reflecting relative input of stars of different ages. To study these relations one needs to combine the model of galactic evolution with the evolutionary scenario code for stars. But the preliminary estimates may be made by a simple model including an initial SF burst during first $\sim 5 \cdot 10^9$ yr of the galaxy life, producing stars with the total mass M_{old} and a consequent $\sim 10^{10}$ yr long stage of the SF with the constant rate $\dot{M}_\star$ (Tutukov and Yungelson 1994). The mass of these stars is $M_{new} = T_H \cdot \dot{M}_\star$. Now, using star formation rate in our Galaxy for normalization, we can estimate the ratio of the rate of certain events connected with the 'old' stars to the frequency of similar events and connected with the 'new' stars:

$$\frac{\nu_{old}}{\nu_{new}} \simeq 3 \, \frac{\nu_{10-15}}{\nu_{tot}} \, \frac{M_{old}}{M_{new}}, \tag{1}$$

where ν_{10-15} is the rate of events from the stars with the age from 10 to 15 billions of years, and ν_{tot} is the rate of events from younger stars in the Galaxy. The average ratio M_{old}/M_{new} for massive galaxies is about four (Gallagher et al. 1994). Therefore, Eq.(1) becomes

$$\frac{\nu_{old}}{\nu_{new}} \simeq 12 \frac{\nu_{10-15}}{\nu_{tot}}. \tag{2}$$

This simple model can be used for the estimates of the influence of the history of SF in the past on the rate of events at the current epoch in massive galaxies like our one.

3. Results: merging white dwarfs and SNe Ia

We assume that the mergers of degenerate components with the total mass exceeding the Chandrasekhar mass evolve into SNe Ia. Figure 1 gives the contribution of stars of different age to the present annual rate of SNe Ia under assumption of constant star formation rate equal to the present galactic rate $\dot{M}_G$. At the same time, Fig. 1 shows the variation of SNe Ia occurence rate for galaxies with 'instantaneous' star formation bursts and mass equal to the product of $\dot{M}_G$ and Hubble time T_H. It is clear that about a half of all SNe Ia explode during $\sim 3 \cdot 10^8$ yr after formation of parent stars. Observations of SNe Ia in spiral galaxies have shown that about two thirds of SNe Ia are located in the arms of the former (Della Valle & Livio 1994). Since arms are shortliving structures (a few of 10^8 yr), this fact shows relative youth of the progenitors of a significant part of SNe Ia.

The model shows that the total mass of merging binary degenerate dwarfs changes from $1.5 - 2.8 M_\odot$ for binaries with ages below $3 \cdot 10^8$ yr to $1.4 - 1.7 M_\odot$ for stars with ages above 10^9 yr (Tutukov and Yungelson 1994). This may be the reason for the change of expansion velocity of SNe Ia envelopes with the age (Branch and van den Bergh 1993).

The model rate of shell He flashes leading to the detonation of carbon in accreting WD in semidetached binaries with nondegenerate helium donors, in which mass exchange is driven by the radiation of gravitational waves, in our Galaxy is about $0.005 \, yr^{-1}$, what is comparable to the total frequency of SNe Ia $\sim 0.003 \, yr^{-1}$. Such supernovae have certainly to be absent among stars of elliptical galaxies. This is a result of relatively short lifetimes of their progenitors ($\lesssim 8 \cdot 10^8$ yr, Iben & Tutukov 1985; Tutukov et al. 1992). Thus, these systems can explain a part of SNe Ia among relatively young stars in spiral and irregular galaxies.

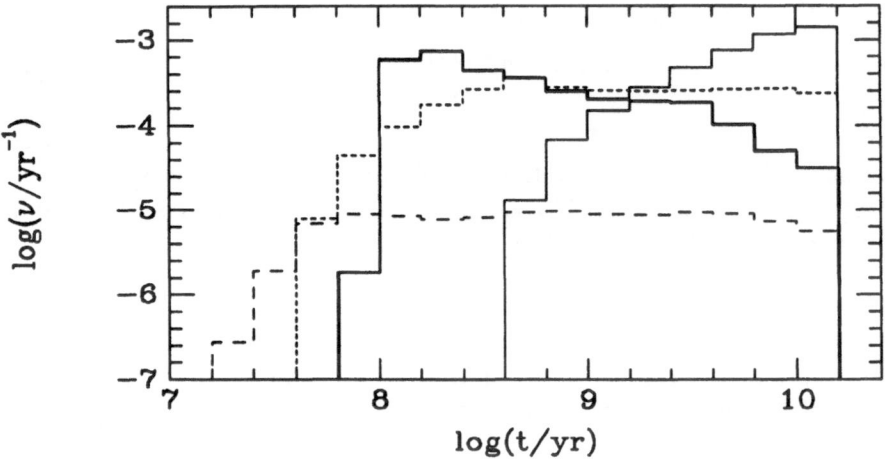

Figure 1. Formation rates of SNe Ia and cataclysmic binaries with accreting degenerate dwarfs of different masses M_d as functions of stellar population age. Thick solid line shows SNe Ia (Tutukov and Yungelson 1994), dashed line marks CB with $M_d > M_\odot$, thin solid line points CB with $M_d < 0.5M_\odot$, and dotted line shows CB with dwarfs of intermediate masses.

Products of mergers of CO white dwarfs and He nondegenerate stars with $M_{CO} \geq 0.7M_\odot$ may become R CrB stars (Iben and Tutukov 1989). Mergers with lower masses of CO dwarfs and mergers of He degenerate dwarfs will be compact hot stars (Iben and Tutukov 1985), occupying in the HR diagram the region of bright OB helium subdwarfs.

4. Results: cataclysmic binaries

The total model formation rate of CB in our Galaxy is $\sim 0.007yr^{-1}$. The whole family of CB in Fig. 1 was divided into three groups according to the masses of accreting dwarfs. The first group, with dwarf masses exceeding $M_\odot$, appears about $3 \cdot 10^7$ years after formation of parent binaries (Fig. 1). Thus, first CB contain only massive dwarfs. The input of CB with low mass dwarfs increases with the age. The oldest CB forming at $t \gtrsim 2 \cdot 10^9$ yr contain mainly helium degenerate dwarfs with $M_d < 0.5M_\odot$. It is evident from Fig. 1, and Eqs. (1,2) that the history of SF cannot change significantly the current rate of formation of CB with CO degenerate dwarfs in our Galaxy (for $M_{old}/M_{new} \sim 4$), since the relative input of stars with ages above 10^{10} years is lower than 10 %. But most of CB with helium dwarfs forming today belong to the oldest stars, therefore the current rate of their formation in our Galaxy may be several times higher than the value estimated for the stationary SF with the current rate.

  Let us now discuss the possible influence of the age on the observed properties of novae. The observed dependence of the novae rate (per unit of luminosity) on galactic color (Della Valle et al. 1994) may be explained as a consequence of the strong decrease of the novae formation rate with the age of stars (Fig. 1). We assume here that novae occur in systems with CO and ONe dwarfs with masses above $0.5M_\odot$. The frequency of SNe Ia (Fig.1) decreases with the age faster than the rate of novae. Thus, the ratio of novae and the supernovae rates grows about ten times in the age range $10^8 - 1.5 \cdot 10^{10}$ years (Fig. 1). Della Valle and Livio (1994) have found about fivefold growth of the observed ratio with transition from 'young' LMC and M33 to 'old' elliptical galaxies in Virgo. Thus, having in mind many remaining uncertainties, model estimates of evolution of this ratio agree with the observed trend.

The observed division of the family of novae into two populations according to masses of accreting dwarfs (Della Valle et al. 1994) can be explained by our model as a result of an overabundance of CB with massive dwarfs among relatively young systems (Fig.1). Disk novae are mainly recently formed products of binaries with initially massive $[(6-11) M_\odot]$ primaries. Their massive progenitors had low ($\sim$ 90 pc) z-coordinates (Scalo 1986). Old 'bulge' novae are descendants of $\sim M_\odot$ stars with average z-coordinates about 300 pc. Thus, the difference in initial masses of primary components may explain distributions of CBs with different dwarf masses over z-coordinate.

However, to give an answer to the question about systematical variation of the mass spectrum of CB dwarfs with the age we have to discuss several additional topics. To get a dependence of the novae rate on the dwarf mass we have to fix the initial mass spectrum of primaries: $dN = aM^{-\alpha}dM$, the relation between the dwarf mass and the initial mass of the star: $M_d = bM^\beta$, and the critical hydrogen-rich shell mass : $M_H = cM_d^{-\gamma}$. The novae frequency ν_N for a fixed donor mass is then

$$\frac{d\nu_N}{dM_d} \sim M_d^{\gamma-1+(1-\alpha)/\beta}. \tag{3}$$

For the usual combination of all powers [$\alpha \sim 2.5$, $\beta \sim 0.4$, $\gamma \sim 10$ (Tutukov and Yungelson 1989)] one gets the power in the last equation equal to 5.25 This distinguishes CB with massive dwarfs among novae. However all numbers in the last equation depend on the chemical composition of stars, which changes with transition from the disk to halo. These dependences are still poorly known. Thus, our conclusions about the age dependence of some CB parameters are valid only if they are limited to the stars belonging to the same chemically homogenious population, e.g., to the galactic disk.

CB with helium dwarfs, acccording to our model, have to be the most common in the elliptical galaxies. Their progenitors had low masses and therefore have to have highest space velocities and z-coordinates. Low dwarf masses ($\sim 0.3 M_\odot$) and low mass exchange rates ($\sim 10^{-10}$ $M_\odot$ yr^{-1}, Tutukov & Yungelson 1979) help to accumulate massive hydrogen shells ($\sim 0.01 M_\odot$, Iben and Tutukov 1992). This mass is about hundred times higher than masses of envelopes of ordinary novae. Thus, because the rates of formation of CB with He and CO dwarfs are comparable (~ 0.003yr^{-1}), novae explosions on He dwarfs have to be rare and probably very powerful events. A possible counterpart of such event may be the bright nova in M31, which has expelled about $0.01 M_\odot$ of hydrogen-rich matter (Mould et al. 1990, see also Iben and Tutukov 1992).

5. Conclusion

1. The average total mass of merging double degenerate dwarfs monotonously decreases with the age of stellar system from $\sim 2.2 M_\odot$ for $t < 3 \cdot 10^8$ yr to about $1.5 M_\odot$ for $t < 10^9$ yr. This may predetermine the change of the expansion velocity of SNe Ia envelopes with the age of a stellar system. An additional possibility to explain the diversity of SNe Ia properties in different galaxies can be attributed to the helium shell explosions as a possible reason for a part of SNe Ia in young populations.

2. The average degenerate dwarf mass in CB decreases with the age. This may explain the difference in the space distribution and other properties of bulge and disk novae.

3. To construct the model of CB population of a galaxy, the star formation history and influence of chemical composition on properties and evolution of close binary stars have to be taken into account.

Acknowledgments

We thank I. Iben, N. Chugai and M. Della Valle for useful discussions, and Mrs. S. Sushko for her help in the preparation of the manuscript. This study was made possible in part by grants MPT 000 from the International Science Foundation, A-01-019 from the ESO C&EE Programme and 93-02-2893 from the Russian Foundation for Fundamental Research. We also thank ISF and SOC for the financial support of participation in this conference.

References

Branch, D. & van den Bergh, S., 1993. *Astron.J.*, **105**, 2231.
Canal, R. & Ruiz-Lapuente, P., 1993. In Proc. IAU Coll.145 'Supernovae and Supernovae Remnants'.
Cappellaro, G.,Turrato, M., Benetti, S., Tsvetkov, D., Bartunov, O. & Makarova, I., 1993. *A&A*, **273**, 383.
Della Valle, M., Bianchini, A., Livio, M. & Orio, M., 1992. *A&A*, **266**, 232.

Della Valle, M., Rosino, L., Bianchini, A. & Livio M., 1994. *Preprint ESO*, No. 977.

Della Valle, M. & Livio, M., 1994. *ApJ*, **423**, L31.

Duerbeck, H.W., 1990. In 'Physics of Classical Novae', eds. A. Cassatella and R. Viotti, (Berlin: Springer), p. 34.

Gallagher, J.S., Hunter, D. & Tutukov, A.V., 1984. *ApJ*, **284**, 544.

Iben, I.,Jr. & Tutukov, A.V., 1989. *ApJ*, **342**, 430.

Iben, I.,Jr. & Tutukov, A.V., 1985. *ApJ SS*, **58**, 661.

Iben, I.,Jr. & Tutukov, A.V., 1992. *ApJ*, **389**, 369.

Iben, I.,Jr., Nomoto, K., Tornambe, A. & Tutukov A.V., 1987. *ApJ*, **317**, 717.

Mould, J., Cohen, J. & Graham, J.R., 1990. *ApJ*, **353**, L35.

Ruiz-Lapuente, P. & Fillipenko, A.V., 1993. In Proc. IAU Coll.145 'Supernovae and Supernovae Remnants'.

Scalo, J.M., 1986. *Fund. Cosmic Phys.*, **11**, 1.

Tutukov, A.V. & Khokhlov, A.M., 1992. *AZh.*, **69**, 754.

Tutukov, A.V. & Yungelson, L.R., 1979. *Acta Astron.*, **29**, 665.

Tutukov, A.V. & Yungelson, L.R., 1989. In 'Physics of Classical Novae', eds. A. Cassatella & R. Viotti, (Berlin: Springer), p. 325.

Tutukov, A.V. & Yungelson, L.R., 1992. *AZh*, **69**, 526.

Tutukov, A.V. & Yungelson, L.R., 1994. *MNRAS*, **268**, 871.

Tutukov, A.V., Yungelson, L.R. & Iben, I.Jr., 1992. *ApJ*, **386**, 197.

Woosley, S.E. & Weaver, T.A., 1994. *ApJ*, **423**, 371.

NOVA POPULATIONS

MASSIMO DELLA VALLE
Dipartimento di Astronomia, Università di Padova
vicolo Osservatorio 5, 35122, Padova, Italy.

Abstract.

Observations of novae in the Milky Way and in extragalactic systems would indicate the existence of two distinct families of novae: fast and bright *disk*-novae, originated by massive white dwarfs, and slow and faint *bulge*-novae, coming from less massive progenitors. A number of observational evidences seems also to suggest that the nova rate varies with the Hubble type of the parent galaxy.

1. Introduction

The search for novae in extragalactic systems has received, in the last decade, a considerable revival of interest after the astronomers realized their potential as distance indicators (see Jacoby et al. 1992 for a review). At the same time, often as a simple by-product, nova surveys enable us to shed light on the properties of the progenitors of the nova populations (e.g. Livio 1992a). In this respect, a valuable piece of information arises from observations of outbursting novae in nearby galaxies, like M31, M33 and the LMC, whose distances have been determined via primary indicators (e.g. Cepheids or RR Lyrae). Due to their closeness, these galaxies provide the best laboratories to study the distribution of the absolute magnitudes of the nova populations and in turn, in view of $M_B(\text{max}) \approx -8.3 - 10 \times \text{Log}(M_{WD}/M_\odot)$ (Livio 1992b), the distribution of the masses of the progenitors. However, systematic observations of novae in extragalactic systems provide another valuable parameter for nova theory: the nova frequency (= number of outbursts per unit of time). The nova frequency is indeed related to the number of progenitors of novae in a given galaxy, N_{CN}, through the simple relation: $N_{CN} = N_{out} \times T_R$, where T_R is the recurrence time between two outbursts and N_{out} the nova rate expressed in terms of number of outbursts per year.

503

A. Bianchini et al. (eds.), Cataclysmic Variables, 503-509.
© *1995 Kluwer Academic Publishers.*

2. *Disk* and *Bulge* Novae

A close investigation of the sites of the nova explosions of M31 novae led Ciardullo et al. (1987) and Capaccioli et al. (1989) to conclude that about 80% of novae of the Andromeda galaxy were produced in the bulge. This result leads to some interesting questions. In the 60's a nova survey of M33 was carried out by Rosino at Asiago Observatory. The analysis of the photographic material (Rosino and Bianchini 1973) led the authors to conclude that the nova rate in M33 had been underestimated until that time (e.g. Hubble and Sandage 1954). To confirm this preliminary result, in the 70's, Rosino started a second survey allowing Della Valle et al. (1994) to derive a nova rate for M33 of ≈ 5 novae per year. Since M33 is a bulgeless galaxy (Bothun 1992) its nova production has to be originated in the disk. All together these facts raise the following question: are there intrinsic differences between *bulge* novae belonging to M31 and *disk* novae originated in M33 or the LMC?

3. Observational Evidences

The existence of two distinct families of novae is supported by a number of observational facts:

a) The Maximum Magnitude *vs.* Rate of Decline relationships for M31 and LMC novae are different (see Fig. 1 and Fig. 2). The plots show that the nova population of the LMC is mainly formed by fast and bright novae, whereas a considerable fraction of novae of M31 is slow and faint.

b) Düerbeck (1990) assumes two different kinds of nova populations to better understand the fact that in the Milky Way the fast and bright novae are located in the neighborhood of the Sun while the slow and faint galactic novae are concentrated toward the galactic bulge.

c) Della Valle et al. (1992) find that the nova population in Galaxy is well represented by assuming two different distributions: the fast and bright 'disk' novae mainly concentrated toward the galactic plane ($z \lesssim 0.1 - 0.2$ kpc), and the slow 'bulge' novae extending up to $z \lesssim 1$ kpc, on average about 1.5–2 magnitudes fainter than the disk ones.

d) Tomaney and Shafter (1992), after monitoring a sample of novae belonging to the bulge of M31, conclude that these novae were spectroscopically different from the galactic novae observed in the neighborhood of the Sun (i.e. disk–novae).

e) The frequency distribution of the apparent magnitude at maximum of novae of M31 (Arp 1956, and Rosino 1964, 1973, 1989) exhibits a bimodal distribution with a peak-to-peak separation of ≈ 1.5 mag. Interestingly enough, such a bimodality was noted by Arp (1956, his Fig. 36), but

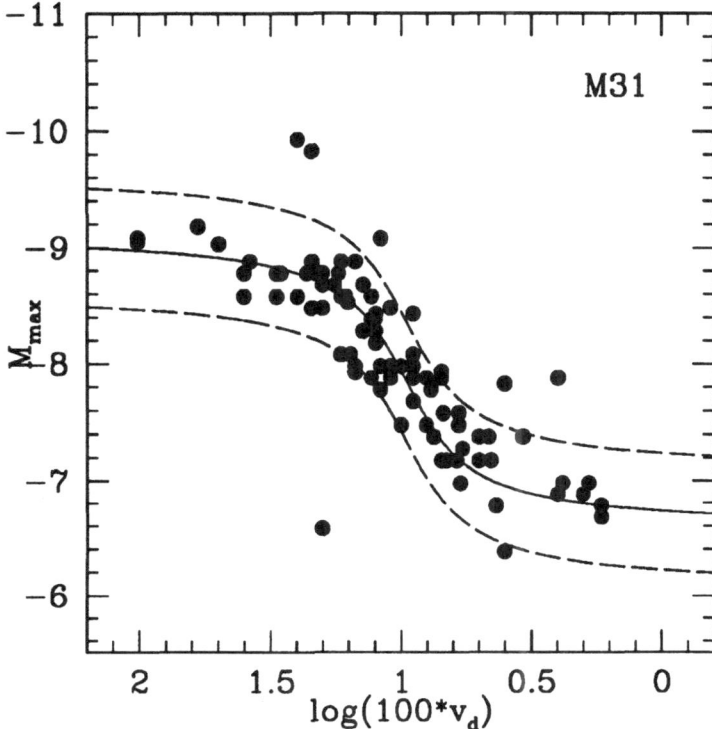

Figure 1. Maximum Magnitude *vs.* Rate of Decline relationships for novae of M31. The magnitudes for each nova have been transformed to absolute magnitudes after assuming $(m-M)_{M31} = 24.30$ and correcting for absorptions. The dashed lines define the 3σ strip of the mean relation.

neglected afterwards. A similar bimodal shape is also found by studying the frequency distribution of the absolute magnitude of the galactic novae. The rate of decline at minimum of the distribution is ≈ 0.13 mag/day (or $\lesssim 15^d$ or $\lesssim 40^d$ in terms of t_2 and t_3 respectively).

f) By using infrared luminosities as describers of the evolved populations Della Valle et al. (1994) find that the nova rate varies with the Hubble type of the parent galaxy (see Fig. 3). The overproduction of nova events in disk dominated systems could be the consequence of selection effects on the nova frequency (Truran and Livio 1986, Ritter et al. 1991). Disk novae are faster and brighter than bulge ones, therefore in view of Livio's (1992b) relationship (see Introduction) one can infer that 'on average' disk and bulge novae have different progenitors. In particular, by assuming $M_B \approx -9$ and $M_B \approx -7.3$ for disk and bulge novae respectively, we find $\langle M_{WD}\rangle_{disk} \approx 1.25\ M_\odot$ and $\langle M_{WD}\rangle_{bulge} \approx 0.8\ M_\odot$ for the masses of the white dwarfs in

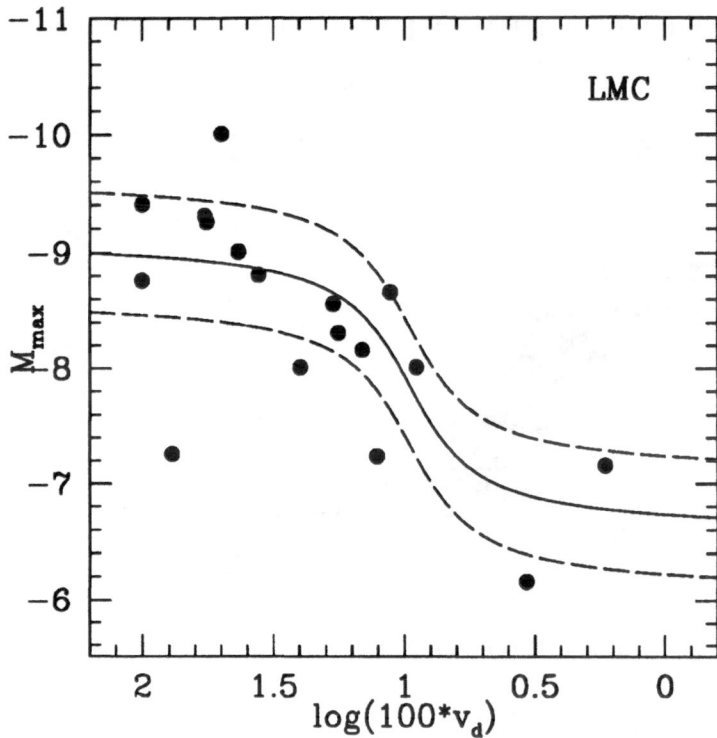

Figure 2. Maximum Magnitude *vs.* Rate of Decline relationships for novae of LMC. The magnitudes for each nova have been transformed to absolute magnitudes after assuming $(m-M)_{LMC} = 18.50$ and correcting for absorptions. The dashed lines define the 3σ strip of the mean relation.

'average' bulge and disk nova. From the results obtained by Ritter et al. (1991) we find that the expected ratio of outbursts between disk and bulge novae is: $N_{disk}/N_{bulge} \approx 4$, which compares well with ≈ 3 obtained from the observations.

The data for the nova populations are presented in Tab. 1.

TABLE 1

Galaxy	H	(B–H)	(m–M)	M_H	$L_{H\odot}$	novae/yr	ν_H
M31	0.86	2.70	24.3	−23.44	7.18	29±4	4.0±0.6
M33	4.36	1.56	24.5	−20.14	0.34	4.6±0.9	13.4±2.8
Virgo	6.09	3.35	31.47	−25.38	42.9	160±57	3.7±1.3
NGC 5128	4.83	2.65	27.96	−23.17	5.4	28±7	5.2±1.3
LMC	−0.9	1.2	18.5	−19.4	0.17	2.5±0.5	14.7±3

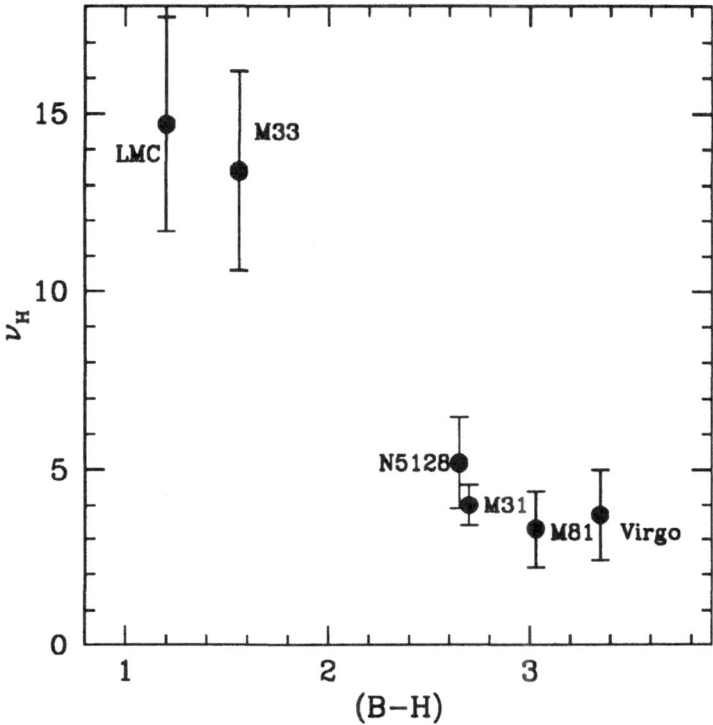

Figure 3. The nova rate per unit of H luminosity as function of the Hubble type of the parent galaxy.

The columns give: the name of the parent galaxy (col. 1), the H magnitude derived from Tully (1988), corrected for foreground and internal absorption (col. 2) and the (B–H) (col. 3). The (B–H) of N5128 and LMC were not directly available from Tully's catalogue. For LMC we have assumed a morphological type Sm and estimated (B–H)≈ 1.2 by extrapolating the color sequence from van den Bergh (1990). However, this value is very close to (B–H)=1.56 of M33 (morphological type Scd) reported in Tully's catalogue. For the peculiar galaxy NGC 5128 we derived (B–H)=2.65 which is bluer by ≈ 0.7 mag than the mean color [(B-H)=3.35] of a normal galaxy having the same Hubble type and according to Ciardullo et al. (1990) slightly bluer than M31. The distance moduli (col. 4) are from van den Bergh (1989). The absolute H magnitude is given in col. 5 and the H luminosity (in 10^{10} $L_\odot$) in col. 6. The nova rate (col. 7) for the LMC is from Graham (1979), M31 rate from Capaccioli et al. (1989), M33 rate from Della Valle et al. (1994) and the NGC 5128 data from Ciardullo et al. (1990). The source for the Virgo data is Pritchet and van den Bergh

(1987). Finally, we report the specific nova rate (novae per year per 10^{10} $L_\odot$) in the H color (col. 8).

g) After studying the spectroscopic evolution of a sample of novae in Galaxy and LMC, Williams (1992) concludes that novae can be divided into two classes of objects: the 'FeII' and the 'He/N' novae. Some preliminary investigations seem to indicate that bulge and disk novae might correlate with the 'FeII' and 'He/N' Williams's classes respectively.

4. Conclusions

Observations of galactic and extragalactic novae seem to indicate that nova systems in disk dominated galaxies are likely to be associated with massive white dwarfs then resulting in fast, intrinsically brighter and more frequent nova events. On the contrary, novae belonging to bulge galaxies are probably related to older and less massive progenitors.

The available data show that the nova rate depends on the Hubble type of the parent galaxy. In particular, the overproduction of novae observed in the late type galaxies may be the consequence of the fact that nova systems containing massive white dwarfs undergo more frequent outbursts.

An invariance in the specific nova rate (nova rate per unit of luminosity of the parent galaxy) across the whole spectrum of Hubble types (Ciardullo et al. 1990) seems unlikely after taking into account that the recurrence time between two consecutive outbursts dramatically depends on the mass of the white dwarf (Truran 1990). Massive white dwarfs in binaries (O–Ne– Mg) form mostly from stars having $\gtrsim 10\ M_\odot$ (Iben and Tutukov 1985) and thus they are probably associated with late type galaxies which have a still active star formation rate (e.g. Kennicutt 1989). In this respect, the study by Tomaney and Shafter (1992), who do not find any signature of outbursts of O–Ne–Mg novae in the bulge of M31, acquires particular relevance. Viceversa, these kinds of nova explosions seem to be relatively frequent in the Milky Way and possibly also in the LMC (e.g. Nova LMC 1991).

Our Galaxy has, probably, a later Hubble type than M31, therefore the use of the nova rate of M31 as unique calibrator to determine the nova rate of the Milky Way could be not adequate. Della Valle and Livio (1994) find a galactic nova rate of ≈ 24 novae per year by using the nova rates of M31, M33 and LMC as calibrators. However, we note the following: if the galactic nova production is disk-dominated (i.e. LMC–like) the galactic nova rate could increase up to 50 novae per year, thus removing part of the discrepancy with the estimate of 73 novae per year obtained by Liller and Mayer (1987).

The most serious deviation from the picture sketched above is the low nova rate occurring in the disk of M31. The rate estimated by Capaccioli et al. (1989) of $\approx$ 6 novae per year is indeed only 25% of what one could expect to find. The most likely explanation of this fact is the extinction within a thick and dusty disk like that of M31, which dramatically affects the pioneering nova surveys carried out in blue or pg colors. A piece of evidence in this direction comes from the nova search in the disk of M31 carried out by Irby and Shafter (in preparation) in H_α color. Since novae remain bright in H_α for much longer than in the B band and these observations are less affected by reddening, incompleteness should not be as much of a problem as in B or pg colors. Preliminary results of this search (Shafter, private communication) seem to suggest that the nova rate in the disk of M31 is likely to be higher than found in previous studies.

References

Arp, H. 1956, A. J., 61, 15.

Bothun, G.D. 1992, A.J., 103, 104.

Capaccioli, M., Della Valle, M., D'Onofrio, M., Rosino, L. 1989, A. J., 97, 1622.

Ciardullo, R., Ford, H., Neill, J.D., Jacoby, G.H., Shafter, A.W. 1987, Ap. J., 318, 520.

Ciardullo, R., Ford, H.C., Williams, R.E., Tamblyn, P., Jacoby, G.H. 1990, A. J., 99, 1079.

Della Valle, M., Bianchini, A., Livio, M., Orio, M. 1992, Astron. Astrophys, 266, 232

Della Valle, M., Livio, M. 1994, Astron. Astrophys, 286, 786

Della Valle, M., Rosino, L., Bianchini, A., Livio, M. 1994, Astron. Astrophys, 287, 403

Graham, J.A. 1979, in Changing Trends in Variable Star Research, IAU Colloquium N. 46, eds. F.M. Bateson, J. Smak, and I.H. Urch, p. 96.

Hubble, E., Sandage, A. 1954, Ann. Rep. Mount Wilson and Palomar Obs., 53, 23.

Iben, I. Jr. and Tutukov, A. V. 1985, Ap. J. Suppl., 58, 661.

Jacoby, G.H., Branch, D., Ciardullo, R., Davies, R., Harris, W., Pierce, M., Pritchet, C., Tonry, J., Welch, D. 1992, PASP, 104, 599.

Kennicutt, R. C. Jr. 1989, Ap. J., 344, 685.

Liller, W., Mayer, B. 1987, PASP, 99, 606

Livio, M. 1992a, in the proceedings of the 22nd 'SAAS FEE' Advanced Course, Interacting Binaries

Livio, M. 1992b, Ap. J., 393, 516.

Pritchet, C.J., van den Bergh, S. 1987, Ap. J., 318, 507.

Ritter, H., Politano, M., Livio, M., Webbink, R. 1991, Ap. J. 376, 177

Rosino, L. 1964, Ann. Astrophys., 27, 498.

Rosino, L. 1973, Astron. Astrophys. Suppl., 9, 347.

Rosino, L., Bianchini, A. 1973, Astron. Astrophys., 22, 461.

Rosino, L., Capaccioli, M., D'Onofrio, M., Della Valle, M. 1989, A.J., 97, 83

Tomaney, A., Shafter, A. 1992, Ap. J. Suppl., 81, 683

Truran, J. 1990, in Physics of Classical Novae, Madrid 1989, eds. A. Cassatella and R. Viotti, p. 373.

Truran, J., Livio, M. 1986, Ap. J., 308, 721.

Tully, B. 1988, Nearby Galaxies Catalog, Cambridge University Press.

van den Bergh, S. 1990a, P.A.S.P., 102, 1318

van den Bergh, S. 1990b, Ann. Rew. Astron. Astrophys., 1, 111.

Williams, R. 1992, A.J., 104, 725

PROPERTIES OF NOVA POPULATION MODELS

U. KOLB
Lick Observatory, UCSC, Santa Cruz CA 95064, USA
and
Max-Planck-Institut für Astrophysik, Garching, Germany

1. Introduction

Classical novae (CNe) are believed to occur in cataclysmic variables (CVs) as a result of a thermonuclear runaway (TNR) in hydrogen-rich layers ontop of the white dwarf (WD) which have accumulated prior to the explosion through mass transfer from the low-mass main-sequence secondary (see e.g. Livio 1994 for a review). The study of collective properties of CNe and the differences to collective properties of CVs in general could provide important clues on physical parameters determining the ignition and appearance of a nova event.

Previous attempts towards a theoretical prediction of the galactic nova population (by Ritter et al. 1991, focusing on the WD mass spectrum of CNe, and by Iben & Fujimoto 1992, aiming at the prediction of the distribution of mass transfer rates and orbital periods in CNe) had to rely on a number of simplifications, both for modeling the secular evolution of CVs and for describing the conditions leading to a TNR.

With the work by Politano (1990), deKool (1992) and Kolb (1993) detailed background models for the galactic population of CVs are now available. Here we report on first steps to transform these to the corresponding nova population models by combining them with recent nova computations by Prialnik et al. (see Prialnik 1994, Schwartzman et al. 1994).

2. Standard nova population models

We begin with numerical experiments to test the sensitivity of the predicted nova population to standard parameters entering the CV background distribution. For that we consider "intrinsic" nova population models obtained from standard CV population models (as in Kolb 1993) by adopting a very

A Bianchini et al (eds) Cataclysmic Variables, 511-515

simple criterion for determining the onset of a TNR. More precisely, we determine the nova outburst density, defined as the distribution of the nova rate $\nu = \dot{M}/\Delta M_{\rm acc}$ (where $\dot{M}$ is the mass transfer rate and $\Delta M_{\rm acc}$ the accreted mass needed to ignite the nova), and neglect any observational selection effects (such as: brighter novae are more likely to be observed). These are discussed in a forthcoming paper.

To compute $\Delta M_{\rm acc}$ we assume that a TNR is triggered when the pressure at the base of the envelope exceeds a critical value (see e.g. Livio 1994) and find

$$\frac{\Delta M_{\rm acc}}{M_\odot} = 2.24\,10^4 \left(\frac{R}{R_\odot}\right)^4 \left(\frac{M_1}{M_\odot}\right)^{-1}, \tag{1}$$

where M_1 and R_1 denote the WD's mass and radius.

Tabel 1 summarizes the essential input parameters for the CV population background models and lists, as characterizing parameters of the resulting nova population, the total nova frequency ν_Σ and intrinsic mean WD mass $\langle M_1 \rangle$ of CNe.

TABLE 1. Total intrinsic nova rate ν_Σ and mean WD mass $\langle M_1 \rangle$ predicted from various nova populations computed with Eq. (1)

model	formation			evolution		$n_{\rm CV}$	ν_Σ	$\langle M_1 \rangle$
	ref	α	$g(q)$	MB law	η	10^{-4} pc^{-3}	10^{-10} pc^{-3}yr^{-1}	$M_\odot$
m056	a	0.3	indep.	VZtotal	0	1.40	1.50	1.32
m052	a	1.0	indep.	VZtotal	0	1.84	1.15	1.29
m055	a	1.0	indep.	VZconv	0	1.84	1.13	1.29
m051	b	1.0	indep.	VZtotal	0	1.12	0.69	1.27
m053	a	1.0	q^{-2}	VZtotal	0	0.59	0.43	1.27
m053	c	1.0	q^{-3}	VZtotal	0	0.12	0.080	1.23
m058	c	1.0	q^{-3}	VZtotal	-0.2		0.049	1.19

Description of table headings. *Formation:* ref: formation rate model from deKool 1992 (a), Kolb & deKool 1993 (b), Politano 1990 (c); α: common envelope efficiency; $g(q)$: mass ratio distribution of ZAMS binaries, $dN \propto g(q)dq$ (indep.: masses independently from the same IMF). *Evolution:* MB law: magnetic braking law (VZtotal: Verbunt & Zwaan 1981; VZconv: see e.g. Kolb & Ritter 1992); η: long-term change of WD mass, $dM_1 = \eta dM_2$. $n_{\rm CV}$: predicted total CV space density.

A typical result for the orbital period and WD mass distribution is shown in Fig. 1 (full line). The strong selection in favour of high-mass WDs found by Ritter et al. (1991) is confirmed. Generally, the rise of the WD mass distribution towards the Chandrasekhar mass is less pronounced (and $\langle M_1 \rangle$ is smaller), if the common envelope efficiency α is large and the initial ZAMS mass ratio distribution $g(q)$ strongly prefers equal masses. Typically,

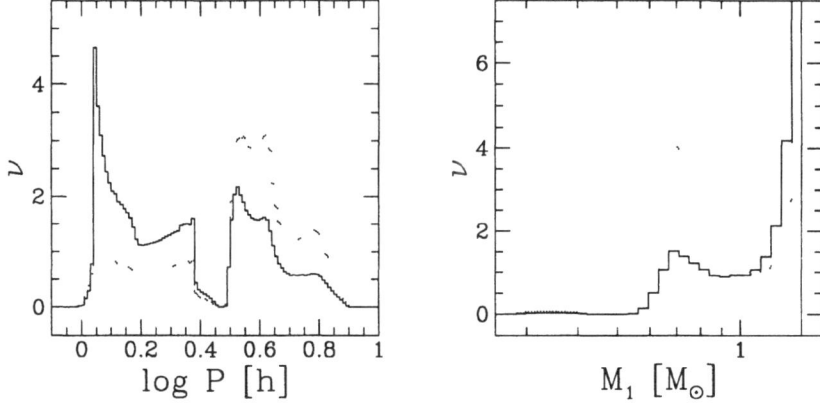

Figure 1. Predicted nova period distribution (left; based on model m052) and WD mass distribution (right; based on model m058). Full line: applying Eq. (1); dotted: for cold nova models according to Eq. (2).

by changing standard parameters in the CV population, the mean WD mass changes by less than $0.05 M_\odot$, in the most extreme case by $\simeq 0.1 M_\odot$. In lowest order the total nova rate scales as the predicted CV space density (and is thus mainly determined by $g(q)$); the ratio r of CNe below to above the period gap varies roughly between $1/3$ and $5/3$.

Although selection effects not yet accounted for in the models and large uncertainties in observational estimates of the WD mass in individual CNe prevent a meaningful direct comparison of the computed distributions with observations, it is interesting to note that the mean observed WD mass in CNe with determined orbital period (6 systems) is $0.79 M_\odot$, and $r \simeq 0.2$ (data from Ritter & Kolb 1995). Observational estimates for the total nova outburst density vary between 10^{-10} and $10^{-9} \mathrm{pc}^{-3} \mathrm{yr}^{-1}$ (see e.g. Della Valle & Duerbeck 1993 and references therein).

3. Cold WD nova population

TNR computations for CNe show that Eq. (1) has to be replaced by a more complicated form; in particular it turns out that ΔM_{acc} depends also on $\dot{M}$ and the WD temperature (see e.g. Kovetz & Prialnik 1985).

Well aware of the fact that the predicted function $\Delta M_{\mathrm{acc}} = \Delta M_{\mathrm{acc}}(M_1, \dot{M}, T_{\mathrm{WD}})$ is still model-dependent, we nevertheless apply here results from one specific set of nova computations where a diffusion-convection mechanism (Prialnik & Kovetz 1984) is assumed to be responsible for the metal enrichment of nova ejecta. These computations constitute the only available consistent set of detailed nova models covering an extended grid over the required range of M_1 and $\dot{M}$ (Prialnik 1994). For simplicity we restrict the discussion here to cold WDs ($T_{WD} = 10^6$K) which prevail in old CVs and thus dominate in the present CV population. The iginition mass.can be approximated by $\Delta M_{acc} = f \Delta M_{acc0}$, where ΔM_{acc0} is the canonical value for ΔM_{acc} given by Eq. (1), and

$$\log f \simeq 2 \log \left(\frac{M_1}{M_\odot} \right) - \frac{1}{3} \log \left(\frac{\dot{M}}{M_\odot \mathrm{yr}^{-1}} \right) - \frac{10}{3} ; \qquad (2)$$

typically $f \simeq 0.1 - 1$. The factor f is large for high-mass WDs (thereby suppressing the corresponding systems in the distribution) and small accretion rates.

Examples of resulting distributions are shown in Fig. 1 (dotted lines). To summarize the main findings: applying (2) reduces $\langle M_1 \rangle$ by $\simeq 0.1 M_\odot$ (a change dominating over the effects discussed in the previous section) and halfs ν_Σ and r, i.e. brings all these values closer to the observed ones. Model m058 (see Tab. 1) produces $r = 0.17$, a value within the observed range.

4. Conclusion and discussion

We have shown that existing standard CV population background models can be combined with theoretical nova models to predict distribution functions for observable properties like orbital period and WD mass of classical novae. The dependence of the ignition mass on binary parameters obtained from recent models for cold (old) WDs turned out to have a significant influence on the resulting nova outburst density. At this point it is interesting to note that for hot (young) WDs $\Delta M_{acc}(M_1, \dot{M})$ has the opposite differential behaviour with respect to $\dot{M}$ and a weaker dependence on M_1, so that the relative weight of high-mass WDs in the corresponding distribution is larger. This suggests that the difference between a halo and disk population of CNe indicated by observations (see e.g. Della Valle et al. 1992, Della Valle 1994) could be related to a different age of the underlying CV population, a possibility yet to be investigated in more detail.

Acknowledgements   The author would like to thank Dina Prialnik and Attay Kovetz for kindly providing results from their nova computations, and Mario Livio, Klaus Schenker and Rudolf Stehle for helpful comments.

References

Della Valle, M., 1994, this volume
Della Valle, M., Bianchini, A., Livio, M., Orio, M., 1992, A&A 266, 232
Della Valle, M., Duerbeck, H.W., 1993, A&A 271, 175
Iben, I., Jr., Fujimoto, M., MacDonald, J., 1992, ApJ 384, 580

Kolb, U., 1993, A&A 271, 149

Kolb, U., deKool, M., 1993, A&A 279, L5

Kolb, U., Ritter, H., 1992, A&A 254, 213

deKool, M., 1992, A&A, 261, 188

Kovetz, A., Prialnik, D., 1985, ApJ 291, 812

Livio, M., 1994, in *Interacting Binaries*, ed. S.N. Shore, M. Livio, E.P.J. van den Heuvel, Springer, Berlin, p. 135

Politano, M., 1990, in *Accretion-Powered Compact Binaries*, ed. C.W. Mauche, Cambridge University Press, Cambridge, p. 421

Prialnik, D., 1994, this volume

Prialnik, D., Kovetz, A., 1984, ApJ 281, 367

Ritter, H., Kolb, U., 1995, to appear in *X-ray Binaries*, W.H.G. Lewin, J. van Paradijs, E.P.J. van den Heuvel (Eds.), Cambridge University Press, Cambridge

Ritter, H., Politano, M., Livio, M., Webbink, R.F., 1991, ApJ 376, 177

Schwartzman, E., Kovetz, A., Prialnik, D., 1994, MNRAS 269, 323

Verbunt, F., Zwaan, C., 1981, A&A 100, L7

HIBERNATION — PROBLEMS AND ALTERNATIVES

K. MUKAI
Code 668, Laboratory for High Energy Astrophysics,
NASA/Goddard Space Flight Center, Greenbelt, MD 20771,
USA.

AND

T. NAYLOR
Department of Physics, Keele University,
Keele, Staffordshire, ST 5 5BG, UK.

Abstract. The hibernation hypothesis has stimulated various avenues of research into what happens to a nova long after its eruption. We highlight some recent observational results: IR light curves of the eclipsing nova WY Sge (1783) and a systematic survey of old nova spectra. Both of them fail to confirm the predictions of hibernation. We also point out that the majority of old novae do not show expected symptoms of an irradiated secondary, which is often invoked as a mechanism to achieve high accretion rate. We therefore explore alternatives in which novae do not hibernate and conclude that available observations are consistent with such models.

1. Introduction

In the pre-hibernation view, old novae were believed to maintain a more-or-less constant appearance in between successive outbursts. Since they tend to be much brighter than dwarf novae, they were treated as a separate sub-class of CVs, without understanding why they may be different. There were two perceived problems with this view: the observed rate of classical nova implied a very high space density of old novae, apparently much higher than the space density of all known CVs; the inferred accretion rates in old novae are higher than the theoretical upper limit for thermo-nuclear runaway (TNR).

A Bianchini et al (eds), Cataclysmic Variables, 517-522

The hibernation hypothesis (see Shara 1989 for a review) tries to solve this by considering a long-term (>100 years) change in the accretion rate, $\dot{M}$, of old novae. In some versions, mass transfer ceases completely; in others, accretion rate simply decreases. A prolonged period of low $\dot{M}$ would solve the theoretical problem of generating TNR in high $\dot{M}$ systems. Some version of hibernation seeks to solve the space density problem by creating many, temporarily detached systems, although very few such systems have been found so far.

In this paper, we summarize observational evidence for or against hibernation, focusing on new observations that have been obtained by our group. We will argue that hibernation is not strongly supported by current observations and present alternatives.

2. Evidence for Hibernation

Hibernation predicts significant changes in the luminosity of old nova centuries after the outburst, and possibly smaller changes decades after the eruption. Observers have attempted to verify or refute these predictions in two different directions.

2.1. ANCIENT NOVAE

One approach is to try to identify and study ancient novae, and recovery of two such systems were claimed by Shara and co-workers: CK Vul (1670) and WY Sge (1783).

The uncertain status of CK Vul is discussed in Naylor et al (1992). There is as yet no firm identification of the stellar remnant of this event; there remains a significant doubt as to whether the 1670 event was a classical nova; most important of all, the available descriptions convincingly show that the 1670 event was not a *typical* classical nova outburst. Regardless of the real nature of this object, we therefore cannot use it to infer the behavior of a typical old nova.

WY Sge (1783) is therefore the oldest securely identified classical nova. Naylor et al (1992) argued, however, that WY Sge is not any fainter than more recent old novae. Apparent faintness is in part due to inclination effect (WY Sge is eclipsing); its optical spectrum resembles those of recent novae, showing that no qualitative change has taken place during the last 200 years. Realizing that the eclipsing nature of this object can be used to derive tight constraints on system parameters, we have obtained high quality IR light curves of WY Sge (Somers et al, in preparation). The eclipse is found to be quite deep in the K band, showing that less than 50% of the light at K comes from the secondary star in this system. This fraction is comparable to that in the nova-like system IX Vel (Haug 1988) and is much smaller than in dwarf novae with similar orbital periods (U Gem and IP Peg, where >90% of K band light is from the secondary: Szkody & Mateo 1986). In other words, WY Sge remains at a brightness similar to old novae and nova-like systems, >200 years after eruption, and has not faded to the level of dwarf novae.

2.2. STATISTICAL APPROACH

The second approach is to try to detect small changes that may be expected within 100 years of eruption by studying a large sample of old novae. A systematic study of pre-eruption light curves and pre- and post-outburst magnitude comparisons had been carried out by Robinson (1975), before the hibernation hypothesis was put forward. In most cases, prenova and postnova magnitudes were found to be the same, usually within a few tenths of a magnitude. Versions of hibernation which invoke different mechanisms (e.g., unspecified long-term fluctuations in $\dot{M}$ before eruption and irradiation of the secondary afterwards) to explain the high luminosity of pre- and post-novae fail to explain this coincidence. This is particularly unlikely in light of the spread in M_v observed among CVs of different subclasses (4–8 for systems with $P_{orb} \sim 4$ hrs; Warner 1987). Robinson (1975) also finds incidences of significant brightening 1–15 years before eruption. However, this is likely to be a red herring: numerical modelling of nova outbursts show that shell nuclear burning has begun by that stage — it builds up slowly over a period of ~ 50 years and only then the explosive stage is reached (see, for example, Prialnik 1986). If prenova brightening is a general feature, then it is more likely to be the result of nuclear burning (i.e., some of the nuclear luminosity leaking out) than the cause. In fact, the entire existing database of pre-nova magnitudes may be irrelevant to the study of hibernation or of conditions that lead to classical nova outbursts.

Studies by Vogt (1990) and Duerbeck (1992) of post-nova magnitudes show secular decline of the order of 1 mag/century. However, the available data are insufficient to predict whether such a decline will continue well after 1 century after the outburst (Naylor et al 1992). Moreover, spectroscopic survey of Ringwald et al (this volume; full report in preparation) shows little sign of qualitative changes in nova spectra after the end of nebular stage. In particular, high excitation lines (non-existent in dwarf novae) are seen in many old novae regardless of age.

3. UV and Optical Spectra of Old Novae

It has been firmly established that old novae are optically bright; this is usually interpreted to imply high accretion rate. However, there is an increasing body of evidence to suggest that we do not understand the disks in old novae. First, in the case of V Per (1887), eclipse mapping shows a very flat temperature profile, leading to a complete failure of the normal steady state disk models (Wood et al. 1992). Of the various patches attempted by Wood et al, only magnetic model (truncated disk) could fit the data. Second, in the case of BT Mon (1938), the UV luminosity as measured with IUE is much smaller than inferred from the optical data using the standard disk model (Selvelli et al 1990). In both these case, the problem is that these systems have flatter UV/optical spectra than the standard model predicts.

Emission lines may provide a clue as to the origin of this problem: Many old novae show prominent emission lines of HeII λ4686 and the Bowen blend. This immediately tells us that there is a hot continuum source with strong flux shortward of the HeII edge at 228Å, and that irradiation is contributing significant line flux. (Note that high excitation lines are not an evidence of magnetic accretion; it simply implies the presence of a hot, ionizing continuum, which magnetic CVs do produce efficiently, but so do old novae.) It is probable that a part of the same EUV flux will be reprocessed into UV/optical continuum photons.

4. Irradiated Secondary?

Versions of hibernation hypothesis invoke irradiation of the secondary to explain temporarily elevated levels of accretion; more generally, evolutionary theories of interacting binaries involving irradiated secondaries have become quite fashionable in recent years. If irradiation plays a significant role in old novae, however, it should have immediate and obvious observational consequences beyond high accretion rate. This follows from the analogy with well-known "reflection effect" in close (but detached) binaries involving a hot sub-dwarf and a late type star, such as MT Ser and BE UMa (Grauer & Bond 1983; Ferguson et al 1987; see also Hilditch 1994 for a review). These systems show prominent photometric and emission line variabilities caused by the changing visibility of the heated face of the secondary.

Light-curve fitting show that most, if not all, the incident flux is radiated away locally in these systems (Hilditch 1994). Since irradiation by the hot white dwarf in an old nova is similar to that in the detached sub-dwarf binaries in terms of the spectrum, we expect same local reprocessing in old novae as well. This is indeed seen in the optical light curves of several old novae: V1500 Cyg, CP Pup, GQ Mus and V2214 Oph. These examples show that irradiated secondary can be a significant contributor of optical light (i.e., $M_v \sim$5). This compares favorably with a simple calculation: Irradiation by a 10^{35} erg s^{-1} white dwarf of a 0.4R$_\odot$ secondary 1R$_\odot$ away would raise the temperature to 10,000K and M_v to 4.7, if there is no accretion disk to shield the secondary from the EUV/FUV flux.

In the majority of old novae, observational evidence for an irradiated secondary is less strong. For example, IR light curves of WY Sge (Sommers et al in preparation) do indicate some irradiation of the secondary, but at a much lower level. This is probably because the accretion disk shields the secondary from irradiating flux near the orbital plane. Instead, irradiation of the accretion disk may be an important process in old novae: an irradiated disk has a flatter temperature distribution than an α-disk and produces high excitation emission lines (§3). If this is the case, then M_v of the disk cannot be used as a good indicator of $\dot{M}$.

We therefore advocate caution in accepting irradiated secondary as a possible cause of high $\dot{M}$ for old novae in general. Theoretical models should be improved: can irradiation still work in the presence of an accretion disk, which would surely shield the L_1 point from direct irradiation (and if so, at what level)? Observationally, how many old novae show evidences of reflection effect, particularly in optical photometry? A sufficient amount of raw observational data may already exist (every failed attempt to determine the orbital period of an old nova through photometry is also an evidence against significant irradiation); if so, a systematic analysis should be carried out.

5. Alternatives to Hibernation

Despite great effort during the last decade, observational evidence for hibernation is still weak. The brightness of old novae may decrease slightly over the period of ~50 years after outbursts, but qualitative differences appear to remain between very old novae and dwarf novae, at least 200 years after outburst.

We propose that CVs have distinct nova-active and nova-inactive stages in their long-term evolution. In a nova-inactive stage, CVs have low accretion rate and likely to undergo dwarf nova outbursts. Such CVs can still have classical nova outbursts, but much less frequently than systems in a nova-active stage. If the frequency of nova outbursts is different by a factor of 100, this would be sufficient to explain the lack of bona fide dwarf novae among old novae. Naylor et al (1992) have already shown that nova rate is a poor indicator of the space density of CVs, when such considerations are taken into account.

Do old novae truly have high $\dot{M}$? We believe this should be a focus of future observational studies. There are indications that accretion disk in old novae reprocess a significant amount of flux from the still-hot white dwarf. This would imply that $\dot{M}$ estimates for old novae may be too high.

This would reduce the apparent spread in $\dot{M}$ for given orbital period, and remove the theoretical problem of generating nova outbursts in high $\dot{M}$ systems.

Old novae may still have somewhat higher $\dot{M}$ than dwarf novae. This in itself may be sufficient to make them nova-active. The different appearances of old novae and dwarf novae are then consequences of different $\dot{M}$ combined with the effect of increased irradiation in old novae.

Alternatively, the effective temperature of the white dwarf (T_{eff}) may be the distinguishing parameter. It is conceivable that the repeated nova outbursts in a nova-active stage can keep the T_{eff} quite high throughout the inter-outburst cycle. This, in turn, could have the effect of speeding up the nova outburst cycle (Prialnik, this volume), although this is an area which needs further study.

6. Conclusion

Extensive observational studies have failed to confirm hibernation of old novae, at least in its more dramatic versions, for the ~200 year period after eruption. Although studies of very old novae should continue, a fresh look at the long-term evolution of CVs seems to be in order.

References

Duerbeck, H. W. 1992, *MNRAS*, **258**, 629.
Ferguson, D.H., Liebert, J., Cutri, R., Green, R.F., Willner, S.P, Steiner, J.E. & Tokarz, S. 1988, *ApJ*, **316**, 399.
Grauer, A.D. & Bond, H.E. 1983, *ApJ*, **271**, 259.
Haug, K. 1988, *MNRAS*, **235**, 1385.
Hilditch, R.W. 1994, *Observatory*, **114**, 212.
Naylor, T., Charles, P.A., Mukai, K & Evans, A. 1992, *MNRAS*, **258**, 449.
Prialnik, D. 1986, *ApJ*, **310**, 222.
Robinson, E.L. 1975, *AJ*, **80**, 515.
Selvelli, P.L., Cassatella, A., Bianchini, A., Friedjung, M. & Gilmozzi, R. 1990, In: *Physics of Classical Novae, IAU Colloq. No. 122*, eds. Cassatella, A. & Viotti, R., Springer-Verlag, Berlin, p65.
Shara, M.M. 1989, *PASP*, **101**, 1.
Szkody, P. & Mateo, M. 1986, *AJ*, **92**, 483.
Vogt, N. 1990, *ApJ*, **356**, 609.
Wood, J.H., Abbott, T.M.C. & Shafter, A.W. 1992, *ApJ*, **393**, 729.

MASS TRANSFER CYCLES IN CV

A.R. KING
Astronomy Group, University of Leicester, Leicester LE1 7RH, U.K.

Abstract. The inferred range of accretion rates in CVs requires them to undergo long–term mass transfer cycles on timescales $\gtrsim 10^5$ y. I show that limit–cycle behaviour can occur in CVs provided that the radius varies in response to the mass transfer rate. This property makes it very likely that CV mass transfer cycles are a result of weak irradiation of the secondary star by the accreting component.

1. Introduction

A striking feature of cataclysmic variables is their great variety. We know of at least two types of novae (classical and recurrent), four types of dwarf novae (U Gem, Z Cam, SU UMa, WZ Sge), and three types of novalikes (UX UMa, VY Scl, RW Sex), and the list could easily be lengthened. An extremely simple model, in which a semidetached low–mass star transfers mass to a white dwarf, is apparently able to describe this vast range of phenomena. This is all the more remarkable since the masses of the two stars are relatively tightly constrained; it is a tribute to the ingenuity of astronomers that is it possible to create plausible models of so many of these systems from such a simple starting point. A key ingredient of this success is a considerable freedom in assigning mass transfer rates. For example, essentially all models of dwarf novae assume that their transfer rates are systematically lower than in steady systems such as novalikes, and there is considerable observational support for this.

This systematic difference is at first sight in conflict with the basic predictions of secular evolution under orbital angular momentum loss. This assigns a nearly unique transfer rate to a CV at a given orbital period, even

A Bianchini et al (eds), Cataclysmic Variables, 523–532
© *1995 Kluwer Academic Publishers*

though observational estimates (e.g. Patterson, 1984; Warner, 1987) show considerable scatter, and systems such as U Gem dwarf novae and various novalikes seem quite happy to inhabit the same period ranges for example. More seriously still, systematic differences in long–term transfer rates would undermine the whole line of explanation of the CV period histogram which has been emerging over the last fifteen years. The essence of this picture is the systematic deviation of the secondary star from its thermal–equilibrium radius as a result of mass transfer. The observed sharpness of the CV minimum period and period gap can only be explained if the secondaries of virtually all systems show almost precisely the same degree of thermal disequilium at a given period. This could not hold if the transfer rates in various CV subtypes were systematically different.

The resolution of this apparent paradox has long been known (e.g. Hameury et al, 1989). The inferred spread in transfer rate $-\dot{M}_2$ versus period P must represent the effect of long–term fluctuations in the transfer rate around its evolutionary (secular) mean. The binary evolves on the angular momentum loss timescale $-J/\dot{J}$, and the evolutionary mean is established on a timescale $t_{\mathrm{evol}} \gtrsim t_{\mathrm{L}}$, where $t_{\mathrm{L}} = H t_J / R_2$ is the timescale for the Roche lobe to move a stellar scaleheight H with respect to a secondary star of radius R_2. As $H/R_2 \simeq 10^{-4}$ and $t_J \gtrsim 10^8$ y we have $t_{\mathrm{evol}} \gtrsim t_{\mathrm{L}} \gtrsim 10^4$ y. Fluctuations on any timescale less than this will not affect the evolutionary mean and the resulting period histogram. Since we do not *observe* dwarf novae becoming novalikes, and so on, the fluctuation time is evidently longer than at least a few centuries: in particular, the fluctuations are a quite distinct phenomenon from the low state behaviour observed in the VY Scl novalike variables.

2. Mass Transfer Fluctuations

The discussion above shows that a knowledge of the properties of the inferred fluctuations is basic to a full understanding of CVs. In particular such insight is required if we are to discover the relationship of the various CV subtypes to each other. Clearly we seek some kind of limit–cycle behaviour; there have been various suggestions as to the likely cause, such as mass or angular momentum losses from accretion discs, or other properties of the accretion flow. All such "external" mechanisms suffer the inherent drawback that their connection to the secondary star may itself be subject to variations such as instabilities, making the condition that the transfer rate should average to the secular mean over many cycles hard to fulfil. Clearly a mechanism involving the properties of this star directly is likely to prove more robust. The most obvious thing to consider is the effect of variations in the radius R_2 of this star. This affects the mass transfer rate $F = -\dot{M}_2$ through a relation of the form

$$F = f(R_2 - R_{\rm L}). \tag{1}$$

(It is convenient to use F instead of $-\dot{M}_2$ both to remove confusion over signs and because the equations are first–order in F.) The function f is very sensitive to its argument (generally exponentially, e.g. Ritter, 1988). In fact we do not need its explicit form, but only the properties (Kolb & Ritter, 1990)

$$\frac{\partial f}{\partial t} = 0, \ \frac{\partial f}{\partial (R_2 - R_{\rm L})} = \frac{f}{H} > 0. \tag{2}$$

These assert that F depends on time only through the difference $R_2 - R_{\rm L}$, and changes significantly if and only if the latter varies by at least a scaleheight. Differentiating (1) wrt t gives

$$\dot{F} = \frac{\partial f}{\partial (R_2 - R_{\rm L})}(\dot{R}_2 - \dot{R}_{\rm L}) \simeq \frac{f}{H}\left(R_2\frac{\dot{R}_2}{R_2} - R_{\rm L}\frac{\dot{R}_{\rm L}}{R_{\rm L}}\right).$$

Using $F = f$ and $R_2 \simeq R_{\rm L}$ gives

$$\dot{F} = F\frac{R_2}{H}\left(\frac{\dot{R}_2}{R_2} - \frac{\dot{R}_{\rm L}}{R_{\rm L}}\right). \tag{3}$$

We see immediately that the usual assumption of *steady* mass transfer ($\dot{F} = 0$) requires $\dot{R}_2/R_2 = \dot{R}_{\rm L}/R_{\rm L}$, i.e. that the stellar radius and Roche lobe move together. As is well known (e.g. King, 1988), in CVs the term $\dot{R}_{\rm L}/R_{\rm L}$ is determined by the orbital angular momentum losses, i.e.

$$\frac{\dot{R}_{\rm L}}{R_{\rm L}} = 2\frac{\dot{J}}{J} + \left(\frac{5}{3} - 2\frac{M_2}{M_1}\right)\frac{F}{M_2}. \tag{4}$$

To fix the radius response $\dot{R}_2/R_2$ one usually assumes some mass-radius relation for the mass-losing star, e.g. $R_2 \propto M_2$ if the star is close to the main sequence. This would give $\dot{R}_2/R_2 = -F/M_2$, and thus the usual relation $F/M_2 = -\dot{J}/J(4/3 - M_2/M_1)$ (cf King 1988) in the steady case. To include the short–term response to radius variations we generalize this relation, and set

$$\frac{\dot{R}_2}{R_2} = K(R_2, F) - \zeta\frac{F}{M_2}. \tag{5}$$

Here ζ is the adiabatic mass–radius index ($\simeq -1/3$ for a low–mass star). Thus $K(R_2, F)$ represents all other forms of radius variation. In particular, it must include the effects of thermal relaxation. For the time being we impose no restrictions on K and simply investigate the effect of different forms of it on F. Using (4) and (5) the equation (3) determining the variation of the transfer rate becomes

$$\dot{F} = 2\frac{R_2}{H}F\left(-\frac{\dot{j}}{J} - D\frac{F}{M_2} + \frac{K}{2}\right),$$
(6)

with

$$D = \frac{5}{6} + \frac{\zeta}{2} - \frac{M_2}{M_1}.$$
(7)

3. Phase-Plane Analysis

The mass–transfer behaviour of the system is completely determined by (5, 6). Since $\dot{J}/J$ is given algebraically in terms of system parameters, which along with ζ are slowly varying, these are two first–order ordinary differential equations for R_2 and F which to a good approximation do not contain t explicitly. Hence in this formulation *the mass transfer behaviour of CVs is determined by a plane autonomous equation system.* It is well known that such systems can be described qualitatively without the need for explicit integration. A future paper will give a full phase–plane analysis of this type, but here I shall use a simpler approach.

We consider the fixed points of the system (5, 6), i.e. the loci in the R_2, F plane specified by the equations $\dot{R}_2 = 0, \dot{F} = 0$. These are

$$F = \frac{M_2 K}{\zeta}$$
(8)

$$F = \frac{M_2}{D}\left(-\frac{\dot{j}}{J} + \frac{K}{2}\right).$$
(9)

The intersection of these curves defines one or more equilibrium points R_0, F_0, which the system will attempt to evolve towards if permitted by the form of the undetermined function K. The evolution near such fixed points thus characterizes the global behaviour of the system. In practice CVs will not reach an equilibrium point: this would require a form of K with the curious property of exactly balancing the adiabatic reaction to mass loss. We therefore expect that reasonable forms of K will prevent evolution towards this point.

To study the behaviour near R_0, F_0 we set $R_2 = R_0 + \Delta R, F = F_0 + \Delta F$

and expand (5, 6) to first order in $\Delta R, \Delta F$. This gives

$$\Delta\dot{F} = \frac{R_2}{H}F_0\left[K_R\Delta R + \left(K_F - \frac{2D}{M_2}\right)\Delta F\right] \tag{10}$$

$$\Delta\dot{R} = R_0\left[K_R\Delta R + \left(K_F - \frac{\zeta}{M_2}\right)\Delta F\right]. \tag{11}$$

Here K_R, K_F are the partial derivatives of K wrt R, F evaluated at the fixed point R_0, F_0.

For any given pair of values K_R, K_F these equations determine the motion of the the system point R_2, F in the neighbourhood of the fixed point, as they give the sign of $\Delta\dot{F}, \Delta\dot{R}$ as this point crosses the curves $\dot{F} = 0, \dot{R} = 0$. The quantities K_R, K_F also fix the slopes of the latter curves as

$$\left(\frac{dF}{dR_2}\right)_{\dot{F}=0} = \frac{(M_2/2D)K_R}{1 - (M_2/2D)K_F}, \tag{12}$$

$$\left(\frac{dF}{dR_2}\right)_{\dot{R}_2=0} = \frac{(M_2/\zeta)K_R}{1 - (M_2/\zeta)K_F}, \tag{13}$$

which follow on differentiating (8, 9).

Figure 1a shows an example for the case

$$K_R < 0, K_F > \frac{2D}{M_2} > 0, \zeta < 0.$$

The curves $\dot{F} = 0, \dot{R}_2 = 0$ cross as shown, dividing the R_2, F plane into four regions; the arrows in the Figure indicate the direction of motion of the system point in each of these regions. As can be seen, the system point must spiral outwards from the fixed point. This is a growing (overstable) oscillation. For different choices of K_R, K_F we get the global behaviours shown in Figures 1b–f, corresponding to damped oscillations, and stable and unstable situations of various types.

4. Limit Cycles

As explained in the Introduction, we are particularly interested in the possibility of limit cycles. In the R_2, F plane, these manifest themselves as closed trajectories. We therefore ask if there exist forms of K which force the system to execute such cycles. Note that this mathematical approach is precisely analogous to that adopted in early discussions of the S–curve giving a local thermal–viscous instability in accretion discs. (e.g. Bath & Pringle

Figure 1

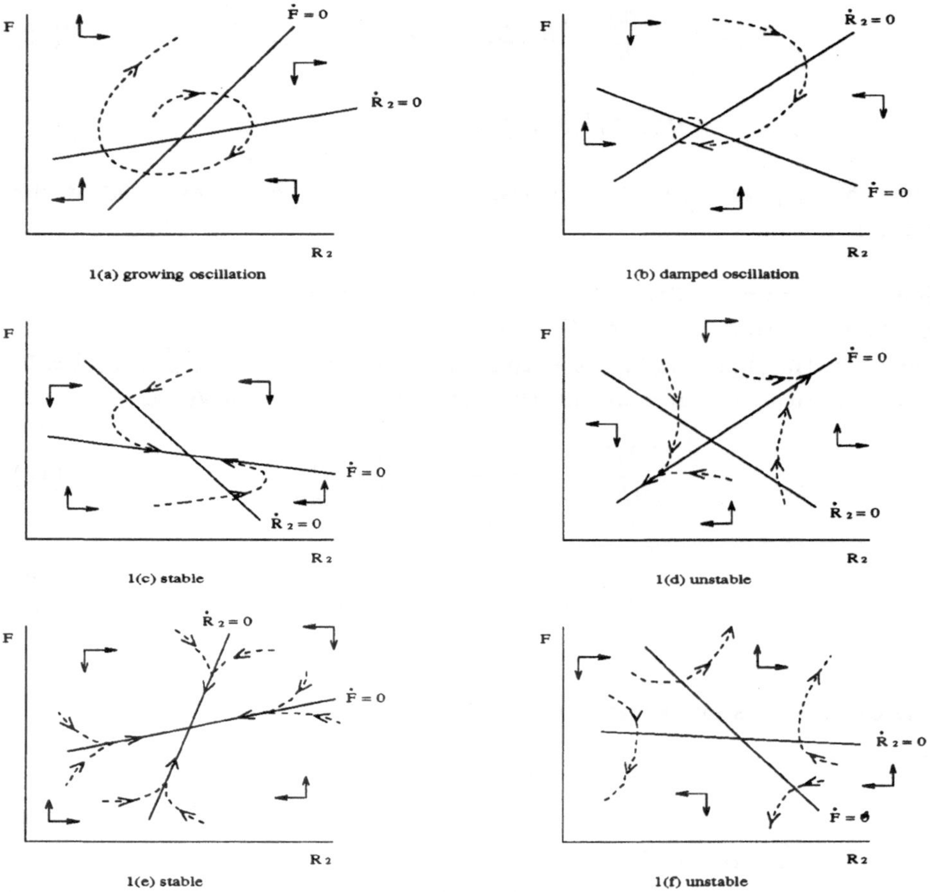

1(a) growing oscillation

1(b) damped oscillation

1(c) stable

1(d) unstable

1(e) stable

1(f) unstable

Figure 1. Possible types of behaviour near a fixed point in the R_2, F plane

1982). The main difference here is that our discussion is already global. A clue is provided by the Poincaré–Bendixson Theorem (e.g. Minorsky 1962): such closed trajectories must surround an unstable fixed point. Also, we require the trajectory to be an attractor for system points outside it. To find a suitable K we thus choose a form giving an unstable fixed point where the curves $\dot{F} = 0$, $\dot{R}_2 = 0$ intersect; away from this point we choose a form which bends these curves in such a way as to produce an attractor (cf Figure 2). A great simplification results from the fact that the timescales in the two governing equations (5, 6) are sharply different; the presence of the factor R_2/H on the rhs of (6) shows that F changes by large factors in response to small changes in R_2. This means that the system will be at-

Figure 2: Limit Cycle

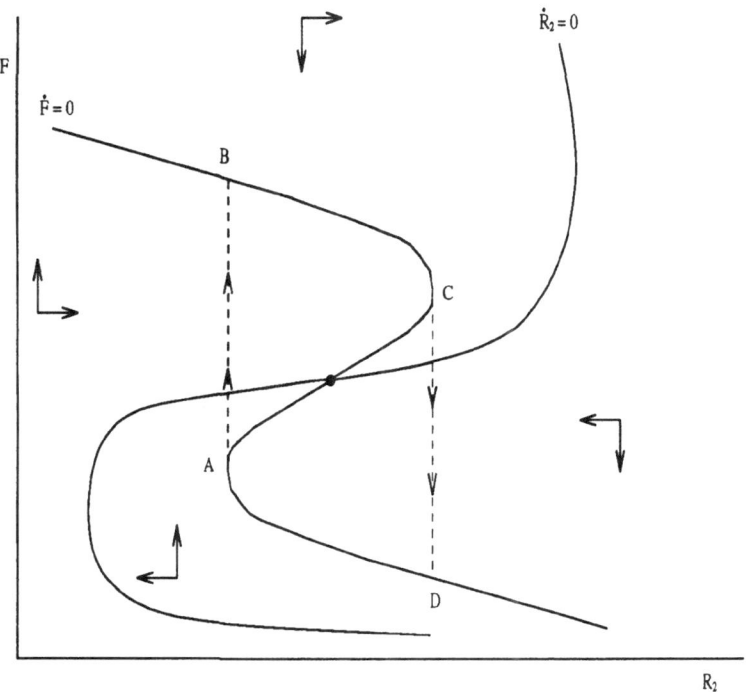

Figure 2. The curve $\dot{F} = 0$ (and $\dot{R}_2 = 0$) for a particular choice of the function $K(R_2, F)$ giving limit cycles

tracted to the curve $\dot{F} = 0$ much more rapidly than to $\dot{R}_2 = 0$; the system will spend most of the time on the $\dot{F} = 0$ curve, attempting to find the intersection with $\dot{R}_2 = 0$ on a much longer timescale $\sim K^{-1}$. Any deviation from the $\dot{F} = 0$ curve is removed on a timescale $\sim (H/R_2 K) \sim 10^{-4} K^{-1}$, corresponding to near–vertical jumps between branches of the $\dot{F} = 0$ curve in the R_2, F plane.

The particular example shown in Fig. 2 is simply the growing oscillation of Fig. 1a, with K (an inverse cubic in F) chosen to deform the curve $\dot{F} = 0$ into a Z–shape in the large. As can be seen from the arrows, the system point can only remain on the sections of the $\dot{F} = 0$ curve having negative slopes; the central section AC with $dF/dR_2 > 0$ is unstable. From a random initial position in the R_2, F plane the system point is rapidly attracted to one of the stable sections of the Z–curve, either to the left of C or to the right of A. Once there, the system point is inevitably attracted into the region bounded by ABCD, and must execute limit cycles as shown. The

Figure 3

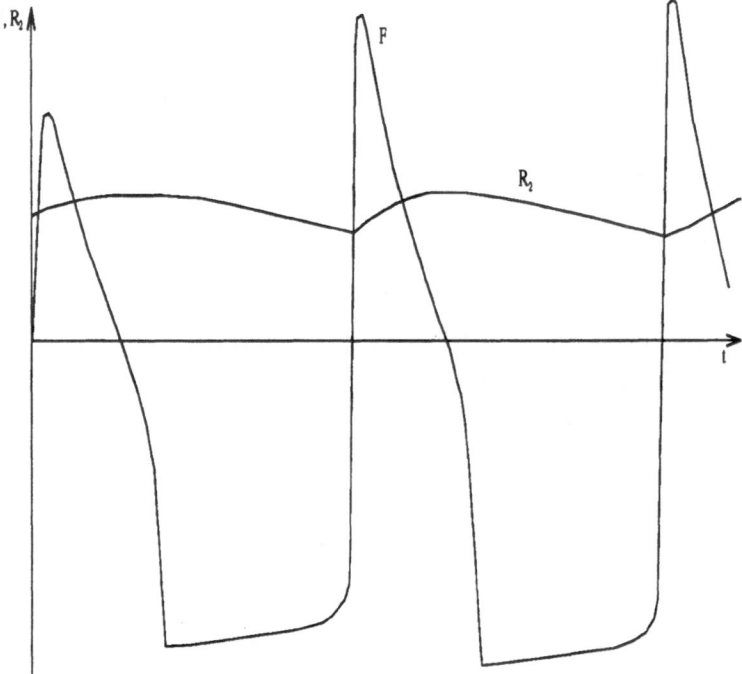

Figure 3. The mass transfer rate $F(t)$ and radius $R_2(t)$ corresponding to the K of Fig. 2

corresponding behaviours of F, R_2 are shown in Fig. 3.

Of course this is for the moment merely a mathematical example: we have still to show that the *real* $\dot{F} = 0$ curve has a form giving a limit cycle. The analogous problem in disc instability theory is not yet solved, as one still has to appeal to large changes in disc viscosity between high and low states, often expressed in terms of discontinuous changes in the Shakura–Sunyaev α–parameter. However constructing examples such as the Z–curve of Fig. 2 does have at least two desirable features:

1. it shows that genuine limit–cycle behaviour is at least possible, and

2. it gives a clue as to the real physical mechanism invoved in producing the cycles.

The latter result follows from noting that K must depend explicitly on F to give us any chance of producing a limit cycle. This is plausible from the discussion here, and can be proved rigorously (King, Frank, Kolb &

Ritter, in prep.). There is only one reasonable physical mechanism that can provide such a dependence of the radius expansion rate on the instantaneous mass transfer rate (remember that the adiabatic response has been subtracted out). *The physical mechanism driving mass transfer cycles is almost certainly irradiation of the secondary star by the accreting component.* We can go further, and make a quantitative estimate of the irradiating luminosity. Clearly limit–cycle behaviour requires the system to be in the neighbourhood of a fixed point, where $\dot{F} \simeq 0, \dot{R}_2 \simeq 0$. From (8, 9) we see that this requires

$$F \simeq -\frac{M_2 \dot{J}}{(D - \zeta/2)J}, K \simeq -\frac{\zeta \dot{J}}{(D - \zeta/2)J}.$$

Since D, ζ are $O(1)$ these relations imply

$$|K| \sim \frac{|\dot{J}|}{J} \sim \frac{F}{M_2} \tag{14}$$

If irradiation is the basic cause of the cycles, K must correspond to the thermal expansion rate given by blocking some part ΔL of the luminosity of the secondary, i.e.

$$K \sim \frac{\Delta L R_2}{G M_2^2}.$$

Combining with (14) we see that

$$\Delta L \sim \frac{G M_2 F}{R_2} \sim \frac{M_2}{M_1} \frac{R_1}{R_2} L_{\rm acc} \sim 10^{-2} L_{\rm acc}, \tag{15}$$

where $L_{\rm acc}$ is the accretion luminosity. Hence the irradiation required to drive the cycles is rather weak, amounting to about 1% of the accretion luminosity only. This required fraction is in very good agreement with what one quite independently expects the secondary to intercept on purely geometrical grounds (the secondary subtends a fractional solid angle of a few percent at the primary, and will not absorb all of the radiation falling on it). Note that for low–mass X–ray binaries the required ΔL is (from (15, with $R_1 \sim 10^6$ cm rather than 10^9 cm) about $10^{-5} L_{\rm acc}$, which is not in good accord with the geometrical cross–section of the secondary.

We conclude that weak irradiation of the secondary star is a very promising candidate for driving mass transfer cycles in CVs. This conclusion, even down to the quantititative estimate (15), is remarkably close to that of Ritter, Kolb & Zhong Zhang (this volume), who give an explicit treatment of the effects of weak irradiation. In the future it should be possible to determine the effective form of the function K in this treatment, and discover under what conditions it leads to limit cycle behaviour.

5. Acknowledgments

I thank Uli Kolb, Hans Ritter and Juhan Frank for many valuable discussions.

6. References

Bath, G.T., Pringle, J.E., 1982, MNRAS 199, 267
Hameury J.-M., King, A.R., Lasota J.-P, 1989, MNRAS 237, 39
King, A.R., 1988, QJRAS 29, 1
Kolb, U., Ritter, H., 1990, A&A 236, 385
Minorsky, N., 1962, *Non-linear Oscillations*, Van Nostrand
Patterson, J., 1984, ApJ Suppl. 54, 443
Ritter, H., 1988 A&A 202, 93
Warner, B. 1987, MNRAS 227, 23

SEARCHING FOR CLOSE DOUBLE DEGENERATES

ANGELA BRAGAGLIA
Osservatorio Astronomico di Bologna
via Zamboni, 33 - I-40126 Bologna, Italy

Despite extensive efforts (Robinson & Shafter 1987; Bragaglia et al. 1990; Foss, Wade & Green 1991; Bragaglia, Greggio & Renzini 1994), not a single "double degenerate" (DD) system with the right properties to be a Supernova Ia precursor has yet been found. We need to increase the number of White Dwarfs examined and to better discriminate in some dubious cases of possible binaries for a decisive observational test. We have now reliable data on more than 80 WDs (2/3 DAs, 1/3 DBs); several of them show marginal detection of radial velocity variations between spectra, a few have a mass so low (Bragaglia, Renzini & Bergeron 1995) to be suggestive of evolution in a close binary system, and three are confirmed binaries (one not-too-close DD, and two WD + red dwarf pairs, Bragaglia et al. 1990). During 1993 we reobserved 6 field DA White Dwarfs, at the ESO 1.5m telescope. From several sky lines and bands we derive a precision of our wavelength scale of ~ 30 kms^{-1}. None of the WDs shows incontrovertible evidence of binarity. We note that all confirmed or suspect binary in our sample show Δv_r's of the order of 100 kms^{-1}, at odds with the expectation (Iben 1990; Yungelson et al. 1994) of a nearly equal number of DDs with the semiamplitude of the radial velocity curve $K \leq 100$ kms^{-1} and with $100 \leq K \leq 500$ kms^{-1}. Further study of all objects showing possible Δv_r's is foreseen.

References

Bragaglia A., Greggio L., Renzini A. & D'Odorico S. 1990, ApJ, 365, L 13
Bragaglia A., Greggio L. & Renzini A. 1994, MemSAIt, 65, 411
Bragaglia A., Renzini A. & Bergeron P. 1995, ApJ, in press
Foss D., Wade R.A. & Green R.F. 1991, ApJ, 374, 281
Iben I. 1990, ApJ, 353, 215
Robinson E.L. & Shafter A.W. 1987, ApJ, 332, 296
Yungelson L.R., Livio M., Tutukov A.V. & Saffer R.A. 1994, ApJ, 420, 336

A Bianchini et al (eds), Cataclysmic Variables 533
© *1995 Kluwer Academic Publishers*

U SCO TYPE RECURRENT NOVA AS A PROGENITOR OF THE HALO TYPE LOW MASS X-RAY BINARIES

KAZUHIRO SEKIGUCHI
South African Astronomical Observatory
P.O. Box 9, Observatory 7935 South Africa

A possible evolutionary link between U Sco subclass recurrent novae (RNe) and low mass X-ray binaries (LMXRBs) has been suggested (Williams *et al.* 1987). The questions were if the white dwarf (WD) of the U Sco subclass can grow to the Chandrasekhar limit and if it undergoes accretion-induced collapse (AIC) to form a neutron star. Indeed, the massive WDs in the U Sco subclass RNe are most likely to be O+Ne+Mg WDs which are increasing mass at high accretion rates ($M > 10^{-8}$ $M_\odot yr^{-1}$) to reach the Chandrasekhar limit (Livio and Truran 1992). Furthermore, such a system may cause AIC by electron capture on ^{24}Mg and ^{20}Ne (Nomoto and Kondo 1991).

From the above arguments, an outline of the evolution of U Sco subclass RNe is as follows. One of the component star of a relatively long orbital period primordial binary is a 8 - 13 $M_\odot$ star. It becomes a helium star of mass $\sim$ 2 - 2.8 $M_\odot$ after tidal mass loss. The star, then, undergoes non-degenerate carbon burning and forms a semi-degenerate O+Ne+Mg core (Nomoto 1984). Roche lobe overflow occurs and common envelope phase, which dissipates the orbital energy of the binary system, follows. The carbon burning shell stops before the O+Ne+Mg core mass reaches to the critical mass of 1.37 $M_\odot$ for neon ignition. A short orbital period pre-RN of an O+Ne+Mg WD and a dwarf binary emerges from the common envelope. Then, the evolution of the nondegenerate secondary fills its Roche lobe to become a U Sco subclass RN. This RN state is relatively short lived ($\sim 10^{-6}$ years). The WD, subsequently, collapse to form a binary system of a neutron star and a slightly evolved secondary. Examples of such a system are the evolved halo X-ray sources Cyg X-2 and 2S 0921-63. Therefore, U Sco subclass RNe may be a progenitor of the halo type LMXRBs. With this scenario, a seemingly too large distance to the U Sco, and also its large distance above the plane, is not a problem, since the system should belongs to the halo population.

References

Livio, M and Truran, J.W. (1992), *Ap. J.*, **389**,695.
Nomoto, K (1984), *Ap. J.*, **277**, 791.
Nomoto, K and Kondo, Y. (1991), *Ap. J.*, **367**, L19.
Williams, R.E., Phillips, M.M. and Heathcote, S.R. (1987), *Ap and SS*, **131**, 681.

A. Bianchini et al. (eds.), Cataclysmic Variables, 534.
© *1995 Kluwer Academic Publishers.*

PRELIMINARY INVESTIGATIONS INTO THE ORBITAL
EVOLUTION OF PRE-CATACLYSMIC VARIABLES

P.B. MARKS
Astronomy Centre, Mathematical and Physical Sciences,
University of Sussex, Falmer, Brighton, BN1 9QH, U.K.

The presently accepted theory of cataclysmic variable formation involves common envelope evolution and subsequent orbital shrinkage due to a loss in the total angular momentum of the system. The mechanisms of angular momentum loss are those of gravitational radiation and magnetic braking via a stellar wind, the efficiency of these processes determining whether the systems will evolve into Roche lobe contact. The results of a numerical calculation of the orbital evolution of a large sample of objects believed to be pre-cataclysmic variables are presented. Two cases are considered: one involving purely gravitational radiation, and one which considers the effects of both gravitational radiation and magnetic braking.

It has been found that when gravitational radiation is considered alone, three of the systems (NN Ser, HW Vir and MT Ser) evolve to become cataclysmic variables within timescales on the order of 10^9 years, the remaining systems taking longer than a Hubble time to reach Roche lobe contact.

When both gravitational radiation and magnetic braking are considered, it is found that, of the ten systems that contain secondaries which can support stellar winds (assumed constant, and of the order of $10^{-10} M_\odot \, \mathrm{yr}^{-1}$), six will evolve into contact within 10^{10} years. However, three of these six systems have mass ratios that are likely to lead to unstable mass transfer, and for a further two systems it is unclear whether the observational data are reliable. This leaves only one system, V471 Tau, from the above wind supporting systems that can be considered to be a pre-cataclysmic variable. With a slight increase in the amount of angular momentum lost, three more systems (HR 8210, Feige 24 and BE UMa) could evolve to Roche lobe contact within a Hubble time and be considered to be pre-cataclysmic variables. However as this is purely speculative, it is concluded that only the four systems mentioned above are likely to be pre-cataclysmic variables.

A Bianchini et al (eds) Cataclysmic Variables 535

LIST OF AUTHORS